PROTEIN INTERACTIONS

Edited by **Jianfeng Cai**
and **Rongsheng E. Wang**

INTECHOPEN.COM

Protein Interactions
Edited by Jianfeng Cai and Rongsheng E. Wang

Published by InTech

Janeza Trdine 9, 51000 Rijeka, Croatia

Edition 2014

Publishing Process Manager Marina Jozipovic

Technical Editor Teodora Smiljanic

Cover Designer InTech Design Team

Additional hard copies can be obtained from orders@intechopen.com

Protein Interactions, Edited by Jianfeng Cai and Rongsheng E. Wang
p. cm.
ISBN 978-953-51-0244-1

INTECH
open science | open minds

Contents

Preface

Protein interactions, including interactions between proteins and proteins, nucleic acids, lipids, carbohydrates, are essential to all aspects of biological processes, such as cell growth, differentiation, and apoptosis. Therefore, investigation and modulation of protein interactions are of significance as it not only reveals the mechanism governing cellular activity, but also leads to potential agents for the treatment of various diseases. In recent years, the development of biochemistry knowledge and instrumentation techniques has greatly facilitated the research in protein interactions. To provide some background information on the protein interactions, and also highlight the examples in the study of protein interactions, this book reviews some latest development in protein interactions, including modulation of protein interactions, applications of analytical techniques, and computer-assisted simulations. It aims to inspire the further development of technologies and methodologies in the understanding and regulation of protein interactions.

Although the chapters included in this book are all addressing protein interactions, we try to separate them into two parts according to their objectives. Chapters in part 1 mainly focus on the investigation of some specific protein-protein or protein-nucleic acid interactions, and try to elucidate the mechanism of specific cellular processes. Part 1 provides some insight of why and how to study protein interactions, and illustrates some approaches to modulate protein interactions. The second part is devoted to the development of various methods for the investigation of protein interactions, including computational modeling. Methods used to study protein interactions often evolve rapidly and many innovative methods or approaches are emerging in this field. The chapters shown in this part would shed light on the further development and application of analytical techniques and computer simulations.

I would like to thank every author because they have devoted their effort and expertise to prepare the outstanding chapters included in this book. I also thank Dr. Rongsheng E. Wang, the co-Editor of this volume, for his tremendous help on the review and editing of the book. Meanwhile, I want to express my deep appreciation to Ms. Marina Jozipovic for her tireless efforts in distributing, organizing, and processing all of the chapters.

Jianfeng Cai
Assistant Professor, Department of Chemistry,
University of South Florida, Tampa, FL
USA

Part 1

Examples of Protein Interactions

1

MOZ-TIF2 Fusion Protein Binds to Histone Chaperon Proteins CAF-1A and ASF1B Through Its MOZ Portion

Hong Yin[1], Jonathan Glass[1] and Kerry L. Blanchard[2]
[1]*Department of Medicine and the Feist-Weiller Cancer Center, LSU Health Sciences Center, Shreveport, LA*
[2]*Eli Lilly & Company, Indianapolis, IN*
USA

1. Introduction

We previously identified a MOZ-TIF2 (transcriptional intermediary factor 2) fusion gene from a young female patient with acute myeloid leukemia (AML) (Liang et al., 1998). MOZ related chromosome translocations include MOZ-CREB-binding protein (MOZ-CBP, t(8;16)(p11;p13)), MOZ-P300(t(8;22)(p11;q13)), MOZ-TIF2(inv(8)(p11q13), and MOZ-NCOA3(t(8;20)(p11;q13)) (Esteyries et al., 2008; Troke et al., 2006). In an animal model, the MOZ-TIF2 fusion product successfully induced the occurrence of AML (Deguchi et al., 2003). Though the mechanisms for leukemogenesis of this fusion protein are poorly understood, analysis of functional domains in the MOZ-TIF2 fusion protein discloses at least two distinct functional domains: 1) the MYST domain containing the C2HC nucleosome recognition motif and the histone acetyltransferase motif in the MOZ portion and 2) the CID domain containing two CBP binding motifs in the TIF2 portion. Together these domains were responsible for AML in mice caused by injecting bone marrow cells transduced with retrovirus containing the MOZ-TIF2 fusion gene. Furthermore, MOZ-TIF2 conferred the properties of leukemic stem cells (Huntly et al., 2004). The MOZ-TIF2 transduced mouse common myeloid progenitors and granulocyte-monocyte progenitors exhibited the ability to serially replated *in vitro*. The cell line derived from transduced progenitors could induce AML in mice. Interestingly, the C543G mutation in C2HC nucleosome recognition motif or in the CBP binding motif (LXXLL) blocked the self-renewal function of MOZ-TIF2 transduced progenitors. More recently, a study using PU.1 deficient mice demonstrated that the interaction between MOZ-TIF2 and PU.1 promoted the expression of macrophage colony-stimulating factor receptor (CSF1R). Cells with high expression of CSF1R are potential leukemia initiating cells(Aikawa et al., 2010). Models suggesting that aberrant transcription by the interaction between MOZ fusion proteins and transcription factors, AML1, p53, PU1, or NF-kB have been well reviewed(Katsumoto et al., 2008).

MOZ as a fusion partner of MOZ-TIF2 is a member of MYST domain family (MOZ/YBF2/SAS2/TIP60) and acetylates histones H2A, H3 and H4 as a histone acetyltransferase (HAT) (Champagne et al., 2001; Kitabayashi et al., 2001). MOZ is a cofactor in the regulation of transcriptional activation of several target genes important to hematopoiesis, such as Runx1 and PU.1 (Bristow and Shore, 2003; Katsumoto et al., 2006; Kitabayashi et al., 2001). $MOZ^{-/-}$ mice died at embryonic day 15 and exhibited a significant decrease of mature erythrocytes (Katsumoto et al., 2006). The histone acetyltransferase activity of MOZ is required to maintain normal functions of hematopoietic stem cells (HSC) (Perez-Campo et al., 2009). Mice with mutation at HAT or MYST domain (G657E) showed a decreased population of HSC in fetal liver. The lineage-committed hematopoietic progenitors from fetal liver cells with $HAT^{-/-}$ mutant had reduced colony formation ability.

In our attempt to find proteins that interact with the fusion protein by using as bait a construct of the MOZ N-terminal fragment, encoding the first 759 amino acids of MOZ-TIF2 fusion gene and containing the H15, PHD, and MYST domains, we were able to isolate two proteins, the p150 subunit or subunit A of the human chromatin assembly factor-1 (p150/CAF-1A) and the human anti-silencing function protein 1 homolog B (ASF1B). Both of these proteins were verified to interact with the MOZ partner of MOZ-TIF2 fusion in the yeast two-hybrid system. The interaction has been further characterized by co-immunoprecipitation, protein pull-down assays, and co-localization by immunohistochemistry. The differences in the interactions of CAF-1A and ASF1B with wild type MOZ and the MOZ-TIF2 fusion proteins may contribute to leukemogenesis.

2.Materials and methods

2.1 The sources of cDNAs and plasmid constructions

The cDNA for MOZ was kindly provided by Julian Borrow (Center for Cancer Research, Massachusetts Institute of Technology, MA) and TIF2 was a kind gift from Hinrich Gronemeyer (Institut de Genetique et de Biologie Moleculaire et Cellulaire, France). A full length MOZ-TIF2 fusion was created by inserting a RT-PCR fragment crossing the MOZ–TIF2 fusion site into the Hind3 site of wild type of human MOZ and the Sac1 site of human TIF2 in pBluescript KS phagemid vector (pBlueKS). The cDNAs for CAF-1A and ASF1B were screened and rescued from Human Bone Marrow MATCHMAKER cDNA Library (BD Biosciences Clontech Palo Alto, CA) by the yeast two-hybrid system using the N-terminal fragment of the MOZ-TIF2 fusion as bait. The cDNAs from the positive clones, which were in the pACT2 vector, were switched into the pBlueKS vector at EcoRI and XhoI sites and sequenced with a T7 primer. The resulting sequences were identified in the NCBI GenBank as the subunit A (p150) of human chromatin assembly factor-1 (GenBank accession No. NM-005483) and human anti-silencing function protein 1 homolog B (GenBank accession No. AF279307). The full length of both cDNAs was confirmed by DNA sequencing with gene specific primers. For the visualization of the expression and localization in mammalian cells, the full length of MOZ, MOZ-TIF2, TIF2, CAF-1A, and ASF1B were subcloned in frame into the C-terminal fluorescent protein Vector, pEGFP or pDsRed2 (BD Biosciences Clontech, Palo Alto, CA) to generate fluorescent fusion proteins. For studies of protein-protein interaction *in vitro*, glutathione S-transferase (GST) fusions of MOZ fragments were constructed in the pGEX vector (Amersham Biosciences, Piscataway, NJ). Briefly, the full length MOZ cDNA was digested with Asp718/BgI2 from pBlueKS-MOZ and was ligated into the pET-30a (EMD Biosciences, Inc. Novagen Madison, WI) plasmid at Asp718/BamH1 site to create the pET-30a-MOZ construct. A PET-30a-MOZ-1/759 (amino acids 1 to 759) construct was generated by removing a Hind3/Hind3 fragment from pET-30a-MOZ and then re-ligating. This fragment was then switched from pET-30a vector to pGEX-4T at a Not1/Xho1 site to construct the pGEX-4T-MOZ-1/759. The pGEX-4T-MOZ-1/313 (amino acids 1 to 313) containing H15 and the PHD domain was generated by the deletion of a 1515 base pair fragment from pGEX-4T-MOZ-1/759 with Hind3 /Blin1 followed by re-ligation. The pGEX-6P-MOZ-488/703 plasmid was constructed by inserting an EcoRV to Eag1 fragment of MOZ (amino acids 488 to 703) containing the C2HC motif and acetyl-CoA binding region to pGEX-6P-2 vector at Sma1/Eag1 sites. To create pET-30a-CAF-1A, the pBlueKS-CAF-1A was first digested with XhoI and then digested partially with NcoI. A 3.1 kb fragment was recovered by agarose electrophoresis and was ligated to NcoI/XhoI sites of pET-30a vector. The pET-30c-ASF1B was constructed by inserting the 1 kilobase EcoR1/Hind3 fragment of pBLueKS-ASF1B into the pET-30c vector at EcoR1 /Hind3 sites.

2.2 Yeast two-hybrid screen

pGBD-MOZ-MYST, a bait plasmid with a fusion of the N-terminal fragment of MOZ-TIF2 to the GAL4 DNA binding domain was constructed by inserting a 2.3 kb fragment encoding amino acids 1 to 759 of human MOZ to BamH1/blunted Bgl 2 sites in the pGBD-C3 vector (James et al., 1996). The bait plasmid was transformed into the yeast host PJ69-2A and mated with pre-transformed Human Bone Marrow MATCHMAKER cDNA Library according to the manufacturer's instruction. The mating culture was plated on 25 x 150 mm triple dropout (TDO) dishes (SD/-His/-Leu/-Trp) and 25 x 150 mm quadruple dropout (QDO) dishes (SD/-Ade/-His/-Leu/-Trp). After incubation for 7 and 14 days, the more than 100 colonies which grew on TDO and QDO dishes were picked for re-screening on SD/-His, SD/-Ade/ and QDO dishes. A total of five colonies were grown from the second screening. The plasmids from each colony were rescued and transformed into KC8 cells. All of the plasmids were re-transformed into the yeast host PJ69-2A and Y187; no auto-transcription activation of any reporter was seen. The pVA3.1 plasmids containing either the murine p53 in PJ69-2A or the PTD1-1 with SV 40 large T antigen in Y187 were used as controls for DNA binding domain and activation domain fusions. The plasmids from positive clones were subjected to restriction enzyme mapping which showed two potential interacting genes which were subsequently sequenced and identified with the NCBI database.

2.3 Co-localization of MOZ or MOZ-TIF2 and CAF-1A or ASF1B

To identify the co-localization of expressed fluorescent fusion proteins, HEK293 cells were grown in DMEM (Mediatech Cellgro, VA) containing 10% fetal bovine serum (FBS) and co-transfected by pEGFP-MOZ or pEGFP-MOZ-TIF2 and pDsRed2-CAF-1A or pDsRed2-ASF1B with Lipofectamine 2000 (Invitrogen, Carlsbad, CA). Briefly, cells were grown on a coverslip in a 12-well plate a day before the transfection in the antibiotic-free medium to reach 80-90% confluence on the next day. 1.6 μg of DNA in 100 μl of Opti-MEM I Reduced Serum Medium (Invitrogen, Carlsbad, CA) was mixed with 100 μl of diluted Lipofectamine 2000 reagent. After incubation for 20 min. at room temperature, the DNA-Lipofectamine 2000 complex was added to the cells and 48 hours later, subcellular location of expressed fluorescent fusion proteins was examined with a Zeiss fluorescent microscope equipped with Axiocam system and by a laser scanning confocal microscope (Bio-Rad Laser Scanning

System Radiance 2000/Nikon Eclipse TE300 microscope). To examine the subcellelular localization of endogenously expressed MOZ and CAF-1A, HEK293 and Hela cells were fixed with 4% paraformaldehyde and then blocked with Ultra V block (Lab Vision Co.CA). For some experiments pre-extraction with 0.3%Triton-X100 was conducted. The fixed cells were then incubated with antibody against MOZ (N-19, Santa Cruz Biotechnology, Inc, Santa Cruz, CA) at 1:100 and /or antibody against CAF-1A (a kind gift from Dr. Bruce Stillman, Cold Spring Harbor, NY). In some experiments, the antibody against CAF-1A and ASF1B were purchased from Cell Signaling Technology, MA. The immunofluorescence of MOZ, CAF-1A, or ASF1B was observed as described above for examination of expressed EGFP fusion proteins.

2.4 Co-immunoprecipitation and immunoblotting

HEK293 cells were transfected with EGFP fusions of MOZ, MOZ-TIF2, or TIF2. After 48 hours of transfection, whole cell lysates was prepared with plastic individual homogenizers in the lysis buffer [50 mM NaCL, 5mM KCL, 1mM EDTA, 20 mM HEPES, pH 7.6, 10% glycerol, 0.5% NP-40, and protease inhibitor cocktails (Roche Applied Science, IN)]. Immunoprecipitation was conducted with an antibody against EGFP (BD Biosciences, Palo Alto, CA). Briefly, 2 μg of anti-EGFP antibody and protein A/G-agarose (Santa Cruz Biotechnology, Santa Cruz, CA) were added to 0.8 ml of cell lysate (about 500 μg protein) and incubated overnight at 4°C with rotation. The precipitate was collected by centrifugation, extensively washed, subjected to SDS-PAGE, transferred onto Hybond-ECL nitrocellulose membrane (Amersham Pharmacia Biotech, Piscataway, NJ), and examined by immunoblotting with the antibody against CAF-1A.

2.5 Expression of GST fusion proteins and GST pull down assay

E. *coli* BL21-CodonPlus®(DE3)-RIL Competent Cells (Stratagene, La Jolla, CA) were transformed with pGEX vectors containing cDNA fragments MOZ-1/759, MOZ-1/313, or MOZ-488/703 and grown in LB medium. To induce protein expression isopropyl β-D-thiogalactopyranoside (IPTG) was added at final concentration of 1mM when the A_{600} of the cultures reached 0.6 to 0.8. After three more hours of growth at 28° C, cells were collected by centrifugation and resuspended in cold PBS containing 1% Triton X-100 and protease inhibitor cocktail and kept on ice for 30 minutes. Cell lysates were prepared by ultrasonication followed by centrifugation at 15,000 rpm for 30 minutes at 4°C. GST fusion proteins were purified with the GST Purification Module (Amersham Pharmacia Biotech, Piscataway, NJ). Purified GST fusion proteins were examined with SDS-PAGE followed by Coomassie Blue staining. To perform GST pull down affinity assays [^{35}S]Methionine-labeled proteins were first produced with Single Tube Protein® System 3 or EcoPro™ T7 system (EMD Biosciences, Inc. Novagen, Madison, WI) from pET 30 vectors carrying full length of CAF-1A or ASF1B. The binding reaction was conducted with 5μl of *in vitro*-translated protein and 3-5 μg of GST alone or GST fusion protein attached to Sepharose 4B beads in 200 μl binding buffer (50mM Tris-HCI , pH 8.0, 100 mM NaCl, 0.3 mM DTT, 10mM MgCl2, 10% glycerol, 0.1% NP40). The reaction was conducted at 4 °C for 1 hour followed by five washes with 400 μl of binding buffer. The final pellet was separated by SDS-PAGE, autoradiography performed, and radioactivity detected with a phosphorimager.

2.6 *In Vitro* protein binding assay with S-tagged fusion protein

The S-tagged fusion of ASF1B was expressed from pET-30c-ASF1B in E. *coli* BL21-CodonPlus® (DE3)-RIL cells after induction with 0.8 mM of IPTG and purification with S-tagged agarose beads. The fusion protein on agarose beads was incubated with 150μl (about 600 μg of protein) of cell extract from HEK293 cells transfected with pEGFP fusion protein. The beads were pelleted, washed, and the "pull-down" proteins examined as described above with the anti-EGFP antibody.

2.7 RNA isolation and microarray analysis

RNA was isolated from stably transfected U937 cells with TRI Reagent® (Molecular Research Center, Inc., Cincinati, OH). The analysis of gene expression profile was conducted on the Human Genome U95A Array (Affymetrix, Inc., Santa Clara, CA). The cRNA was synthesized from 10μg of total RNA. The hybridization and signal detection was completed in the Core Facility at LSUHSC-Shreveport according to the standard Affymetrix protocol. The human U95A array represents 12,256 oligonucleotides of known genes or expression tags. The expression profile was analyzed with GeneSifter software. In pairwise analysis, the quality was set as 0.5 for at least one group in order to minimize the effect of low intensity or poor quality spots. Genes with a $\geq$ 2-fold change and with $P<0.05$ in a student T-test were considered as either significantly up or down regulated genes. To find genes either commonly or differentially expressed in the gene list, we set the quality as 1 to obtain positive expressed genes in pattern navigation analysis. The analysis results were exported for Venn Diagram analysis using the GeneSifter intersector tool.

3.Results

3.1 Screening for MOZ interacting proteins by the yeast two-hybrid system

A MOZ cDNA fragment encoding amino acids 1 to 759 cloned into pGBD was used as the bait in the yeast two-hybrid system in which the prey was a human cDNA bone marrow library. After a second screening five β-galactosidase positive clones grew on SD/-His plates. To eliminate any of these clones as representing false positive clones, plasmid DNA from each clone was rescued using KC8 cells and transformed into PJ69-2A cells carrying pGDB-MOZ-MYST. The transformants were then selected on five different media: -Trp/ -Leu, -His, -His+5mM 3-amino-1,2,4-triazole (3-AT), -His+10mM 3-AT, and -Ade and interaction with the MOZ fragment was verified in all five of the clones (Figure 1). Clone 3.1 grew on -His, -His+10mM 3-AT, and -Ade medium indicative of a strong physical interaction; the other clones only grew on -His and -His + 5 mM 3-AT, but not on -Ade, indicating a weaker interaction. DNA sequencing of the putatively strongly MOZ interacting protein demonstrated that the cDNA encoded the full length CAF-1A. The more weakly interacting cDNAs represented the entire coding region of ASF1B.

3.2 Identify the interaction between MOZ and CAF-1A in human cells

In yeast, the MYST family member Sas2 was found to interact with Cac1, the largest subunit of Saccharomyces cerevisiae chromatin assembly factor-I (CAF-1) (Meijsing and Ehrenhofer-Murray, 2001) but it is not known if the interaction between the homologous proteins in mammalian cells, MOZ and CAF-1A, takes place in human cells and if any interaction occurs between the MOZ-TIF2 fusion protein and CAF-1A. To address these areas we looked for interactions by co-immunoprecipitation using transfections with the

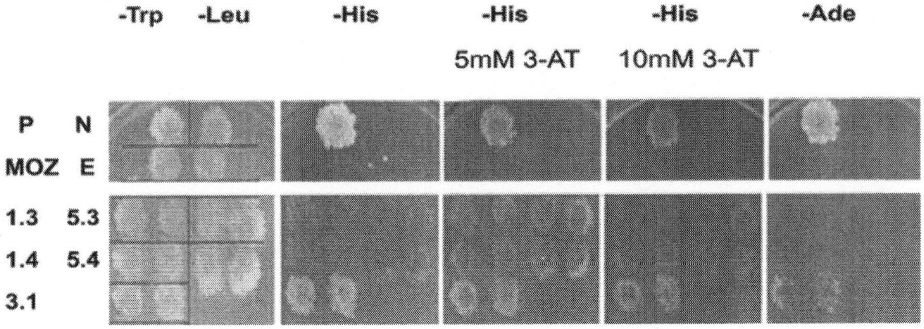

Fig. 1. Protein interaction between MOZ and CAF-1A or ASF1B in the yeast two-hybrid system. The yeast two-hybrid system was used with pretransformed Matchmaker libraries as detailed in the Methods. The bait was the fragment encoding amino acids 1 to 759 of the human MOZ gene in the pGAL 4 DNA-BD vector. In the upper panel controls are plated on 5 different selection media: **P**, positive control diploid with plasmid pDT1-1 encoding an AD/SV40 large T-antigen fusion protein and pVA3-1 carrying DNA-BD/murine P53 fusion protein. **N**, negative control diploid. MOZ, a diploid with GAL4 DNA-BD+ MOZ fragment of amino acids 1 to 759. **E**, a diploid with GAL4 DNA-BD vector only. In the lower panel the five clones (1.3, 1.4, 3.1, 5.3, and 5.4) that were positive after a second screening were plated in duplicate on the same media. Clones 1.3, 1.4, 5.3, and 5.4 show an interaction between MOZ and ASF1B; clone 3.1 shows an interaction between the MOZ and CAF-1A. Trp, tryptophan, Leu, leucine, His, histidine, Ade, adenine, 3-AT, 3-amino-1,2,4,triazole.

MOZ and MOZ-TIF2 fusion constructs into HEK293 cells which express CAF-1A (Figure 2). In these experiments the HEK293 cells were transfected with EGFP fusions of MOZ, MOZ-TIF2 and TIF2, the expressed fusion proteins precipitated with anti-EGFP antibody and the presence of co-precipitated CAF-1A assayed by western blot analysis. Only with the product of the EGFP-MOZ construct was a significant amount of CAF-1A precipitated (Figure 2A); a far smaller amount was precipitated with MOZ-TIF2. By comparison to the intensity of the CAF-1A band in the input lane, which represents 10% of the amount of lysate subjected to immunoprecipitation, approximately 35-40% of the HEK293 cell CAF-1A was estimated to be co-precipitated with the transfected MOZ. In contrast, less than 10% of the CAF-1A co-precipitated with MOZ-TIF2 (Figure 2A). The differences in the amount of CAF-1A precipitated were not a result of altered expression of CAF-1A or of differences in expression levels of the transfectants as the expression of CAF-1A was not affected by any of the three transfectants (Figure 2B) and the EGFP-tagged MOZ and MOZ-TIF2 proteins showed similar levels of expression, while TIF2 showed a 2-3 fold higher expression than MOZ and MOZ-TIF2 (Figure 2C).

3.3 The MOZ portion of MOZ-TIF2 fusion interacts physically with CAF-1A through the N-terminal of MOZ

Using the yeast two-hybrid system we have shown that CAF-1A interacted with a MOZ fragment extending from amino acids 1 to 759. Within this region are PHD (amino acids 195-320) and MYST (amino acids 562-750) domains that are potential sites for the interaction with (Figure 3A) (Champagne et al., 1999).

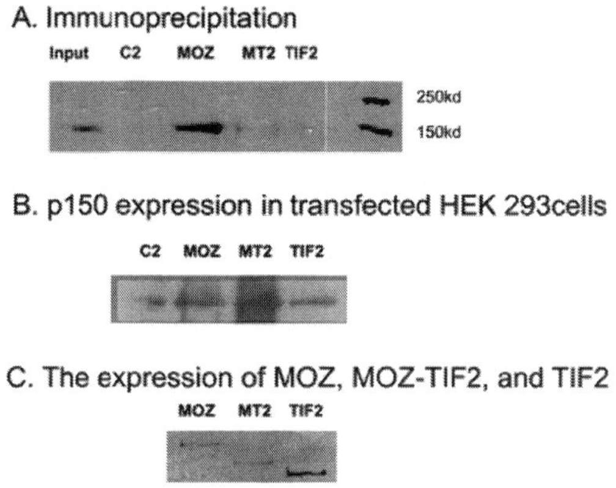

Fig. 2. **Co-precipitation of CAF-1A (p150) with EGFP-tagged MOZ, MOZ-TIF2, and TIF2.** The EGFP constructs of MOZ, MOZ-TIF2, and TIF2 were transfected into HEK293 cells. **Panel A.** After 48 hours, whole cell extracts were prepared in lysis buffer and subjected to immunoprecipitation with anti-EGFP antibody, followed by SDS-PAGE, and western blot analysis with anti-p150 antibodies. The input lane corresponds to 10% of the amount of lysate subjected to immunoprecipitation. Lane C2 represents the pEGFP-C2 vector alone and MT2 represents MOZ-TIF2. **Panel B**. The lysates of the various transfectants were subjected to SDS-PAGE followed by western blot analysis with anti- p150 antibody to demonstrate the expression level of p150 in the transfected cells. **Panel C**. The same lysates used in Panel B were subjected to a western blot analysis with anti-EGFP antibody to demonstrate the expression of EGFP-tagged MOZ, MOZ-TIF2 and TIF2.

To further define the region containing the binding domain, a pull down assay using GST fusion proteins was established. First, a GST-tagged MOZ fragment from amino acids 1 to 759 was used to pull down CAF-1A and to demonstrate that the GST did not interfere with the MOZ-CAF-1A interactions shown earlier (Figure 3B). We then generated two GST-tagged MOZ fragments, one encompassing amino acids 1-313 (MOZ-1/313) containing the H15 and PHD domains and the other from amino acids 488-703 (MOZ-488/703) including the C2HC motif and acetyl-CoA binding region (Figure 3 C, left panel). These peptides were used with [^{35}S]methionine labeled CAF-1A synthesized in an *in vitro* translation system and interactions detected with a GST pull down assay (Figure 3C). For equivalent amounts of fusion peptides more MOZ-1/313 was bound to CAF-1A than MOZ-488/703 (Figure 3C). As a percentage of the input radioactivity, MOZ-1/313 pulled down about 30 % of the [^{35}S]methionine labeled CAF-1A while MOZ-488/703 pulled down only 14%. Further analysis of domain interactions showed that strongest binding was seen between MOZ-1/313 and CAF-1A-176/327 among all peptides (Figure 3D). CAF-1A-176/327 pulled down about 328% of [^{35}S]methionine labeled MOZ-1/313 and pulled down only 76% of MOZ-488/703 while CAF-1A-620/938 pulled down 20% and 28% of MOZ-1/313 and MOZ-488/703, respectively.

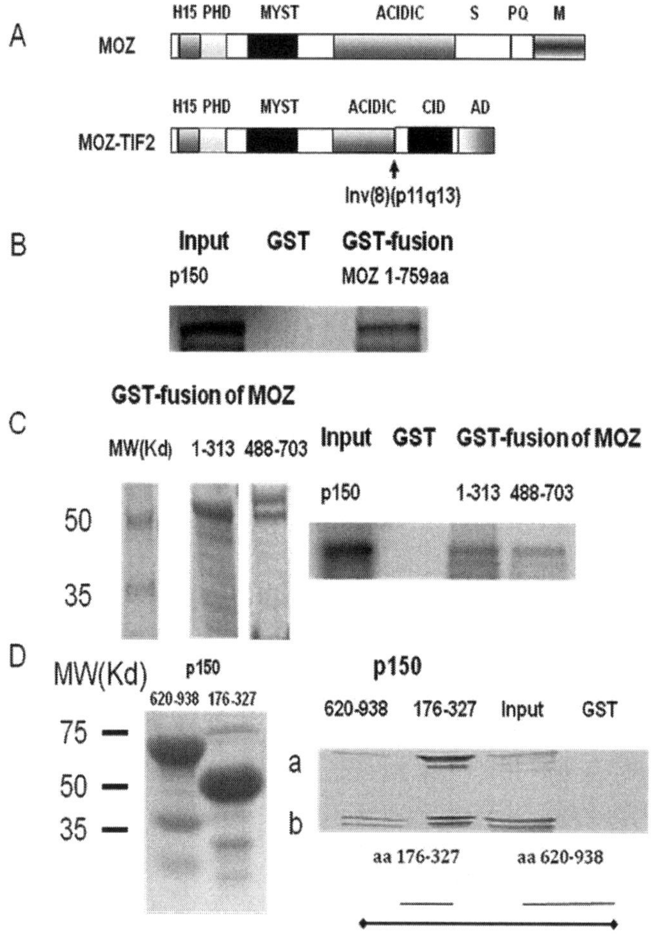

Fig. 3. The interaction between MOZ fragments and CAF-1A (p150). GST-tagged MOZ fragments were expressed and purified with glutathione Sepharose 4B as described in Materials and Methods. [^{35}S]-methionine labeled p150 protein was produced from a T7-driven pET-30 plasmid with an *in vitro* translation system. A, binding assay was conducted with [^{35}S]-methionine labeled p150 and the GST-tagged MOZ fragments. The input lane is 10% of the [^{35}S] methionine p150 protein added to the binding assay. A, schematic structure

of MOZ and MOZ-TIF2. B, interaction between p150 and the MOZ fragment from amino acids 1 to 759 using the binding assay as described in the Materials and Methods. C, left panel, SDS-PAGE of the purified GST-MOZ-1/313 and GST-MOZ-488/703 peptides to demonstrate that the peptides were of the expected molecular weights; right panel, as described in Materials and Methods [^{35}S]-methionine labeled p150 synthesized in a cell-free translation system was incubated *in vitro* with equivalent amounts of GST fusions with MOZ-1/313 or MOZ-488/703, the resulting complexes isolated by GST-pull down assay, and the amount of [^{35}S]-methionine labeled p150 detected by radioautography following SDS-PAGE. D, left panel, SDS-PAGE of the purified GST-p150-176/327 and GST-p150-620/938 fusion peptides; right panel, GST pulldown assays as described in C with [^{35}S]-methionine labeled MOZ-1/313 (a) and MOZ-488/703(b) peptides. The bottom line indicates the full length p150 protein.

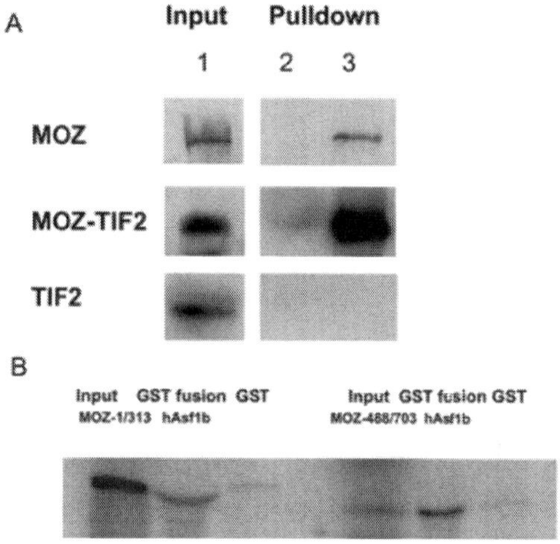

Fig. 4. ASF1B interacts with MOZ and MOZ-TIF2. **Panel A**. HEK293 cells were transfected with EGFP-MOZ, EGFP-MOZ-TIF2, and EGFP-TIF2 as detailed in the Materials and Methods. At 48 hours after transfection cell lysates were incubated with S-tagged ASF1B absorbed to S-tag agarose beads and after extensive washing the proteins bound to ASF1B were analyzed by SDS-PAGE with subsequent western blot analysis with anti-GFP antibody. Lane 1, 10% of input; lane 2, S-tag protein alone; lane 3, S-tagged ASF1B. **Panel B**. GST pull down assays were performed as detailed above incubating GST- ASF1B with [^{35}S]-methionine labeled MOZ-1/313 or MOZ-488/703 peptides synthesized in a cell-free translation system as described in the Material and Methods.

3.4 Confirmation of ASF1B as an interacting protein of MOZ and MOZ-TIF2

The yeast two-hybrid system also revealed a cDNA encoding another protein, ASF1B, which interacted with the MOZ-1/759 fragment. To verify the interaction between MOZ and ASF1B and to examine if the MOZ-TIF2 fusion protein also interacts with ASF1B, we conducted pull down assays and examined co-localization of proteins similar to the studies with CAF-1A. A S-tag fusion cDNA with ASF1B was created in the pET-30c vector and the fusion protein was labeled with [^{35}S]methionine by an in vitro transcription/translation system. The expressed fusion protein was purified with S-tag agarose beads and incubated with cell lysates containing expressed EGFP fusions of MOZ, MOZ-TIF2 and TIF2. Subsequently, EGFP proteins that interacted with ASF1B were identified by western blot analysis with an anti-EGFP antibody (Figure 4A). Both EGFP-MOZ and EGFP-MOZ-TIF2 could be demonstrated to interact with ASF1B. MOZ-TIF2 appeared to interact more strongly with the percentage of EGFP fusion protein bound to ASF1B approximately 240% over the input for MOZ-TIF2 and 70% for MOZ, respectively. TIF2 showed no binding to ASF1B. To further identify the ASF1B binding domain in MOZ, the GST-tagged ASF1B was incubated with [^{35}S]methionine labeled MOZ-1/313 and MOZ-488/703 (Figure 4B). The MOZ-488/703 fragment showed stronger binding to ASF1B than MOZ-1/313. The percentage of ASF1B bound to the MOZ-1/313 fragment represented about 25% of the input while the percentage of ASF1B bound to the MOZ-488/703 fragment was 150% of the input.

3.5 The co-localization of MOZ and MOZ-TIF2 with CAF-1A and ASF1B

To further verify the interaction of MOZ with CAF-1A, we first examined by indirect immunohistochemistry the localization of endogenous MOZ and CAF-1A in Hela cells to determine if the subcellular distribution was similar by confocal immunofluorescence microscopy (Figure 5A). In Hela cells both MOZ and CAF-1A were predominately localized in interphase nuclei (Figure 5A-a). As the chromatin condensed in metaphase MOZ distributed dominantly in cytoplasm and disassociated from the spindle-chromosome in some cells (Figure 5A-b and 5A-c). CAF-1A was observed either to disassociate from (Figure 5A-b) or bind to spindle-chromosomes (Figure 5A-c). However, cytoplasmic co-localization of MOZ and CAF-1A was still seen as detected by the persistence of yellow by confocal microscopy. In anaphase, with paired chromosome separation, CAF-1A was still bound to the spindle-chromosome but MOZ was fully dissociated (Figure 5A-d) but with persistent co-localization of both in the cytoplasm. To determine if the MOZ-TIF2 fusion protein has similar localization as MOZ and co-localized with CAF-1A, HEK293 cells were transfected with EGFP-MOZ or EGFP-MOZ-TIF2 and DsRed2-CAF-1A (Figure 5B). Both EGFP-MOZ and EGFP-MOZ-TIF2 showed a predominantly nuclear localization in HEK293 cells in interphase. However, the, EGFP-MOZ-TIF2 fusion protein appeared in larger aggregates compared to the more homogenously distributed MOZ. In the merged image the MOZ co-localization with CAF-1A appeared stronger than the MOZ-TIF2-CAF-1A co-localization (Figure 5B, top panel, merge). To examine the binding of MOZ, MOZ-TIF2, and CAF-1A to the interphase chromatin we conducted pre-extraction with Triton-X100 in EGFP-MOZ and EGFP-MOZ-TIF2 transfected HEK293 cells (Figure 5C). In the interphase, all three proteins, EGFP-MOZ, EGFP-MOZ-TIF2, and CAF-1A showed resistance to pre-extraction and the co-localization with DAPI-stained DNA. Similarly, the co-localization of EGFP-MOZ-TIF2 with ASF1B was shown in transfected HEK293 cells (Figure 6A). Interestingly, EGFP-MOZ-TIF2 exhibited stronger co-localization with DsRed2-ASF1B than EGFP-MOZ in pre-extracted HEK293 cells (Figure 6B, merge).

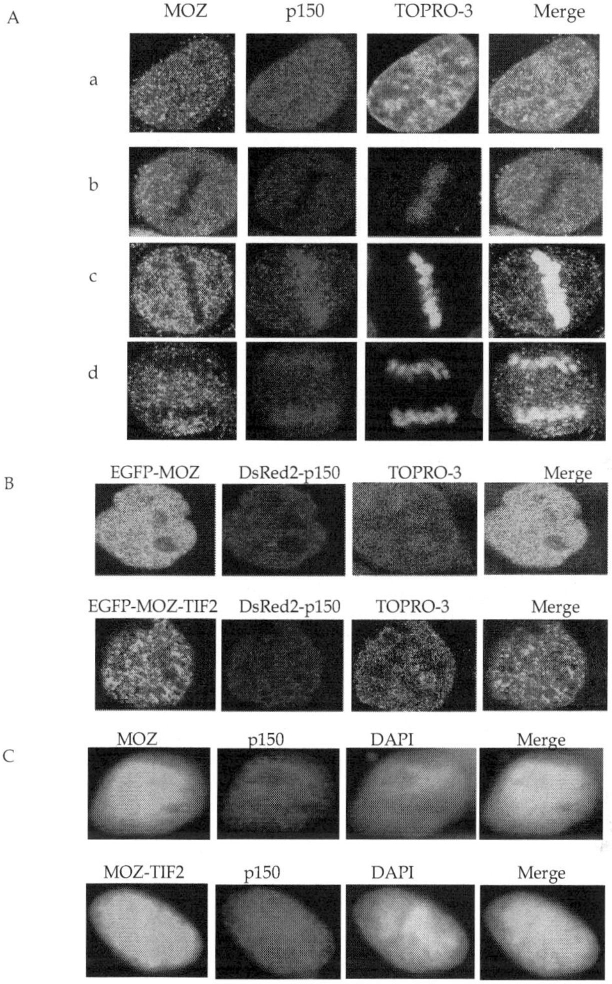

Fig. 5. Subcellular localization of MOZ, MOZ-TIF2, CAF-1A (p150). **A**. Indirect immunofluorescence of MOZ (green) and p150 (red) in HeLa cells at interphase and metaphase observed by confocal microscopy with the nuclei stained with Topro-3. **B**. Confocal microscope images were obtained of HEK293 cells co-transfected with EGFP-MOZ and DsRed2-p150 or EGFP-MOZ-TIF2 and DsRed2-p150 as detailed in the Materials and Methods and nuclei stained with Topro-3. **C**. HEK293 cells transfected with EGFP-MOZ (green) and EGFP-MOZ-TIF2 (green) and stained with anti-p150 antibody after pre-extraction with Triton-X100. The fluorescent images were obtained at x100 with a Zeiss fluorescent microscope.

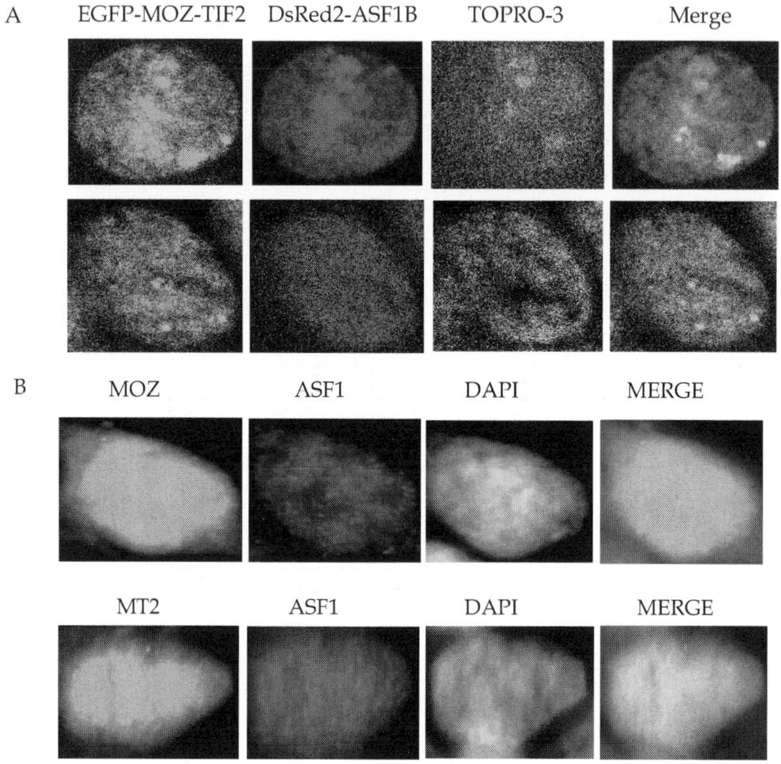

Fig. 6. **A**. Confocal microscope images were obtained of HEK293 cells co-transfected with EGFP-MOZ-TIF2 and DsRed2-ASF1B. The nuclei were stained with Topro-3. **B**. HEK293 cells were transfected with EGFP-MOZ or EGFP-MOZ-TIF2. 48 hours later, cells were pre-extracted, fixed, and immune-stained with anti-ASF1B antibody. Fluorescent images were photographed at x100 with a Zeiss fluorescent microscope.

3.6 Altered gene expression profile in U937 cells stably transfected with MOZ-TIF2

CAF-1 and ASF1, as histone chaperon proteins are essential in maintaining the nucleosome structure after DNA replica and in DNA repair. In yeast, CAF-1 and ASF1 are regulators of global gene expression (Zabaronick and Tyler, 2005). However, if MOZ and MOZ-TIF2, as proteins that associate with CAF-1 and ASF1, affect global gene expression is not known. We established stable transfection clones from U937 cells with forced expression of MOZ and MOZ-TIF2 and analyzed global gene expression of these cell clones. Compared to the expression profile of control cells stably transfected with pcDNA3 vector alone, MT2 caused a > 2-fold change in expression with 181 genes increasing and 106 genes decreasing expression (p = 0.01). Over expression of wild type MOZ also altered gene expression (>2-fold increase in 132 genes and >2-fold decrease in 88 genes, p=0.01). In addition, a differential gene expression signature was seen between MOZ and MOZ-TIF2 in a Venn diagram analysis (Figure 7). The signature-expressed genes are 189 with pcDNA3, 84 with MOZ, and 427 with MOZ-TIF2, respectively. Further pairwise analysis of differential expression of genes between MOZ and MOZ-TIF2 indicated that there 28 genes increasing over 2 fold (Table 1) and 34 genes decreasing over 2 fold (Table 2) in MOZ-TIF2 compared with that in MOZ. The altered genes between MOZ and MOZ-TIF2 are involved in multiple cell functions such as signal transduction, cell response to stimulus, cell cycle, chromosome structure, development, and tumor progression.

Ratio	p-value	Gene Name
6.41	0.002777	Transcribed locus, weakly similar to XP_537423.2 PREDICTED: similar to LINE-1 reverse transcriptase homolog [Canis familiaris
5.78	0.001076	Malic enzyme 3, NADP(+)-dependent, mitochondrial
4.8	0.042772	Testis derived transcript (3 LIM domains)
4.01	0.021676	Bone morphogenetic protein 1
3.71	0.031657	Interleukin 8 receptor, beta
3.63	0.012968	NADH dehydrogenase (ubiquinone) 1 beta subcomplex, 6, 17kDa
3.09	0.042154	calreticulin
2.73	0.047444	Actin binding LIM protein 1
2.55	0.040271	Ribosome binding protein 1 homolog 180kDa (dog)
2.43	0.016348	vesicle amine transport protein 1 homolog (T. californica)
2.41	0.004052	Calreticulin
2.39	0.039279	histone cluster 1, H2bi
2.37	0.039548	insulin-like growth factor binding protein 2, 36kDa
2.36	0.012714	Inversin
2.36	0.048093	PTK2B protein tyrosine kinase 2 beta
2.29	0.010907	RAP1 interacting factor homolog (yeast)
2.22	0.00445	CD160 molecule
2.13	0.042832	Carnitine palmitoyltransferase 1B (muscle)
2.13	0.010751	PCTAIRE protein kinase 1
2.12	0.006671	Neutrophil cytosolic factor 4, 40kDa
2.11	0.039481	neurogranin (protein kinase C substrate, RC3)
2.1	0.036043	Tyrosine 3-monooxygenase/tryptophan 5-monooxygenase activation protein, epsilon polypeptide
2.09	0.002837	Mediator complex subunit 21
2.06	0.008281	Insulin-like growth factor binding protein 2, 36kDa
2.04	0.032872	acylphosphatase 2, muscle type
2.03	0.006805	FK506 binding protein 1A, 12kDa
2.02	0.006539	Neurochondrin
2	0.001062	Syntaxin 5

Table 1. Up-regulated genes in MOZ-TIF2 vs MOZ.

Ratio	p-value	Gene Name
14.58	0.034066	Fibroblast growth factor receptor 2
10.51	0.048971	Transcribed locus
5.7	0.020122	Spectrin, beta, non-erythrocytic 1
5.3	0.007972	Sulfotransferase (Sulfokinase) like gene, a putative GS2 like gene
5.17	0.005226	Defensin, beta 1
3.92	0.00956	chorionic somatomammotropin hormone-like 1
3.68	0.002866	elongation factor, RNA polymerase II, 2
3.64	0.04087	RAP2A, member of RAS oncogene family
3.57	0.040904	CD2 molecule
3.57	0.039775	Met proto-oncogene (hepatocyte growth factor receptor)
3.34	6.34E-05	Adipose differentiation-related protein
2.99	0.049188	ATPase, Ca++ transporting, plasma membrane 4
2.61	0.013043	X-ray repair complementing defective repair in Chinese hamster cells 2
2.49	0.002229	regulatory solute carrier protein, family 1, member 1
2.47	0.018837	CMP-N-acetylneuraminate monooxygenase) pseudogene
2.45	0.049139	Angiogenic factor with G patch and FHA domains 1
2.45	0.026164	SCY1-like 3 (S. cerevisiae)
2.36	0.003507	spermidine/spermine N1-acetyltransferase 1
2.34	0.046867	ATPase, class VI, type 11A
2.27	0.027531	ecotropic viral integration site 2A
2.22	0.045457	Ubiquitin specific peptidase like 1
2.21	0.035349	Cyclin-dependent kinase 6
2.17	0.000434	CDC14 cell division cycle 14 homolog B (S. cerevisiae)
2.17	0.032088	Kruppel-like factor 10
2.17	0.049619	Starch binding domain 1
2.16	0.023854	Homeodomain interacting protein kinase 3
2.15	0.030337	Ectodermal-neural cortex (with BTB-like domain)
2.14	0.010869	Angiogenic factor with G patch and FHA domains 1
2.12	0.049079	Reversion-inducing-cysteine-rich protein with kazal motifs
2.11	0.042149	suppressor of Ty 3 homolog (S. cerevisiae)
2.08	0.037371	Nuclear receptor subfamily 1, group D, member 2
2.08	0.022954	cytochrome P450, family 1, subfamily A, polypeptide 1
2.06	0.004142	Peroxisomal biogenesis factor 5
2.06	0.033184	Fem-1 homolog c (C. elegans)

Table 2. Down-regulated genes in MOZ-TIF2 vs MOZ.

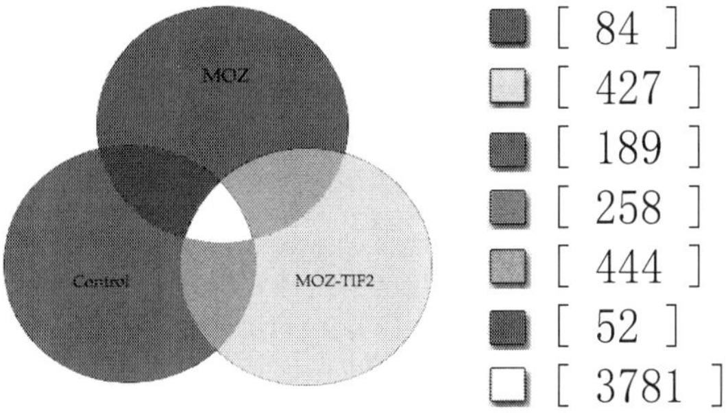

Fig. 7. The Venn diagram of signature gene expression among pcDNA3, MOZ, and MOZ-TIF2. The positive expressed genes were picked up as described in Materials and Methods. The number in brackets indicates the signature genes.

4. Discussion

In order to gain understanding of the function of the MOZ-TIF2 fusion protein we used the yeast two-hybrid system to screen a human bone marrow cDNA library and identified two proteins, CAF-1A and ASF1B, that interacted with the MOZ partner of MOZ-TIF2. The CAF-1A is the largest subunit of CAF-1 which is responsible for bringing histones H3 and H4 to newly synthesized DNA to constitute a nucleosome during DNA replication and DNA repair (Moggs et al., 2000; Shibahara and Stillman, 1999; Smith and Stillman, 1989). CAF-1 controls S-phase progression in euchromatic DNA replication (Klapholz et al., 2009). During chromatin assembly CAF-1 is localized at the replication loci through the association with the proliferation cell nuclear antigen (PCNA), interacting with the N-terminal PCNA binding motif in the CAF-1A. CAF-1 has also been shown to have a role in transcription regulation and epigenetic control of gene expression by interacting with methyl-CpG binding protein and by contributing non-methylation dependent gene silencing (Reese et al., 2003; Sarraf and Stancheva, 2004; Tchenio et al., 2001). A dominant-negative mutant of CAF-1A arrests cell cycle in S-phase (Ye et al., 2003). The loss of CAF-1 is lethal in human cells and increases the sensitivity of cells to UV and other DNA damaging reagents (Game and Kaufman, 1999; Nabatiyan and Krude, 2004). In addition, CAF-1 has been suggested as a clinical marker to distinguish quiescent from proliferating cells (Polo et al., 2004). ASF1B, the other MOZ-TIF2 interacting protein identified, is one of two human ASF1 proteins and participates in chromatin assembly by interacting with the p60 unit of CAF-1 (Mello et al., 2002). The function of ASF1 overlaps with CAF-1 but contributes mainly to chromatin-mediated gene silencing (Meijsing and Ehrenhofer-Murray, 2001; Mello et al., 2002; Osada et al., 2001). In the process of nucleosome formation during DNA replication, ASF1 synergizes functionally with CAF-1 by binding histone H3/H4 and delivers histone H3 and H4 dimers to CAF-1 (Tyler et al., 1999; Tyler et al., 2001). As with CAF-1 mutations, mutations in ASF1 raise the sensitivity of cells to DNA damage (Daganzo et al., 2003; Emili et al., 2001; Le et al., 1997). In yeast, the absence of ASF1 leads to enhanced genetic instability and sister chromatid exchange (Prado et al., 2004). Recent study revealed that the expression of ASF1B, like CAF-1A, was proliferation-dependent (Corpet et al., 2011). Both CAF-1 and ASF1 are important in maintaining genetic stability and hence mutations or aberrant expression in either may contribute to carcinogenesis.

Our initial results demonstrated that the MOZ portion of the MOZ-TIF2 fusion protein interacted with the human CAF-1A and ASF1B. These associations are consistent with previous findings that a MYST family member in yeast, SAS (something about silencing) protein, interacts with Cac1, a yeast homologue of human CAF-1A, and yeast ASF1 and that the interaction contributed to the silencing of the ribosomal DNA locus (Meijsing and Ehrenhofer-Murray, 2001). However, in our experiments with the yeast two-hybrid system, the association between the MOZ-1/759 fragment and CAF-1A was stronger than the interaction of the MOZ-1/759 fragment and ASF1B. The clones of MOZ-1/759 and CAF-1A grew in both -His and -Ade selection media while the clones of MOZ-1/759 and ASF1B grew only in the -His medium. These results suggest that the intensity of interaction of the MOZ fragment with each chaperone is different and the interactions may involve different domains of MOZ. With the GST pull-down assays, we were able to verify the physical interactions using purified proteins and to begin probing the regions of MOZ involved in the interactions. Our results demonstrated that CAF-1A bound primarily to the N-terminus of MOZ (MOZ-1/313) while ASF1B bound to the domain containing C2HC motif and acetyl-CoA binding region (MOZ-488/703). To exclude possible indirect interactions caused by using a mammalian transcription/ translation system, the pull-down assay was also conducted using an E. coli translation system (EcoPro™ T7 System, EMD Biosciences, Novagen, San Diego, CA) with the same interactions being seen again (data not shown). The binding of CAF-1A and ASF1B to two distinct regions within the MOZ fragment involved in the MOZ-TIF2 fusion protein suggests that MOZ-TIF2 positively influences participation in chromatin assembly.

The experiments reported here also begin to shed some light on aberrant function of the MOZ-TIF2 fusion protein by comparing semi-quantatively the strength of association of CAF-1A and ASF1B with MOZ and MOZ-TIF2. In the co-immunopreciptiation and S-tagged pull down experiments, CAF-1A appeared to interact more strongly with MOZ than MOZ-TIF2. These observations were confirmed by the increased co-localization seen in confocal microscopy of the co-transfected cells at interphase. The converse was seen in the interactions of ASF1B with an apparent greater intensity of interaction of ASF1B with MOZ-TIF2 than MOZ alone. Again, this interaction was confirmed in pre-extracted HEK293 cells. It seems that MOZ-TIF2 fusion protein changed the binding priorities of MOZ. These differences may occur because of the necessity of appropriate folding or other higher order structural changes in the full-length MOZ, which are obviated in the fusion protein. In addition, we noticed that the localization of MOZ and CAF-1A was altered in mitotic cells, suggesting that the function of interactions in chromatin assembly and modification depend on cell division cycle. Previously, CAF-1 has been observed to disassociate from chromosomes during the M phase and to be inactivated in mitosis (Marheineke and Krude, 1998). However, we have seen the binding of CAF-1A to the spindle-like chromosome during the metaphase and anaphase in immune-stained Hela cells. It is not clear if the altered association of CAF-1A with chromosome indicates a physiological process during the mitosis or is the artificial results either of fixation and stain process or the limitation of the antibody. A further investigation is necessary to determine the dynamic change of the association.

Using stably transfected U937 cells, we were able to find MOZ-TIF2-correlated changes in the global expression profile of genes and identify a signature-expression profile for MOZ-TIF2. However, as MOZ and TIF2 function as transcription co-factors and as CAF-1 and ASF1 are regulators of global transcription the altered gene expression by MOZ-TIF2 cannot be ascribed to the interaction of MOZ-TIF2 with CAF-1A and ASF1B alone. Interestingly, inspite of 427 expressed signature genes of MOZ-TIF2, only 62 genes were found with over two-fold significant change between MOZ-TIF2 and MOZ, suggesting that differences in expression level between MOZ and MOZ-TIF2 of most most signature genes signature genes could be relatively small.

We are currently examining the hypothesis that the association of MOZ-TIF2 with chromatin assembly factors affects the nucleosome structure and/or histone modification such that histone acetylation status would contribute to leukemogenesis. This hypothesis assumes that the MOZ-TIF2 fusion protein may alter constitution of the chromatin assembly factor complex and then change global gene expression. A possible target for this type of altered function would be that the fusion protein could alter the recruitment of CBP to the complex via LXXLL motifs in TIF2 portion (Voegel et al., 1998; Yin et al., 2007).

5. Conclusions

We demonstrate that both MOZ and MOZ-TIF2 interacts with ASF1B via its MYST domain and interacts with CAF-1A via its zinc finger domain. MOZ and MOZ–TIF2 co-localize with CAF-1A and ASF1B in interphase nuclei. MOZ-TIF2, compared to MOZ, preferentially binds to ASF1B rather than to CAF-1A. MOZ-TIF2 interferes with the function of wild type MOZ and alters global gene expression in U937 cells.

6. References

Aikawa, Y., Katsumoto, T., Zhang, P., Shima, H., Shino, M., Terui, K., Ito, E., Ohno, H., Stanley, E. R., Singh, H., *et al.* (2010). PU.1-mediated upregulation of CSF1R is crucial for leukemia stem cell potential induced by MOZ-TIF2. Nat Med *16*, 580-585, 581p following 585.

Bristow, C. A., and Shore, P. (2003). Transcriptional regulation of the human MIP-1alpha promoter by RUNX1 and MOZ. Nucleic Acids Res *31*, 2735-2744.

Champagne, N., Bertos, N. R., Pelletier, N., Wang, A. H., Vezmar, M., Yang, Y., Heng, H. H., and Yang, X. J. (1999). Identification of a human histone acetyltransferase related to monocytic leukemia zinc finger protein. J Biol Chem *274*, 28528-28536.

Champagne, N., Pelletier, N., and Yang, X. J. (2001). The monocytic leukemia zinc finger protein MOZ is a histone acetyltransferase. Oncogene *20*, 404-409.

Corpet, A., De Koning, L., Toedling, J., Savignoni, A., Berger, F., Lemaitre, C., O'Sullivan, R. J., Karlseder, J., Barillot, E., Asselain, B., *et al.* (2011). Asf1b, the necessary Asf1 isoform for proliferation, is predictive of outcome in breast cancer. Embo J *30*, 480-493.

Daganzo, S. M., Erzberger, J. P., Lam, W. M., Skordalakes, E., Zhang, R., Franco, A. A., Brill, S. J., Adams, P. D., Berger, J. M., and Kaufman, P. D. (2003). Structure and function of the conserved core of histone deposition protein Asf1. Curr Biol *13*, 2148-2158.

Deguchi, K., Ayton, P. M., Carapeti, M., Kutok, J. L., Snyder, C. S., Williams, I. R., Cross, N. C., Glass, C. K., Cleary, M. L., and Gilliland, D. G. (2003). MOZ-TIF2-induced acute myeloid leukemia requires the MOZ nucleosome binding motif and TIF2-mediated recruitment of CBP. Cancer Cell *3*, 259-271.

Emili, A., Schieltz, D. M., Yates, J. R., 3rd, and Hartwell, L. H. (2001). Dynamic interaction of DNA damage checkpoint protein Rad53 with chromatin assembly factor Asf1. Mol Cell *7*, 13-20.

Esteyries, S., Perot, C., Adelaide, J., Imbert, M., Lagarde, A., Pautas, C., Olschwang, S., Birnbaum, D., Chaffanet, M., and Mozziconacci, M. J. (2008). NCOA3, a new fusion partner for MOZ/MYST3 in M5 acute myeloid leukemia. Leukemia *22*, 663-665.

Game, J. C., and Kaufman, P. D. (1999). Role of Saccharomyces cerevisiae chromatin assembly factor-I in repair of ultraviolet radiation damage in vivo. Genetics *151*, 485-497.

Huntly, B. J., Shigematsu, H., Deguchi, K., Lee, B. H., Mizuno, S., Duclos, N., Rowan, R., Amaral, S., Curley, D., Williams, I. R., *et al.* (2004). MOZ-TIF2, but not BCR-ABL, confers properties of leukemic stem cells to committed murine hematopoietic progenitors. Cancer Cell *6*, 587-596.

James, P., Halladay, J., and Craig, E. A. (1996). Genomic libraries and a host strain designed for highly efficient two-hybrid selection in yeast. Genetics *144*, 1425-1436.

Katsumoto, T., Aikawa, Y., Iwama, A., Ueda, S., Ichikawa, H., Ochiya, T., and Kitabayashi, I. (2006). MOZ is essential for maintenance of hematopoietic stem cells. Genes Dev *20*, 1321-1330.

Katsumoto, T., Yoshida, N., and Kitabayashi, I. (2008). Roles of the histone acetyltransferase monocytic leukemia zinc finger protein in normal and malignant hematopoiesis. Cancer Sci *99*, 1523-1527.

Kitabayashi, I., Aikawa, Y., Nguyen, L. A., Yokoyama, A., and Ohki, M. (2001). Activation of AML1-mediated transcription by MOZ and inhibition by the MOZ-CBP fusion protein. Embo J *20*, 7184-7196.

Klapholz, B., Dietrich, B. H., Schaffner, C., Heredia, F., Quivy, J. P., Almouzni, G., and Dostatni, N. (2009). CAF-1 is required for efficient replication of euchromatic DNA in Drosophila larval endocycling cells. Chromosoma *118*, 235-248.

Le, S., Davis, C., Konopka, J. B., and Sternglanz, R. (1997). Two new S-phase-specific genes from Saccharomyces cerevisiae. Yeast *13*, 1029-1042.

Liang, J., Prouty, L., Williams, B. J., Dayton, M. A., and Blanchard, K. L. (1998). Acute mixed lineage leukemia with an inv(8)(p11q13) resulting in fusion of the genes for MOZ and TIF2. Blood *92*, 2118-2122.

Marheineke, K., and Krude, T. (1998). Nucleosome assembly activity and intracellular localization of human CAF-1 changes during the cell division cycle. J Biol Chem *273*, 15279-15286.

Meijsing, S. H., and Ehrenhofer-Murray, A. E. (2001). The silencing complex SAS-I links histone acetylation to the assembly of repressed chromatin by CAF-I and Asf1 in Saccharomyces cerevisiae. Genes Dev *15*, 3169-3182.

Mello, J. A., Sillje, H. H., Roche, D. M., Kirschner, D. B., Nigg, E. A., and Almouzni, G. (2002). Human Asf1 and CAF-1 interact and synergize in a repair-coupled nucleosome assembly pathway. EMBO Rep *3*, 329-334.

Moggs, J. G., Grandi, P., Quivy, J. P., Jonsson, Z. O., Hubscher, U., Becker, P. B., and Almouzni, G. (2000). A CAF-1-PCNA-mediated chromatin assembly pathway triggered by sensing DNA damage. Mol Cell Biol *20*, 1206-1218.

Nabatiyan, A., and Krude, T. (2004). Silencing of chromatin assembly factor 1 in human cells leads to cell death and loss of chromatin assembly during DNA synthesis. Mol Cell Biol *24*, 2853-2862.

Osada, S., Sutton, A., Muster, N., Brown, C. E., Yates, J. R., 3rd, Sternglanz, R., and Workman, J. L. (2001). The yeast SAS (something about silencing) protein complex contains a MYST-type putative acetyltransferase and functions with chromatin assembly factor ASF1. Genes Dev *15*, 3155-3168.

Perez-Campo, F. M., Borrow, J., Kouskoff, V., and Lacaud, G. (2009). The histone acetyl transferase activity of monocytic leukemia zinc finger is critical for the proliferation of hematopoietic precursors. Blood *113*, 4866-4874.

Polo, S. E., Theocharis, S. E., Klijanienko, J., Savignoni, A., Asselain, B., Vielh, P., and Almouzni, G. (2004). Chromatin assembly factor-1, a marker of clinical value to distinguish quiescent from proliferating cells. Cancer Res *64*, 2371-2381.

Prado, F., Cortes-Ledesma, F., and Aguilera, A. (2004). The absence of the yeast chromatin assembly factor Asf1 increases genomic instability and sister chromatid exchange. EMBO Rep *5*, 497-502.

Reese, B. E., Bachman, K. E., Baylin, S. B., and Rountree, M. R. (2003). The methyl-CpG binding protein MBD1 interacts with the p150 subunit of chromatin assembly factor 1. Mol Cell Biol *23*, 3226-3236.

Sarraf, S. A., and Stancheva, I. (2004). Methyl-CpG binding protein MBD1 couples histone H3 methylation at lysine 9 by SETDB1 to DNA replication and chromatin assembly. Mol Cell *15*, 595-605.

Shibahara, K., and Stillman, B. (1999). Replication-dependent marking of DNA by PCNA facilitates CAF-1-coupled inheritance of chromatin. Cell *96*, 575-585.

Smith, S., and Stillman, B. (1989). Purification and characterization of CAF-I, a human cell factor required for chromatin assembly during DNA replication in vitro. Cell *58*, 15-25.

Tchenio, T., Casella, J. F., and Heidmann, T. (2001). A truncated form of the human CAF-1 p150 subunit impairs the maintenance of transcriptional gene silencing in mammalian cells. Mol Cell Biol *21*, 1953-1961.

Troke, P. J., Kindle, K. B., Collins, H. M., and Heery, D. M. (2006). MOZ fusion proteins in acute myeloid leukaemia. Biochem Soc Symp, 23-39.

Tyler, J. K., Adams, C. R., Chen, S. R., Kobayashi, R., Kamakaka, R. T., and Kadonaga, J. T. (1999). The RCAF complex mediates chromatin assembly during DNA replication and repair. Nature *402*, 555-560.

Tyler, J. K., Collins, K. A., Prasad-Sinha, J., Amiott, E., Bulger, M., Harte, P. J., Kobayashi, R., and Kadonaga, J. T. (2001). Interaction between the Drosophila CAF-1 and ASF1 chromatin assembly factors. Mol Cell Biol *21*, 6574-6584.

Voegel, J. J., Heine, M. J., Tini, M., Vivat, V., Chambon, P., and Gronemeyer, H. (1998). The coactivator TIF2 contains three nuclear receptor-binding motifs and mediates transactivation through CBP binding-dependent and -independent pathways. Embo J *17*, 507-519.

Ye, X., Franco, A. A., Santos, H., Nelson, D. M., Kaufman, P. D., and Adams, P. D. (2003). Defective S phase chromatin assembly causes DNA damage, activation of the S phase checkpoint, and S phase arrest. Mol Cell *11*, 341-351.

Yin, H., Glass, J., and Blanchard, K. L. (2007). MOZ-TIF2 repression of nuclear receptor-mediated transcription requires multiple domains in MOZ and in the CID domain of TIF2. Mol Cancer *6*, 51.

Zabaronick, S. R., and Tyler, J. K. (2005). The histone chaperone anti-silencing function 1 is a global regulator of transcription independent of passage through S phase. Mol Cell Biol *25*, 652-660.

2

Autophagy-Mediated Defense Response of Mouse Mesenchymal Stromal Cells (MSCs) to Challenge with *Escherichia coli*

N.V. Gorbunov[1,*], B.R. Garrison[1], M. Zhai[1], D.P. McDaniel[2], G.D. Ledney[3], T.B. Elliott[3] and J.G. Kiang[3,*]
[1]*The Henry M. Jackson Foundation for the Advancement of Military Medicine, Inc.*
[2]*The Department of Microbiology and Immunology, School of Medicine,*
[3]*Radiation Combined Injury Program, Armed Forces Radiobiology Research Institute, Uniformed Services University of the Health Sciences, Bethesda, Maryland, USA*

1. Introduction

Symbiotic microorganisms are spatially separated from their animal host, e.g., in the intestine and skin, in a manner enabling nutrient metabolism as well as evolutionary development of protective physiologic features in the host such as innate and adaptive immunity, immune tolerance, and function of tissue barriers (1,2). The major interface barrier between the microbiota and host tissue is constituted by epithelium, reticuloendothelial tissue, and mucosa-associated lymphoid tissue (MALT) (2,3).

Traumatic damage to skin and the internal epithelium in soft tissues can cause infections that account for 7% to 10% of hospitalizations in the United States (4). Moreover, wound infections and sepsis are an increasing cause of death in severely ill patients, especially those with immunosupression due to exposure to cytotoxic agents and chronic inflammation (4). It is well accepted that breakdown of the host-bacterial symbiotic homeostasis and associated infections are the major consequences of impairment of the "first line" of anti-microbial defense barriers such as the mucosal layers, MALT and reticuloendothelium (1-3). Under these impairment conditions of particular interest then is the role of sub-mucosal structures, such as connective tissue stroma, in the innate defense compensatory responses to infections.

The mesenchymal connective tissue of different origins is a major source of multipotent mesenchymal stromal cells (i.e., colony-forming-unit fibroblasts) (5, 6). Recent discovery of immunomodulatory function of mesenchymal stromal cells (MSCs) suggests that they are essential constituents that control inflammatory responses (6-7).

Recent *in vivo* experiments demonstrate promising results of MSC transfusion for treatment of acute sepsis and penetrating wounds (7-9). The molecular mechanisms underlying MSC

* Corresponding Authors

action in septic conditions are currently under investigation. It is known to date that (i) Gram-negative bacteria can induce an inflammatory response in MSCs *via* cascades of Toll-like receptor (type 4) and the nucleotide-binding oligomerization domain-containing protein 2 (NOD2) complexes recognizing the conserved pathogen-associated molecular patterns; (ii) activated MSCs can modulate the septic response of resident myeloid cells; and (iii) activated MSCs can directly suppress bacterial proliferation by releasing antimicrobial factors (10, 11).

Considering all of the above factors including the fact that MSCs are ubiquitously present in the sub-mucosal structures and conjunctive tissue, one would expect involvement of these cells in formation of antibacterial barriers and host-microbiota homeostasis. From this perspective our attention was attracted by the phagocytic properties of mesenchymal fibroblastic stromal cells documented in an early period of their investigation (5, 12). The phagocytosis mechanism is closely and synchronously connected with the cellular mechanisms of biodegradation mediated by the macroautophagy-lysosomal (autolysosomal) system (13-15). The last one decomposes proteins and organelles as well as bacteria and viruses inside cells and, therefore, is considered as a part of the innate defense mechanism (13- 15).

Macroautophagy (hereafter referred to as autophagy) is a catabolic process of bulk lysosomal degradation of cell constituents and phagocytized particles (16). Autophagy dynamics in mammalian cells are well described in recent reviews (14, 17-20). Thus, it was proposed that autophagy is initiated by the formation of the phagophore, followed by a series of steps, including the elongation and expansion of the phagophore, closure and completion of a double-membrane autophagosome (which surrounds a portion of the cytoplasm), autophagosome maturation through docking and fusion with an endosome (the product of fusion is defined as an amphisome) and/or lysosome (the product of fusion is defined as an autolysosome), breakdown and degradation of the autophagosome inner membrane and cargo through acid hydrolases inside the autolysosome, and recycling of the resulting macromolecules through permeases (14). These processes, along with the drastic membrane traffic, are mediated by factors known as autophagy-related proteins (i.e., ATG-proteins) and the lysosome-associated membrane proteins (LAMPs) that are conserved in evolution (21). The autophagic pathway is complex. To date there are over 30 ATG genes identified in mammalian cells as regulators of various steps of autophagy, e.g., cargo recognition, autophagosome formation, etc. (14, 22). The core molecular machinery is comprised of (i) components of signaling cascades, such as the ULK1 and ULK2 complexes and class III PtdIns3K complexes, (ii) autophagy membrane processing components, such as mammalian Atg9 (mAtg9) that contributes to the delivery of membrane to the autophagosome as it forms, and two conjugation systems: the microtubule-associated protein 1 (MAP1) light chain 3 (i.e., LC3) and the Atg12–Atg5–Atg16L complex. The two conjugation systems are proposed to function during elongation and expansion of the phagophore membrane (14, 19, 22, 23). A conservative estimate of the autophagy network counts over 400 proteins, which, besides the ATG-proteins, also include stress-response factors, cargo adaptors, and chaperones such as p62/SQSTM1 and heat shock protein 70 (HSP70) (15, 19, 22, 24, 26-28).

Autophagy is considered as a cytoprotective process leading to tissue remodeling, recovery, and rejuvenation. However, under circumstances leading to mis-regulation of the autolysosomal pathway, autophagy can eventually cause cell death, either as a precursor of apoptosis in apoptosis-sensitive cells or as a result of destructive self-digestion (29).

Based on this information we hypothesized that challenge of MSCs with *Escherichia coli* (*E. coli*) can induce a complex process where bacterial phagocytosis is accompanied by activation of autolysosomal pathway and stress-adaptive responses in MSCs. The objective of this current chapter is to provide evidence of this hypothesis.

2. Hypothesis test: Experimental procedures and technical approach

2.1 Bone marrow stromal cells

Bone marrow stromal cells were obtained from 3- to 4-month-old B6D2F1/J female mice using a protocol adapted from STEMCELL Technologies, Inc., and were expanded and cultivated in hypoxic conditions (5% O_2, 10% CO_2, 85% N_2) for approximately 30 days in MESENCULT medium (STEMCELL Technologies, Inc.) in the presence of antibiotics. Phenotype, proliferative activity, and colony-forming ability of the cells were analyzed by flow cytometry and immunofluorescence imaging using positive markers for mesenchymal stromal cells: CD44, CD105, and Sca1. The results of these analyses showed that the cultivated cells displayed properties of mesenchymal stromal clonogenic fibroblasts.

The experiments were performed in a facility accredited by the Association for the Assessment and Accreditation of Laboratory Animal Care-International (AAALAC-I). All animals used in this study received humane care in compliance with the Animal Welfare Act and other federal statutes and regulations relating to animals and experiments involving animals and adhered to principles stated in the Guide for the Care and Use of Laboratory Animals, NRC Publication, 1996 edition.

2.2 Challenge of MSCs with *Escherichia coli* bacteria

MSC cultures of approximate 80% confluency were challenged with proliferating *E. coli* ($1x10^7$ microorganisms/ml) for 1-5 h in antibiotic-free media. For assessment of the cellular alteration $\geq$ 5 h the incubation medium was replaced with fresh medium containing penicillin and streptavidin antibiotics. Bacteria-cell interaction was monitored with time-lapse microscopy using DIC imaging of MSCs and fluorescence imaging of *E. coli* labeled with PSVue® 480, a fluorescent cell tracking reagent (www.mtarget.com). At the end of the experiments the cells were either (i) harvested, washed, and lysed for qRT-PCR and immunoblot analyses, (ii) fixed for transmission electron microscopy and fluorescence confocal imaging, or (iii) used live for imaging of Annexin V reactivity, dihydrorhodamine 123, a sensitive indicator of peroxynitrite reactivity, and colony formation. With this protocol the cells were tested for (i) phagocytic activity; (ii) autolysosomal activity; (iii) production of reactive oxygen (ROS) and nitrogen species, (iii) stress responses to *E. coli*; (iv) genomic DNA damage and pro-apoptotic alterations; and (v) colony-forming ability. The results of observations indicated that challenge with *E. coli* did not diminish viability and colony forming ability of the cells under the selected conditions (Fig.1). Stimulation of MSCs with *E. coli* resulted in expression of the proinflammatory genes, IL-1α, IL-1β, IL-6, and iNOS, as determined with qRT-PCR analysis.

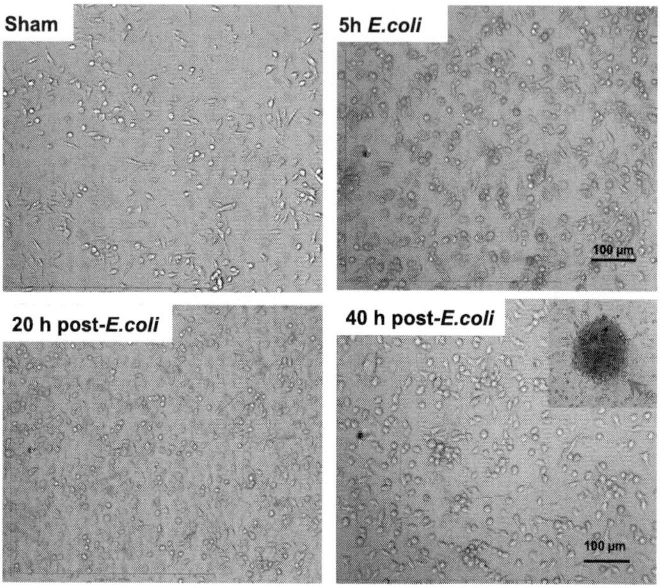

Conditions: MSCs were incubated with ~1x10⁷/ml *E. coli* for 5 h in medium (without antibiotics). After 5 h the medium was replaced with fresh medium (with antibiotics) and MSCs were incubated for another 40 h. Inset: formation of colonies (red arrowhead) occurred at 72 h post-exposure to *E. coli*.

Fig. 1. Bright field microscopy of MSCs challenged with *E. coli*. Images presented in the panels are MSCs at different time-points following exposure of MSCs to *E. coli*.

2.3 Analysis of the cell proteins

Proteins from MSCs were extracted in accordance with the protocol described previously (30). The aliquoted proteins (20 µg total protein per gel well) were separated on SDS-polyacrylamide slab gels (NuPAGE 4-12% Bis-Tris; Invitrogen, Carlsbad, CA). After electrophoresis, proteins were blotted onto a PDVF membrane and the blots were incubated with antibodies (1 µg/ml) raised against MAP LC3, Lamp-1, p62/SQSTM1, p65(NFκB), Nrf2, HSP70, iNOS, and actin (Abcam, Santa Cruz Biotechnology Inc., LifeSpan Biosciences, Inc., eBiosciences) followed by incubation with species-specific IgG peroxidase conjugate. IgG amounts did not alter after radiation. IgG, therefore, was used as a control for protein loading.

2.4 Immunofluorescent staining and image analysis

MSCs (5 specimens per group) were fixed in 2% paraformaldehyde and analyzed with fluorescence confocal microscopy following labeling (30). Normal donkey serum and antibody were diluted in phosphate-buffered saline (PBS) containing 0.5% BSA and 0.15% glycine. Any nonspecific binding was blocked by incubating the samples with purified normal donkey serum (Santa Cruz Biotechnology, Inc., Santa Cruz, CA) diluted 1:20. Primary antibodies were raised against MAP LC3, Lamp-1, p62/SQSTM1, p65(NFκB),

Nrf2, Tom 20, and iNOS. That was followed by incubation with secondary fluorochrome-conjugated antibody and/or streptavidin-AlexaFluor 610 conjugate (Molecular Probes, Inc., Eugene OR), and with Hoechst 33342 (Molecular Probes, Inc., Eugene OR) diluted 1:3000. Secondary antibodies used were AlexaFluor 488 and AlexaFluor 594 conjugated donkey IgG (Molecular Probes Inc., Eugene OR). Negative controls for nonspecific binding included normal goat serum without primary antibody or with secondary antibody alone. Five confocal fluorescence and DIC images of crypts (per specimen) were captured with a Zeiss LSM 7100 confocal microscope. The immunofluorescence image analysis was conducted as described previously (30).

2.5 Transmission Electron Microscopy (TEM)

MSCs in cultures were fixed in 4% formaldehyde and 4% glutaraldehyde in PBS overnight, post-fixed in 2% osmium tetroxide in PBS, dehydrated in a graduated series of ethanol solutions, and embedded in Spurr's epoxy resin. Blocks were processed as described previously (30). The sections of embedded specimens were analyzed with a Philips CM100 electron microscope.

2.6 RNA isolation and qRT-PCR

Total cellular RNA was isolated from MSC pellets using the Qiagen RNeasy miniprep kit, quantified by measuring the absorbance at 260nm on a Nanodrop, and qualified by electrophoresis on a 1.2% agarose gel. cDNA was synthesized using Superscript II (Invitrogen) and qRT-PCR was performed using SYBR Green iQ Supermix (Bio-Rad), each according to the manufacturers' instructions. The quality of qRT-PCR data were verified by melt curve analysis, efficiency determination, agarose gel electrophoresis, and sequencing. Relative gene expression was calculated by the method of Pfaffl using the formula $2^{-\Delta\Delta Ct}$(31).

2.7 Statistical analysis

Statistical significance was determined using one-way ANOVA followed by post-hoc analysis with pair-wise comparison by Tukey-Kramer test. Significance is reported at a level of $p<0.05$.

3. Response of MSCs to challenge with *E. coli*

3.1 Phagocytosis and autolysosomal degradation of *E. coli* bacteria by MSCs

TEM images presented in Fig. 2 show different stages of cell-bacterium interaction. The uptake of microorganisms occurred in at least two independent events. The first event encompassed engulfing and taking in particles by the cell membrane extrusions (Fig. 2A1). The second event was tethering and "zipping" of adhered particles by the cell plasma membrane (Fig. 2A2 - 2A5). The time-lapse fluorescence microscopy observation indicated that these events proceeded quickly and the uptake process required a few minutes (not shown). Thereafter, a significant amount of bacteria in MSCs was observed within 1 h of co-incubation of the cells. The phagocytized bacteria were subjected to autolysosomal degradation (Fig 2B). Formation of the double-membrane autophagosomes, which incorporated bacteria, was observable in MSCs at 3 h of co-incubation and during a further period of observation. Fusion of autophagosomes with lysosomes also occurred at this period. Fragmentation of bacterial constituents was observed at 5 h of co-incubation and appearance of bacterial "ghosts" at 24 h (Fig. 2B).

Various cells eliminate bacterial microorganisms by autophagy, and this elimination is in many cases crucial for host resistance to bacterial translocation. Although autophagy is a non-selective degradation process, autophagosomes do not form randomly in the cytoplasm, but rather sequester the bacteria selectively (32, 33). Therefore, autophagosomes that engulf microbes are sometimes much larger than those formed during degradation of cellular organelles, suggesting that the elongation step of the autophagosome membrane is involved in bacteria-surrounding autophagy (33). The mechanism underlying selective induction of autophagy at the site of microbe phagocytosis remains unknown. However, it is likely mediated by pattern recognition receptors, stress-response elements, and adaptor proteins, e.g., p62/SQSTM1, which target bacteria and ultimately recruit factors essential for formation of autophagosomes (13,14, 33, 34).

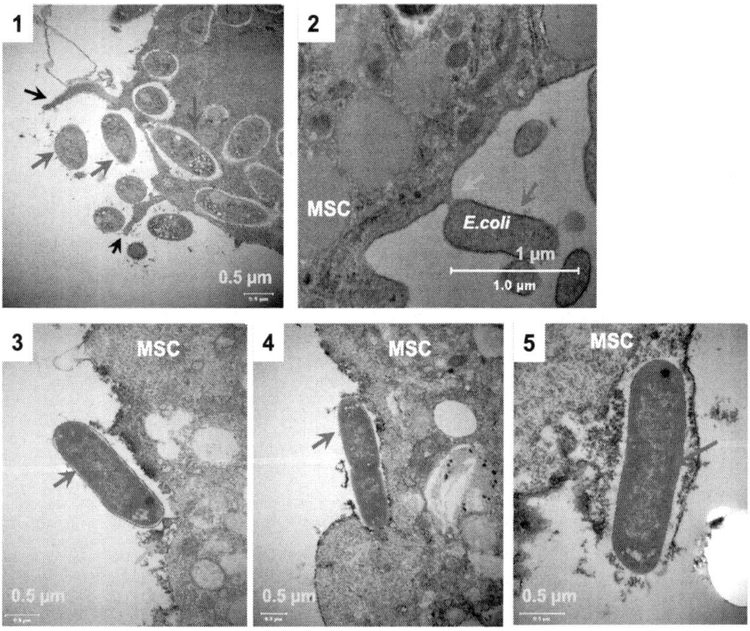

A

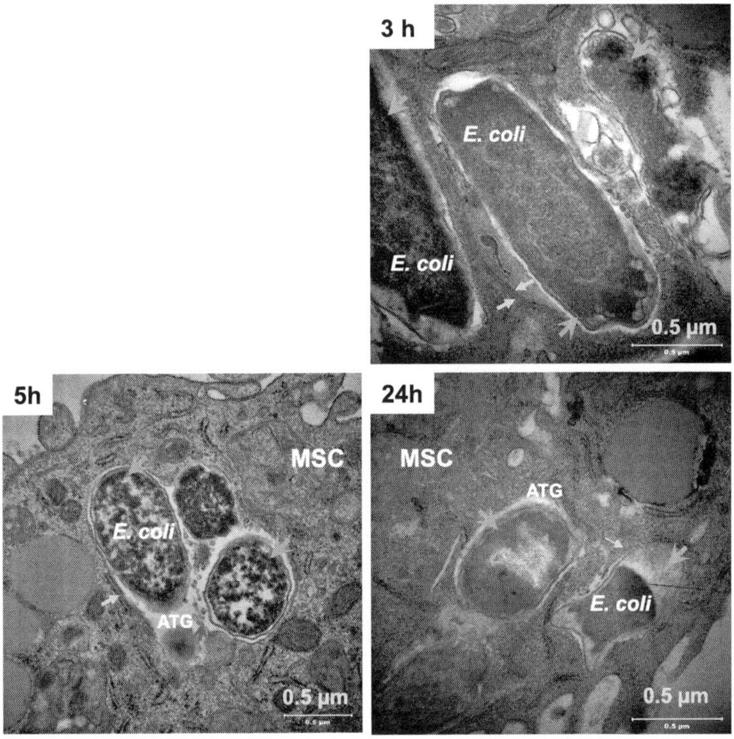

B

Conditions: MSCs were incubated with ~1×10^7/ml *E. coli* either for 3 h or 5 h in MesenCult Medium (without antibiotics). After 5 h the medium was replaced with fresh medium (with antibiotics) and MSCs were incubated for another 19 h.

Fig. 2. Transmission electron micrographs (TEM) of *E. coli* phagocytosis by MSCs and autolysosomal degradation of phagocytized bacteria.
A) Panel A1: Engulfing and up-take of bacteria (red arrows) by the cell plasma membrane extrusions (black arrows). Panels A2-A5: Tethering and zipping (green arrows) and up-take of bacteria (red arrows) by the cell plasma membrane. Specimens were fixed at 3 h co-incubation of MSCs with bacteria.
B) Autolysosomal degradation of phagocytized bacteria at different time-points after exposure of MSCs to *E. coli* (green arrows). Autophagosome (ATG) membranes are indicated with yellow arrows. Lysosome fusion with autophagosomes is indicated with red arrows.

The results of TEM were corroborated by the data obtained with immunoblotting and immunofluorescence confocal imaging of autophagy MAP (LC3) protein, lysosomal LAMP1 and the ubiquitin-associated target adaptor p62. A key step in the autophagosome biogenesis is the conversion of light-chain protein 3 type I (LC3-I, also known as ubiqitin-like protein, Atg8) to type II (LC3-II). The conversion occurs via the cleavage of the LC3-I carboxyl terminus by a redox-sensitive Atg4 cysteine protease. The subsequent binding of the modified LC3-I to phosphatidylethanolamine, i.e., process of lipidation of LC3-I, on the isolation membrane, as it forms, is mediated by E-1- and E-2-like enzymes Atg7 and Atg3 (14). Therefore, conversion of LC3-I to LC3-II and formation of LC3-positive vesicles are considered to be a marker of activation of autophagy (14). A growing body of evidence suggests involvement of chaperone HSP70 in regulation of LC3-translocation. The results of immunoblot analysis of the proteins indicated an increase in the LC3-I to LC3-II - transition in the *E. coli* -challenged MSCs (Fig. 3).

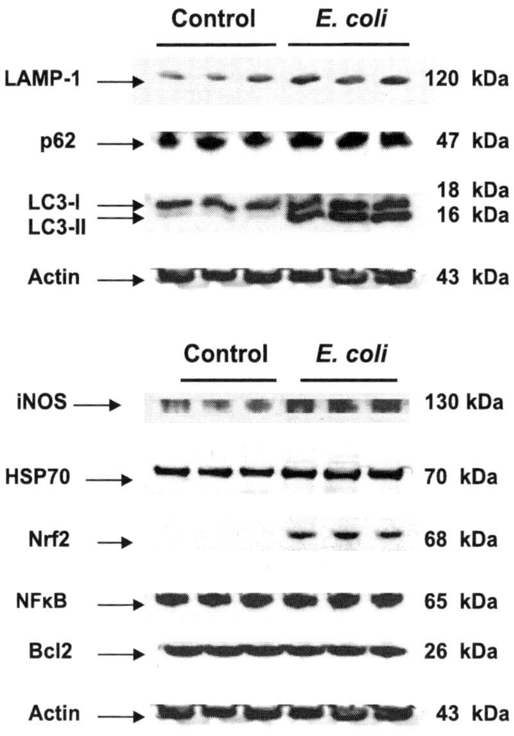

Conditions: MSCs were incubated with ~1x10^7/ml *E. coli* for 3 h in MesenCult Medium (without antibiotics). After 3 h the medium was replaced with fresh medium (with antibiotics) and MSCs were further incubated for another 21 h.

Fig. 3. Immunoblotting analysis of LC3, LAMP1 autolysosomal proteins, p62 adaptor protein, and stress-response elements: NF-κB(p65), Nrf2, HSP70 in MSCs challenged with *E. coli.*

The images presented in Fig. 4A indicate an increase of formation of LC3-positive vesicles in MSCs challenged with *E. coli*. The LC3 immunoreactivity co-localized with immunoreactivity to LAMP1, a marker of lysosomes, indicating presence of fusion of autophagosomes with lysosomes, i.e., formation of autolysosomes (Fig. 4A). This effect

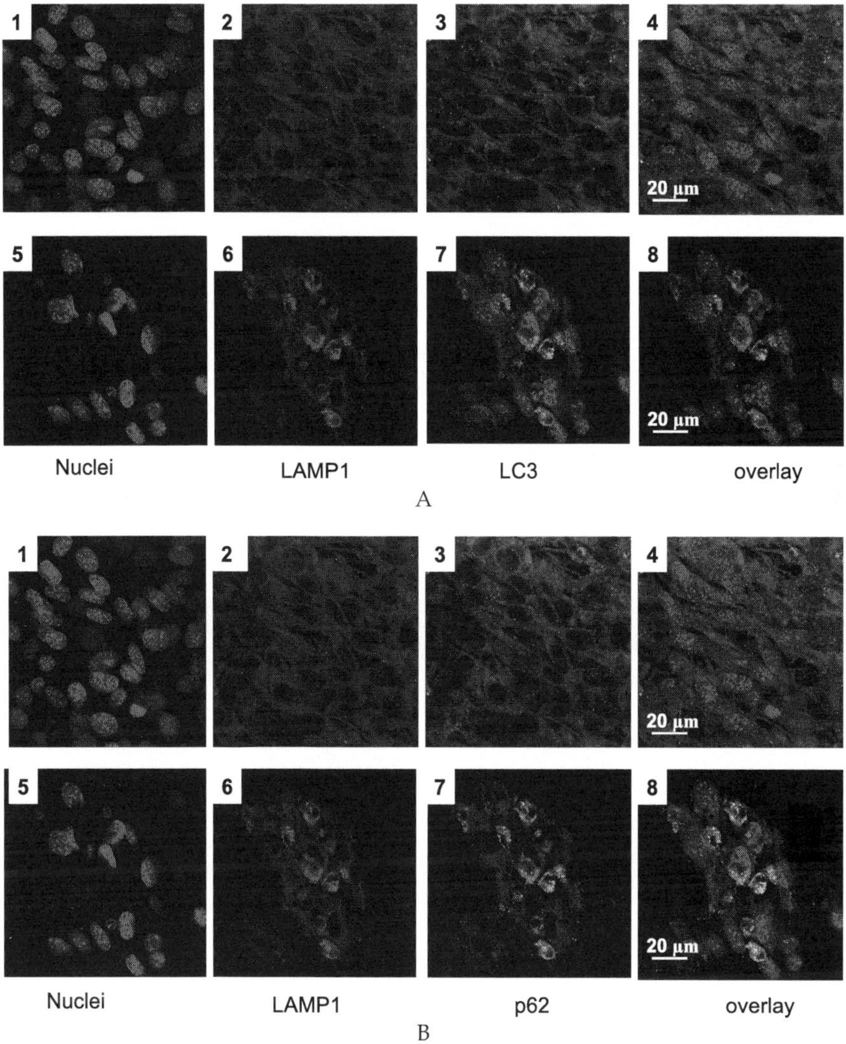

Conditions: MSCs were incubated with ~$1x10^7$/ml *E. coli* for 3 h in MesenCult Medium (without antibiotics). After 3 h the medium was replaced with fresh medium (with antibiotics) and MSCs were further incubated for 21 h. Projections of LAMP1 protein (red channel) are shown in panels A2, A6, B2, and B6. Projections of LC3 protein (green channel) are shown in panels A3 and A7. Projections of p62 protein (green channel) are shown in panels B3 and B7. Counterstaining of nuclei was with Hoechst 33342 (blue channel). Panels A4, A8, B4, and B8 are overlay of signals acquired in the red, green, and blue channels. The confocal images were taken with pinhole setup to obtain 0.5 μm Z-sections.

Fig. 4. Immunofluorescence confocal imaging of the LC3, LAMP1, and p62 protein in MSCs challenged with *E. coli*. Panels A1-A4 and B1-B4 are control specimens. Panels A5-A8 and B5-B8 are challenged with *E. coli*.

was accompanied by the presence of immunoreactivity to p62, a marker of ubiquitin-dependent target transport, in autolysosomes that was associated with autophagy of *E. coli* (Fig. 4B, Fig. 5). The image analysis of autophagy was supported by results of immunoblotting of the proteins (Fig. 3). It should be noted that pre-incubation of cell cultures with wortmannin, an autophagy inhibitor, resulted in apoptotic transformations and ultimately loss of confluency approximately 3 h after challenge with *E. coli* (not shown).

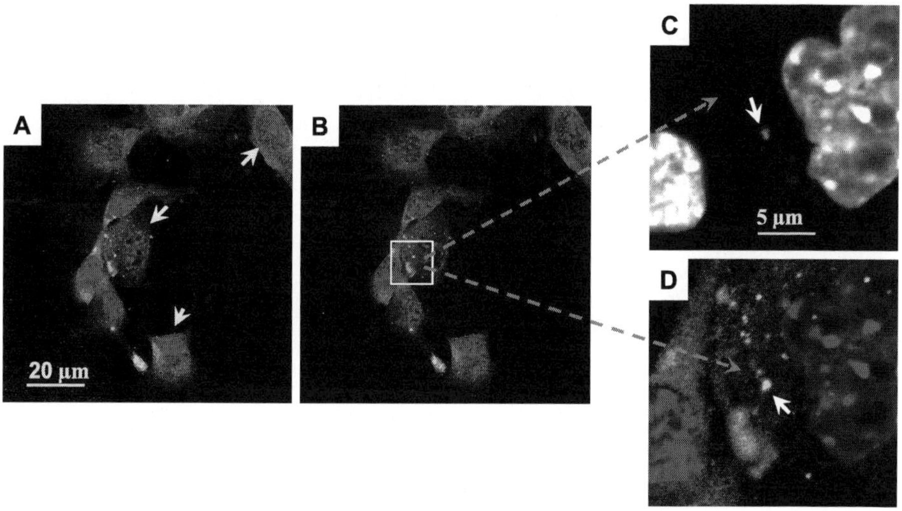

Panel A: Projection of FOXO3a (red channel; nuclear FOXO3a is indicated with yellow arrows) and p62 (green channel). Panel B: Projection of LC3 (red channel) and p62 (green channel). Counterstaining of nuclei with Hoechst 33342 appears in blue color. Panels C and D – selected area indicated in panel B.
Panel C: Signal acquired in the blue channel; bacterial DNA is indicated with white arrow.
Panel D: Signals acquired in the blue, red and green channels; co-localization of bacterial nucleus with p62 and LC3 proteins is indicated with white arrow.

Conditions: MSCs were incubated with ~1x10^7/ml *E. coli* for 3 h in MesenCult Medium (without antibiotics). After 3 h the medium was replaced with fresh medium (with antibiotics) and MSCs were incubated for further 21 h. The confocal images were taken with pinhole setup to obtain 0.5 μm Z-sections.

Fig. 5. Immunofluorescence confocal imaging of LC3, p62, phagocytized bacteria, and nuclear fraction of FOXO3a in MSCs challenged by *E. coli*.

Autolysosomal degradation of phagocytized bacteria can involve reactive oxygen and nitrogen species ultimately leading to up-regulation of stress-adaptive elements (13). Confocal fluorescence imaging of formation of reactive nitrogen species in autolysosomes was conducted using dihydrorhodamine 123, a sensitive indicator of peroxynitrite. The results of assessment of oxidative environment in the MSC autolysosomes containing *E. coli* are presented in Fig. 6. The appearance of reactivity to dihydrorhodamine 123 was likely due to up-regulation of nitric oxide synthase induced in MSCs in response to challenge with *E. coli*. It was hypothesized this increase in redox events in MSCs could at least in part contribute to degradation of the phagocytized bacteria. Indeed, as shown in Fig. 7 bacterial nuclei present in autolysosomes were positive to terminal deoxynucleotidyl transferase dUTP nick-end labeling (TUNEL).

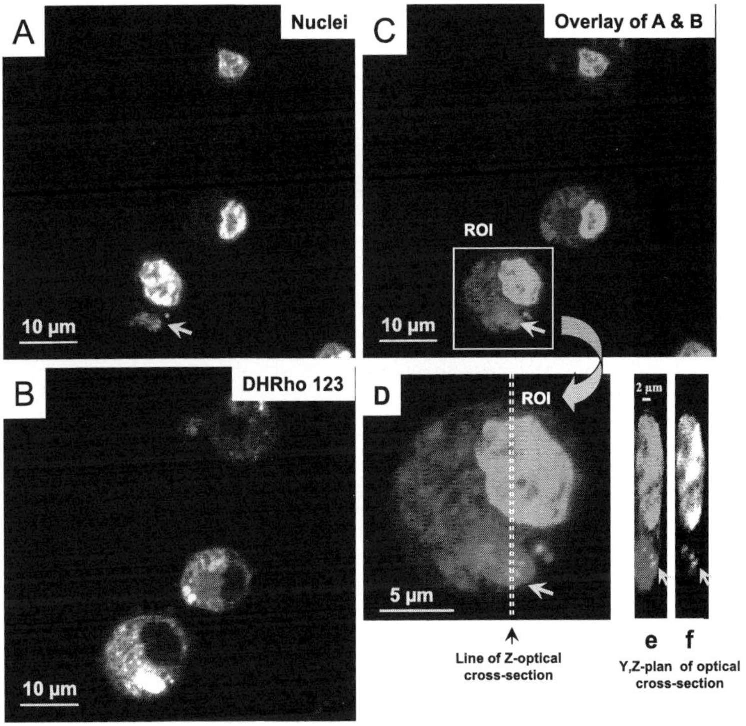

Panels A-C are projections of nuclei and oxidized fluorescent product of dihydrorhodamine 123. The images acquired in the blue (panel A) and green (panel B) are shown in grayscale; then, the images were overlaid in panel C in pseudo-colors that are "red" and "green", respectively. Panel D is the selected area indicated in panel C, where nuclei are green, oxidized dihydrorhodamine 123 (DHRho 123) is red, and co-localization of nuclei and DHRho 123 is in yellow colors. The presence of bacterial genomic DNA in the autolysosome appears in yellow as result of interference of red and green colors. Experimental conditions were the same as indicated in Fig. 5.

Fig. 6. Assessment of production of peroxynitrite in *E. coli*-challenged MSCs using dihydrorhodamine 123 probe.

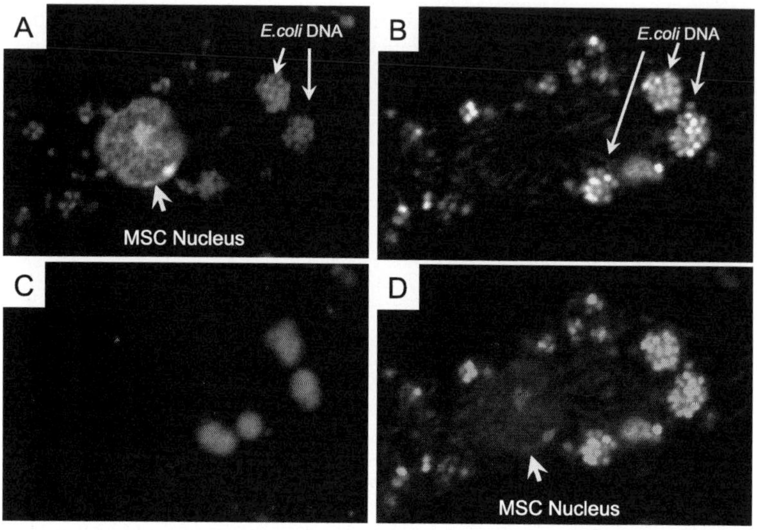

Panel A: Projection the nuclear DNA is indicated with yellow arrows (blue channel, counterstaining of nuclei with Hoechst 33342). Panel B: Projection of TUNEL-positive DNA (green channel).
Panel C: Projection of tyrosine-phosphorylated caveolin-1 (red channel). Panel D: Overlay of the images presented in panels A, B, and C. TUNEL-positive bacterial nuclei appear in yellow as result of interference of blue and green. TUNEL - positive staining of bacterial DNA occurred in autolysosomes. Experimental conditions were the same as indicated in Fig. 5.

Fig. 7. Assessment of bacterial DNA damage in *E. coli*-challenged MSCs using terminal deoxynucleotidyl transferase dUTP nick-end labeling (TUNEL).

3.2 Stress-response of MSCs following challenge with *E. coli* bacteria

General stress responses are characterized by conserved signaling modules that are interconnected to the cellular adaptive mechanisms. It is proposed that stress induced by inflammatory factors, microorganisms, and oxidants triggers a cascade of responses attributed to specific sensitive transcriptional and post-transcriptional mechanisms mediating inflammation, antioxidant response, adaptation, and remodeling (36-42). The components of the oxidative stress response employ a battery of redox-sensitive thiol-containing molecules, such as glutathione (GSH), thioredoxin 1 (TRX1)/thioredoxin reductase, apurinic/apyrimidinic endonuclease/redox effector factor-1 (APE/Ref-1), and transcription factors (such as nuclear factor-kappa B (NF-κB) and nuclear factor (erythroid-derived 2)-like 2 (Nrf2). Overall, these effector proteins play a major role in maintaining the steady-state intracellular balance between pro-oxidant production, antioxidant capacity, and repair of oxidative damage (39, 43). While NF-κB and Nrf2 are normally sequestered in the cytoplasm bound to their native inhibitors, i.e., IκB and Keap-1 respectively, bacterial products, pro-inflammatory factors, and oxidative stress can stimulate their translocation to the nucleus (38, 41, 44). NF-κB and Nrf2 are known to regulate numerous genes that play a crucial role in the host response to sepsis (40, 45) and therefore, have relevance to the current study. Regulation of Nrf2 function is controlled by numerous factors among which Nrf2 conjugates with Keap-1.

Dissociation of the Nrf2/Keap-1 complex results from a modification of cysteine residues in Keap-1 through either their conjugation or oxidation (40, 43, 45).

Two major redox systems, the GSH and TRX1 systems, control intracellular thiol/disulfide redox environments. While the GSH/GSSG couple provides a major cellular redox buffer, TRXs serve a more specific function in regulating redox-sensitive proteins (46). These two redox systems function at different sites in the Nrf2 signaling pathway: first, the cytoplasmic dissociation of Nrf2 is primarily regulated by cytoplasmic GSH concentrations, and second, the nuclear reduction of Nrf2 cysteine 506 (required for Nrf2 binding of DNA) is primarily regulated by TRX1 (45). Redox dependence of DNA-binding activity of NF-κB has been broadly discussed (39, 47). DNA-binding activity of NF-κB can drastically increase in the presence of the reduced form of the redox factor-1 (Ref-1) redox-converted by TRX (39, 47). It should be noted that up-regulation of Nrf2 and NF-κB via autophagy-dependent mechanisms can also occur *via* lysosomal degradation of IκB and Keap-1, (48). Therefore, we do not exclude autophagy-dependent activation of these transcriptional factors in *E. coli*-treated cells. Taking into consideration all of the above, one would assume that a battery of stress-sensitive mechanisms mediated by survival transcription factors such as NF-κB, Nrf2, and FOXO3a are involved in adaptive response of MSCs challenged with *E. coli*.

Immunoblot analysis of stress-response proteins indicated that control MSCs had relatively high amounts of constitutively present NF-κB. Challenge of cells with *E. coli* resulted in prompt (within 1 h) increases in the nuclear fraction of NF-κB as determined with confocal immunofluorescence imaging (not shown). But, we did not observe a similar pattern when we assessed nuclear Nrf2. That could be due to an extremely low level of constitutive Nrf2 in the cells (Fig. 3). A drastic increase in the nuclear fraction of NF-κB occurred during the period of the observation, i.e., 24 h post-exposure (Fig. 8). This effect was accompanied

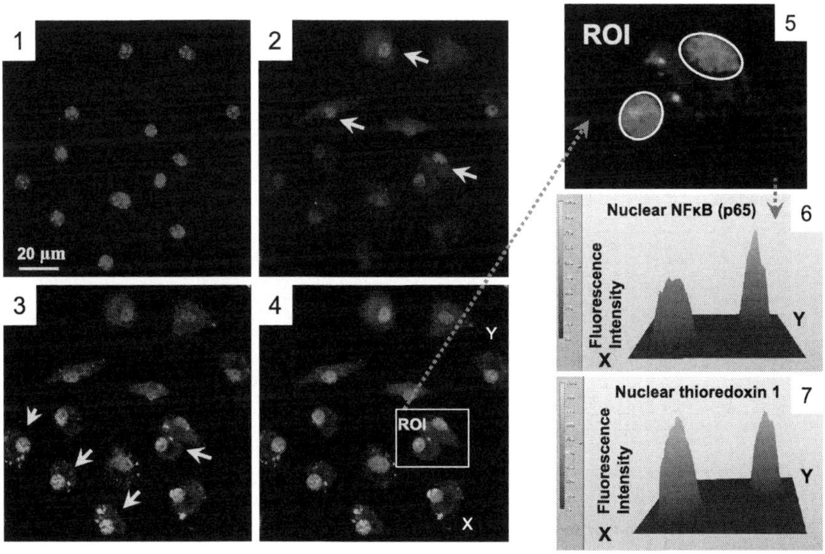

A

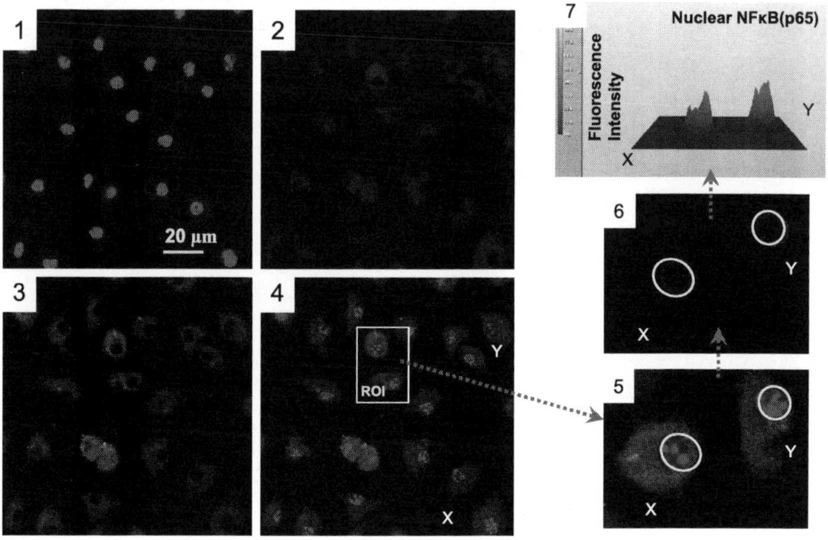

B

Panel 1: Projection of the nuclear DNA (blue channel, counterstaining of nuclei with Hoechst 33342). Panel 2: Projection of NFκB(p65) (red channel, nuclear localization is indicated with yellow arrows). Panel 3: Projection of thioredoxin 1 (green channel, nuclear localization is indicated with yellow arrows). Panel 4: Overlay of the images presented in panels 1, 2, and 3. Panels 5-7: analysis of nuclear fractions of NFκB(p65) and thioredoxin 1 in ROI indicated in panel 4. Experimental conditions were the same as indicated in Fig. 5.

Fig. 8. Assessment of nuclear fractions of NF-κB(p65) and thioredoxin 1 in MSCs challenged with *E. coli*. (A) Challenge with *E. coli.*; (B) Control.

by transactivation of NF-κB-dependent proinflammatory factors such as IL-1α, IL-1β, IL-6, and iNOS (Fig. 9). Interestingly, pre-incubation of the cells with pyrrolidine dithiocarbamate, an inhibitor of NF-κB translocation, resulted in development of pro-apoptotic alterations and loss of confluency in *E. coli*-treated MSCs (not shown). The response to *E. coli*–induced stress was also associated with increases in nuclear fractions of Ref-1 and TRX-1 (Figs. 8 and 10); these reducing agents appeared in close proximity with the nuclear NF-κB (Figs. 8 and 10). Moreover, the MSC stress-response at 24 h was characterized by significant expression of Nrf2 protein (Fig. 3) that accumulated in cell nuclei (Fig. 11). Based on these observations we concluded that the MSC response to challenge with *E. coli* activates complex molecular machinery designed to eliminate environmental microorganisms and increase adaptive capacity to stress. That conclusion contributes to a broad perspective on the role of stromal cells in the host innate defense and on the cell molecular mechanisms mediating resistance of cells to damage. Considering that the cell can employ a battery of stress-response factors operating synchronously, we focused our attention on other cellular components that are crucial for cell survival, e.g., mitochondria, the caveolae vesicular system, and signaling cascades mediated by transcriptional factor FOXO3a.

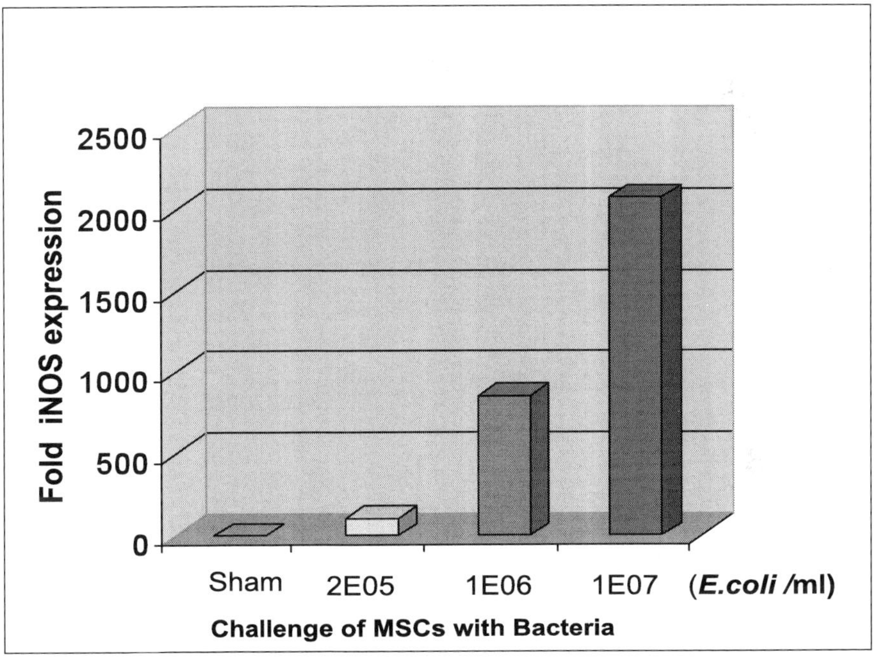

Fig. 9. qRT-PCR assessment of iNOS transactivation in MSCs challenged with *E. coli*. Conditions: MSCs were incubated with bacteria for 3 h in MesenCult Medium (without antibiotics). After 3 h the cells were harvested and lysed for extraction of RNA.

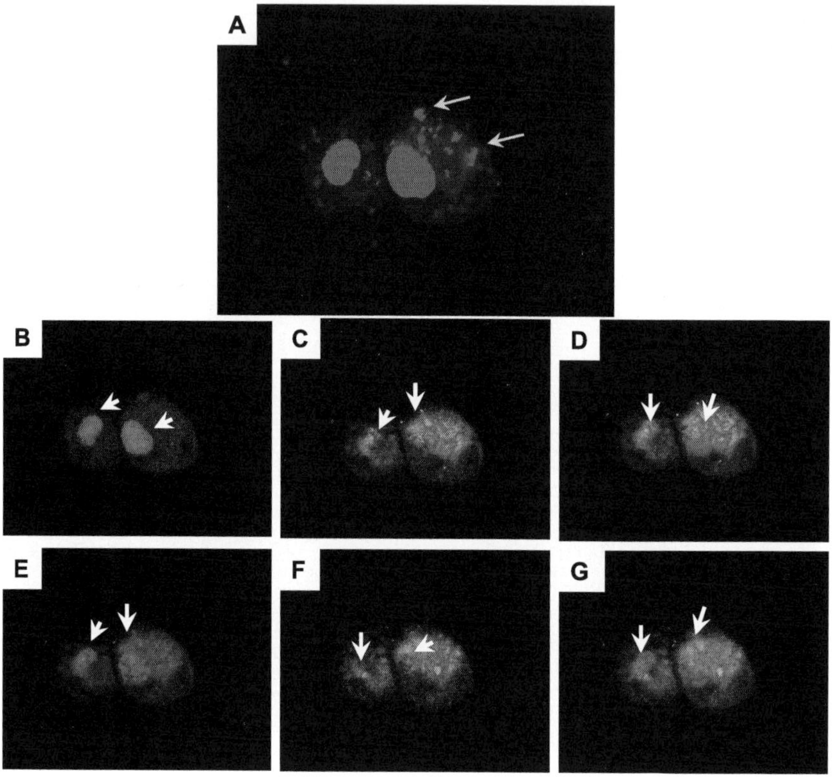

Panel A: Projection of the nuclear DNA (blue channel, high intensity; counterstaining of nuclei with Hoechst 33342). Bacterial nuclei are indicated with yellow arrow. Panel B: Projection of NFκB(p65) (red channel) and nuclear DNA (blue channel); nuclear co-localization of NFκB(p65) is indicated with white arrows. Panel C: Projection of Ref1 protein (green channel, nuclear localization is indicated with white arrows). Panel D: Overlay of the images presented in panels B and C. Nuclear co-localization of Ref1 and NFκB(p65) is indicated with white arrows. Panel E: Projection of Ref1 (red channel) and nuclear DNA (blue channel); nuclear co-localization of Ref1 is indicated with white arrows.
Panel F: Projection of thioredoxin 1 protein (green channel, nuclear localization is indicated with white arrows). Panel G: Overlay of the images presented in panels E and F. Nuclear co-localization of Ref1 and thioredoxin 1 is indicated with white arrows. Experimental conditions were the same as indicated in Fig. 5.

Fig. 10. Assessment of nuclear co-localization of NF-κB, thioredoxin 1, and Ref1 in MSCs challenged with *E. coli*.

FOXO3a, a member of a family of mammalian forkhead transcription factors of the class O, was recently proposed as mediator of diverse physiologic processes, including regulation of stress resistance and survival (49, 50). Thus, it is shown in our study that in response to oxidative stress, FOXO3a along with Nrf2 can promote cell survival by inducing the expression of antioxidant enzymes and factors involved in cell cycle withdrawal, such as the cyclin-dependent kinase inhibitor (CKI) p27 (50). We analyzed FOXO3a transcriptional factor in MSCs responding to *E. coli* challenge. Fig. 12 shows that the presence of *E. coli* increased FOXO3a protein in MSCs. The data suggest that, indeed, this FOXO3a transcriptional factor is also implicated in the stress-response to *E. coli* challenge.

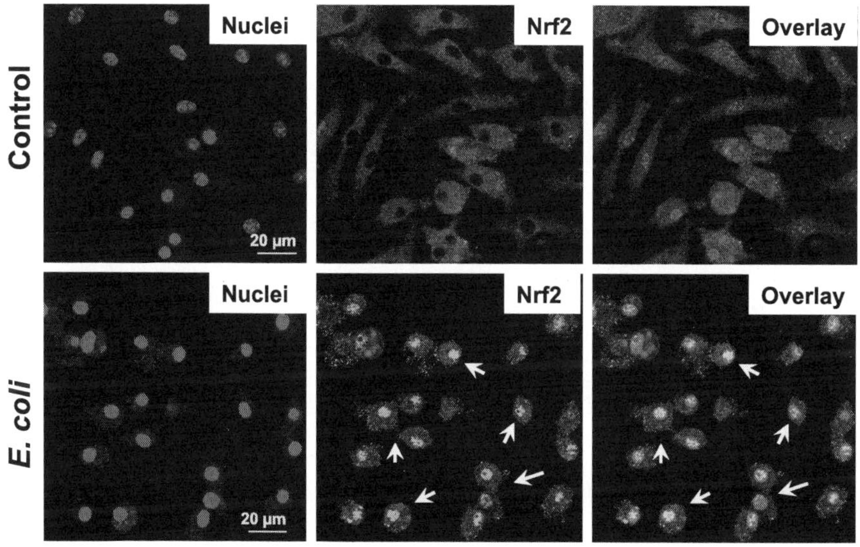

Fig. 11. Assessment of nuclear fractions of Nrf2 in MSCs challenged with *E. coli*. Counterstaining of nuclear DNA was with Hoechst 33342 (blue channel). Nrf2 staining is in green. Nrf2 localized in nuclei appears in turquoise/green color due to interference of "green" and "blue" (indicated with arrows). Experimental conditions were the same as indicated in Fig. 5.

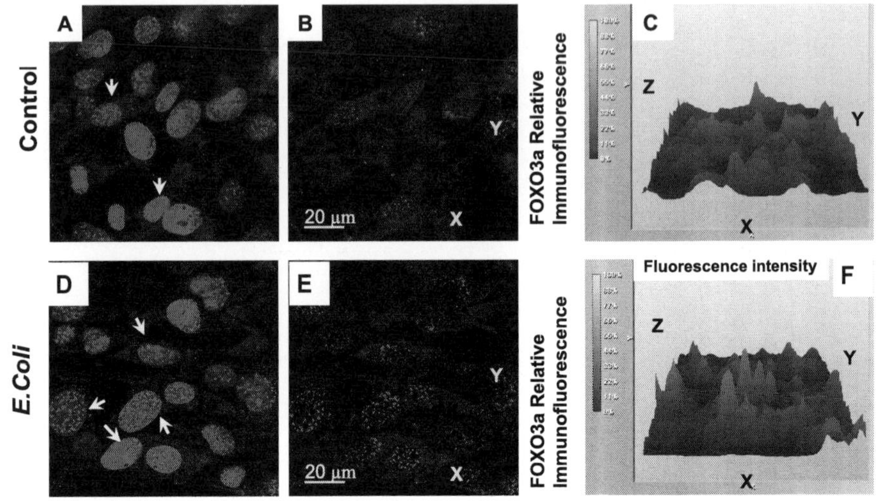

Control: Panel A-C. Challenged with *E. coli.*: Panels D-F.
Panels A and D: Projection of FOXO3a (red channel) and nuclear DNA (blue channel); nuclear localization of FOXO3a is indicated with white arrows. Panels B and E: Projection of FOXO3a protein (red channel only). Panel C and F: Relative intensity of the FOXO3a immunofluorescence shown in panels B and E, respectively. Experimental conditions were the same as indicated in Fig. 5.

Fig. 12. Immunofluorescence assessment of nuclear fraction of FOXO3a in MSCs challenged with *E. coli.*

4. Conclusion

Multipotent fibroblast-type mesenchymal cells are the essential components of the stroma, which supports tissue barriers and integrity (51). Disturbance in the stroma, composed of endothelial, fibroblastic and myofibroblastic cells as well as macrophages and other inflammatory cells - can be a critical step triggering bacterial translocation and sepsis - exacerbating a variety of injury types. This chapter aims to define whether MSCs can contribute to antibacterial innate defense mechanisms.

The antibacterial defense response of MSCs was characterized by extensive phagocytosis and inactivation of *E. coli* mediated by autolysosome mechanisms. *E. coli*-challenged MSCs showed increased transactivation of NF-κB, Nrf2, and FOXO3a stress-response transcriptional factors and associated expression of proinflammatory mediators. These observations were accompanied by a compensatory antioxidant response of MSCs mediated by nuclear translocation of Nrf2, Ref-1 and thioredoxin 1.

Taken together our data support the hypothesis that (i) MSCs contribute to the innate defense response to bacterial infection; (ii) the mechanism of MSC responses involves specific macroautophagy and nitroxidation mediated by iNOS; and (iii) MSCs are armed against self-injury by the mechanisms degrading phagocytized *E. coli.*

5. Acknowledgements

The authors thank HM1 Neil Agravante and Ms. Dilber Nurmemet for their technical support.

5.1 Grants

This work was supported by AFRRI Intramural RAB2CF (to JGK) and NIAID YI-AI-5045-04 (To JGK).

There are no ethical and financial conflicts in the presented work.

5.2 Disclaimer

The opinions or assertions contained herein are the authors' private views and are not to be construed as official or reflecting the views of the Uniformed Services University of the Health Sciences, AFRRI, the United States Department of Defense, or the National Institutes of Health.

6. References

[1] Turner HL, Turner JR. Good fences make good neighbors: Gastrointestinal mucosal structure. Gut Microbes. 2010; 1(1):22-9.

[2] Hooper LV, Macpherson AJ. Immune adaptations that maintain homeostasis with the intestinal microbiota. Nat Rev Immunol. 2010; 10(3):159-69.

[3] Wershil BK, Furuta GT. Gastrointestinal mucosal immunity. J Allergy Clin Immunol. 2008; 121(2 Suppl):S380-3.

[4] Vinh DC, Embil JM. Rapidly progressive soft tissue infections. Lancet Infect Dis. 2005; 5(8):501-13.

[5] Friedenstein A. Stromal-hematopoietic interrelationships: Maximov's ideas and modern models. Haematol Blood Transfus. 1989; 32:159-67.

[6] Sethe S, Scutt A, Stolzing A. Aging of mesenchymal stem cells. Ageing Res Rev. 2006; 5(1):91-116.

[7] Mezey E, Mayer B, Németh K. Unexpected roles for bone marrow stromal cells (or MSCs): a real promise for cellular, but not replacement, therapy. Oral Dis. 2010; 16(2):129-35.

[8] English K, French A, Wood KJ. Mesenchymal stromal cells: facilitators of successful transplantation? Cell Stem Cell. 2010; 7(4):431-42.

[9] McFarlin K, Gao X, Liu YB, Dulchavsky DS, Kwon D, Arbab AS, Bansal M, Li Y, Chopp M, Dulchavsky SA, Gautam SC. Bone marrow-derived mesenchymal stromal cells accelerate wound healing in the rat. Wound Repair Regen. 2006; 14(4):471-8.

[10] Tyndall A and Pistoia V. Mesenchymal stem cells combat sepsis. Nat Med. 2009; 15: 18 - 20.

[11] Krasnodembskaya A, Song Y, Fang X, Gupta N, Serikov V, Lee JW, Matthay MA. Antibacterial effect of human mesenchymal stem cells is mediated in part from secretion of the antimicrobial peptide LL-37. Stem Cells. 2010; 28(12):2229-38.

[12] Hall SE, Savill JS, Henson PM, Haslett C. Apoptotic neutrophils are phagocytosed by fibroblasts with participation of the fibroblast vitronectin receptor and involvement of a mannose/fucose-specific lectin. J Immunol. 1994; 153(7):3218-27.

[13] Levine B, Mizushima N, Virgin HW. Autophagy in immunity and inflammation. Nature. 2011; 469(7330):323-35.

[14] Yang Z, Klionsky DJ. Eaten alive: a history of macroautophagy. Nat Cell Biol. 2010; 12(9):814-22;

[15] Yano T, Kurata S. Intracellular recognition of pathogens and autophagy as an innate immune host defence. J Biochem. 2011; 150(2):143-9.

[16] Mizushima N, Levine B, Cuervo AM, Klionsky DJ. Nature. Autophagy fights disease through cellular self-digestion. 2008; 451:1069–75.

[17] Klionsky DJ. The Autophagy Connection. Dev Cell. Author manuscript; available in PMC 2011 July 20.

[18] Tooze SA, Yoshimori T. The origin of the autophagosomal membrane. Nat Cell Biol. 2010; 12(9):831-5.

[19] Weidberg H, Shvets E, Elazar Z. Biogenesis and cargo selectivity of autophagosomes. Annu Rev Biochem. 2011; 80:125-56.

[20] Eskelinen EL. New insights into the mechanisms of macroautophagy in mammalian cells. Int Rev Cell Mol Biol. 2008; 266:207-47.

[21] Eskelinen EL, Saftig P. Autophagy: a lysosomal degradation pathway with a central role in health and disease. Biochim Biophys Acta. 2009; 1793(4):664-73.

[22] Mizushima N, Levine B. Autophagy in mammalian development and differentiation. Nat Cell Biol. 2010; 12(9):823-30.

[23] Kabeya Y, Mizushima N, Yamamoto A, Oshitani-Okamoto S, Ohsumi Y, Yoshimori T. LC3, GABARAP and GATE16 localize to autophagosomal membrane depending on form-II formation. J Cell Sci. 2004; 117(Pt 13):2805-12.

[24] Behrends C, Sowa ME, Gygi SP, Harper JW. Network organization of the human autophagy system. Nature. 2010; 466(7302):68-76.

[25] Lipinski MM, Hoffman G, Ng A, Zhou W, Py BF, Hsu E, Liu X, Eisenberg J, Liu J, Blenis J, Xavier RJ, Yuan J. A genome-wide siRNA screen reveals multiple mTORC1 independent signaling pathways regulating autophagy under normal nutritional conditions. Dev Cell. 2010; 18(6):1041-52.

[26] Viiri J, Hyttinen JM, Ryhänen T, Rilla K, Paimela T, Kuusisto E, Siitonen A, Urtti A, Salminen A, Kaarniranta K. p62/sequestosome 1 as a regulator of proteasome inhibitor-induced autophagy in human retinal pigment epithelial cells. Mol Vis. 2010; 16:1399-414.

[27] Ryhänen T, Hyttinen JM, Kopitz J, Rilla K, Kuusisto E, Mannermaa E, Viiri J, Holmberg CI, Immonen I, Meri S, Parkkinen J, Eskelinen EL, Uusitalo H, Salminen A, Kaarniranta K. Crosstalk between Hsp70 molecular chaperone, lysosomes and proteasomes in autophagy-mediated proteolysis in human retinal pigment epithelial cells. J Cell Mol Med. 2009; 13(9B):3616-31.

[28] Behl C. BAG3 and friends: co-chaperones in selective autophagy during aging and disease. Autophagy. 2011; 7(7):795-8.

[29] Sridhar S, Botbol Y, Macian F, Cuervo AM. Autophagy and Disease: always two sides to a problem. J Pathol. 2011 Oct 12. doi: 10.1002/path.3025. [Epub ahead of print];

[30] Gorbunov NV, Garrison BR, Kiang JG. Response of crypt Paneth cells in the small intestine following total-body gamma-irradiation. Int J Immunopathol Pharmacol. 2010; 23(4):1111-23.

[31] Pfaffl MW. A new mathematical model for relative quantification in real-time RT-PCR. Nucleic Acids Res. 2001 29(9):2002-7.

[32] Nakagawa I, Amano A, Mizushima N, Yamamoto A, Yamaguxhi H, Kamimoto T, Nara A, Funao J, Nakata M, Tsuda K, Hamada S, Yoshimori T. Autophagy defends cells against invading Group A *Streptococcus*. Science. 2004; 306:1037-40.

[33] Yano T, Kurata S. Intracellular recognition of pathogens and autophagy as an innate immune host defence. J Biochem. 2011; 150(2):143-9.

[34] Xiao G. Autophagy and NF-kappaB: fight for fate. Cytokine Growth Factor Rev. 2007; 18(3-4):233-43.

[35] Ichimura Y, Komatsu M. Selective degradation of p62 by autophagy. Semin Immunopathol. 2010; 32(4):431-6.

[36] Kültz D. Molecular and evolutionary basis of the cellular stress response. Annu Rev Physiol. 2005; 67: 225-57.

[37] Burhans WC, Heintz NH. The cell cycle is a redox cycle: linking phase-specific targets to cell fate. Free Radic Biol Med. 2009; 47(9):1282-93.

[38] Sebban H, Courtois G. NF-kappaB and inflammation in genetic disease. Biochem Pharmacol. 2006; 72(9):1153-60.

[39] Janssen-Heininger YM, Mossman BT, Heintz NH, Forman HJ, Kalyanaraman B, Finkel T, Stamler JS, Rhee SG, van der Vliet A. Redox-based regulation of signal transduction: principles, pitfalls, and promises. Free Radic Biol Med. 2008; 45(1):1-17.

[40] Nguyen T, Nioi P, Pickett CB. The Nrf2-Antioxidant Response Element Signaling Pathway and Its Activation by Oxidative Stress J Biol Chem. 2009; 284(20):13291-5.

[41] Thimmulappa RK, Lee H, Rangasamy T, Reddy SP, Yamamoto M, Kensler TW, Biswal S. Nrf2 is a critical regulator of the innate immune response and survival during experimental sepsis. J Clin Invest. 2006; 116(4):984-95.

[42] Nivon M, Richet E, Codogno P, Arrigo AP, Kretz-Remy C. Autophagy activation by NFkappaB is essential for cell survival after heat shock. Autophagy. 2009; 5(6):766-83.

[43] Maher J, Yamamoto M. The rise of antioxidant signaling--the evolution and hormetic actions of Nrf2. Toxicol Appl Pharmacol. 2010; 244(1):4-15.

[44] Cronin JG, Turner ML, Goetze L, Bryant CE, Sheldon IM. Toll-Like Receptor 4 and MYD88-Dependent Signaling Mechanisms of the Innate Immune System Are Essential for the Response to Lipopolysaccharide by Epithelial and Stromal Cells of the Bovine Endometrium. Biol Reprod. 2011 Nov 2. [Epub ahead of print].

[45] Hansen JM, Watson WH, Jones DP. Compartmentation of Nrf-2 redox control: regulation of cytoplasmic activation by glutathione and DNA binding by thioredoxin-1. Toxicol Sci. 2004; 82(1):308-17.

[46] Jones DP. Methods Enzymol. 2002; 348:93-112. Redox potential of GSH/GSSG couple: assay and biological significance.

[47] Nishi T, Shimizu N, Hiramoto M, Sato I, Yamaguchi Y, Hasegawa M, Aizawa S, Tanaka H, Kataoka K, Watanabe H, Handa H. Spatial redox regulation of a critical cysteine residue of NF-kappa B in vivo. J Biol Chem. 2002; 277(46):44548-56.

[48] Weidberg H, Shvets E, Elazar Z. Biogenesis and cargo selectivity of autophagosomes. Annu Rev Biochem. 2011; 80:125-56.

[49] Miyamoto K, Araki KY, Naka K, Arai F, Takubo K, Yamazaki S, Matsuoka S, Miyamoto T, Ito K, Ohmura M, Chen C, Hosokawa K, Nakauchi H, Nakayama K, Nakayama KI, Harada M, Motoyama N, Suda T, Hirao A. Foxo3a is essential for maintenance of the hematopoietic stem cell pool. Cell Stem Cell. 2007; 1(1):101-12.

[50] Burhans WC, Heintz NH. The cell cycle is a redox cycle: linking phase-specific targets to cell fate. Free Radic Biol Med. 2009; 47(9):1282-93.

[51] Powell DW, Pinchuk IV, Saada JI, Chen X, Mifflin RC. Mesenchymal cells of the intestinal lamina propria. Annu Rev Physiol. 2011; 73:213-37.

3

The Use of Reductive Methylation of Lysine Residues to Study Protein-Protein Interactions in High Molecular Weight Complexes by Solution NMR

Youngshim Lee[1], Sherwin J. Abraham[2] and Vadim Gaponenko[1,*]
[1]*Department of Biochemistry and Molecular Genetics, University of Illinois at Chicago, Chicago, IL*
[2]*Department of Molecular and Cellular Physiology, Stanford University, School of Medicine, Beckman Center, Stanford, CA*
USA

1. Introduction

While solution state NMR is very well suited for analysis of protein-protein interactions occurring with a wide range of affinities, it suffers from one significant weakness, known as the molecular weight limitation. This limitation stems from the efficient nuclear relaxation processes in macromolecules larger than 30 kDa (Wider & Wüthrich, 1999). These relaxation processes cause rapid decay of NMR signals. Although the use of transverse relaxation optimized spectroscopy (TROSY) approaches has made solution state NMR of large proteins and protein-protein complexes more feasible, it is still limited by the ability to produce isotope enriched proteins (Pervushin et al., 1997). However, there is a significant number of proteins for which no convenient system for stable isotope incorporation exists. We recently utilized reductive methylation methodology to demonstrate that it is possible to introduce ^{13}C-enriched methyl groups into lysine residues in otherwise unlabeled proteins with the purpose of studying protein-ligand and protein-protein interactions by NMR (Abraham et al., 2008).

Reductive methylation is commonly used to improve crystallization of proteins (Schubot & Waugh, 2004). Studies show that success of protein crystallization improves significantly through reductive methylation of solvent exposed lysines due to a reduction in surface entropy. Reductive methylation does not alter significantly protein structures and native protein-protein interactions (Gerken et al., 1982; Kurinov et al., 2000; Rayment, 1997; Walter et al., 2006). Despite clear advantages offered by reductive methylation, this technique remains underutilized in solution NMR. Here we show that reductive methylation allows characterization of high molecular weight protein-protein complexes that is not achievable using traditional NMR approaches.

* Corresponding Author

For reductive methylation of NMR protein samples, ^{13}C-enriched carbonyl compound (e.g. ^{13}C-formaldehyde) and reducing agents are required. The primary amine of lysine in polypeptide molecules acting as a nucleophile attacks the carbonyl group of formaldehyde. This reaction results in formation of an intermediate imine through the carbonyl-condensation process. The intermediate imine subsequently reacts with a proton donor to give rise to the higher order amine (Scheme 1). The solvent exposed lysine residues are frequently dimethylated when a sufficient amount of formaldehyde is present.

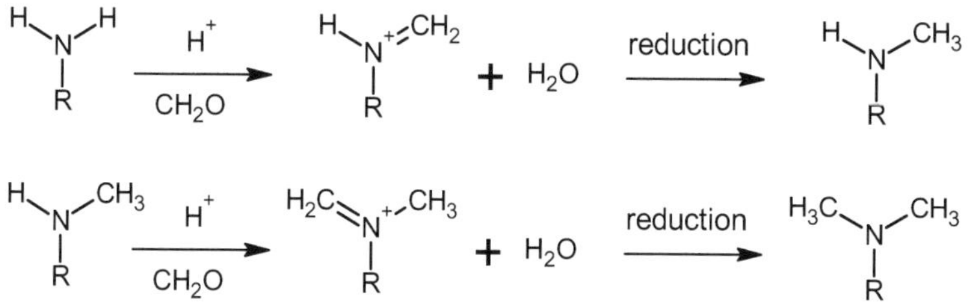

Scheme 1.

The reductive methylation technique offers several advantages. First, proteins purified from their native hosts can be directly used for enrichment with stable isotopes. In this way, the protein molecules are likely to retain their correct fold and post-translational modifications. Second, since only a small amount of ^{13}C-labeled formaldehyde is used in the reaction the reductive methylation procedure is significantly more economical than the traditional isotope enrichment protocols. Finally, the use of ^{13}C-labeled methyl groups in lysines offers an opportunity to observe NMR signals with favorable relaxation properties in large molecular weight proteins due to reduced order parameters for lysine side-chains (Abraham et al., 2009). In this report we not only demonstrate that observation of NMR signals in high molecular weight non-isotope enriched proteins is possible but also that investigation of conformational changes due to binding in protein-protein complexes is amenable to solution state NMR through reductive methylation.

2. Cardiac muscle proteins: Actin, tropomyosin, and troponin complex

Muscle contraction is caused by cyclic interaction between myosin and actin filaments. In cardiac muscle, regulation of contraction is controlled by the troponin complex and tropomyosin which bind to the actin filament (Galińska-Rakoczy et al., 2008; Kobayashi et al., 2008; Kobayashi & Solaro, 2005). The actin filaments consist of polymerized actin (F-actin) molecules which contain myosin binding sites. At rest, the myosin binding site is concealed by tropomyosin forming a coiled-coil dimer that lies in the two grooves of actin. Seven actin molecules interact with one tropomyosin dimer. Each tropomyosin dimer also binds one troponin complex composed of three subunits: troponin C, troponin I, and troponin T. The N-terminal domain of troponin C has a calcium binding pocket. The troponin complex, together with tropomyosin, regulate muscle contraction in a Ca^{2+}-dependent manner. This is accomplished by altering accessibility of actin binding sites to myosin. Being a Ca^{2+} sensor, troponin functions as an on/off switch for muscle contraction. Muscle contraction occurs when Ca^{2+} binds to the regulatory site in troponin C. Conversely, the muscle relaxes when Ca^{2+} dissociates. When Ca^{2+} concentration is high, Ca^{2+} binding to troponin C induces a structural change in the troponin complex that causes relocation of tropomyosin away from the actin groove. Due to tropomyosin relocation, the myosin binding site on actin is exposed and cross-bridge formation is initiated between actin and myosin. Troponin I is known to inhibit myosin cross-bridge formation by inducing relocation of tropomyosin. Troponin T associates with troponin C and I to form the complete troponin complex. Troponin T also binds to tropomyosin and actin to inhibit myosin binding to thin filaments.

Alpha-helical coiled-coil tropomyosin assembles into filaments in the end-to-end configuration and interacts with actin polymers. When bound to polymerized actin, tropomyosin filament spans seven consecutive actin monomers forming a 369 kDa complex. One troponin binds to each tropomyosin coiled-coil dimer such that the molecular ratio for actin, Tm, and troponin is 7:2:1. There are two kinds of interaction between the troponin complex and actin-tropomyosin. One is Ca^{2+}-independent binding through troponin T, anchoring the troponin complex to actin-tropomyosin. The other is Ca^{2+}-dependent regulatory interactions through inhibitory C-terminal half of troponin I, turning muscle contraction "on" and "off". The cytoplasmic Ca^{2+} concentration is essential for muscle contraction. However, allosteric regulation of the troponin complex is also known to be an important contributor. Solution NMR can detect conformational changes in protein molecules and thus is a good tool to study the allosteric regulation. We utilize the reductive methylation technique because the thin fiber is a large protein-protein complex containing molecules that are difficult to produce as recombinant proteins for enrichment with stable isotopes.

2.1 Conformation of reductively methylated cardiac troponin C free and as part of the cardiac troponin complex in the presence and absence of Ca^{2+}

2.1.1 Methods of preparation of reductively methylated troponin complex

Reductive methylation of troponin C for NMR experiments was performed using ^{13}C-enriched formaldehyde and borane-ammonia complex ($NH_3.BH_3$) as a reducing agent. Briefly, 20 µL of 1 M borane-ammonia complex and 40 µL of ^{13}C formaldehyde were added to 1 mL of troponin C in methylation buffer (10 mM HEPES pH 7.6, 50 mM $MgCl_2$, 50 mM $CaCl_2$, and 1 mM β-mercaptoethanol). The reaction mix was incubated at 4 °C with stirring for 2 h. The procedure was repeated one more time with a final addition of 10 µL of ^{13}C formaldehyde and was incubated at 4 °C with stirring overnight. The reaction was stopped by adding 200 mM glycine and the undesired reaction products and excess reagents were removed by extensive dialysis against 10 mM Tris/HCl pH 7.6, 50 mM $MgCl_2$, 50 mM $CaCl_2$, and 1 mM β-mercaptoethanol. To obtain the troponin complex, troponin I and troponin T were added to methylated troponin C in the 1:1:1 molar ratio. To obtain larger molecular weight complexes tropomyosin was added to the troponin complex containing methylated troponin C in the 2:1 molar ratio.

The NMR experiments were performed on samples containing 20 μM troponin C (either alone or in complex) in NMR buffer containing 40 mM Tris-HCl (pH 10.0), 50 mM KCl, 1 mM β-mercaptoethanol, and either 50 mM $CaCl_2$ or 50 mM $MgCl_2$. All ^1H-^{13}C heteronuclear single-quantum correlation (HSQC) spectra were acquired on the 600 MHz Bruker Avance spectrometer fitted with a cryoprobe using 128 indirect points at 25 °C. The data were processed using NMRPipe software (Delaglio et al., 1995).

2.1.2 Results

NMR ^1H-^{13}C HSQC experiments were performed on free reductively methylated troponin C in the presence and absence of Ca^{2+} and on the cardiac troponin complex containing reductively methylated troponin C in the presence and absence of Ca^{2+}. The results of these experiments are shown in Figure 1. All of the acquired spectra display the expected 12 signals representing methyl groups on eleven lysines in troponin C and one on the N-terminal

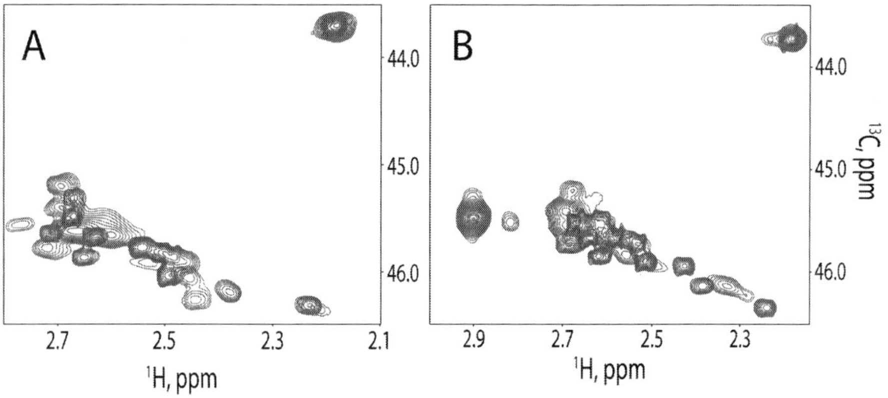

Fig. 1. An overlay wof ^{13}C-^1H HSQC spectra of reductively methylated 20 μM troponin C (blue) and the troponin complex consisting of full length troponin C, troponin I, and troponin T (red). The spectra in (A) were recorded in the presence of 50 mM Ca^{2+}. The spectra in (B) were recorded in the absence of Ca^{2+} and the presence of 50 mM Mg^{2+}. The spectra were acquired at 600 MHz at 25 °C with 256 indirect points. The buffer conditions are 40 mM Tris-HCl (pH 10.0), 50 mM KCl, 1 mM β-mercaptoethanol, 50 mM $CaCl_2$ (A) or 50 mM $MgCl_2$ (B).

primary amine. Comparison of spectra of free troponin C in the presence and absence of Ca^{2+} (Fig. 1A and 1B) reveals significant differences in methyl chemical shift values. These chemical shift perturbations indicate expected structural rearrangements in the N-terminal domain of troponin C caused by Ca^{2+} binding. Comparison of NMR spectra of free Ca^{2+}-bound troponin C with Ca^{2+}-bound troponin C in the troponin complex reveals significant perturbations in nine out of twelve methyl chemical shifts (Fig. 1A). This observation suggests involvement of troponin C in intermolecular interactions with components of the troponin complex. In the absence of Ca^{2+} only five out of twelve signals experience significant chemical shift perturbations (Fig. 1B). One possible explanation of this is that in the absence of Ca^{2+}, troponin C is less extensively engaged in protein-protein interactions within the troponin complex. Together, we demonstrate that using reductive methylation it is possible to characterize protein-protein interactions within the troponin complex by NMR despite the high molecular weight of the protein system.

2.2 NMR experiments with reductively methylated 369kDa actin-tropomyosin complex suggest that a global conformational rearrangement is induced in polymerized actin upon tropomyosin binding

2.2.1 Methods of preparation of reductively methylated actin-tropomyosin complex

Globular actin was dialyzed into 10mM phosphate buffered saline, pH 7.4, 0.1mM $MgCl_2$, 1mM dithiothreitol, 0.1mM ATP, and 0.01% NaN_3, to make actin filaments. Initially 20mM borane ammonia complex and 40mM ^{13}C-formaldehyde (20% w/w in H_2O) were added into 0.7mL of 60 μM F-actin and the mixture was stirred for 2 hours at 4°C. Addition of borane ammonia complex and ^{13}C-formaldehyde was repeated and mixture was incubated for another 2 hours at 4°C. After incubation, 10 mM borane ammonia complex was added to the mixture. The mixture was incubated at 4°C with stirring overnight. To quench the reaction, the 50 μL of 2M Tris-HCl was added. To study the change in actin structures upon binding of Tm, Tm is added into ^{13}C methylated F-actin to make 7.5 μM of final Tm concentration whereas the concentration of F-actin is 37μM. The molar ratio of actin and tropomyosin was 5 to 1. The samples were dialyzed against 10mM phosphate buffered saline, pH 7.4, with 1mM $MgCl_2$, 0.1mM ATP, 0.01% NaN_3 and 10% D_2O was added for further NMR experiments. All NMR experminents were carried out on Bruker Avance 600 or 900 NMR spectrometers equipped with cryogenic probes. The 2D ^{1}H-^{13}C edited HSQC experiments were processed with NMRPipe software (Delaglio et al., 1995).

2.2.2 Results

To assess conformational changes occurring in polymerized actin upon binding tropomyosin we performed a reductive methylation reaction on actin and carried out ^{1}H-^{13}C HSQC experiments on actin alone and on actin in the presence of tropomyosin (Fig. 2). In the spectrum of polymerized actin seven out of nineteen expected signals were observable. Significant chemical shift changes in four out of seven signals in actin were detected upon Tm binding. Lysines are evenly distributed in the actin structure with no accumulation in any one particular area. Therefore, the data shown here indicates that binding of tropomyosin causes a global conformational change in the structure of polymerized actin.

This observation is contrary to many computational models that propose that tropomyosin binding sites in actin are small and global changes do not occur in the actin-tropomyosin complex.

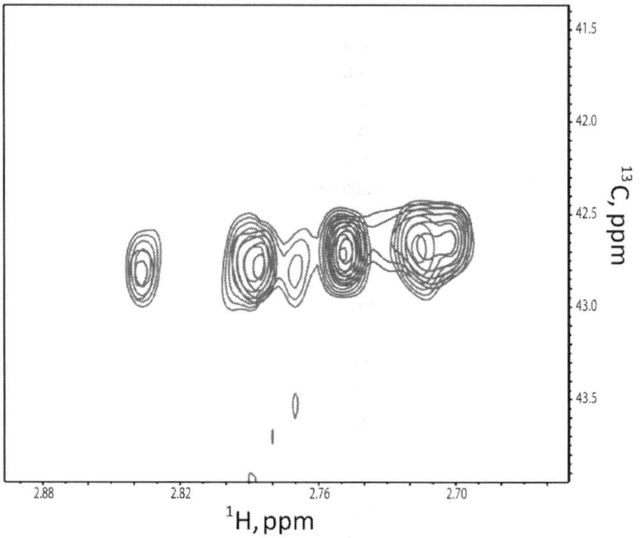

Fig. 2. An overlay of ^{1}H-^{13}C HSQC spectra of reductively methylated polymerized actin (red) and actin-tropomyosin complex (blue). The spectra were acquired at 900 MHz at 25 °C with 256 indirect points. The buffer conditions are 10 mM phosphate buffered saline (pH 7.4), 150 mM KCl, 50 mM $MgCl_2$, and 1 mM ATP.

3. Conclusion

In conclusion, we have described an important novel application of the reductive methylation methodology to observation of conformational changes in high molecular weight protein-protein complexes by NMR. Using cardiac troponin C as a model system, for which structural information is available, we confirmed that the proposed methodology allows detection of conformational rearrangements in cardiac troponin C upon Ca^{2+} binding. This was done in the context of the full-length troponin complex. Similar experiments would have been very difficult to perform using conventional NMR approaches due to the high molecular weight limitation. We also show that reductive methylation can be used to discover novel conformational changes in a 369 kDa actin-tropomyosin complex. For the first time we show that actin undergoes a global conformational change upon tropomyosin binding. This appears to be the only way such molecular events can be observed. The available computational models were unable to predict this phenomenon. Electron microscopy images of the cardiac thin fiber are too low resolution to detect a conformational change in actin. Crystallization of polymerized actin is not feasible due to heterogeneity of actin fibers. In addition, there is no good procedure for production of recombinant actin that would allow traditional approaches for stable isotope enrichment for NMR. The functional significance of actin conformational rearrangements upon binding of tropomyosin is still under investigation. However, the discovery that these conformational changes occur in actin is a significant step forward.

4. Acknowledgment

We acknowledge Dr. Tomoyoshi Kobayashi at the University of Illinois at Chicago for providing purified actin, tropomyosin, and protein samples for the troponin complex. We also acknowledge support from the National Cancer Institute grant R01CA135341 to Vadim Gaponenko.

5. References

Abraham, S. J.; Hoheisel, S. & Gaponenko, V. (2008). Detection of protein-ligand interaction by NMR using reductive methylation of lysine residues. *J. Biomol. NMR*, Vol. 42, No. 2, pp. 143-148.

Abraham, S. J.; Kobayashi, T.; Solaro, R. J. & Gaponenko, V. (2009). Differences in lysine pKa values may be used to improve NMR signal dispersion in reductively methylated proteins. *J. Biomol. NMR*, Vol. 43, No. 4,, pp. 239-246.

Delaglio F.; Grzesiek S., Vuister G. W., Zhu G., Pfeifer J. & Bax A. (1995) NMRPipe: a multidimensional spectral processing system based on UNIX pipes. *J. Biomol. NMR*, Vol. 6, No 3, pp. 277-293.

Galińska-Rakoczy A.; Engel P., Xu C., Jung H., Craig R., Tobacman L. S, & Lehman W. (2008). Structural basis for the regulation of muscle contraction by troponin and tropomyosin. *J. Mol. Biol.*, Vol. 379, No. 5, pp. 929–935.

Gerken T. A.; Jentoft J. E., Jentoft N. & Dearborn D. G. (1982) Intramolecular interactions of amino groups in ^{13}C reductively methylated hen egg-white lysozyme. *J. Biol. Chem.*, Vol. 257, No. 6, pp. 2894–2900.

Kobayashi T.; Jin L. & de Tombe P. P. (2008) Cardiac thin filament regulation. *Pflugers Arch.* Vol. 457 No. 1, pp. 37–46.

Kobayashi T. & Solaro R. J. (2005) Calcium, thin filaments and the integrative biology of cardiac contractility. *Annu. Rev. Physiol.* Vol. 67, pp. 39-67.

Kurinov I. V.; Mao C., Irvin J. D. & Uckun F. M. (2000) X-ray crystallographic analysis of pokeweed antiviral protein-II after reductive methylation of lysine residues. *Biochem. Biophys. Res. Commun.*, Vol. 275, No. 2, pp. 549–552.

Pervushin K.; Riek R., Wider G. & Wüthrich K. (1997) Attenuated T_2 relaxation by mutual cancellation of dipole-dipole coupling and chemical shift anisotropy indicates an avenue to NMR structures of very large biological macromolecules in solution. *Proc. Natl. Acad. Sci. USA* Vol. 94, No. 23, pp. 12366-12371.

Rayment I. (1997) Reductive alkylation of lysine residues to alter crystallization properties of proteins. *Methods Enzymol.* Vol. 276, pp. 171–179.

Schubot F. D. & Waugh D. S. (2004) A pivotal role for reductive methylation in the *de novo* crystallization of a ternary complex composed of *Yersinia pestis* virulence factors YopN, SycN and YscB. *Acta Crystallogr., Sect D; Biol. Crystallogr.* Vol. 60, pp. 1981-1986.

Walter T. S.; Meier C., Assenberg R., Au K. F., Ren J., Verma A., Nettleship J. E., Owens R. J., Stuart D. I. & Grimes J. M. (2006) Lysine methylation as a routine rescue strategy for protein crystallization. *Structure* Vol. 14, No. 11, pp. 1617–1622.

Wider G. & Wüthrich K. (1999) NMR spectroscopy of large molecules and multimolecular assemblies in solution. *Curr. Opin. Struct. Biol.* Vol. 9, No. 5, pp. 594-601.

4

Regulation of Protein-Protein Interactions by the SUMO and Ubiquitin Pathways

Yifat Yanku and Amir Orian
Technion-Israel Institute of Technology
Israel

1. Introduction

Post-transcriptional modifications of proteins by ubiquitin and ubiquitin-like proteins (UBLs) such as SUMO (Small Ubiquitin-related Modifier) regulate the function of protein-networks, enable cells to respond to signaling cues during development and to cope with the changing environment during adult life. The ubiquitin and SUMO pathways have profound impacts on protein stability, localization, protein-protein interactions and function.

In this chapter we will review mechanistic and biological aspects of protein-protein interactions that are regulated by ubiquitin and SUMO. We will describe the covalent tagging of proteins by ubiquitin and SUMO, and the enzymatic machineries that regulate these modifications. Subsequently, we will discuss how ubiquitylated or SUMOylated proteins are recognized by ubiquitin and SUMO recognition motifs present on interacting proteins. We will also illuminate how these non-covalent interactions regulate diverse cellular processes such as DNA repair, transcription, signaling, and autophagy in health and disease. Finally, we will address the crosstalk between the ubiquitin and SUMO pathways by SUMO-Targeted Ubiquitin Ligases (STUbLs).

2. Covalent modification of proteins by ubiquitin and SUMO

Ubiquitylation is a post-transcriptional modification where ubiquitin, a 76 amino acids long polypeptide, is covalently attached to proteins. Originally, ubiquitylation was considered as a "death-tag" targeting proteins for degradation by the 26S proteasome. However, over the last two decades non-proteolytic roles of ubiquitylation have also been found to impact protein function, cellular localization and protein-protein interactions. Furthermore, in addition to ubiquitin, ubiquitin-like proteins were identified and collectively termed UBLs. These proteins include among others SUMO (Small Ubiquitin Like Modifier), Nedd8 (Neural precursor cell expressed developmentally down-regulated 8), ISG15 (interferon stimulated gene 15) and FAT10, and all share at least one of the ubiquitin canonical folds (Hershko, 1983; Hochstrasser, 2009).

Among UBLs, the most studied modifier is SUMO. Vertebrates possess four different SUMO isoforms, termed SUMO1-4. While SUMO1-3 are efficiently conjugated to target proteins by specific SUMO enzymes, it is less clear if SUMO4 is conjugated to proteins *in vivo*. SUMO conjugation of target proteins results in a change in their activity, affecting protein localization, and modulates composition of protein complexes. In some cases SUMOylation may act as a priming modification that promotes ubiquitylation via a specific specialized sub-type of E3 ubiquitin ligases termed SUMO-Targeted Ubiquitin Ligases (Abed et al., 2011b; Praefcke et al., 2011). Within the scope of this chapter we will focus solely on the interactions mediated by ubiquitin and SUMO.

2.1 SUMO and Ubiquitin pathways

Ubiquitylation or SUMOylation are both mediated by a tripartite enzymatic cascade comprised of specific sets of enzymes. Ubiquitylation is mediated by the E1 ubiquitin-activating enzyme, E2-ubiquitin-conjugating enzyme (Ubc), and an E3 ubiquitin ligase (Hershko 1983; Pickart 2001). SUMOylation is similarly carried out by a different set of specialized SUMO specific E1, E2 and E3 enzymes. Both Ubiquitin and SUMO are covalently conjugated via their C-terminus Gly residue to free NH_2 amine group of the target protein that may reside on the ε-amino group of a Lys residue along the protein sequence, or on the amino terminus of the protein. Post-transcriptional regulation by ubiquitin and UBLs are tightly regulated, as covalent modification by ubiquitin or SUMO requires ATP. In the human genome only two genes coding for E1, ubiquitin-activating enzymes have been found. It is estimated that about one hundred E2s (ubiquitin-activating enzymes), and hundreds or more E3 ubiquitin-protein ligases exist. Interestingly, in plants the ubiquitin and SUMO pathways are greatly expanded. Collectively, these observations probably reflect the high degree of specificity and regulatory role of these pathways (Hershko & Ciechanover, 1998; Weissman et al., 2011).

E3 ligases are the "match makers" that directly recognize the targeted protein substrate. Ubiquitin E3 ligases can be classified into two main functional classes. The first is the HECT (Homologous to the E6-AP Carboxyl Terminus) domain ligases that accept ubiquitin from an E2 enzyme in the form of a thio-ester via a Cys residue in their catalytic domain, thus forming a thio-ester bond directly with the ubiquitin molecule. The second and most abundant class is the RING (Really Interesting New Gene) finger E3 ligases which utilizes a metal binding domain harbouring Zn+2 ions to facilitate ubiquitylation (Deshaies & Joazeiro, 2009). RING ligases are a diverse subclass, encompassing several hundreds of proteins in the human genome. This large family of ligases is further divided into modular sub-classes: 1. Single subunit ligases such as c-Cbl and parkin, that directly bind to both the E2 enzymes and the targeted ubiquitylation protein substrate, and 2. RING E3 ligases that function as a multi-protein-complex that recruits substrates via separate subunits. Examples for this subclass are the well-characterized APC/C (Anaphase promoting complex; McLean et al., 2011) and, the Cullin-based RING E3 ligases (SCF). A subclass of RING-like ligases is the U-box ligases, that contain a modified RING motif lacking canonical cysteine residues for Zn2+ coordination (Hindley et al., 2001, Patterson, 2002).

While only one ubiquitin isoform exists, three functional SUMO isoforms have been characterized: SUMO1 and SUMO2/3 that share 97% sequence identity. SUMO pathway components are much less diverse, comprising of only one known E2 conjugating enzyme termed Ubc9 and a few known E3 ligases. A key difference between SUMO1 and SUMO2/3 is the lack of a SUMOylation motif in SUMO1. Therefore, SUMO2/3 can form high molecular weight SUMO polymers with greater affinity than SUMO1. Similarly to ubiquitin ligation, SUMOylation is mostly facilitated via one of two main mechanisms: by recruiting an E2-SUMO to the substrate or by enhancing the conjugation of SUMO to a substrate already bound to the E2 (Ulrich, 2009). Like E3 ubiquitin ligases, SUMO E3 ligases are diverse and have been categorized into three families. The first class is PIAS like (Protein Inhibitor of Activated STAT–signal transducer and activator of transcription) proteins which posses an SP-RING domain functioning similarly to the ubiquitin RING domain and interacting with Ubc9. The second class is the RanBP2/Nup358 (Nuclear pore proteins Ran binding protein 2 and nucleoporin 358) like proteins, which harbour tandem elements that are capable of binding both Ubc9 and SUMO. A third group includes proteins like polycomb group protein Pc2, TOPORS (Topoisomerase I-binding RING finger protein) and likely HDAC4 (histone deacetylase 4) whose molecular mechanism of SUMO ligation is not fully understood (Gareau & Lima, 2010; Hannoun et al., 2010).

2.1.1 Mono ubiquitylation and SUMOylation

Mono-ubiquitylation is formed in most cases by a covalent attachment of the carboxy terminal glycine of the ubiquitin polypeptide to the ε-amino group within the side chain of Lys residues on the target protein. In some cases the attachment site could be the free NH2-group of the protein's first amino acid. Mono-ubiquitylation or mono-SUMOylation are mediated by ubiquitin-conjugating enzymes (Ubcs, E2s) or in the case of SUMO by the single SUMO conjugating enzyme, Ubc9. A single protein can be modified by several mono-ubiquitin monomers, resulting in multi-mono-ubiquitylation.

Mono-ubiquitylation most commonly has a regulatory function as a cellular signal, marking transmembrane receptors for recycling in the lysosome, or alternatively marking specific histone tails, thereby impacting chromatin structure. Yet, it can also serve as a degradation signal. As for SUMO, the three SUMO isoforms differ in their conjugation ability, but all three forms can be conjugated to form mono-SUMOylated substrates. Like ubiquitin, SUMO is bound to the target protein via a covalent attachment of its C-terminal carboxyl group to the ε-amino group on the Lys residue of the modified protein. Subsequently, SUMOylated and ubiquitylated proteins are recognized by a specific interaction motif that functions as a 'receptor' for these proteins on their interaction partners. In both cases multiple Lys residues within the target substrate can undergo ubiquitin/UBL modifications, generating multi-ubiquitylated or SUMOylated proteins (Hurley, 2006; Gareau & Lima, 2010).

2.1.2 Poly ubiquitylation and SUMOylation

Successive rounds of ubiquitylation or SUMOylation of a single covalently attached mono-ubiquitin or mono-SUMO molecules will generate poly-ubiquitin or poly-SUMO chains. Ubiquitin- or SUMO-protein ligase enzymes (E3), together with distinct E2s are essential for catalyzing poly-ubiquitylation, SUMOylation, and determine substrate specificity as well as govern chain structure. Following mono-ubiquitiylation, E2 and E3 ligases conjugate subsequent ubiquitin units, forming additional iso-peptide bonds between the carboxyl group of the newly added ubiquitin molecule and the ε-amino group of the Lys residue of the already covalently attached ubiquitin molecule. Recent work suggests that these chains serve as a versatile "code" that regulates protein fate, and that the internal structure and length of the chain directly impacts its recognition by "reader" proteins (Weismann et al., 2011).

Poly-ubiquitylation may be linked through any of the seven Lys residues present in the ubiquitin molecule - Lys^{6}, Lys^{11}, Lys^{27}, Lys^{29}, Lys^{33}, Lys^{48} and Lys^{63} or the ubiquitin N-terminal Met (Met^{1}). Additional ubiquitin chains can be linked through the same Lys residue within ubiquitin, forming homotypic chains. One unique form are linear poly-ubiquitin chains where the C-terminal Gly of the ubiquitin molecule is sequentially linked to the N-terminal Met of the next ubiquitin molecule. In addition, chains linked through different Lys side chains can form mixed poly-ubiquitin chains (i.e. harbouring alternating Lys linkage types) or even branched trees of ubiquitin molecules in which more than one Lys residue is extended. Thus, while only one isoform of ubiquitin exists, diverse arrays of ubiquitin chains are formed, dictating different globular structures. Aside from the role in proteasomal degradation mediated by chains with a Lys^{48} linkage, other types of poly-ubiquitylation chains regulate protein-protein interactions in processes such as DNA repair or immune signalling. For example, linear poly-ubiquitin chains are important for the regulation of NFκB signalling (as shortly described below, and Harper & Schulman, 2006; Ikeda & Dikic, 2008; Iwai & Tokunaga, 2009; Weismann et al., 2011; Kim et al., 2011; Behrends & Harper, 2011).

K63-linked ubiquitin chains have been reported to function as scaffolds for the recruitment of other signaling proteins upon cytokine stimulation, and recently an emerging unique role for ubiquitin conjugation in TNF-R signaling was characterized. TNF receptor-1 (TNF-R) activation results in K^{63}-linked ubiquitin chains. These chains are specifically generated by two ubiquitin RING E3 ligases named Haeme-Oxidized-IRP2-ubiquitin-Ligase-1 (HOIL-1) and Haeme-Oxidized-IRP2-ubiquitin-Ligase-1-Interacting-Protein (HOIP). HOIL-1 and HOIP along with a third protein, SHARPIN, form the linear ubiquitin chain assembly complex (LUBAC) that catalyzes linear head-to-tail ubiquitylation by ligating the N-terminal Met^{1} residue of the ubiquitin molecule to the C-terminal Gly residue of another ubiquitin molecule. TNF- induced LUBAC complex binds to an activator of the NFκB pathway named NFκB Essential Modifier (NEMO). LUBAC conjugates linear poly-ubiquitin chains to NEMO and enhances the interaction between NEMO and the TNF-R signaling complex. Since NEMO is required for efficient activation of the TNF-R signaling complex, LUBAC activity influences activation of the NFκB pathway. Taken together, linear ubiquitylation in this case serves as a survival machinery required for the proper activation of the TNF-R signaling complex, NFκB gene induction, and protection from TNF-induced cell death (Haas et al., 2009; Gerlech et al., 2011; Iwai, 2011; Niu et al., 2011).

As for SUMO, three SUMO1-3 genes encode for proteins that differ from one another in their ability to form SUMO chains. Only SUMO2/3 possess a Lys residue within a consensus motif ΨKXE that facilitates the formation of SUMO chains. SUMO-1 lacks this consensus site and therefore formation of poly-SUMO chains is less favorable. Yet, *in vitro* it can form polymeric chains and can serve as the chain terminator of SUMO chains. Of the eight Lys residues encoded in the SUMO molecule, SUMO chains are predominantly formed via Lys^{11}. SUMOylation of proteins was thought to enhance transcriptional repression; however, new findings suggest a more diverse function for SUMOylation. SUMOylation was found to affect sub-cellular localization and is also involved in intra-nuclear localization, transport and apoptosis, as well as in targeting proteins for ubiquitin-mediated degradation (Ulrich 2008; Matic et al., 2008; Ulrich, 2009).

2.1.3 Ubiquitin and SUMO chain editing

Covalent ubiquitylation or SUMOylation is a reversible process in which de-ubiquitylating enzymes (DUBs) or sentrins/SUMO specific proteases (SENPs), promote the cleavage of the iso-peptide bond and release ubiquitin or SUMO molecules, respectively. About 80 known DUBs, and less than 10 SENPs, are devoted to removing covalent ubiquitin or SUMO modifications. DUBs serve to perform three distinct roles in the cell; First, DUBs can cleave some of the ubiquitin molecules that are transcribed as a linear fusion chain for future conjugation processes. Second, DUBs mediate the removal of ubiquitin from tagged proteins prior to their degradation, allowing for recycling of ubiquitin molecules for future conjugation processes. Third, DUBs can trim ubiquitin chains, subsequently changing their length and structure (Komander, et al., 2009a). Editing of ubiquitin chains by specific ubiquitin peptidases may impact their recognition by the proteasome, or affect protein-protein interaction. Interaction between DUBs and ubiquitylated proteins is mediated through ubiquitin interacting motifs within the DUBs (UIMs and UBD, ubiquitin binding domains, see below and chapter 3.1.1). DUB-mediated chain editing is essential for regulation of chromatin structure, and is involved in DNA damage repair pathways as well as endosomal targeting of membrane bound receptors (Katz, 2010).

One of the well-studied DUBs is the tumor suppressor CYLD (cylindromatosis associated DUB). CYLD is a negative regulator of Wnt, NFκB and JNK signaling pathways in immunity and inflammation. Among CYLD's targets are substrates with Lys^{63}-linkage ubiquitylation chains like TRAF6, BCL3, PLK1 that regulate cell cycle proliferation and apoptosis (Massoumi, 2010). CYLD also forms a protein complex with the ubiquitin ligase Itch. Together this editing complex inhibits TAK1, which is required for termination of the immune response (Wertz, 2011). Importantly, mutations in CYLD are associated with cylindromatosis, a predisposition to benign tumors of the hair follicle in the skin and other secretory glands. Another work established that CYLD functions as a tumor suppressor and its loss is associated with cancer (Bignell et al., 2000). In this regard, DUBs are emerging as excellent targets for small molecule inhibitors. For example, recent work from the Dixit group revealed that genetically or chemically targeting USP1 induces muscle stem-cell differentiation, and can serve as a molecular target for therapy of osteosarcoma that is highly resistant to conventional chemotherapy (Williams, 2011).

Like ubiquitiylation, SUMOylation is also a reversible process, and a family of seven sentrin-specific proteases (SENPs) catalyzes de-SUMOylation. Family members differ from one another based on SUMO chain specificity and cellular localization that is determined by distinct N-terminal domains. Biochemical and genetic experiments revealed that SENPs show high degrees of specificity. Towards example, the SENP6 and SENP7 have greater affinity de-conjugation of di- and poly-SUMO2/3 chains than SUMO1 (Lima & Reverter, 2008). SENPs harbor a Ulp domain at their C-terminus which facilitates the cleavage of the isopeptidic bond between the SUMO molecule and the Lys group on the modified protein. SENP-mediated de-conjugation plays an important role in the regulation of developmental and signaling processes. It is also important for tightly regulating the levels of free SUMO in the cell. The regulatory function of SENPs is biologically relevant in development. SENP3/5 are required for ribosomal biogenesis, and targeting SENP2 during cardiac development impairs the expression of key cardiac factors Gata2 and Gata6 (Yun et al, 2008; Kang et al., 2010). Specifically, in SENP2 null embryos, SUMOylated polycomb group Pc2/CBX4 complex accumulates on the promoters of PcG target genes, leading to repression of Gata4 and Gata6 transcription. SENPs also play a key role in tumorigenesis; expression of SENP1 transforms prostate cancer cells and activates Androgen Receptor (AR) signaling. In addition, SENP3 regulates angiogenesis via its impact on HIF1α-associated coactivator p300, and elevated mRNA levels of SENP6, 7 are linked to breast cancer (Cheng, 2010; Bawa-Khalfe & Yeh, 2010).

3. Recognition motifs for ubiquitin and SUMO ligases

Are there preferred (consensus) sites for ubiquitylation? While recognition motifs for recruitment of ubiquitin ligases ("Degrons") exist, it appears that site-specific ubiquitylation is more promiscuous. This variability may stem from the different structural requirements for the diverse interactions of heterogeneous substrates with a variety of E2-E3 complexes. In contrast, the acceptor Lys residues in many SUMOylated proteins reside within a consensus motif ΨKXE (Ψ- hydrophobic residue; X, any amino acids; E, glutamic or aspartic), forming a unique conformation, which interacts directly with the specific hydrophobic groove on the Ubc9 enzyme. In addition to the canonical SUMOylation consensus sequence, longer consensus sequences with specific characteristics have been identified. Among them are the inverted form of the canonical sequence, a hydrophobic cluster motif enriched with a consecutive sequence of large hydrophobic residues, a phosphorylation dependent motif – PSDM (**ΨKXEXXS^{P}**) and a negatively charged amino-acid motif - NDSM **(ΨKXEXXEEEE).** The existence of this highly characterized consensus motif correlates with to the existence of a single unique E2-SUMO conjugating enzyme. Yet, some proteins can be SUMOylated without the presence of the characterized consensus sequence (Ulrich, 2009).

3.1 Recognition of ubiquitylated and SUMOylated proteins

Regardless of the final fate of the modified proteins, covalent tagging by ubiquitin or SUMO is "sensed" by protein motifs that subsequently mediate protein-protein interactions involved in numerous cellular processes. In this section we will discuss the currently known domains that recognize mono and poly-ubiquitin/UBL chains. While our understanding of these interactions is in its infancy, it is the focus of intensive research. For simplicity we will focus on "sensing " ubiquitin and SUMO monomers and polymers.

Proteins modified by ubiquitylation or SUMOylation interact non-covalently with other proteins via ubiquitin binding domains (UBD), or SUMO interacting motifs (SIM), respectively. These motifs are present in many proteins and thereby mediate multiple interactions, and have the potential to induce conformational changes and to impact the avidity of existing protein complexes or to form new protein complexes (see Fig 1).

3.2 Ubiquitin binding domains

UBDs are motifs that enable the association of proteins with either mono ubiquitin or ubiquitin polymers. More than 20 domains have already been identified, and more than 150 different human proteins harbor a versatile combinations of UBDs. Most UBDs interact with ubiquitylated proteins via a hydrophobic patch that include Leu8, Ile44 and Val44 within the ubiquitin molecule on one hand, and a α-helix motif of the UBD on the other hand.

Intensive structure analysis identified that UBD includes, among others, zinc finger interacting domains (ZnF/PAZ/UBZ), plekstrin homology fold (PH fold), and ubiquitin association domain (UBA) as well as ubiquitin-conjugating-like domains (Dikic et al., 2009).

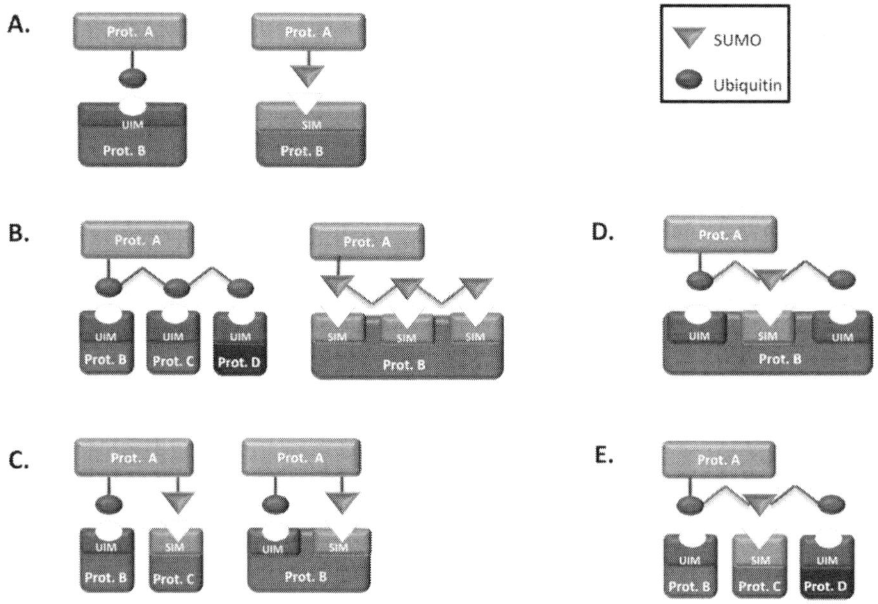

Fig. 1. **Regulation of protein-protein interactions by Ub/SUMO signalling**
(A). Mono-ubiquitin or mono-SUMO modified substrates are recognized by Ubiquitin Interacting Motif (UIM/UBDs), or by SUMO interacting motifs (SIMs). **(B).** Poly-ubiquitin or poly-SUMO chains can mediate interaction with several proteins in tandem, or with a single protein harboring several UIM/SIM domains. **(C).** Combinatorial modifications by mono ubiquitin or SUMO can be sensed by multiple proteins each harboring a discrete interaction motif, or via multiple domains within a single protein. **(D-E).** Similar to mono ubiquitylation and SUMOylation, modification by UB/SUMO mixed-chains can be recognized by multiple domains either within a single protein (D), or in the context of several proteins within a protein complex (E).

Different UBDs recognize and interact with different ubiquitin polymers. Some UBDs bind specifically to mono-ubiquitylated proteins. For example the mammalian Eap45 subunit of the endosomal sorting complex harbors a UBD domain termed GLUE domain that interacts with the hydrophobic patch of the ubiquitin molecule (Hirano et al., 2006).

Interestingly, the classical α-helix ubiquitin-interacting motif UIM, such as the one that is present in the proteasome S5a subunit and yeast Rpn10 as well as RAP80, can be found in some cases in an "inverted" orientation to form "inverted UIM" (IUIM/MIU). In most cases, each UBD interacts with a single ubiquitin molecule. Yet, ZnF binding domains, such as the one found in the guanine nucleotide exchange factor RABEX; interact with multiple surfaces on ubiquitin. Hence, in this case a single ubiquitin molecule can interact with three ZnF motifs simultaneously (Penengo et al., 2006). Tandem repeats of UBDs may dictate chain specific interaction. For example, RAP80, which is recruited together with BRCA1 to damage sites, harbors two adjacent UIMs that binds Lys^{63} but not of Lys^{48} Linked-poly-ubiquitin chains (Hicke et al., 2005; Harper & Schulman, 2006; Komander 2009b; Dikic et al. 2009). In contrast, the ubiquitin receptor RAD23A, that is required for targeting proteins to the proteasome, has a C-terminal UBA domain. RAD23 UBA domain has a 6.3 fold higher affinity to Lys^{48} than to Lys^{63} chains and a 70 fold higher affinity to Lys^{48} chains than to free ubiquitin (Raasi et al., 2005).

The specificity of different UBDs toward chain linkage is greatly dependent on UBDs present in DUBs. For example, while isoT is dedicated for the de-conjugation of Lys^{48} chains, CYLD is specific for Lys^{63}. Recently a novel UBD that recognizes linear poly-ubiquitin chains (UBAN) was characterized. Importantly, the ability of a cell to respond to TNFα and to activate the IKK kinase complex is compromised in cells that have a mutated UBAN domain within the NEMO protein that is required for activation of the IKK complex (Rahighi et al., 2009; Lo et al., 2009). Likewise, recent work has shown that the ESCRT sorting complex that is involved in targeting the EGF receptor for lysosomal degradation is based on a combination of various UBDs on different ubiquitin receptors. Together these ubiquitin receptors form a large protein complex harboring a high avidity interaction surface with an ubiquitylated cargo (Raiborg & Stenmark, 2009).

An interesting case where binding to ubiquitin chains via ubiquitin receptors plays a key role is autophagy. Autophagy is used by macrophages as a defense mechanism against infection by invading intracellular bacteria. A molecular link between autophagy and ubiquitylation was established following the identification of autophagy receptors, which simultaneously bind both ubiquitin and autophagy-specific ubiquitin-like modifiers (Atg8). Several ubiquitin-related autophagy pathways have already been characterized such as the ubiquitin-NDP52-LC3 pathway, which targets group-A-Streptococcus, *Salmonella typhimurium*, or the ubiquitin-p62-LC3 pathway, which targets *Mycobacterium tuberculosis* and *Listeria monocytogenes*. *Listeria* is a gram-positive pathogen that expresses several virulence proteins including a hemolysin (listeriolysin O, LLO). LLO bacterial proteins in macrophages infected with *Listeria* were found to form small aggregates associated with poly-ubiquitin chains and to undergo selective autophagy by the p62-LC3 pathway (Ogawa et al., 2011). A recent report determined that targeting the SUMO conjugating E2, Ubc9, for degradation and subsequently inhibiting SUMOylation mediate part of the virulence of *Listeria*. Furthermore the Dikic lab recently showed that invading *Salmonella* are coated with poly-ubiquitin chains ligated by a yet to be identified ligase. Subsequently, the ubiquitin chain binding proteins p62 and NDP52 bind to the poly-ubiquitin chains and recruit the protein Optineurin (OPTN) that upon its phosphorylation by Tank Binding Kinase (TBK) connects the coated pathogen to LC3 autophagy receptors allowing the engulfment and autophagy of the pathogen. Thus, an emerging network of interactions between the SUMO pathway, ubiquitin chains, ubiquitin binding proteins and ubiquitin like proteins (Atg8) plays a key role in elimination of bacteria and innate immunity (Wild et al., 2011; Weidberg et al., 2011).

3.3 SUMO interacting motifs

In analogy to ubiquitylation, SUMOylation is also "sensed" by a specific protein motif, termed SUMO Interacting Motif (SIM). SIM motifs are characterized by a sequence motif of hydrophobic amino acids (V/I) X (V/I) (V/I). The SIM domain interacts with the hydrophobic patch on SUMO. This hydrophobic interaction is re-enforced with other weak non-covalent interactions that are formed between basic residues on the SUMO and the acidic residues flanking the SIM domain (Gareau & Lima, 2010). SIM–mediated recruitment plays a key role in transcriptional repression. SUMO/SIM-mediated binding is required for the recruitment of histone de-acetylases and histone de-methylases to co-repressor complexes, as well as impacts the activities of chromatin remodeling factors. Examples for these interactions are the SIM/SUMO dependent recruitment of HDAC2 to SUMOylated Elk2, and the SUMO2/3 and SIM dependent recruitment of the Lys demethylase (LSD) to the CoRest co-repressor complex. In this case, CoREST1 binds directly and non-covalently SUMOylated REST (NRSF) to bridge HDAC2 and LSD (Yang & Sharrocks, 2004; Gill, 2005; Ouyang & Gill, 2009). Furthermore, SUMO/SIM interactions are likely to impact nucleosome remodeling as the recruitment of the ATP-remodeling complex protein Mi-2 to the transcription factor SP3 is SIM/SUMO dependent (Stielow et al., 2006).

SIMs and UBDs are targets for posttranscriptional modification (PTMs). These PTMs impact the structural properties of SIMs, UBDs or their immediate vicinity, resulting in a change in the binding properties of the modified domains. For example a CKII-mediated phosphorylation site near the SIM domain in the co-repressor DAXX shifts specificity between SUMO prologs. The DAXX (Fas death domain associated protein) is a transcriptional co-repressor that binds to a variety of transcription factors at the promoter sites of anti-apoptotic genes. The SIM domain within DAXX facilitates a non-covalent interaction with other SUMOylated proteins. DAXX binding to SUMO-1 but not SUMO2/3 is enhanced by CKII phosphorylation of DAXX $Ser^{737,\ 739}$ surrounding its SIM domain, enhancing its recruitment and subsequent transcriptional repression of these anti-apoptotic genes (Chang et al., 2011; Mukhopadhyay & Matunis 2011). Thus, the observations that SIMs and likely UBDs are targeted to PTMs by signaling pathways provide evidence for another layer of regulation that establish a direct crosstalk between signaling pathways and ubiquitin/SUMO signals.

4. Cross talk between ubiquitylation, SUMOylation, and the function of SUMO-targeted ubiquitin ligases

While both ubiquitin and SUMO pathways have been well studied individually, the long-speculated nature of the crosstalk between SUMO and ubiquitin pathways has been molecularly enigmatic. Importantly the interplay between SUMOylation and ubiquitylation can be a critical determinant in signaling, transcription, and cancer (Karscher 2006). For example, the equilibrium between SUMOylation and ubiquitylation can influence the balance between p53 nuclear localization and stabilization, cytoplasmic export and degradation, as well as regulating the activity and stability of Hypoxia Induced Factor (HIF; Lee et al., 2006; Carter et al., 2007; Carbia-Nagashima et al., 2007).

The crosstalk between Ubiquitin and SUMO is mediated at several levels. First, SUMOylation or ubiquitylation on the same Lys residue can differentially regulate the activity and fate of several proteins. For example, ubiquitylation of Lys164 of the Proliferating Small Nuclear Antigen (PCNA), which is required for replication and DNA damage response, enhances the recruitment of translesion error-prone DNA polymerases. Yet, genetic evidence suggests that SUMOylation at this site by the SUMO ligase Siz-1 at S-phase promotes PCNA association with the Srs2 helicase and restricts the helicase activity (Bergink & Jentch 2009). Second, enzymes within each pathway are targets for modification by the other pathway. For example, the ubiquitin ligase E2-25k undergoes SUMOylation at its core domain that inhibits its activity. Another example is the DUB USP25 that is regulated by both SUMOylation and ubiquitylation. In this case SUMOylation inhibits its function, and ubiquitylation at the same site enhances its enzymatic activity. An interesting case is the ubiquitin/SUMO ligase TOPORS. TOPORS is unique as it can catalyze the formation of either ubiquitin or SUMO chains. Importantly, a phospho-switch induced by the polo like kinase, PLK1 results site-specific phosphorylation of TOPORS, inhibiting its ability to SUMOylate, but enhancing its ubiquitylation activity (Yang et al., 2009).

However, until recently it was not clear how does the cell directly "sense" and integrate the ubiquitin and SUMO signals at the single protein level. A first direct and enzymatic link between the two pathways was established by the identification of SUMO-targeted ubiquitin ligases (STUbLs). STUbLs are a unique group of RING proteins: they bind non-covalently to the SUMO moiety of SUMOylated proteins via several SIM motifs, and subsequently target the SUMOylated protein for ubiquitylation via a RING domain [Sun et al., 2007; Geoffroy &Hay, 2009; Abed et al 2011b]. STUbLs impact protein stability, localization, and are required for the maintenance of genomic integrity, transcription and are involved cancer. Thus, STUbLs integrate the SUMO and ubiquitin pathway and generate a SUMO/ubiquitin dual signal that may serve as an additional level of regulation of protein-protein interactions.

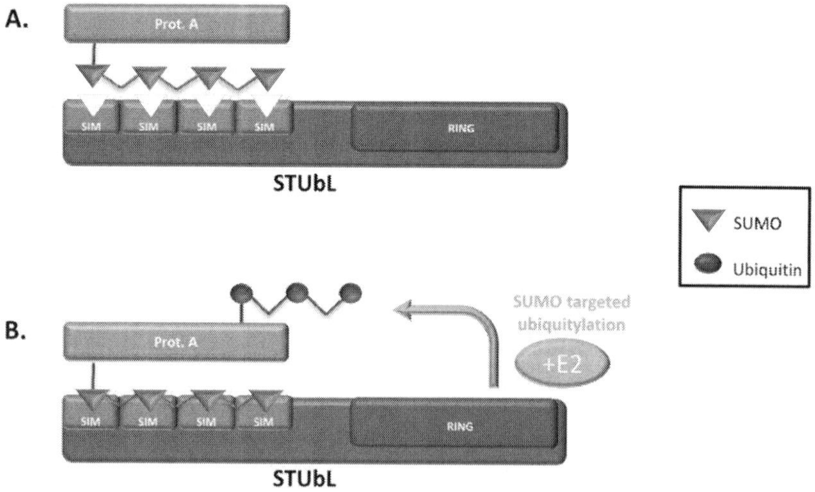

Fig. 2. Classical mode of action of STUbL: STUbL interact with SUMOylated-proteins via its SUMO Interacting Motifs (SIM). Subsequently, dimers of STUbL proteins interact with charged E2 -Ub complex, and catalyze poly-ubiquitylation via the RING domain.

4.1 Characterization of STUbL proteins

STUbLs are highly conserved in eukaryotes. Members of the STUbL family include for example: the yeast *S. pombe* Slx8-Rfp; *S. cerevisiae* Slx5–Slx8; *H. sapiens* RNF4; *D. discoideum* MIP1; and the *D. melanogaster* Degringolade (Dgrn). Yet, no clear STUbL orthologs exist in the worm *C. elegans*. These members are structurally and functionally conserved, as RNF4 protein can substitute for the yeast and fly genes in functional assays (Abed et al., 2011a Abed et al, 2011b; Barry et al., 2011; Praefcke et al., 2011). Recent structural work from the Hay lab uncovered that RNF4, and likely other STUbLs, function as dimers and that dimer formation is actively required to facilitate SUMO-dependent ubiquitin conjugation (Plechanovova' et al. 2011).

Several observations link STUBLs to protein degradation: 1. STUbLs bind and ubiquitylate SUMO chains, 2. STUbLs enhance the degradation of SUMOylated proteins 3. Genetic ablation of STUbL genes in yeast, flies, and cancer cells results in accumulation of poly-SUMOylated proteins. Among the most studied substrates of RNF4 are the promyelocytic leukaemia protein, PML and its derived oncogene PML-RAR. An elegant set of experiments by several groups established that arsenic-induced phosphorylation enhances poly SUMOylation of PML and recruitment of RNF4. Subsequently, RNF4-dependent ubiquitylation targets SUMOylated and ubiquitylated PML for degradation via the 26S proteasome (Lallemand-Breitenbach et al., 2008; Tathem et al., 2008). This is highly relevant to the treatment of acute promyelocytic leukemia, where a combination of Retinoic Acid (RA) with arsenic treatment can result in 90% cure (Lallemand, 2011). Other bona-fide human substrates of RNF4 are the kinetochore proteins CENP-I and VHL. In addition, proteomic analysis using RNF4 as a bait identified a wide spectrum of proteins that are bound by RNF4 (Makhopadhyay et al., 2010; Tatham et al., 2011). While the exact nature of these interactions requires further characterization, GO analysis already points out that SUMO-dependent ubiquitylation by RNF4 is relevant to a large verity of protein complexes and involves diverse cellular process.

An interesting issue is the recognition of proteins by STUbLs. While by their definition STUbLs recognize SUMOylated substrates via their SIM motifs, recent reports suggest a more complex picture. For example, the recognition of the Mat2α repressor by Slx5: Slx8 does not require substrate SUMOylation, but does require intact SIM motifs (Xia et al., 2010) Furthermore, the binding of the *Drosophila* STUbL Dgrn to the Notch-related HES repressor proteins is independent of SUMOylation, and is mediated by Dgrn's RING domain and not the SIM motif. Yet, the SIM motifs are required for ubiquitylation *in vitro* and for Dgrn's impact *in vivo* (Abed et al., 2011a, Barry et al., 2011). Thus, it is highly likely that specific interactions are determined *in vivo* by a dual recognition machinery, where the SIM motif interacts with the poly-SUMO chain and the RING domain interacts with other non-SUMO determinants in the target protein or adjacent proteins. In addition, recent work suggests that RNF4 can interact with proteins such as Nip45 that harbour two SUMO Like Domains (SLDs), but are not SUMOylated, thus expanding the spectrum of RNF4 targets (Sekiyama et al., 2010).

4.2 Cellular processes regulated by STUbLs

STUbLs are required for normal development, for the cell's ability to cope with genotoxic stress, and to maintain genome stability (Prudden et al., 2007; Nagi et al. 2008; Nagi et al.,

2011and Barry et al., 2011). During mouse development, RNF4 is highly expressed in the stem cell compartment of the developing gonads and brain (43). RNF4 was also identified as a gene that is specifically expressed in hematopoietic, embryonic, and neural progenitor cells, likely representing its role in "stemness" (Galili et al., 2000; Ramalho-Santos et al., 2002). During early *Drosophila* development Dgrn localizes to centrosomes, and *dgrn* null embryos accumulate SUMOylated proteins. *dgrn* null embryos show genomic instability phenotypes; they fail to incorporate DNA into centrosomes, assemble aberrant mitotic spindles and exhibit chromosomal bridges at anaphase. Cells in *dgrn* null embryos show high SUMO content and fall from the embryo surface into the center of the syncytium (Barry et al., 2011). These findings fit well with those reported for the yeast STUbLs, as yeast lacking *Slx5:Slx8* display genomic instability, and are hypersensitive to replication stress. For example, yeast deficient in *Slx5: Slx8* fail to replicate DNA upon hydroxy-urea treatment (Prudden 2007, Rouse, 2009). While the protein substrates of STUbLs in this context are still unknown, the observations that many of the proteins involved in the DNA damage response are ubiquitylated, SUMOylated, or contain SLD motifs suggest that STUbLs are targeting specific regulatory "nodes" in the DNA response network. Since genomic instability is a hallmark of cancer cells, proteins such as STUbLs may be the "Achilles heel " in specific cancerous settings. Therefore we predict that STUbLs inhibition will results in collapse of the tumorigenic network and cancer elimination.

Interestingly, and prior to the identification of RNF4 as a dedicated STUbL protein, RNF4 was identified by the Palvimo lab as a potent transcriptional regulator that functions both as a co-activator or a co-repressor depending on the cellular context. For example RNF4 was shown to be essential for androgen and steroid receptor-mediated target gene activation (Moilanen et al., 1998; Poukka et al., 2000;). In addition, we found that in transcription the consequences of Dgrn activity are not strictly limited to targeting SUMOylated proteins for degradation (Abed et al., 2011a; Barry et al., 2011). Importantly, Dgrn/RNF4-mediated ubiquitylation impacts the affinity between proteins, inhibiting the interaction of a given protein with one protein but not affecting its ability to bind other protein. Specifically, we found that during fly development Dgrn serves as a molecular selector that determines co-repressor recruitment as described below. Dgrn-mediated ubiquitylation of the HES-related repressor Hairy inhibits its ability to interact with its co-repressor Groucho but not with other Hairy co-repressors such as dCtBP. In addition, we find that Dgrn specifically targets SUMOylated Gro for sequestration. Yet, the exact cellular and molecular details surrounding sequestration require further exploration. Accordingly, Dgrn antagonize Hairy/Groucho-mediated repression in transcription and function in cells and *in vivo*. Genome wide association studies using DamID profiling unveiled that the activity of Dgrn is relevant genome wide. Thus, Dgrn serves as a "molecular selector" that determines protein-protein interactions. We found that this "selector" activity of Dgrn/RNF4 is likely relevant also to HES independent processes and in other types of co-factor switches, and may directly impact chromatin structure (Abed et al., 2011a; Hu et al., 2010; Orian unpublished). We speculate that this activity of STUbLs will be highly relevant not only in transcription, but in the regulation of protein-protein interactions in other cellular process such as the selective recruitment of proteins to DNA repair foci.

5. Conclusion and future challenges

We focused on ubiquitin and SUMO signalling as a mode to regulate protein-protein interactions. We predict that the lessons learned during the last decades regarding ubiquitin and SUMO will be highly relevant for other UBLs and non UBL modifications. An important concept that emerges from these studies is that combinatorial posttranscriptional modifications by ubiquitin/UBLs serves as a molecular tool to regulate and establish diverse and selective, signal-induced protein-protein interactions. Furthermore, proteins that have ubiquitin like or SUMO like domains, and the ability of specific enzymes to catalyse different and distinct chains of ubiquitin/UBL proteins, further add to this diversity. This complexity is also reflected at the level of "reader" proteins that contain several UBD/UBL recognition motifs and that bind only a discrete combinatorial ubiquitin/UBL signal. Thus, we can envision how a relatively small number of signalling pathways and a limited pool of ubiquitin/UBLs can generate discrete protein-protein interactions. We predict that a key objective for future studies will be to understand the enzymatic machinery that dictates selective recruitment In distinct cellular processes. The identification of these enzymes has direct implications beyond basic research. It will pave the way to design highly selective inhibitors that will impact specific pathways with minimal side effects, features that are desired in the clinic such as in the case of cancer treatments.

6. Acknowledgment

We thank Tom Schultheiss and members of the Orian lab for discussions and comments on the manuscript. This work was supported by; ISF grants (F.I.R.ST 1215/07 and 418/09), ICRF grant (2011-3075-PG), a special ICA concert in the name of Menashe Mani, and the Rappaport Research Fund to AO.

7. References

Abed, M., Barry, K. C., Kenyagin, D., Koltun, B., Phippen, T. M., Delrow, J. J., Parkhurst, S. M. & Orian, A. (2011a). Degringolade, a SUMO-targeted ubiquitin ligase, inhibits Hairy/Groucho-mediated repression. *EMBO J*, Vol. 30, No. 7, (April 2011), pp. 1289-1130, ISSN 1460-2075

Abed, M., Bitman-Lotan, E. & Orian, A. (2011b). A fly view of a SUMO-targeted ubiquitin ligase. *Fly (Austin)*, Vol. 5, No. 4, (October 2011), PMID 21857164, ISSN 1933-6942

Barry, K. C., Abed, M., Kenyagin, D., Werwie, T. R., Boico, O., Orian, A. & Parkhurst, S. M. (2011). The Drosophila STUbL protein Degringolade limits HES functions during embryogenesis. *Development*, Vol. 138, No. 9, (April 2011), pp. 1759-1769, ISSN 1477-9129

Bawa-Khalfe, T. & Yeh, E. T. (2010). SUMO Losing Balance: SUMO Proteases Disrupt SUMO Homeostasis to Facilitate Cancer Development and Progression. *Genes Cancer*, Vol. 1, No. 7, (June 2010), pp. 748-752, ISSN 1947-6027

Behrends, C. & Harper, J. W. (2011). Constructing and decoding unconventional ubiquitin chains. *Nat Struct Mol Biol*, Vol. 18, No. 5, (May 2011), pp. 520-528, ISSN 1545-9985

Bergink, S. & Jentsch, S. (2009). Principles of ubiquitin and SUMO modifications in DNA repair. *Nature*, Vol. 458, No. 7237, (March 2009), pp. 461-467, ISSN 1476-4687

Bignell, G. R., Warren, W., Seal, S., Takahashi, M., Rapley, E., Barfoot, R., Green, H., Brown, C., Biggs, P. J., & Lakhani, S. R. (2000). Identification of the familial cylindromatosis tumour-suppressor gene. *Nat Genet*, Vol. 25, No. 2. (June 2000), pp. 160-165, ISSN 1061-4036

Carbia-Nagashima, A., Gerez, J., Perez-Castro, C., Paez-Pereda, M., Silberstein, S., Stalla, G. K., Holsboer, F. & Arzt, E. (2007). RSUME, a small RWD-containing protein, enhances SUMO conjugation and stabilizes HIF-1alpha during hypoxia. *Cell*, Vol. 131, No. 2, (October 2007), pp. 309-323, ISSN 0092-8674

Carter, S., Bischof, O., Dejean, A. & Vousden, K. H. (2007). C-terminal modifications regulate MDM2 dissociation and nuclear export of p53. *Nat Cell Biol*, Vol. 9, No. 4, (April 2007), pp. 428-435, ISSN 1465-7392

Chang, C. C., Naik, M. T., Huang, Y. S., Jeng, J. C., Liao, P. H., Kuo, H. Y., Ho, C. C., Hsieh, Y. L., Lin, C. H., & Huang, N. J. (2011). Structural and functional roles of Daxx SIM phosphorylation in SUMO paralog-selective binding and apoptosis modulation. *Mol Cell*, Vol. 42, No. 1, (April 2011), pp. 62-74, ISSN 1097-4167

Cheng, J., Kang, X., Zhang, S. & Yeh, E. T. (2007). SUMO-specific protease 1 is essential for stabilization of HIF1alpha during hypoxia. *Cell*, Vol. 131, No. 3, (November 2007), pp. 584-595, ISSN 0092-8674

Deshaies, R. J. & Joazeiro, C. A. (2009). RING domain E3 ubiquitin ligases. *Annu Rev Biochem*, Vol. 78, pp. 399-434, ISSN 1545-4509

Dikic, I., Wakatsuki, S. & Walters, K. J. (2009). Ubiquitin-binding domains - from structures to functions. *Nat Rev Mol Cell Biol*, Vol. 10, No. 10, (October 2009), pp. 659-671, ISSN 1471-0080

Galili, N., Nayak, S., Epstein, J. A. & Buck, C. A. (2000). Rnf4, a RING protein expressed in the developing nervous and reproductive systems, interacts with Gscl, a gene within the DiGeorge critical region. *Dev Dyn*, Vol. 218, No. 1, (May 2000), pp. 102-111, ISSN 1058-8388

Gareau, J. R. & Lima, C. D. (2010). The SUMO pathway: emerging mechanisms that shape specificity, conjugation and recognition. *Nat Rev Mol Cell Biol*, Vol. 11, No. 12, (December 2010). pp. 861-871, ISSN 1471-0080

Geoffroy, M. C. & Hay, R. T. (2009). An additional role for SUMO in ubiquitin-mediated proteolysis. *Nat Rev Mol Cell Biol*, Vol. 10, No. 8, (August 2009), pp. 564-568, ISSN 1471-0080

Gerlach, B., Cordier, S. M., Schmukle, A. C., Emmerich, C. H., Rieser, E., Haas, T. L., Webb, A. I., Rickard, J. A., Anderton, H., & Wong, W. W. (2011). Linear ubiquitination preventsinflammation and regulates immune signalling. *Nature*, Vol. 471, No. 7340, (March 2011), pp. 591-596, ISSN 1476-4687

Gill, G. (2005). Something about SUMO inhibits transcription. *Curr Opin Genet Dev*, Vol. 15, No.5, (October 2005), pp. 536-541, ISSN 0959-437X

Haas, T. L., Emmerich, C. H., Gerlach, B., Schmukle, A. C., Cordier, S. M., Rieser, E., Feltham, R., Vince, J., Warnken, U., & Wenger, T. (2009). Recruitment of the linear ubiquitin chain assembly complex stabilizes the TNF-R1 signaling complex and is required for TNF-mediated gene induction. *Mol Cell*, Vol. 36, No. 5, (December 2009), pp. 831-844, ISSN 1097-4164

Hannoun, Z., Greenhough, S., Jaffray, E., Hay, R. T. & Hay, D. C. (2010). Post-translational modification by SUMO. *Toxicology*, Vol. 278, No. 3, (December 2010), pp. 288-293, ISSN 1879-3185

Harper, J. W. & Schulman, B. A. (2006). Structural complexity in ubiquitin recognition. *Cell* Vol. 124, No. 6, (March 2006), pp. 1133-1136, ISSN 0092-8674

Hershko, A. (1983). Ubiquitin: roles in protein modification and breakdown. *Cell*, Vol. 34, No. 1, (August 1983), pp. 11-12, ISSN 0092-8674

Hershko, A. & Ciechanover, A. (1998). The ubiquitin system. *Annu Rev Biochem*, Vol. 67, pp. 425-479, ISSN 0066-4154

Hicke, L., Schubert, H. L. & Hill, C. P. (2005). Ubiquitin-binding domains. *Nat Rev Mol Cell Biol*, Vol. 6, No. 8, (August 2005), pp. 610-621, ISSN 1471-0072

Hindley, C. J., McDowell, G. S., Wise, H. & Philpott, A. (2011). Regulation of cell fate determination by Skp1-Cullin1-F-box (SCF) E3 ubiquitin ligases. *Int J Dev Biol*, Vol. 55, No. 3, pp. 249-260, ISSN 1696-3547

Hirano, S., Suzuki, N., Slagsvold, T., Kawasaki, M., Trambaiolo, D., Kato, R., Stenmark, H. & Wakatsuki, S. (2006). Structural basis of ubiquitin recognition by mammalian Eap45 GLUE domain. *Nat Struct Mol Biol*, Vol. 13, No. 11, (November 2006), pp. 1031-1032, ISSN 1545-9993

Hochstrasser, M. (2009). Origin and function of ubiquitin-like proteins. *Nature*, Vol. 458, No. 7237, (March 2009), pp. 422-429, ISSN 1476-4687

Hu, X. V., Rodrigues, T. M., Tao, H., Baker, R. K., Miraglia, L., Orth, A. P., Lyons, G. E., Schultz, P. G. & Wu, X. (2010). Identification of RING finger protein 4 (RNF4) as a modulator of DNA demethylation through a functional genomics screen. *Proc Natl Acad Sci USA*, Vol. 107, No. 34, (August 2010), pp. 15087-15092, ISSN 1091-6490

Hurley, J. H., Lee, S. & Prag, G. (2006). Ubiquitin-binding domains. *Biochem J*, Vol. 399, No. 3, (November 2006), pp. 361-372, ISSN 1470-8728

Ikeda, F. & Dikic, I. (2008). Atypical ubiquitin chains: new molecular signals. 'Protein Modifications: Beyond the Usual Suspects' review series. *EMBO Rep*, Vol. 9, No. 6, (June 2008), pp. 536-542, ISSN 1469-3178

Iwai, K. (2011). Linear polyubiquitin chains: a new modifier involved in NFkappaB activation and chronic inflammation, including dermatitis. *Cell Cycle*, Vol. 10, No. 18, (September 2011), pp. 3095-3104, ISSN 1551-4005

Iwai, K. & Tokunaga, F. (2009). Linear polyubiquitination: a new regulator of NF-kappaB activation. *EMBO Rep*, Vol. 10, No. 7, (July 2009), pp. 706-713, ISSN 1469-3178

Kang, X., Qi, Y., Zuo, Y., Wang, Q., Zou, Y., Schwartz, R. J., Cheng, J. & Yeh, E. T. (2010). SUMO-specific protease 2 is essential for suppression of polycomb group protein-mediated gene silencing during embryonic development. *Mol Cell*, Vol. 38, No. 2, (April 2010), pp. 191-201, ISSN 1097-4164

Katz, E. J., Isasa, M. & Crosas, B. (2010). A new map to understand deubiquitination. *Biochem Soc Trans*, Vol. 38, No. Pt1, (February 2010), pp. 8-21, ISSN 1470-8752

Kerscher, O., Felberbaum, R. & Hochstrasser, M. (2006). Modification of proteins by ubiquitin and ubiquitin-like proteins. *Annu Rev Cell Dev Biol*, Vol. 22, pp. 159-180, ISSN 1081-0706

Kim, W., Bennett, E. J., Huttlin, E. L., Guo, A., Li, J., Possemato, A., Sowa, M. E., Rad, R., Rush, J., & Comb, M. J. (2011). Systematic and quantitative assessment of the ubiquitin-modified proteome. *Mol Cell*, Vol. 44, No. 2, (October 2011), pp. 325-340, ISSN 1097-4164

Komander, D., Clague, M. J. & Urbe, S. (2009a). Breaking the chains: structure and function of the deubiquitinases. *Nat Rev Mol Cell Biol*, Vol. 10, No. 8, (August 2009), pp. 550-563, ISSN 1471-0080

Komander, D., Reyes-Turcu, F., Licchesi, J. D., Odenwaelder, P., Wilkinson, K. D. & Barford, D. (2009b). Molecular discrimination of structurally equivalent Lys 63-linked and linear polyubiquitin chains. *EMBO Rep*, Vol. 10. No. 5, (April 2009), pp. 466-473, ISSN 1469-3178

Lallemand-Breitenbach, V., Jeanne, M., Benhenda, S., Nasr, R., Lei, M., Peres, L., Zhou, J., Zhu, J., Raught, B. & de The, H. (2008). Arsenic degrades PML or PML-RARalpha through a SUMO-triggered RNF4/ubiquitin-mediated pathway. *Nat Cell Biol*, Vol. 10, No. 5, (May 2008), pp. 547-555, ISSN 1476-4679

Lallemand-Breitenbach, V., Zhu, J., Chen, Z. & de The, H. (2011). Curing APL through PML/RARA degradation by As(2)O(3). *Trends Mol Med*. PMID 22056243, ISSN 1471-499X

Lee, M. H., Lee, S. W., Lee, E. J., Choi, S. J., Chung, S. S., Lee, J. I., Cho, J. M., Seol, J. H., Baek, S. H., & Kim, K. I. (2006). SUMO-specific protease SUSP4 positively regulates p53 by promoting Mdm2 self-ubiquitination. *Nat Cell Biol*, Vol. 8, No. 12, (December 2006), pp. 1424-31, ISSN 1465-7392

Lima, C. D. & Reverter, D. (2008). Structure of the human SENP7 catalytic domain and poly-SUMO deconjugation activities for SENP6 and SENP7. *J Biol Chem*, Vol. 283, No. 46, (November 2008), pp. 32045-32055, ISSN 0021-9258

Lo, Y. C., Lin, S. C., Rospigliosi, C. C., Conze, D. B., Wu, C. J., Ashwell, J. D., Eliezer, D. & Wu, H. (2009). Structural basis for recognition of diubiquitins by NEMO. *Mol Cell*, Vol. 33, No. 5, (February 2009), pp. 602-615, ISSN 1097-4164

Massoumi, R. (2010). Ubiquitin chain cleavage: CYLD at work. *Trends Biochem Sci*, Vol. 35, No. 7, (July 2010), pp. 392-399, ISSN 0968-0004

Matic, I., van Hagen, M., Schimmel, J., Macek, B., Ogg, S. C., Tatham, M. H., Hay, R. T., Lamond, A. I., Mann, M. & Vertegaal, A. C. (2008). In vivo identification of human small ubiquitin-like modifier polymerization sites by high accuracy mass spectrometry and an in vitro to in vivo strategy. *Mol Cell Proteomics*, Vol. 7, No, 1 (January 2007), pp. 132-144, ISSN 1535-9476

McLean, J. R., Chaix, D., Ohi, M. D. & Gould, K. L. (2011). State of the APC/C: organization, function, and structure. *Crit Rev Biochem Mol Biol*, Vol. 46, No. 2, (April 2011), pp. 118-136, ISSN 1549-7798

Moilanen, A. M., Poukka, H., Karvonen, U., Hakli, M., Janne, O. A. & Palvimo, J. J. (1998). Identification of a novel RING finger protein as a coregulator in steroid receptor-mediated gene transcription. *Mol Cell Biol*, Vol. 18, No. 9, (September 1998), pp. 5128-5139, ISSN 0270-7306

Mukhopadhyay, D., Arnaoutov, A. & Dasso, M. (2010). The SUMO protease SENP6 is essential for inner kinetochore assembly. *J Cell Biol*, Vol. 188, No. 5, (March 2010), pp. 681-692, ISSN 1540-8140

Mukhopadhyay, D. & Matunis, M. J. (2011). SUMmOning Daxx-mediated repression. *Mol Cell*, Vol. 42, No. 1, (April 2009), pp. 4-5, ISSN 1097-4164

Nagai, S., Davoodi, N. & Gasser, S. M. (2011). Nuclear organization in genome stability: SUMO connections. *Cell Res*, Vol. 21, No. 3, (March 2011), pp. 474-485, ISSN 1748-7838

Nagai, S., Dubrana, K., Tsai-Pflugfelder, M., Davidson, M. B., Roberts, T. M., Brown, G. W., Varela, E., Hediger, F., Gasser, S. M. & Krogan, N. J. (2008). Functional targeting of DNA damage to a nuclear pore-associated SUMO-dependent ubiquitin ligase. *Science*, Vol. 322, No. 5901, (October 2008), pp. 597-602, ISSN 1095-9203

Niu, J., Shi, Y., Iwai, K. & Wu, Z. H. (2011). LUBAC regulates NF-kappaB activation upon genotoxic stress by promoting linear ubiquitination of NEMO. *EMBO J*, Vol. 30, No. 18), pp. 3741-3753, ISSN 1460-2075

Ogawa, M., Yoshikawa, Y., Mimuro, H., Hain, T., Chakraborty, T. & Sasakawa, C. (2011). Autophagy targeting of Listeria monocytogenes and the bacterial countermeasure. *Autophagy*, Vol. 7, No. 3, (March 2011), pp. 310-314, ISSN 1554-8635

Ouyang, J. & Gill, G. (2009). SUMO engages multiple corepressors to regulate chromatin structure and transcription. *Epigenetics*, Vol. 4, No. 7, (October 2010), pp. 440-444, ISSN 1559-2308

Patterson, C. (2002). A new gun in town: the U box is a ubiquitin ligase domain. *Sci STKE*, Vol. 2002, No. 116, (January 2002), pp. pe4, ISSN 1525-8882

Penengo, L., Mapelli, M., Murachelli, A. G., Confalonieri, S., Magri, L., Musacchio, A., Di Fiore, P. P., Polo, S. & Schneider, T. R. (2006). Crystal structure of the ubiquitin binding domains of rabex-5 reveals two modes of interaction with ubiquitin. *Cell*, Vol. 124, No. 6. (March 2006), pp. 1183-1195, ISSN 0092-8674

Pickart, C. M. (2001). Mechanisms underlying ubiquitination. *Annu Rev Biochem*, Vol. 70, pp. 503-533, ISSN 0066-4154

Plechanovova, A., Jaffray, E. G., McMahon, S. A., Johnson, K. A., Navratilova, I., Naismith, J. H. & Hay, R. T. (2011). Mechanism of ubiquitylation by dimeric RING ligase RNF4. *Nat Struct Mol Biol*, Vol. 18, No. 9, (August 2011), pp. 1052-1059, ISSN 1545-9985

Poukka, H., Aarnisalo, P., Santti, H., Janne, O. A. & Palvimo, J. J. (2000). Coregulator small nuclear RING finger protein (SNURF) enhances Sp1- and steroid receptor-mediated transcription by different mechanisms. *J Biol Chem*, Vol. 275, No. 1, (January 2000), pp. 571-579, ISSN 0021-9258

Praefcke, G. J., Hofmann, K. & Dohmen, R. J. (2011). SUMO playing tag with ubiquitin. *Trends Biochem Sci.*, (October 2011), ISSN 0968-0004

Prudden, J., Pebernard, S., Raffa, G., Slavin, D. A., Perry, J. J., Tainer, J. A., McGowan, C. H. & Boddy, M. N. (2007). SUMO-targeted ubiquitin ligases in genome stability. *EMBO J*, Vol. 26, No. 18, (September 2007), pp. 4089-4101, ISSN 0261-4189

Raasi, S., Varadan, R., Fushman, D. & Pickart, C. M. (2005). Diverse polyubiquitin interaction properties of ubiquitin-associated domains. *Nat Struct Mol Biol*, Vol. 12, No. 8, pp. 708-714, ISSN 1545-9993

Rahighi, S., Ikeda, F., Kawasaki, M., Akutsu, M., Suzuki, N., Kato, R., Kensche, T., Uejima, T., Bloor, S., & Komander, D. (2009). Specific recognition of linear ubiquitin chains by NEMO is important for NF-kappaB activation. *Cell*, Vol. 136, No. 6, (March 2009), pp. 1098-1109, ISSN 1097-4172

Raiborg, C. & Stenmark, H. (2009). The ESCRT machinery in endosomal sorting of ubiquitylated membrane proteins. *Nature*, Vol. 458, No. 7237, (March 2009), pp. 445-452, ISSN 1476-4687

Ramalho-Santos, M., Yoon, S., Matsuzaki, Y., Mulligan, R. C. & Melton, D. A. (2002). "Stemness": transcriptional profiling of embryonic and adult stem cells. *Science*, Vol. 298, No. 5593, (October 18), pp. 597-600, ISSN 1095-9203

Rouse, J. (2009). Control of genome stability by SLX protein complexes. *Biochem Soc Trans*, Vol. 37, No. Pt3, pp. 495-510, ISSN 1470-8752

Sekiyama, N., Arita, K., Ikeda, Y., Hashiguchi, K., Ariyoshi, M., Tochio, H., Saitoh, H. & Shirakawa, M. (2010). Structural basis for regulation of poly-SUMO chain by a SUMO-like domain of Nip45. *Proteins*, Vol. 78, No. 6, pp. 1491-1502, ISSN 1097-0134

Stielow, B., Sapetschnig, A., Kruger, I., Kunert, N., Brehm, A., Boutros, M. & Suske, G. (2008). Identification of SUMO-dependent chromatin-associated transcriptional repression components by a genome-wide RNAi screen. *Mol Cell*, Vol. 29, No. 6, (April 2008), pp. 742-754, ISSN 1097-4164

Sun, H., Leverson, J. D. & Hunter, T. (2007). Conserved function of RNF4 family proteins in eukaryotes: targeting a ubiquitin ligase to SUMOylated proteins. *EMBO J*, Vol. 26, No. 18, (September 2007), pp. 4102-4112, ISSN 0261-4189

Tatham, M. H., Geoffroy, M. C., Shen, L., Plechanovova, A., Hattersley, N., Jaffray, E. G., Palvimo, J. J. & Hay, R. T. (2008). RNF4 is a poly-SUMO-specific E3 ubiquitin ligase required for arsenic-induced PML degradation. *Nat Cell Biol*, Vol. 10, No. 5, (May 2008), pp. 538-546, ISSN 1476-4679

Tatham, M. H., Matic, I., Mann, M. & Hay, R. T. (2011). Comparative proteomic analysis identifies a role for SUMO in protein quality control. *Sci Signal*, Vol. 4, No. 178, pp. rs4, ISSN 1937-9145

Ulrich, H. D. (2008). The fast-growing business of SUMO chains. *Mol Cell*, Vol. 32, No. 3, (November 2008), pp. 301-305, ISSN 1097-4164

Ulrich, H. D. (2009). The SUMO system: an overview. *Methods Mol Biol*, Vol. 497, pp. 3-16, ISSN 1064-3745

Weidberg, H. & Elazar, Z. (2011). TBK1 mediates crosstalk between the innate immune response and autophagy. *Sci Signal*, Vol. 4, No. 187, (August 2008), pp. pe39, ISSN 1937-9145

Weissman, A. M., Shabek, N. & Ciechanover, A. (2011). The predator becomes the prey: regulating the ubiquitin system by ubiquitylation and degradation. *Nat Rev Mol Cell Biol*, Vol. 12, No. 9, (August 2011), pp. 605-620, ISSN 1471-0080

Wertz, I. E. (2011). It takes two to tango: a new couple in the family of ubiquitin-editing complexes. *Nat Immunol*, Vol. 12, No. 12, (November 2011), pp. 1133-1135, ISSN 1529-2916

Wild, P., Farhan, H., McEwan, D. G., Wagner, S., Rogov, V. V., Brady, N. R., Richter, B., Korac, J., Waidmann, O., & Choudhary, C. (2011). Phosphorylation of the autophagy receptor optineurin restricts Salmonella growth. *Science*, Vol. 333, No. 6039, (May 2011), pp. 228-233, ISSN 1095-9203

Williams, S. A., Maecker, H. L., French, D. M., Liu, J., Gregg, A., Silverstein, L. B., Cao, T. C., Carano, R. A. & Dixit, V. M. (2011). USP1 deubiquitinates ID proteins to preserve a mesenchymal stem cell program in osteosarcoma. *Cell*, Vol. 146, No. 6, (September 2011), pp. 918-930, ISSN 1097-4172

Xie, Y., Rubenstein, E. M., Matt, T. & Hochstrasser, M. (2010). SUMO-independent in vivo activity of a SUMO-targeted ubiquitin ligase toward a short-lived transcription factor. *Genes Dev*, Vol. 24, No. 9, (May 2010), pp. 893-903, ISSN 1549-5477

Yang, S. H. & Sharrocks, A. D. (2004). SUMO promotes HDAC-mediated transcriptional repression. *Mol Cell*, Vol. 13, No. 4, (February 2004), pp. 611-617, ISSN 1097-2765

Yang, X., Li, H., Zhou, Z., Wang, W. H., Deng, A., Andrisani, O. & Liu, X. (2009). Plk1-mediated phosphorylation of Topors regulates p53 stability. *J Biol Chem*, Vol. 284, No. 28, (July 2009), pp. 18588-18592, ISSN 0021-9258

Yun, C., Wang, Y., Mukhopadhyay, D., Backlund, P., Kolli, N., Yergey, A., Wilkinson, K. D. & Dasso, M. (2008). Nucleolar protein B23/nucleophosmin regulates the vertebrate SUMO pathway through SENP3 and SENP5 proteases. *J Cell Biol*, Vol. 183, No. 4, (November 2008), pp. 589-595, ISSN 1540-8140

Zhang, X. D., Goeres, J., Zhang, H., Yen, T. J., Porter, A. C. & Matunis, M. J. (2008). SUMO-2/3 modification and binding regulate the association of CENP-E with kinetochores and progression through mitosis. *Mol Cell*, Vol. 29, No. 6, (March 2008), pp. 729-741, ISSN 1097-4164

5

Functional Protein Interactions in Steroid Receptor-Chaperone Complexes

Thomas Ratajczak[1,2,*], Rudi K. Allan[1,2], Carmel Cluning[1,2] and Bryan K. Ward[1,2]
[1]*Laboratory for Molecular Endocrinology, Western Australian Institute for Medical Research and the UWA Centre for Medical Research, The University of Western Australia, Nedlands WA,*
[2]*Department of Endocrinology & Diabetes, Sir Charles Gairdner Hospital, Hospital Avenue, Nedlands WA, Australia*

1. Introduction

Heat shock protein 90 (Hsp90) is unique in that it chaperones a select group of client proteins and assists their folding in preparation for key regulatory roles in cellular signalling. Steroid receptors are among the most extensively studied Hsp90 chaperone substrates and belong to the large nuclear receptor superfamily of hormone-activated transcription factors that respond to hormonal cues through conformational changes induced by hormone binding within the ligand-binding domain (LBD). In an ATP-dependent assembly process, high affinity hormone binding is achieved through the direct interaction of the steroid receptor LBD with Hsp90 and specific Hsp90-associated chaperones. After synthesis, steroid receptors enter the Hsp90 chaperoning pathway by initial assembly with Hsp40, followed by incorporation of Hsp70 and Hip. The binding of Hop and Hsp90 then generates an intermediate receptor complex which is further modified by the release of Hsp70 and Hop, allowing a transition of the receptor to hormone-binding competency. Recruitment of p23 leads to formation of mature receptor complexes capable of binding hormone with high affinity and characterized by the additional presence of one of the immunophilin cochaperones, FKBP51, FKBP52, CyP40 and PP5. This dynamic assembly of receptors to a hormone-activatable state, together with a selective functionality of receptors associated with specific Hsp90-immunophilin complexes provides mechanisms through which Hsp90 and the immunophilin cochaperones may regulate hormone-induced signalling events. This may occur directly by enhancing hormone binding as has been observed for AR, GR and PR associated with Hsp90-FKBP52 complexes or indirectly by facilitating nuclear import of receptor as seen

[*] Corresponding Author

subsequent to the hormone-induced exchange of FKBP51 by FKBP52 in GR-Hsp90 complexes. For more in depth summaries related to the mechanism and functional consequences of steroid receptor assembly with the Hsp90 chaperone machine, readers are referred to recent reviews (Echeverria & Picard, 2010; Picard, 2006; Pratt & Toft, 2003; Ratajczak *et al.*, 2003; Riggs *et al.*, 2004; Smith & Toft, 2008).

It is understood that ligand binding induces conformational changes within the steroid receptor LBD, facilitating release of Hsp90 and its cochaperones and exposing elements required for homodimerization, nuclear translocation and DNA binding. The mechanisms through which Hsp90 chaperone machinery regulates the physiological response to steroid hormones mediated by steroid receptors remain unclear. In early work, multiple approaches that included deletion analyses, peptide competition studies and use of the *in vitro* receptor-Hsp90 heterocomplex assembly system present in rabbit reticulocyte lysate were aimed at defining the regions within steroid receptors and Hsp90 responsible for interaction (Pratt & Toft, 1997). These revealed that the GR LBD was essential for formation of apo-GR-Hsp90 heterocomplexes and defined a ~100-amino acid minimal segment (human GR residues 550-653) required for high-affinity Hsp90-binding. The region contains the so-called signature sequence (human GR residues 577-596) that is conserved among steroid receptors, and may contribute to the stability of receptor-Hsp90 interaction. Despite the identification of this core Hsp90 interaction domain, other results suggested a role for nearly all of the LBD in GR association with Hsp90. Similar studies with PR and ERα also concluded that several regions throughout the LBD participate in the assembly of receptor-Hsp90 complexes, although for ERα the much less stable association of the LBD with Hsp90 requires a short upstream sequence (human ERα residues 251-71), located at the C-terminal end of the DNA-binding domain, to confer increased stability. Since Hsp90 has not been shown to bind directly to this upstream sequence, it has been proposed that the region may alternatively serve as a contact site for Hsp90 cochaperones (e.g. FKBP52) (Pratt & Toft, 1997).

Studies by the Toft laboratory, with mutants of chicken Hsp90α translated *in vitro* in reticulocyte lysate, have shown that the PR-Hsp90 interaction can tolerate deletion of the first 380-residues within the 728-amino acid chicken Hsp90α sequence to produce a hormone-activatable receptor. By contrast, selected regions (amino acids 381-441 and 601-677) in the C-terminal half of chicken Hsp90α were shown to be particularly important for PR-Hsp90 binding, with their deletion also interfering with receptor hormone responsiveness (Sullivan & Toft, 1993). An alternate approach by Baulieu and coworkers, in which human GR was coexpressed in baculovirus-infected insect cells with wild type or mutant chicken Hsp90α containing selective internal deletions (ΔA: 221-290; ΔB: 530-581; ΔZ: 392-419), revealed a loss of GR-Hsp90 interaction upon deletion of region A within the N-terminal domain, whereas deletions of regions B and Z afforded aggregated receptor-Hsp90 complexes in which receptor was unable to bind hormone (Cadepond *et al.*, 1993). An extension of these studies by the same laboratory, to chicken ERα and human MR, also concluded that deletion of the A domain in chicken Hsp90α negates interaction with both receptors (Binart *et al.*, 1995). None of the deletions affected ERα hormone-binding capacity, but MR failed to bind aldosterone with removal of region B. Although these investigations led to conflicting conclusions in relation to the role of the Hsp90 N-terminal domain in receptor interaction, it is appreciated that the introduced modifications may have caused structural perturbations leading to a disruption of Hsp90 functions elsewhere in the protein, possibly hampering valid interpretation of the results (Pratt & Toft, 1997).

Recent developments have led to the crystallographic analysis of steroid receptors, as well as Hsp90 and several of its cochaperones. At the same time, the use of the yeast two-hybrid system has revealed novel interactions between specific steroid receptors and selected cochaperones involved in the Hsp90 chaperoning pathway. Additionally, further insight is now available into the mechanism(s) that underlie the potentiation of AR, GR and PR by FKBP52. This review provides a summary of this recent progress with a focus on steroid receptor, Hsp90 and cochaperone contact domains that mediate interactions important for steroid receptor function.

2. Hsp90-steroid receptor interactions

2.1 GR LBD sub-regions required for assembly of apo-GR-Hsp90 complexes; GR structure

Further endeavours to identify sequences within the GR LBD critical for Hsp90 recognition were undertaken jointly by the Simons and Pratt laboratories. In initial studies using COS-7 cell-expressed receptor chimeras comprising glutathione S-transferase (GST) fused to the N-terminal end of an intact rat GR LBD and testing for recovered Hsp90, they found that a 7-residue amino-terminal truncation of the LBD eliminated both Hsp90 and steroid binding (Xu *et al.*, 1998). This allowed them to determine the 7-amino acid sequence, TPTLVSL (equivalent to amino acids 547-553 in rat GR and residues 529-535 in human GR, see Fig. 4), to be essential for the GR-Hsp90 interaction. Alignment of this sequence with the corresponding region in other steroid receptors revealed a conserved hydrophobic domain contained within helix 1 of the receptor LBD structure. It was proposed that the sequence defined a structure important for the unfolding of the hormone binding pocket, permitting steroid access and resulting in the exposure of a hydrophobic contact domain for stable Hsp90 interaction (Xu *et al.*, 1998). Extending the 7-amino acid sequence to include Leu554 in rat GR (Leu536 in human GR), gave the sequence TPTLVSLL and led to the recognition of the LXXLL protein-protein interaction motif within helix 1 (Giannoukos *et al.*, 1999). Such motifs have previously been shown to mediate interactions between transcriptional coactivators and members of the steroid/nuclear receptor super family (Ratajczak, 2001). Mutation of the first two leucine residues within the motif (L550S/L553S in rat GR) caused an increased rate of steroid dissociation, resulting in a dramatic loss of transcriptional activity. From a predicted GR structure, the GR LBD was seen as a "hinged pocket" with helices 1-6 comprising one side of the steroid-binding domain. In this model, the LXXLL motif within helix 1 was proposed to function as a hydrophobic clasp, helping to close one end of the steroid binding pocket by forming intramolecular contacts with residues in helices 8 and 9 on the opposite arm of the pocket, as well as residues in helix 3 and the intervening loop between helices 3 and 4 (Giannoukos *et al.*, 1999). The LXXLL motif was proposed then to play a key role in stabilizing GR LBD tertiary structure and would, as a consequence, make important contributions to steroid binding activity.

A mutational study of specific rat GR LBD residues within the previously defined minimal high affinity binding segment for Hsp90 revealed that alanine substitution of the conserved

Pro643 (analogous to human GR Pro625) profoundly reduced both the stability of the GR-Hsp90 heterocomplex, as well as transcriptional activity, despite retaining almost normal hormone-binding affinity (Caamano *et al.*, 1998). The negative effect on transcriptional function was related to a defect in nuclear translocation for the mutated receptor. Together the results strengthened the case for the requirement of Hsp90 as a critical component of steroid receptor signalling and identified an essential role for proline residue 643, located within an exposed hydrophobic loop between helices 5 and 6 in the receptor, in maintaining the apo-GR-Hsp90 interaction.

The x-ray structure of the human GR LBD, liganded to dexamethasone, resembles those for AR and PR, bound to their respective agonists and confirmed a helical sandwich arrangement for the steroid binding pocket (Bledsoe *et al.*, 2002). Pro625 was shown to be a key residue of a novel receptor dimerization interface involving reciprocal hydrophobic interactions between the helix 5-6 loop residues, Pro625 and Ile628 from each LBD and a hydrophobic bond network between the LBDs involving residues within the helix 1-3 loops (see Fig. 4). Since Pro625 is also central to the stability of GR-Hsp90 heterocomplexes, the finding suggested an overlap between the interface for receptor dimerization and an important contact domain for Hsp90. Indeed, this may form part of the mechanism that allows the Hsp90 chaperone complex to restrict transactivation of receptor in the absence of hormone (Picard, 2006). In comparison to GR, studies have revealed that ERα is less reliant on Hsp90 regulatory control over its hormone-dependent function (Picard *et al.*, 1990), allowing the ERα LBD to mediate dimerization in the absence of hormone *in vivo* (Aumais *et al.*, 1997). ERα homodimer formation in the LBD is mediated through helix 10, thus differing in configuration to that of GR (Bledsoe *et al.*, 2002). It is of interest that for ERα, substitution of a valine residue for Gly400, also within the helix 5-6 loop of the ERα LBD, induces a conformational change that destabilizes the receptor LBD, promoting a stronger, more stable association with Hsp90, similar to that for GR and rendering receptor transactivation more hormone-dependent (Aumais *et al.*, 1997).

2.2 Hsp90 structure; Amphipathic helices 1 and 2 in the Hsp90 C-terminal domain with potential for GR-binding

The x-ray crystal structure of the C-terminal dimerization domain of htpG, the *Escherichia coli* Hsp90, was recently solved by Agard and coworkers, revealing a dimerization motif defined by a four-helix bundle interface derived from the interaction of helices 4 and 5 of one monomer with equivalent helices from a second monomer (Harris *et al.*, 2004). The structure also identified helix 2, a flexible, solvent exposed amphipathic helix, as a potential chaperone substrate-binding site. Hydrophobic residues within helix 2 are strongly conserved in Hsp90 homologues across species, suggesting an important underlying function. This was supported by other studies in which deletion of a region encompassing the corresponding helix 2 sequence in yeast Hsp82 impaired viability (Louvion *et al.*, 1996), while the point mutation, A587T, which defines the start of the helix, compromised the ability of Hsp82 to promote GR activity and caused a general reduction in Hsp90 function (Nathan &Lindquist, 1995). Core hydrophobic residues within the helix 2 sequence were observed to share sequence similarity with helix 12 of steroid receptors, leading to a proposal that Hsp90 helix 2 acts as a receptor helix 12 mimic in apo-receptor-Hsp90 complexes, occupying the normal activation function 2 (AF2) position of helix 12 following hormone binding (Jackson *et al.*, 2004). Structural elucidation of full-length yeast Hsp90 (Ali *et al.*, 2006) allowed the recognition of helix 1, also consisting of a solvent-exposed, hydrophobic surface within the Hsp90 C-terminal domain, as a possible contact site for protein-protein interactions (Fang *et al.*, 2006). The highly conserved hydrophobic sequence of this helix closely matches the LXXLL recognition motif of the Steroid Receptor Coactivator/p160 family of coactivators that modulate receptor transcriptional activity by interacting with the AF2 agonist-induced hydrophobic groove of nuclear receptors (Ratajczak, 2001).

2.3 Flexible positioning of receptor LBD helix 12; Hsp90 helix 2 induces apo-GR helix 12 to adopt the GR-RU486 antagonist conformation

Recent studies by Darimont and coworkers have confirmed that Hsp90 helix 2 stabilizes unliganded GR by engaging apo-GR at the position normally occupied by receptor helix 12 in response to hormonal activation and forcing the flexible helix 12 to bind to the hydrophobic groove, at the same time preventing receptor interaction with coactivators (Fang *et al.*, 2006). The resulting structure corresponds to the native conformation of unliganded GR, with an orientation of helix 12 similar to that in antagonist (RU486)-bound GR (Fang *et al.*, 2006; Kauppi *et al.*, 2003). On agonist binding, hormone-induced conformational changes within the LBD of holo-GR promote the replacement of Hsp90 helix 2 by receptor helix 12, causing loss of Hsp90 chaperone machinery and establishing the AF2 contact domain for coactivator interaction. Alternatively, the new structure might facilitate Hsp90 helix 1 binding to the hydrophobic groove. Since Hsp90 helices 1 and 2 are proximally located at the Hsp90 C-terminus, this exchange of receptor-Hsp90 interactions, which is partly determined by the dynamics of receptor helix 12, may likely be achieved within the one receptor-Hsp90 complex (Fig. 1).

The hormone-induced progression from apo- to holo-GR-Hsp90 complexes, through changes in the mode of receptor-Hsp90 interaction resulting from altered receptor LBD conformation, provides a suitable model for visualising the transition between inactive and active receptor that may also involve the participation of Hsp90 cochaperones such as FKBP51 and FKBP52. Although FKBP51 is the preferred cochaperone in mature GR-Hsp90 complexes (Barent *et al.*, 1998; Nair *et al.*, 1997), FKBP52 has been shown to promote increased GR hormone binding affinity and to potentiate the transcriptional activity of the receptor (Riggs *et al.*, 2003). It is possible that the observed hormone-induced interchange of FKBP51 by FKBP52 in GR-Hsp90 complexes, resulting in the favoured nuclear translocation of receptor complexes (Davies *et al.*, 2002), might be initiated by a change in GR LBD conformation elicited by the transfer of receptor interaction from Hsp90 helix 2 to helix 1, both helices being close to the common TPR binding site for immunophilin cochaperones in the C-terminal region of Hsp90. Unique steroid receptor LBD conformations then might be an important determinant of receptor preferences for specific immunophilin cochaperones within receptor-Hsp90 complexes (e.g. FKBP51 in GR, PR and MR complexes (Barent *et al.*, 1998; Nair *et al.*, 1997); PP5, the major cochaperone in GR complexes (Silverstein *et al.*, 1997) and CyP40, the prevalent immunophilin in ER complexes (Ratajczak *et al.*, 1990)), allowing these cochaperones to potentially modulate receptor function (Ratajczak *et al.*, 2003; Smith & Toft, 2008).

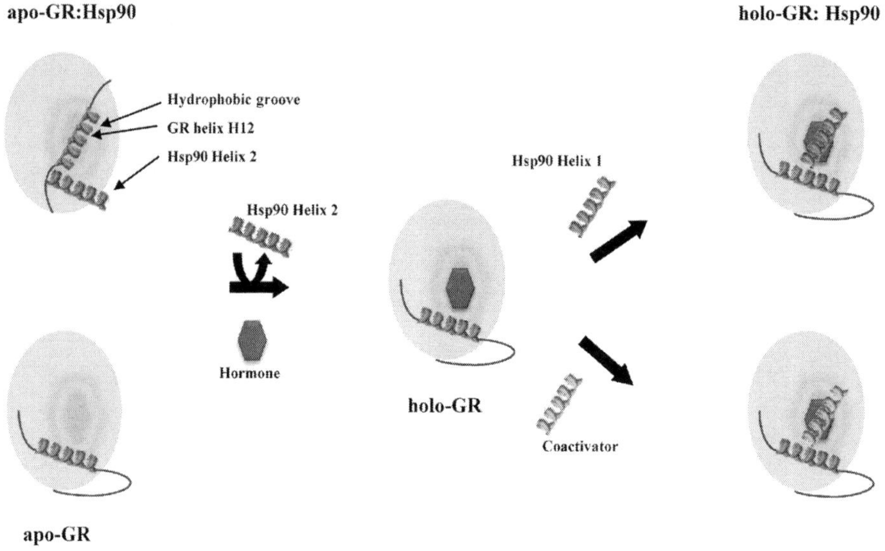

Hsp90 helix 2 binds apo-GR at the position normally occupied by GR helix H12, forcing H12 to dock within the hydrophobic groove, thus stabilizing the unliganded hormone-binding pocket. With hormone binding, GR H12 replaces Hsp90 helix 2 providing contacts for AF2-interacting coactivators or for Hsp90 helix 1.

Fig. 1. Hsp90 interactions with apo-GR and holo-GR.

3. Hsp90/Hsp70-cochaperone interactions

3.1 TPR cochaperones

Folding of newly synthesized peptides to functionally mature proteins, such as steroid receptors, is actively regulated by Hsp70 and Hsp90 with their cochaperones in what is known as the Hsp70/Hsp90-based chaperone machinery (Pratt & Toft, 2003). Cochaperones can regulate the nucleotide status, and thus function, of Hsp70 and Hsp90, and deliver non-native proteins to their respective polypeptide-binding domains for folding. Those cochaperones that regulate Hsp70 include Hsp40, Hsc70-interacting protein (Hip), Hsp-organizing protein (Hop) and small glutamine-rich TPR protein (SGT), while Hsp90 is regulated by cochaperones that include Hop, p23, PP5, CyP40, FKBP51 and FKBP52. C-terminal of Hsp70-interacting protein (CHIP) is another cochaperone that regulates both Hsp70 and Hsp90. Fig. 2 shows the domain architecture of the immunophilin and other TPR cochaperones with an established role in Hsp70 and/or Hsp90 chaperone function.

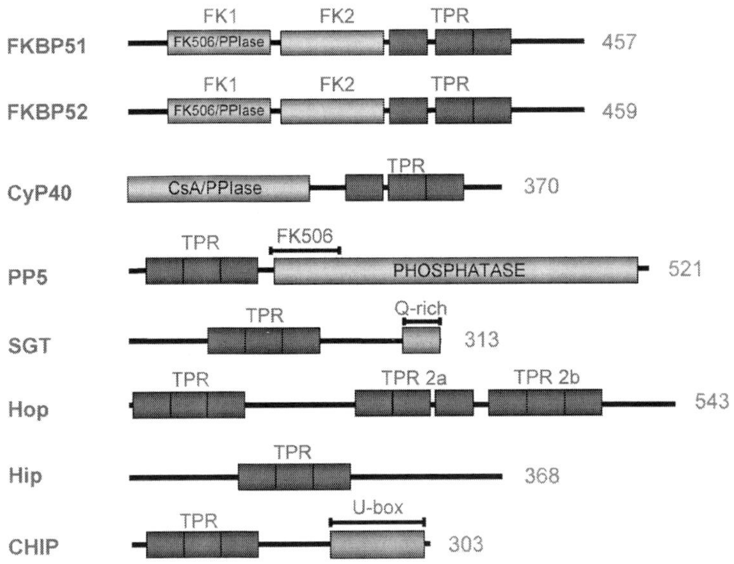

TPR domains are depicted in red whilst other specialized functional domains are highlighted in other various colours and labelled accordingly. Abbreviations: FKBP, FK506-binding protein; PPIase, peptidylprolyl isomerase; TPR, tetratricopeptide repeat; CyP40, cyclophilin 40; CsA, cyclosporin A; PP5, protein phosphatase 5; SGT, small glutamine-rich TPR protein; Hop, Hsp-organizing protein; Hip, Hsc70-interacting protein; CHIP, C-terminal of Hsp70-interacting protein.

Fig. 2. Schematic presentation of the domain structures of TPR-containing proteins associated with the Hsp70/Hsp90 chaperone machinery.

Since the crystallization of the PP5 TPR domain, the structures of several other steroid receptor-associated TPR-containing proteins have been solved. There are now full-length structures available for bovine CyP40, human FKBP52, PP5 and Hop, human and squirrel monkey FKBP51, and mouse CHIP, as well as the structure of the human SGT TPR domain. It is known that TPR domains in these proteins can mediate interactions with Hsp70 and/or Hsp90 (Angeletti *et al.*, 2002; Smith, 2004), but in addition to their Hsp-recognition domains, each also possesses other localized functional domains important for their own conformation and/or the regulation of associated proteins.

3.1.1 CyP40, FKBP51 and FKBP52

CyP40 and the two FKBPs have a similar structural arrangement, each possessing an N-terminal binding site for the immunosuppressants cyclosporin A or FK506, respectively, and a C-terminal TPR domain (Sinars *et al.*, 2003; Taylor *et al.*, 2001; Wu *et al.*, 2004). The cyclophilin domain of CyP40 is similar to other single-domain cyclophilins (Kallen *et al.*, 1998). In FKBP51 and FKBP52, FK506 binds to the first of two FKBP domains, termed FK1, while the second domain, called FK2, lacks drug-binding activity. Bound immunosuppressants inhibit the peptidylprolyl isomerase (PPIase) activity of the cyclophilin and FK1 domains, which may be important for target protein regulation by direct or indirect association. Fig. 3 provides a structural comparison between CyP40, FKBP51 and FKBP52 immunophilin cochaperones.

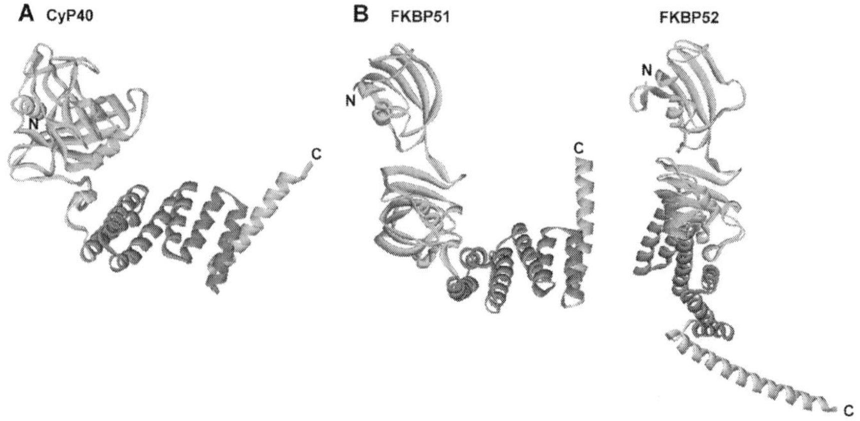

A, CyP40 and **B,** FKBP51, FKBP52. The CsA-binding domain (CyP40) and FK regions (FKBP51 and FKBP52) are shown in green. Core TPR domains for CyP40, FKBP51 and FKBP52 are depicted in red, with the final extended helices, at the C-terminal ends of each protein, shown in yellow.

Fig. 3. Ribbon representations of molecular structures of TPR-containing proteins.

3.1.2 PP5

PP5 is a phosphatase that dephosphorylates serine and threonine residues on target proteins (Barford, 1996; Cohen, 1997). Crystallisation of the full-length phosphatase in the absence of ligands or binding partners revealed the structural organization of the autoinhibited form of PP5 (Yang *et al.*, 2005). The TPR domain in PP5 is oriented to the N-terminus and is linked to a C-terminal phosphatase catalytic domain followed by a short C-terminal subdomain. In this inactive conformation, the TPR domain engages with the catalytic domain in such a way as to restrict target protein access to the enzymatic site, and this structure is stabilized by the C-terminal subdomain. Suppression of catalytic activity can be abolished by an allosteric conformational change that disrupts the TPR-catalytic domain interface, and this can be induced upon binding of polyunsaturated fatty acids or Hsp90 to the TPR domain (Chen & Cohen, 1997; Ramsey & Chinkers, 2002; Skinner *et al.*, 1997).

3.1.3 Hop

Hop plays a dual role in mature steroid receptor complex assembly by recruiting Hsp90 to preformed Hsp70-receptor complexes and inhibiting the ATPase of Hsp90 for client loading onto the chaperone for subsequent folding (Chen *et al.*, 1996b; Chen & Smith, 1998; Dittmar *et al.*, 1996; Kosano *et al.*, 1998; Prodromou *et al.*, 1999; Siligardi *et al.*, 2004). Hop has an N-terminal TPR domain (TPR1) followed by an aspartic acid/proline (DP)-rich region, and two more adjacent TPR domains (TPR2a and TPR2b) followed by a second DP-rich region.

3.1.4 Hip

Hip functions as a transient component of native steroid receptor complexes and enters the assembly cycle once Hsp70 ATPase activity has been stimulated by Hsp40 (Frydman & Höhfeld, 1997; Höhfeld *et al.*, 1995). Hip acts to stabilize the ADP-bound state of Hsp70 that is necessary for high affinity interaction with unfolded substrates (Frydman &Höhfeld, 1997; Höhfeld *et al.*, 1995). Structurally, Hip consists of an N-terminal oligomerization domain that is important for the functional maturation of GR in yeast (Nelson *et al.*, 2004), a central TPR domain and an adjacent highly charged region which are both required for Hsp70 binding (Prapapanich *et al.*, 1996b) and a C-terminal DP-rich domain that helps direct the intermediate stage recruitment of Hop-Hsp90 during assembly of steroid receptor complexes (Prapapanich *et al.*, 1998).

3.1.5 CHIP

The cochaperones described above are involved in maintaining an activatable conformation of Hsp70/Hsp90-dependent "clients", but TPR proteins also function to mediate the degradation of misfolded proteins, indicating a role in quality control (Cyr *et al.*, 2002). Selection of proteins for degradation is mediated by E3 ubiquitin ligases, and CHIP is a member of this enzymatic class (Jiang *et al.*, 2001; Murata *et al.*, 2001). CHIP has an N-terminal TPR domain and a C-terminal U-box domain that mediates its ligase activity, which promotes ubiquitylation of target substrates prior to their degradation by the proteasome.

3.1.6 SGT

Human SGT binds to viral protein U (Vpu) and Group specific Antigen, 2 proteins associated with human immunodeficiency virus-1, and the rat homologue was identified as an interactor of the non-structural protein NS-1 of the parvovirus H-1. The central TPR domain in SGT is flanked by an N-terminal dimerization domain and a C-terminal glutamine-rich domain involved in association with type 1 glucose transporter (Callahan *et al.*, 1998; Cziepluch *et al.*, 1998; Liou & Wang, 2005).

3.2 Regulation of Hsp70 and Hsp90 ATPases by TPR cochaperones

Both Hsp70 and Hsp90 require ATP for their functional association with substrates (Pratt & Toft, 2003). In the case of a steroid receptor, Hip binding to the N-terminal ATPase domain of Hsp70, possibly through a unique TPR binding site located within this region (see below), stabilizes the Hsp70-receptor complex (Frydman & Höhfeld, 1997; Höhfeld *et al.*, 1995) in a step that may be important for recognition by Hop and loading of the receptor onto Hsp90 for further processing. Hop contains three distinct TPR domains (TPR1, TPR2a, TPR2b) (Fig. 2), with TPR1 and TPR2a providing anchor points for the C-terminal EEVD peptides of Hsp70 and Hsp90, respectively. These specific interactions, coupled with domain-domain interactions, also involving its TPR domains, allow Hop to play a key role in coordinating the actions of Hsp70 and Hsp90 (Carrigan *et al.*, 2006; Chen *et al.*, 1996b; Chen & Smith, 1998; Odunuga *et al.*, 2003; Prodromou *et al.*, 1999; Ramsey *et al.*, 2009; Scheufler *et al.*, 2000). While the TPR acceptor site for Hop in the C-terminal region of Hsp90 serves to anchor the cochaperone, studies have shown that Sti1, the yeast homologue of Hop, markedly inhibits the ATPase activity of yeast Hsp90 through secondary interactions that block the ATP-binding pocket in the Hsp90 N-terminal domain (Prodromou *et al.*, 1999). By directly competing with Sti1 for binding to Hsp90, the CyP40 yeast homologue Cpr6 can negate the Sti1-mediated blockade of Hsp90 ATPase activity following TPR protein exchange (Prodromou *et al.*, 1999). In contrast, in vitro studies with human Hop determined that the cochaperone had no influence on the weak basal ATPase activity of human Hsp90, but significantly inhibited the increased rate of ATP hydrolysis by Hsp90 in response to interaction with the ligand binding domain of GR, an established Hsp90 client protein (McLaughlin *et al.*, 2002). On the other hand, FKBP52, which like CyP40 binds competitively with Hop to the C-terminal TPR interaction site of Hsp90, was shown to enhance Hsp90 ATPase activity stimulated by GR (McLaughlin *et al.*, 2002). This control over ATP utilization is important for the functional activity of newly synthesized substrates, but ATPase regulation is also required for the degradation of improperly folded substrates. CHIP can bind Hsp70 and inhibit Hsp40-stimulated Hsp70 ATPase activity, and has been reported to deplete cellular GR levels (Ballinger *et al.*, 1999; Connell *et al.*, 2001). Therefore, CHIP can be regarded as a degradatory cochaperone of Hsp70 and Hsp90. SGT negatively regulates Hsp70 such that the chaperone has a reduced ability to refold denatured luciferase (Angeletti *et al.*, 2002).

3.3 Determinants of Hsp70 and Hsp90 interaction with TPR cochaperones

Deletion studies were the first to demonstrate that TPR domains mediated binding to Hsp90 (Barent *et al.*, 1998; Chen *et al.*, 1996a; Radanyi *et al.*, 1994; Ratajczak & Carrello, 1996). Determination of the TPR domain structure of PP5 revealed that the packing of adjacent TPR units generated an exposed groove capable of accepting a target protein peptide (Das *et al.*, 1998). Although TPR motifs are highly degenerate, they display a consistent pattern of key residues important for structural integrity. The two α-helical sub-domains in each TPR motif are arranged such that the groove is mainly composed of residues from the A helix of each repeat, while B helix residues are buried to form the structural backbone of the superhelix, and this groove forms a critical Hsp recognition surface.

In a PP5 mutagenesis study, Russell and coworkers carefully selected A helix residues with side-chains extended into the groove and identified four basic residues important for PP5-Hsp90 interaction (Russell *et al.*, 1999). These amino acids are highly conserved in other Hsp90-binding TPR proteins, and mutation of aligned residues in CyP40 confirmed their importance in Hsp90 recognition (Ward *et al.*, 2002). The key recognition sequence for the TPR domain in these proteins is the EEVD peptide located at the extreme C-terminus of Hsp90 (Carrello *et al.*, 1999; Chen *et al.*, 1998; Young *et al.*, 1998), which is conserved in Hsp70. Crystallization of individual Hop TPR domains with Hsp70 and Hsp90 N-terminally extended EEVD peptides has defined the mechanism of TPR domain-peptide interaction (Scheufler *et al.*, 2000). The TPR1 domain of Hop binds to Hsp70, while the TPR2a domain mediates Hsp90 recognition (Chen *et al.*, 1996b; Lassle *et al.*, 1997). The groove in each TPR domain accommodates their respective peptide in an extended conformation where the ultimate aspartate residue is tightly held by electrostatic interactions with TPR residue side-chains in a two-carboxylate clamp. Additional EEVD contacts involve hydrogen-bonding, while amino acids upstream of the EEVD enhance the affinity of the peptides for TPR domains and mediate specificity of Hsp70 and Hsp90 to TPR1 and TPR2a, respectively. Notably, Hop TPR2a provides an example of where an additional sequence within the TPR domain doesn't disrupt the overall structure. TPR2a contains an insertion between units 2 and 3 that extends the helices by a single turn but does not impact Hsp90 peptide recognition (Scheufler *et al.*, 2000).

The Hsp90 dimerization domain, located in the C-terminal region upstream of the MEEVD peptide, contributes to TPR cochaperone recognition (Chen *et al.*, 1998) and contains the putative binding site for novobiocin, a coumarin-based Hsp90 inhibitor (Marcu *et al.*, 2000). *In vitro* studies demonstrated that novobiocin had a differential effect on Hsp90-immunophilin cochaperone interaction, suggesting that the TPR cochaperones modulate Hsp90 function through distinct contacts within the Hsp90 C-terminal domain (Allan *et al.*, 2006).

Although EEVD interactions with the TPR domain groove are critical for Hsp binding, regions outside of the TPR domains are also important in mediating recognition. TPR domains are typically followed by a seventh α-helix that packs against and extends beyond the TPR domain and has been shown to be involved in binding Hsp90 in addition to the TPR domain. FKBP51 and FKBP52 have different affinities for Hsp90 and are assembled differentially with specific receptor complexes, and these differences map in part to sequences C-terminal of their respective TPR domains (Barent *et al.*, 1998; Cheung-Flynn *et al.*, 2003; Pirkl & Buchner, 2001). The charge-Y motif was identified and found to be essential for FKBP-Hsp90 interaction, which was also confirmed for CyP40, but sequences further downstream in FKBP51 and FKBP52 differentially regulated Hsp90 binding (Allan *et al.*, 2006; Cheung-Flynn *et al.*, 2003; Ratajczak & Carrello, 1996). The acidic linker flanking the N-terminus of the CyP40 TPR domain was also shown to be important for efficient interaction (Mok *et al.*, 2006; Ratajczak & Carrello, 1996). Although an interaction partner for Hop TPR2b has yet to be identified, mutations in TPR2b reduced Hop interaction with both Hsp70 and Hsp90, while mutations in the C-terminal DP-rich region inhibited Hop binding to Hsp70 (Chen & Smith, 1998; Nelson *et al.*, 2003).

3.4 Alternative modes of Hsp70 and Hsp90 recognition by TPR cochaperones

Like Hop, CHIP binds to both Hsp70 and Hsp90 (Ballinger *et al.*, 1999; Connell *et al.*, 2001), but CHIP interacts with either of these major chaperones through a single TPR domain. Recent elucidation of the binding of Hsp90 C-terminal peptide (NH_2-DDTSRMEEVD) with the CHIP TPR domain has revealed that the peptide sequence is not accommodated in an extended conformation as for Hop, but turns at the methionine residue and becomes buried within a hydrophobic pocket (Zhang *et al.*, 2005). This pocket can accommodate either the methionine or isoleucine that lies immediately upstream of the EEVD sequence in Hsp90 and Hsp70, respectively, and the peptide is twisted, negating the role of upstream residues in conferring the same specificity seen in binding Hop TPR domains. SGT also recognizes Hsp70 and Hsp90 via its single TPR domain, but possibly through a different mechanism to that described for CHIP as SGT lacks the residues that form the hydrophobic pocket which allows the respective C-terminal peptides in the chaperones to twist (Dutta & Tan, 2008).

Hydrophobic pockets themselves may also be important structural features within TPR domains that confer Hsp specificity, as the crystal structure of Hop TPR2a with the non-cognate Hsp70 peptide shows the hydrophobic pocket to be less accommodating for the Ile

(-5) residue in the extended Hsp70 peptide than Met (-5) in the extended Hsp90 peptide, with the notable feature of a lack of bending by the Hsp70 peptide, such as with CHIP, to perhaps enhance affinity for TPR2A (Kajander *et al.*, 2009).

General cell UNC-45 (GCUNC-45), a member of the UNC-45/Cro1/She4p (UCS) protein family, is a TPR protein that regulates PR chaperoning by Hsp90 by preventing activation of Hsp90 ATPase activity (Chadli *et al.*, 2006). Hsp90-binding experiments in the presence of Hop revealed a novel GCUNC-45 TPR recognition site in the N-terminal domain of Hsp90, which also bound FKBP52 (Chadli *et al.*, 2008a). Further analysis defined a non-contiguous EEVD-like motif, centered in and around the Hsp90 N-terminal ATP-binding pocket, arranged in a structural conformation that can recognize TPR domains. Nucleotide binding negatively regulates the interaction. These authors also alluded to CyP40 binding to the N-terminal interaction motif, although Onuoha and coworkers have recently confirmed CyP40 interaction only with the C-terminal domain of Hsp90 (Onuoha *et al.*, 2008). GCUNC-45 is the first cochaperone to display a preferential association with Hsp90β over the Hsp90α isoform, resulting in functional Hsp90β-GCUNC-45 interactions that more efficiently block progression of PR chaperoning than seen with Hsp90α-GCUNC-45 complexes (Chadli *et al.*, 2008b). An EEVD-like motif interaction with a TPR domain has also been described for androgen receptor recognition by SGT, where binding is mediated by the first 2 TPR motifs of the SGT TPR domain and the hinge region located between the DNA-binding and ligand-binding domains in the receptor (Buchanan *et al.*, 2007).

Hip has similarly been reported to bind the Hsp70 N-terminal ATPase domain via its TPR domain (Höhfeld *et al.*, 1995). Through this interaction, Hip, originally identified in progesterone receptor complex assembly (Prapapanich *et al.*, 1996a; Smith, 1993), can stabilize substrate-Hsp70 binding and competitively counteract the destabilizing effects of the non-TPR cochaperone BAG1 (Bimston *et al.*, 1998; Gebauer *et al.*, 1997; Höhfeld & Jentsch, 1997; Takayama *et al.*, 1997). The Hip-Hsp70 interaction also allows for the simultaneous association of Hip with Hsp70-Hop complexes (Gebauer *et al.*, 1997; Prapapanich *et al.*, 1996a). By analogy with the mode of GCUNC-45 interaction with Hsp90, there is the possibility that Hip targets a similar TPR recognition site in the N-terminal region of Hsp70. However, Hip is unique among the steroid receptor-associated TPR proteins in terms of Hsp recognition in that it binds Hsp70 independently of EEVD interactions (Höhfeld *et al.*, 1995), and that efficient binding may be due to a greater requirement for additional Hsp-interaction determinants, such as the adjacent highly charged region and a C-terminal DP-repeat domain (Prapapanich *et al.*, 1998). It is possible the mechanism of Hsp70 recognition by Hip is not unique, but may be utilized by some of the steroid-receptor TPR cochaperones to interact with binding partners in distinct cellular pathways. Dutta and Tan (2008) reported the SGT TPR domain is sufficient to bind Vpu and identified the sequence ^{31}KILRQ^{35} in Vpu as being important for this interaction.

4. p23 and Cdc37 interaction with Hsp90

p23 is an essential component involved in stabilizing mature steroid receptor-Hsp90 complexes and binds to the ATP-bound conformation of a Hsp90 dimer characterised by high affinity for client proteins (Ali *et al.*, 2006; Felts & Toft, 2003; McLaughlin *et al.*, 2006;

Richter *et al.*, 2004). Conformational changes that accompany ATP binding promote dimeric interaction between the N-terminal domains of the Hsp90 C-terminal dimer to form distinct binding surfaces for separate p23 molecules, thus further underpinning the ATP-bound conformation (Ali *et al.*, 2006; Karagöz *et al.*, 2010). In a recent model proposed for the Hsp90 cochaperone cycle, entry of an immunophilin cochaperone into an existing client protein-Hsp90-Sti1/Hop-Hsp70 complex forms an intermediate complex important for cycle progression. Conversion of Hsp90 to the closed conformation on ATP and subsequent p23 binding then favours the release of Sti1/Hop (Li *et al.*, 2011).

Cdc37 serves as an adaptor predominantly facilitating protein kinase interaction with Hsp90, although additional client proteins, including steroid receptors have been identified (MacLean & Picard, 2003). Similar to Hop, Cdc37 arrests the Hsp90 ATPase cycle and functions as an "early" cochaperone for the recruitment of protein kinase clients to the Hsp90 machinery. Hsp90 binding maps to the Cdc37 C-terminal region, while kinase interaction occurs via the N-terminal domain (Roe *et al.*, 2004). Hsp90 ATPase activity is coupled to an opening and closing of a molecular clamp generated by the constitutive C-terminal Hsp90 dimer at one end in combination with the ATP-dependent association of the N-terminal domains at the other (Prodromou *et al.*, 2000). A structural view of the Hsp90-Cdc37 complex shows Cdc37 located as a dimer between the N-terminal domains of the clamp, thus preventing their interaction (Roe *et al.*, 2004). With cycle progression, loss of one Cdc37 monomer leads to the formation of a stable $(Hsp90)_2$-Cdc37-kinase complex (Vaughan *et al.*, 2006; Vaughan *et al.*, 2008).

5. Receptor-cochaperone interactions

5.1 Cortisol resistance in New World primates; The key role of FKBP51; Structures of FKBP51 and FKBP52

Analysis of glucocorticoid resistance in New World primates, such as squirrel monkey, has demonstrated that the high circulating cortisol levels result from elevated expression and greatly increased incorporation of FKBP51 into GR-Hsp90 complexes, causing a significant decrease in GR hormone binding affinity (Denny *et al.*, 2000; Reynolds *et al.*, 1999; Scammell *et al.*, 2001). FKBP51 then appears to have a major role in stabilizing an inactive receptor conformation. The FK506 drug-binding pocket of FKBP51 is inaccessible to FK506 in low affinity hormone-binding GR heterocomplexes. However, incubation of receptor cytosols from squirrel monkey lymphocytes with FK506 prevented assembly of FKBP51 with GR-Hsp90 complexes, correlating with a sharp increase in receptor hormone binding and affinity. On the other hand, recognition of FK506 by FKBP52 appeared unaffected by whether the immunophilin exists as a component of mature, high affinity hormone-binding GR complexes or not (Denny *et al.*, 2000; Tai *et al.*, 1992). Furthermore, the immunosuppressant blocks FKBP52-mediated potentiation of GR activity (Riggs *et al.*, 2003). The inhibitory influence of FKBP51 on GR activity requires both FK domains, as well as Hsp90 binding, but is not reliant on FKBP51 PPIase activity (Denny *et al.*, 2005). FK506 may likely serve to sterically hinder receptor LBD interactions with the FK1 domain of FKBP51 and FKBP52 essential for inhibitory and activation effects on receptor, respectively. This differential action of FK506 may arise from distinct domain orientations that have been defined from recent structures of the two immunophilins (Sinars *et al.*, 2003; Wu *et al.*, 2004). Unique interactions between receptor and the FKBP51 and FKBP52 cochaperones have been further highlighted by results showing that deletion of the Asp195, His196, Asp197 insertion within the FK2 domain of FKBP51 compromised assembly of the immunophilin into PR complexes, whereas removal of the corresponding FK2 insertion loop from FKBP52 had no affect on receptor association (Sinars *et al.*, 2003). This raises the possibility that direct interaction of FK2 in FKBP51 with PR might favour the preferred association of FKBP51 over FKBP52 with this receptor.

5.2 Cortisol resistance in the guinea-pig; Do guinea pig GR LBD changes favour FKBP51 binding over FKBP52?

In contrast to the New World primates, the cause of glucocorticoid resistance in the guinea pig, a New World hystricomorph, has been delineated to an unstructured loop between helix 1 and helix 3 of the guinea pig GR LBD. Five amino acid substitutions in this region differentiate guinea pig GR from the human receptor, with at least four contributing to the low binding affinity phenotype (Fuller *et al.*, 2004). It has been predicted that these crucial residues (Ile538, His539, Ser540, Thr545 and Ser546) lying on the surface of the guinea pig GR LBD, disrupt a contact domain for FKBP52, favouring increased association with FKBP51 and conformational changes that compromise high affinity cortisol binding. Using a yeast-based assay (Riggs *et al.*, 2003) with rat GR substituted in the helix 1 to helix 3 loop with the guinea pig GR-specific residues, we have recently confirmed that FKBP52 can efficiently potentiate the transcriptional activity of the mutated GR, thus discounting a central role of this region in receptor-FKBP52 interaction [Cluning C and Ratajczak T, unpublished observations].

5.3 FKBP52 potentiation of AR, GR and PR

Direct interaction studies between bacterially expressed FKBP52 and GST-tagged, wild type human GR and C-terminal truncation mutants of the receptor purified from Sf9 cell extracts, identified a 35-amino acid region (hGR 465-500), between the DNA-binding domain and the LBD, to be sufficient for FKBP52 binding, with optimal interaction requiring involvement of the LBD (Silverstein *et al.*, 1999). However, recent demonstration of FKBP52 potentiation of GR activity in association with increased receptor hormone binding affinity has definitively localized the FKBP52 effect to the GR LBD (hGR 521-777) and at the same time pointed to a requirement of FKBP52 PPIase activity residing in the FK1 domain (Riggs *et al.*, 2003). Studies with FKBP52 knockout mouse strains have extended the critical physiological role of FKBP52 to cellular responses controlled by both AR (Cheung-Flynn *et al.*, 2005) and PR (Tranguch *et al.*, 2005; Yang *et al.*, 2006), while similar influences of this immunophilin cochaperone on ERα (Riggs *et al.*, 2003) and MR (Gallo *et al.*, 2007) activity have not been observed, despite the assembly of FKBP52 with Hsp90 complexes containing these receptors.

5.4 Molecular basis of FKBP52 action; Potential interaction of FKBP52 with the BF3 regulatory site

An initial understanding that FKBP52 potentiation of AR, GR and PR activity was dependent on the FK1-mediated PPIase function of the immunophilin, prompted speculation that FKBP52 might target a key proline likely to be conserved among these receptors and that this critical residue would be located on the surface of the LBD, accessible to the cochaperone and in a position where it might influence the shape of the ligand binding pocket (Cheung-Flynn *et al.*, 2005). Although several such candidate prolines exist in the intervening loops between receptor LBD helices, a more extensive mutational analysis of the FK1 catalytic site has excluded a role for the FKBP52 PPIase activity in receptor potentiation (Riggs *et al.*, 2007). Rather, recent evidence has identified a loop overhanging the FK1 catalytic pocket in FKBP52 that is responsible for the functional difference between FKBP52 and FKBP51 relating to AR (and GR/PR) potentiation (Riggs *et al.*, 2007). It is proposed that a critical proline within this loop (human FKBP52 Pro119) allows specific contact with a region of the AR LBD (a structural feature that is also common to GR and PR), thus helping to stabilize an LBD conformation favourable for high affinity hormone binding and leading to efficient transcriptional activation (Riggs *et al.*, 2007). It is speculated that a leucine substitution within the corresponding FK1 sequence of FKBP51 alters the loop conformation sufficiently to disrupt this functionally important contact. The possibility exists that in the hormone-induced transition from inactive to active states of AR-Hsp90 complexes associated with FKBP51 and FKBP52, respectively, Hsp90 orients FKBP52 to achieve unique interactions with the receptor LBD, allowing Hsp90 to facilitate optimal hormone binding and to further fine-tune the hormonal response.

Prior to investigations establishing a noncatalytic involvement of the FKBP52 PPIase domain in the modulation of receptor function, an early attempt to identify the putative proline substrate for FKBP52 isomerase activity within the AR LBD utilized AR-P723S, a proline mutant associated with androgen insensitivity syndrome (Cheung-Flynn *et al.*, 2005). Although predicted to display basal activity, coupled with a lack of response to hormone in the presence of FKBP52, this mutant was characterized by subnormal activity in the absence of FKBP52, showing full restoration to wild type receptor activity levels with the cochaperone on exposure to hormone (Cheung-Flynn *et al.*, 2005). Such a favoured response reflects a greater dependence of the AR-Pro723S mutant on FKBP52 for normal activity. Pro723 lies within the signature sequence conserved among all steroid receptors (Brelivet *et al.*, 2004), close to a region directly involved in ligand binding and is situated in a solvent exposed loop between helices 3 and 4, which combine together with the mobile helix 12 to form the AF2 coactivator binding pocket (He *et al.*, 2004; Matias *et al.*, 2000b). For AR, AF2 initially has a preferred interaction with the AR N-terminal domain, resulting in an intramolecular fold that precedes receptor dimerization and appears critical for AR function (He *et al.*, 2001; He *et al.*, 2004; Schaufele *et al.*, 2005). Pro723 also forms part of the recently identified BF-3 surface that has the ability to allosterically alter the AF2 binding pocket of AR (Estébanez-Perpiñá *et al.*, 2007) (Fig. 4). BF-3 residues altered through natural mutations linked to androgen insensitivity and those associated with prostate cancer, either diminish or enhance AR AF2 activity, respectively, underlining the importance of the BF-3 surface for AR function (Estébanez-Perpiñá *et al.*, 2007). FKBP52 rescue of AR-Pro723S activity might signify FKBP52 influence over some part of the BF-3 allosteric regulatory site leading to conformational changes that allow full recovery of AR activity. Indeed, Cox and coworkers have recently identified small-molecule inhibitors of FKBP52-enhanced AR function in prostate cancer cells that target a region of the AR LBD overlapping the BF3 surface (De Leon *et al.*, 2011) (Fig. 4). Multiple residues that contribute to the FKBP52 sensitivity of AR, some of which form part of the binding site for MJC13, the lead compound, have been identified (De Leon *et al.*, 2011) (Fig. 4). Since MJC13 helps to maintain an intact AR-Hsp90-FKBP52 complex at low hormone concentrations, it is possible that the inhibitor interferes with a critical next step - a hormone-induced, FKBP52-dependent transitory change in AR conformation necessary for nuclear translocation. Sequence comparisons have revealed some conservation of BF-3 residues within the LBDs for AR, GR, MR and PR, suggesting the presence of BF-3-like regulatory domains in each receptor (Estébanez-Perpiñá *et al.*, 2007) (Fig. 4). A very limited conservation of these residues is apparent in ERα, suggesting the formation of a BF-3 type surface that is unique to this receptor (Estébanez-Perpiñá *et al.*, 2007) (Fig. 4). Both ERα and MR behave differently to AR, GR and PR, through their inability to respond to FKBP52. Certain structural differences within their LBDs distinguish these two receptors from the other members of this subfamily (De Leon *et al.*, 2011) (Fig. 4). Since FKBP52 also regulates GR and PR activity, most likely through specific BF3 surfaces, there is the potential for the development of FKBP52-specific inhibitors targeting GR and PR function to treat a range of steroid hormone-based diseases (Moore *et al.*, 2010). The BF-3 pocket is a potential target for second-site modulators that can allosterically block agonist-activated AR function to inhibit prostate cancer cell growth (Joseph *et al.*, 2009).

5.4.1 FKBP51 is an androgen-regulated gene that promotes assembly of mature AR-Hsp90 complexes

FKBP51 is recognised as a highly sensitive AR-regulated gene that functions as an important component of a feed-forward mechanism linked to the partial reactivation of AR-signalling pathways in the absence of androgens, leading to the outgrowth of androgen-independent tumours (Amler *et al.*, 2000; Febbo *et al.*, 2005; Magee *et al.*, 2006; Mousses *et al.*, 2001; Tomlins *et al.*, 2007). Sanchez and coworkers have confirmed a significantly increased expression of FKBP51, but not that of FKBP52, in most prostate cancer tissues and in androgen-dependent and androgen-independent cell lines (Periyasamy *et al.*, 2010), suggesting that FKBP51 might have a critical role in prostate cancer growth and progression. FKBP51 overexpression was found to increase the AR transcriptional response by facilitating hormone-binding competence through the assembly of the AR LBD with mature FKBP51-Hsp90-p23 complexes (Ni *et al.*, 2010), resulting in higher levels of androgen-liganded receptor and providing a pathway for AR-dependent signalling and growth in a low-androgen environment. The ability of FKBP51 to enhance AR transcription and chaperone complex assembly appears to be dependent on FKBP51 PPIase activity mediated by the FK1 domain and requires Hsp90 binding through its TPR domain (Ni *et al.*, 2010).

6. Receptor LBD contacts with other Hsp90 cochaperones

6.1 PP5; GCUNC-45; SGT

The domain structure of the Hsp90 cochaperone, PP5, a serine/threonine protein phosphatase (Chen *et al.*, 1994; Chinkers, 1994), is characterised by a C-terminal phosphatase catalytic domain and an N-terminal TPR domain that competes with FKBP51, FKBP52 and CyP40 for the TPR binding site at the Hsp90 C-terminus during assembly into mature steroid receptor-Hsp90 complexes (Banerjee *et al.*, 2008; Chen *et al.*, 1996a; Hinds Jr & Sanchez, 2008). Through its TPR domain, PP5 has also been shown to bind directly to ERα and ERβ, an interaction that targets the LBDs of these receptors, but does not require the C-terminal region incorporating

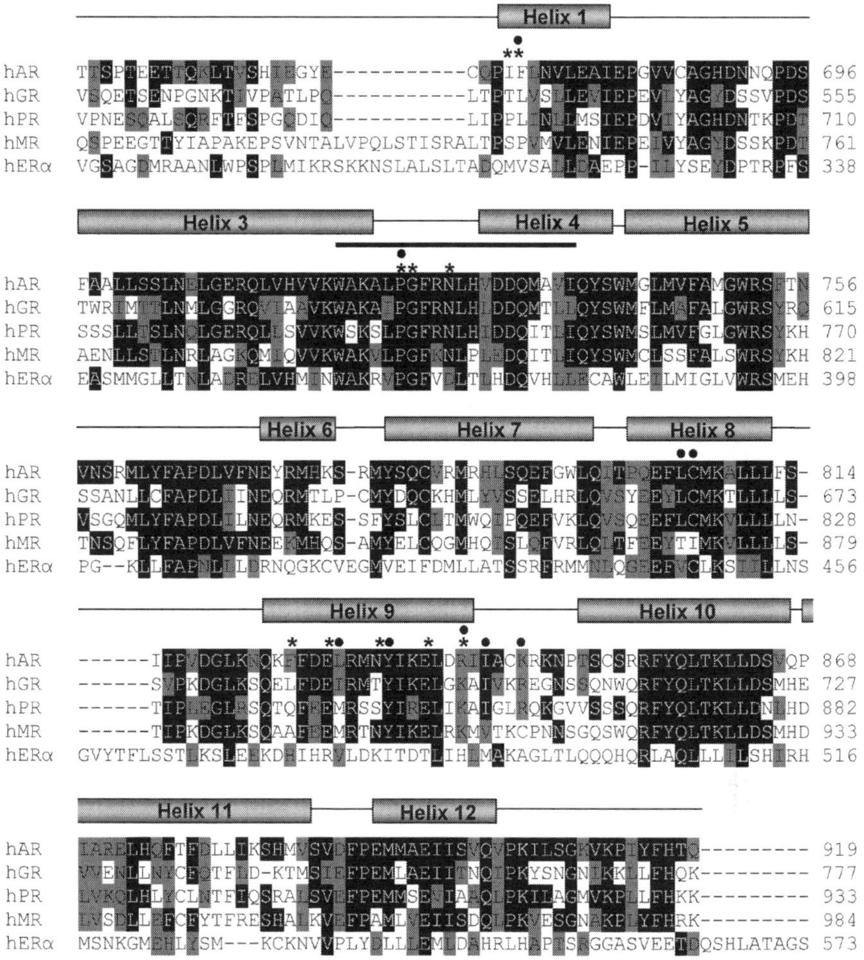

NCBI accession numbers for receptor sequences are: AR - NP000035, ERα - NP000116, GR - NP001018087, MR - NP000892, PR - NP000917. The ERα sequence has 595 amino acids and is shown terminated at residue 573. LBD helices are based on the structure of AR liganded to R1881 (Matias *et al.*, 2000a) (PDB ID 1E3G). The nuclear receptor signature sequence is indicated (thick black line). Residues that map to the BF-3 allosteric regulatory site defined for AR are highlighted with an asterisk (*). Multiple residues that contribute to the FKBP52 sensitivity of AR and form the putative binding site for MJC13 (De Leon *et al.*, 2011) are highlighted with a black circle (•). Identical residues are shown white against black; conserved residues (black on grey) are based on the following scheme: (P, G), (M, C), (Y, W, F, H), (L, V, I, A), (K, R), (E, Q, N, D) and (S, T).

Fig. 4. Multiple sequence alignment of human steroid receptor LBDs.

the helix 11-12 loop and helix 12 central to AF2 function (Ikeda *et al.*, 2004). PP5 was found to function as a negative regulator of ERα transcription *in vivo* by inhibiting epidermal growth factor (EGF)-dependent phosphorylation of Ser118 in the receptor N-terminal domain. Although demonstration of a direct PP5-ERα interaction was consistent with a non-involvement of Hsp90, a role for this major molecular chaperone in the *in vivo* effects of PP5 on ERα function cannot be discounted. Similar observations have been reported for GR with evidence suggesting that PP5-dependent modulation of receptor N-terminal phosphorylation within the GR-Hsp90 apo-receptor complex is mediated through contacts between the phosphatase and receptor LBD (Wang *et al.*, 2007).

A yeast two-hybrid screen, using bait encompassing both the hinge region and LBD of human PR, liganded with the mixed antagonist RU486, identified GCUNC-45 as a PR-binding protein (Chadli *et al.*, 2006). Presence of two LXXLL motifs (similar to NR boxes of known transcriptional coregulatory proteins) within the interacting clone, corresponding to the C-terminal end of GCUNC-45, suggested a mode of interaction similar to that for receptor recognition of transcription coactivators (Ratajczak, 2001), although this remains to be confirmed. Both FKBP52 and CyP40 compete with GCUNC-45 for the N-terminal TPR site, with nucleotides causing a reduction in Hsp90 binding affinity for these cochaperones in this region and favouring their interaction with the Hsp90 C-terminus during progression of receptor to a hormone-binding state (Chadli *et al.*, 2008b). GCUNC-45 therefore, appears to have a role upstream of FKBP52 and CyP40, at an intermediate stage of the receptor activation pathway.

The Hsp70/Hsp90 cochaperone, SGT, has been shown to interact through its TPR domain with the hinge region of human AR, which contains a peptide sequence structurally resembling the EEVD binding site for TPR proteins at the extreme C-terminus of Hsp70 and Hsp90 (Buchanan *et al.*, 2007). It has been proposed that, as a component of AR-Hsp90 complexes, SGT regulates the ligand sensitivity of AR signalling by limiting receptor trafficking to the nucleus at low hormone concentrations and maintaining the receptor within the cytoplasm of the cell.

6.2 p23; Cdc37

Disruption of the p23 gene in mice has revealed that although p23 is not essential for overall perinatal development its absolute requirement for perinatal survival is linked to impaired GR function arising most likely from instability of GR-Hsp90 complexes in the absence of p23 (Grad *et al.*, 2006; Picard, 2006). These findings suggest that GR might be a key molecular target for p23. Overexpression experiments with p23 in tissue culture cells have revealed both positive and negative influences on GR function (Freeman *et al.*, 2000; Wochnik *et al.*, 2004), as well as differential effects on other steroid receptors - increasing PR activity, while decreasing the activities of AR, ERα and MR (Freeman *et al.*, 2000). In yeast, p23 has been shown to be a positive regulator of ERα transcriptional activation, being most effective at low ERα levels and hormone concentrations, consistent with the proposed role for p23 as a component of mature ERα-Hsp90 complexes (Knoblauch & Garabedian, 1999). Ectopic expression of p23 in MCF-7 breast cancer cells increased both hormone-dependent and hormone-independent ERα transcriptional activity (Knoblauch & Garabedian, 1999).

Thus, while the major impact of p23 on ERα is likely to be through an Hsp90-dependent effect on estradiol binding, p23 overexpression may also influence receptor activity independent of ligand binding and may participate in the disassembly of receptors at cognate response elements (Freeman *et al.*, 2000; Freeman & Yamamoto, 2001; Freeman & Yamamoto, 2002). It is of interest that although p23 increases AR transcriptional activity in a variety of mammalian cell lines, partly by increasing ligand binding competence of the receptor, Hsp90 inhibitors could not abolish the AR coactivation potential of p23, consistent with an Hsp90-independent role of p23 in AR function (Querol Cano L and Bevan CL, unpublished observations).

Genetic studies in yeast have revealed that Cdc37 plays a role in AR hormone-dependent transactivation through functional interactions with the AR LBD, although the hormone-binding properties of the receptor appear to be unaffected (Fliss *et al.*, 1997). The association with Cdc37 is specific to AR since it does not occur with closely related nuclear receptors such as GR (Rao *et al.*, 2001). Depletion of Cdc37 using RNA interference caused growth arrest in both AR-positive and AR-negative prostate cancer cells, and in the former led to a loss of AR transcriptional activity with a concomitant decrease in androgen-dependent gene expression (Gray *et al.*, 2007). The targeting of Cdc37 in prostate cancer causes growth inhibition that correlates with decreased signalling through multiple pathways - the extracellular signal-regulated kinase (ERK) and Akt kinase cascades, as well as reduced AR-dependent signalling (Gray *et al.*, 2008).

7. Conclusions

We have arrived at a better understanding of the molecular mechanisms that allow the Hsp90 chaperone to modulate steroid receptor function through direct contact with receptor LBDs. Critical to this regulation is the ability of Hsp90 to coordinate and bring to receptor-Hsp90 complexes a selection of cochaperones whose specialized influences target receptor LBDs and combine, at various stages of the receptor activation pathway, to alter receptor hormone-binding status, cellular location and transcriptional activity. A number of these cochaperones may impact on steroid receptor function independently of Hsp90. Substantial gaps still remain, however in our knowledge of how the interplay between Hsp90 and its cochaperones affects receptor function. For example, while it is known the CyP40 yeast homologue, Cpr6, regulates Hsp90 ATPase activity during receptor assembly (Prodromou *et al.*, 1999) and studies of a second yeast homologue, Cpr7, have provided some insight into the role of this immunophilin in Hsp90-dependent signalling by steroid receptors (Duina *et al.*, 1996; Duina *et al.*, 1998), a coherent mechanism at the molecular level has yet to be defined. From the structural similarity between CyP40 and FKBP52, both being characterized by N-terminal PPIase and C-terminal TPR domains, it is tempting to draw parallels for their mechanism of action. Within steroid receptor-Hsp90 complexes it is possible that, as for FKBP52, the CyP40 PPIase domain forms productive interactions with the receptor LBD, serving to modulate receptor conformation and function. This may be of relevance for the function of ERα, purification of which led to the isolation of CyP40 in ERα-Hsp90 complexes (Ratajczak *et al.*, 1993) and for the regulation of AR in prostate cancer where CyP40 appears to be overexpressed (Periyasamy *et al.*, 2010).

Hsp90 is required for the proper function of several key regulatory proteins including multiple tyrosine and serine/threonine kinases and steroid receptors, many of which are involved in promoting malignancy (Calderwood *et al.*, 2006; Pearl, 2005; Whitesell & Lindquist, 2005). The aim of targeting and pharmacological manipulation of the Hsp90 chaperoning system has led to the ongoing development and clinical evaluation of novel Hsp90 and chaperone inhibitors for potential application in therapies against selected malignancies (Donnelly *et al.*, 2010; Kim *et al.*, 2009), syndromes arising from dysfunctional protein folding and neurodegenerative diseases (Jinwal *et al.*, 2010). With growing understanding of the novel mechanisms through which Hsp90 cochaperones modulate the function of specific clients, strategies are now evolving for the targeting of chaperone-client interactions in a wide range of human diseases (De Leon *et al.*, 2011; Gray *et al.*, 2008).

8. Abbreviations

Hsp, heat shock protein; TPR, tetratricopeptide repeat; PPIase, prolylpeptidyl isomerase; FKBP, FK506-binding protein; CyP40, cyclophilin 40; PP5, serine/threonine protein phosphatase type 5; GCUNC-45, general cell UNC-45; αSGT, small glutamine-rich tetratricopeptide repeat containing protein α; AR, androgen receptor; ERα, estrogen receptor α; ERβ, estrogen receptor β; GR, glucocorticoid receptor; MR, mineralocorticoid receptor; PR, progesterone receptor; LBD, ligand-binding domain; AF2, activation function 2; GST, glutathione S-transferase.

9. Acknowledgments

The authors wish to acknowledge support from the National Health & Medical Research Council of Australia, the National Breast Cancer Foundation and the Sir Charles Gairdner Hospital Research Fund. The authors also thank colleagues for permitting citation of their data prior to publication.

10. References

Ali M. M. U., Roe S. M., Vaughan C. K., Meyer P., Panaretou B., Piper P. W., Prodromou C. & Pearl L. H. (2006). Crystal structure of an Hsp90-nucleotide-p23/Sba1 closed chaperone complex. *Nature*, Vol.440, No.7087, pp. 1013-1017.

Allan R. K., Mok D., Ward B. K. & Ratajczak T. (2006). Modulation of chaperone function and cochaperone interaction by novobiocin in the C-terminal domain of Hsp90: evidence that coumarin antibiotics disrupt Hsp90 dimerization. *J Biol Chem*, Vol.281, No.11, pp. 7161-7171.

Amler L. C., Agus D. B., LeDuc C., Sapinoso M. L., Fox W. D., Kern S., Lee D., Wang V., Leysens M., Higgins B., Martin J., Gerald W., Dracopoli N., Cordon-Cardo C., Scher H. I. & Hampton G. M. (2000). Dysregulated expression of androgen-responsive and nonresponsive genes in the androgen-independent prostate cancer xenograft model CWR22-R. *Cancer Res*, Vol.60, No.21, pp. 6134-6141.

Angeletti P. C., Walker D. & Panganiban A. T. (2002). Small glutamine-rich protein/viral protein U-binding protein is a novel cochaperone that affects heat shock protein 70 activity. *Cell Stress Chaperones*, Vol.7, No.3, pp. 258-268

Aumais J. P., Lee H. S., Lin R. & White J. H. (1997). Selective interaction of Hsp90 with an estrogen receptor ligand-binding domain containing a point mutation. *J Biol Chem*, Vol.272, No.18, pp. 12229-12235.

Ballinger C. A., Connell P., Wu Y., Hu Z., Thompson L. J., Yin L.-Y. & Patterson C. (1999). Identification of CHIP, a novel tetratricopeptide repeat-containing protein that interacts with heat shock proteins and negatively regulates chaperone functions. *Mol Cell Biol*, Vol.19, No.6, pp. 4535-4545.

Banerjee A., Periyasamy S., Wolf I. M., Hinds T. D., Yong W., Shou W. & Sanchez E. R. (2008). Control of glucocorticoid and progesterone receptor subcellular localization by the ligand-binding domain is mediated by distinct interactions with tetratricopeptide repeat proteins. *Biochemistry*, Vol.47, No.39, pp. 10471-10480.

Barent R. L., Nair S. C., Carr D. C., Ruan Y., Rimerman R. A., Fulton J., Zhang Y. & Smith D. F. (1998). Analysis of FKBP51/FKBP52 chimeras and mutants for Hsp90 binding and association with progesterone receptor complexes. *Mol Endocrinol*, Vol.12, No.3, pp. 342-354.

Barford D. (1996). Molecular mechanisms of the protein serine/threonine phosphatases. *Trends Biochem Sci*, Vol.21, No.11, pp. 407-412.

Bimston D., Song J., Winchester D., Takayama S., Reed J. C. & Morimoto R. I. (1998). BAG-1, a negative regulator of Hsp70 chaperone activity, uncouples nucleotide hydrolysis from substrate release. *EMBO J*, Vol.17, No.23, pp. 6871-6878.

Binart N., Lombès M. & Baulieu E. E. (1995). Distinct functions of the 90 kDa heat-shock protein (hsp90) in oestrogen and mineralocorticosteroid receptor activity: effects of hsp90 deletion mutants. *Biochem J*, Vol.311, 797-804.

Bledsoe R. K., Montana V. G., Stanley T. B., Delves C. J., Apolito C. J., McKee D. D., Consler T. G., Parks D. J., Stewart E. L., Willson T. M., Lambert M. H., Moore J. T., Pearce K. H. & Xu H. E. (2002). Crystal structure of the glucocorticoid receptor ligand binding domain reveals a novel mode of receptor dimerization and coactivator recognition. *Cell*, Vol.110, 93-105.

Brelivet Y., Kammerer S., Rochel N., Poch O. & Moras D. (2004). Signature of the oligomeric behaviour of nuclear receptors at the sequence and structural level *EMBO Rep*, Vol.5, No.4, pp. 423-429.

Buchanan G., Ricciardelli C., Harris J. M., Prescott J., Yu Z. C.-L., Jia L., Butler L. M., Marshall V. R., Scher H. I., Gerald W. L., Coetzee G. A. & Tilley W. D. (2007). Control of androgen receptor signaling in prostate cancer by the cochaperone small glutamine rich tetratricopeptide repeat containing protein α. *Cancer Res*, Vol.67, No.20, pp. 10087-10096.

Caamano C. A., Morano M. I., Dalman F. C., Pratt W. B. & Akil H. (1998). A conserved proline in the Hsp90 binding region of the glucocorticoid receptor is required for Hsp90 heterocomplex stabilization and receptor signaling. *J Biol Chem*, Vol.273, No.32, pp. 20473-20480.

Cadepond F., Binart N., Chambraud B., Jibard N., Schweizer-Groyer G., Segard-Maurel I. & Baulieu E. E. (1993). Interaction of glucocorticosteroid receptor and wild-type or mutated 90-kDa heat shock protein coexpressed in baculovirus-infected Sf9 cells. *PNAS*, Vol.90, No.22, pp. 10434-10438.

Calderwood S. K., Khaleque M. A., Sawyer D. B. & Ciocca D. R. (2006). Heat shock proteins in cancer: chaperones of tumorigenesis. *Trends Biochem Sci*, Vol.31, No.3, pp. 164-172.

Callahan M. A., Handley M. A., Lee Y.-H., Talbot K. J., Harper J. W. & Panganiban A. T. (1998). Functional interaction of human immunodeficiency virus type 1 Vpu and Gag with a novel member of the tetratricopeptide repeat protein family. *J Virol*, Vol.72, No.6, pp. 5189-5197.

Carrello A., Ingley E., Minchin R. F., Tsai S. & Ratajczak T. (1999). The common tetratricopeptide repeat acceptor site for steroid receptor-associated immunophilins and Hop is located in the dimerization domain of Hsp90. *J Biol Chem*, Vol.274, No.5, pp. 2682-2689.

Carrigan P. E., Sikkink L. A., Smith D. F. & Ramirez-Alvarado M. (2006). Domain:domain interactions within Hop, the Hsp70/Hsp90 organizing protein, are required for protein stability and structure. *Protein Sci*, Vol.15, No.3, pp. 522-532.

Chadli A., Bruinsma E. S., Stensgard B. & Toft D. (2008a). Analysis of Hsp90 cochaperone interactions reveals a novel mechanism for TPR protein recognition. *Biochemistry*, Vol.47, No.9, pp. 2850-2857.

Chadli A., Felts S. J. & Toft D. O. (2008b). GCUNC-45 is the first Hsp90 co-chaperone to show α/β isoform specificity *J Biol Chem*, Vol.283, No.15, pp. 9509-9512.

Chadli A., Graham J. D., Abel M. G., Jackson T. A., Gordon D. F., Wood W. M., Felts S. J., Horwitz K. B. & Toft D. (2006). GCUNC-45 is a novel regulator for the progesterone receptor/Hsp90 chaperoning pathway. *Mol Cell Biol*, Vol.26, No.5, pp. 1722-1730.

Chen M.-S., Silverstein A. M., Pratt W. B. & Chinkers M. (1996a). The tetratricopeptide repeat domain of protein phosphatase 5 mediates binding to glucocorticoid receptor heterocomplexes and acts as a dominant negative mutant. *J Biol Chem*, Vol.271, No.50, pp. 32315-32320.

Chen M. X. & Cohen P. T. W. (1997). Activation of protein phosphatase 5 by limited proteolysis or the binding of polyunsaturated fatty acids to the TPR domain. *FEBS Lett*, Vol.400, No.1, pp. 136-140.

Chen M. X., McPartlin A. E., Brown L., Chen Y. H., Barker H. M. & Cohen P. T. W. (1994). A novel human protein serine/ threonine phosphatase, which possesses four tetretricopeptide repeat motifs and localizes to the nucleus. *EMBO J*, Vol.13, No.18, pp. 4278-4290.

Chen S., Prapapanich V., Rimerman R. A., Honore B. & Smith D. F. (1996b). Interactions of p60, a mediator of progesterone receptor assembly, with heat shock proteins Hsp90 and Hsp70. *Mol Endocrinol*, Vol.10, No.6, pp. 682-693.

Chen S. & Smith D. F. (1998). Hop as an adaptor in the heat shock protein 70 (Hsp70) and Hsp90 chaperone machinery. *J Biol Chem*, Vol.273, No.52, pp. 35194-35200.

Chen S., Sullivan W. P., Toft D. O. & Smith D. F. (1998). Differential interactions of p23 and the TPR-containing proteins Hop, Cyp40, FKBP52 and FKBP51 with Hsp90 mutants. *Cell Stress Chaperones*, Vol.3, No.2, pp. 118-129.

Cheung-Flynn J., Prapapanich V., Cox M. B., Riggs D. L., Suarez-Quian C. & Smith D. F. (2005). Physiological role for the cochaperone FKBP52 in androgen receptor signaling. *Mol Endocrinol*, Vol.19, No.6, pp. 1654-1666.

Cheung-Flynn J., Roberts P. J., Riggs D. L. & Smith D. F. (2003). C-terminal sequences outside the tetratricopeptide repeat domain of FKBP51 and FKBP52 cause differential binding to Hsp90. *J Biol Chem*, Vol.278, No.19, pp. 17388-17394.

Chinkers M. (1994). Targeting of a distinctive protein-serine phosphatase to the protein kinase-like domain of the atrial natriuretic peptide receptor. *PNAS*, Vol.91, No.23, pp. 11075-11079.

Cohen P. T. W. (1997). Novel protein serine/threonine phosphatases: variety is the spice of life. *Trends Biochem Sci*, Vol.22, No.7, pp. 245-251.

Connell P., Ballinger C. A., Jiang J., Wu Y., Thompson L. J., Hohfeld J. & Patterson C. (2001). The co-chaperone CHIP regulates protein triage decisions mediated by heat-shock proteins. *Nat Cell Biol*, Vol.3, No.1, pp. 93-96.

Cyr D. M., Höhfeld J. & Patterson C. (2002). Protein quality control: U-box-containing E3 ubiquitin ligases join the fold. *Trends Biochem Sci*, Vol.27, No.7, pp. 368-375.

Cziepluch C., Kordes E., Poirey R., Grewenig A., Rommelaere J. & Jauniaux J.-C. (1998). Identification of a novel cellular TPR-containing protein, SGT, that interacts with the nonstructural protein NS1 of parvovirus H-1. *J Virol*, Vol.72, No.5, pp. 4149-4156.

Das A. K., Cohen P. T. W. & Barford D. (1998). The structure of the tetratricopeptide repeats of protein phosphatase 5: implications for TPR-mediated protein-protein interactions. *EMBO J*, Vol.17, No.5, pp. 1192-1199.

Davies T. H., Ning Y.-M. & Sanchez E. R. (2002). A new first step in activation of steroid receptors: hormone-induced switching of FKBP51 and FKBP52 immunophilins. *J Biol Chem*, Vol.277, No.7, pp. 4597-4600.

De Leon J. T., Iwai A., Feau C., Garcia Y., Balsiger H. A., Storer C. L., Suro R. M., Garza K. M., Lee S., Sang Kim Y., Chen Y., Ning Y.-M., Riggs D. L., Fletterick R. J., Guy R. K., Trepel J. B., Neckers L. M. & Cox M. B. (2011). Targeting the regulation of androgen receptor signaling by the heat shock protein 90 cochaperone FKBP52 in prostate cancer cells. *PNAS*, Vol.108, No.29, pp. 11878-11883.

Denny W. B., Prapapanich V., Smith D. F. & Scammell J. G. (2005). Structure-function analysis of Squirrel Monkey FK506-binding protein 51, a potent inhibitor of glucocorticoid receptor activity. *Endocrinology*, Vol.146, No.7, pp. 3194-3201.

Denny W. B., Valentine D. L., Reynolds P. D., Smith D. F. & Scammell J. G. (2000). Squirrel monkey immunophilin FKBP51 is a potent inhibitor of glucocorticoid receptor binding. *Endocrinology*, Vol.141, No.11, pp. 4107-4113.

Dittmar K. D., Hutchison K. A., Owens-Grillo J. K. & Pratt W. B. (1996). Reconstitution of the steroid receptor·Hsp90 heterocomplex assembly system of rabbit reticulocyte lysate. *J Biol Chem*, Vol.271, No.22, pp. 12833-12839.

Donnelly A. C., Zhao H., Reddy Kusuma B. & Blagg B. S. J. (2010). Cytotoxic sugar analogues of an optimized novobiocin scaffold. *Med Chem Commun*, Vol.1, No.2, pp. 165-170.

Duina A. A., Chang H.-C. J., Marsh J. A., Lindquist S. & Gaber R. F. (1996). A cyclophilin function on Hsp90-dependent signal transduction. *Science*, Vol.274, 1713-1715.

Duina A. A., Marsh J. A., Kurtz R. B., Chang H.-C. J., Lindquist S. & Gaber R. F. (1998). The peptidyl-prolyl isomerase domain of the CyP-40 cyclophilin homolog Cpr7 is not required to support growth or glucocorticoid receptor activity in *Saccharomyces cerevisiae*. *J Biol Chem*, Vol.273, No.18, pp. 10819-10822.

Dutta S. & Tan Y.-J. (2008). Structural and functional characterization of human SGT and its interaction with Vpu of the human immunodeficiency virus type 1. *Biochemistry*, Vol.47, No.38, pp. 10123-10131.

Echeverria P. C. & Picard D. (2010). Molecular chaperones, essential partners of steroid hormone receptors for activity and mobility. *Biochim Biophys Acta - Mol Cell Res*, Vol.1803, No.6, pp. 641-649.

Estébanez-Perpiñá E., Arnold L. A., Nguyen P., Rodrigues E. D., Mar E., Bateman R., Pallai P., Shokat K. M., Baxter J. D., Guy R. K., Webb P. & Fletterick R. J. (2007). A surface on the androgen receptor that allosterically regulates coactivator binding. *PNAS*, Vol.104, No.41, pp. 16074-16079.

Fang L., Ricketson D., Getubig L. & Darimont B. (2006). Unliganded and hormone-bound glucocorticoid receptors interact with distinct hydrophobic sites in the Hsp90 C-terminal domain. *PNAS*, Vol.103, No.49, pp. 18487-18492.

Febbo P. G., Lowenberg M., Thorner A. R., Brown M., Loda M. & Golub T. R. (2005). Androgen mediated regulation and functional implications of FKBP51 expression in prostate cancer. *J Urol*, Vol.173, No.5, pp. 1772-1777.

Felts S. J. & Toft D. O. (2003). p23, a simple protein with complex activities. *Cell Stress Chaperones*, Vol.8, No.2, pp. 108-113.

Fliss A. E., Fang Y., Boschelli F. & Caplan A. J. (1997). Differential *in vivo* regulation of steroid hormone receptor activation by Cdc37p. *Mol Biol Cell*, Vol.8, No.12, pp. 2501-2509.

Freeman B. C., Felts S. J., Toft D. O. & Yamamoto K. R. (2000). The p23 molecular chaperones act at a late step in intracellular receptor action to differentially affect ligand efficacies. *Genes Dev*, Vol.14, 422-434.

Freeman B. C. & Yamamoto K. R. (2001). Continuous recycling: a mechanism for modulatory signal transduction. *Trends Biochem Sci*, Vol.26, No.5, pp. 285-290.

Freeman B. C. & Yamamoto K. R. (2002). Disassembly of transcriptional regulatory complexes by molecular chaperones. *Science*, Vol.296, No.5576, pp. 2232-2235.

Frydman J. & Höhfeld J. (1997). Chaperones get in touch: the Hip-Hop connection. *Trends Biochem Sci*, Vol.22, No.3, pp. 87-92.

Fuller P. J., Smith B. J. & Rogerson F. M. (2004). Cortisol resistance in the New World revisited. *Trends Endocrinol Metab*, Vol.15, No.7, pp. 296-299.

Gallo L. I., Ghini A. A., Piwien Pilipuk G. & Galigniana M. D. (2007). Differential recruitment of tetratricopeptide repeat domain immunophilins to the mineralocorticoid receptor influences both heat-shock protein 90-dependent retrotransport and hormone-dependent transcriptional activity. *Biochemistry*, Vol.46, No.49, pp. 14044-14057.

Gebauer M., Zeiner M. & Gehring U. (1997). Proteins interacting with the molecular chaperone Hsp70/Hsc70: physical associations and effects on refolding activity. *FEBS Lett*, Vol.417, No.1, pp. 109-113.

Giannoukos G., Silverstein A. M., Pratt W. B. & Simons Jr S. S. (1999). The seven amino acids (547-553) of rat glucocorticoid receptor required for steroid and Hsp90 binding contain a functionally independent LXXLL motif that is critical for steroid binding. *J Biol Chem*, Vol.274, No.51, pp. 36527-36536.

Grad I., McKee T. A., Ludwig S. M., Hoyle G. W., Ruiz P., Wurst W., Floss T., Miller C. A., III & Picard D. (2006). The Hsp90 cochaperone p23 is essential for perinatal survival. *Mol Cell Biol*, Vol.26, No.23, pp. 8976-8983.

Gray P. J., Prince T., Cheng J., Stevenson M. A. & Calderwood S. K. (2008). Targeting the oncogene and kinome chaperone CDC37. *Nat Rev Cancer*, Vol.8, No.7, pp. 491-495.

Gray P. J., Stevenson M. A. & Calderwood S. K. (2007). Targeting Cdc37 inhibits multiple signaling pathways and induces growth arrest in prostate cancer cells. *Cancer Res*, Vol.67, No.24, pp. 11942-11950.

Harris S. F., Shiau A. K. & Agard D. A. (2004). The crystal structure of the carboxy-terminal dimerization domain of htpG, the *Escherichia coli* Hsp90, reveals a potential substrate binding site. *Structure*, Vol.12, No.6, pp. 1087-1097.

He B., Bowen N. T., Minges J. T. & Wilson E. M. (2001). Androgen-induced NH_2- and COOH-terminal interaction inhibits p160 coactivator recruitment by activation function 2. *J Biol Chem*, Vol.276, No.45, pp. 42293-42301.

He B., Gampe Jr R. T., Kole A. J., Hnat A. T., Stanley T. B., An G., Stewart E. L., Kalman R. I., Minges J. T. & Wilson E. M. (2004). Structural basis for androgen receptor interdomain and coactivator interactions suggests a transition in nuclear receptor activation function dominance. *Mol Cell*, Vol.16, No.3, pp. 425-438.

Hinds Jr T. D. & Sánchez E. R. (2008). Protein phosphatase 5. *Int J Biochem Cell Biol*, Vol.40, No.11, pp. 2358-2362.

Höhfeld J. & Jentsch S. (1997). GrpE-like regulation of the Hsc70 chaperone by the anti-apoptotic protein BAG-1. *EMBO J*, Vol.16, No.20, pp. 6209-6216.

Höhfeld J., Minami Y. & Hartl F.-U. (1995). Hip, a novel cochaperone involved in the eukaryotic Hsc70/Hsp40 reaction cycle. *Cell*, Vol.83, No.4, pp. 589-598.

Ikeda K., Ogawa S., Tsukui T., Horie-Inoue K., Ouchi Y., Kato S., Muramatsu M. & Inoue S. (2004). Protein phosphatase 5 is a negative regulator of estrogen receptor-mediated transcription. *Mol Endocrinol*, Vol.18, No.5, pp. 1131-1143.

Jackson S. E., Queitsch C. & Toft D. (2004). Hsp90: from structure to phenotype. *Nat Struct Mol Biol*, Vol.11, No.12, pp. 1152-1155.

Jiang J., Ballinger C. A., Wu Y., Dai Q., Cyr D. M., Höhfeld J. & Patterson C. (2001). CHIP is a U-box-dependent E3 ubiquitin ligase. Identification of Hsc70 as a target for ubiquitylation. *J Biol Chem*, Vol.276, 42938-42944.

Jinwal U. K., Koren J., Borysov S. I., Schmid A. B., Abisambra J. F., Blair L. J., Johnson A. G., Jones J. R., Shults C. L., O'Leary J. C., Jin Y., Buchner J., Cox M. B. & Dickey C. A. (2010). The Hsp90 cochaperone, FKBP51, increases Tau stability and polymerizes microtubules. *J Neurosci*, Vol.30, No.2, pp. 591-599.

Joseph J. D., Wittmann B. M., Dwyer M. A., Cui H., Dye D. A., McDonnell D. P. & Norris J. D. (2009). Inhibition of prostate cancer cell growth by second-site androgen receptor antagonists. *PNAS*, Vol.106, No.29, pp. 12178-12183.

Kajander T., Sachs J. N., Goldman A. & Regan L. (2009). Electrostatic interactions of Hsp-organizing protein tetratricopeptide domains with Hsp70 and Hsp90: computational analysis and protein engineering. *J Biol Chem*, Vol.284, 25364-25374.

Kallen J., Mikol V., Taylor P. & D.Walkinshaw M. (1998). X-ray structures and analysis of 11 cyclosporin derivatives complexed with cyclophilin A. *J Mol Biol*, Vol.283, No.2, pp. 435-449.

Karagöz G. E., Duarte A. M. S., Ippel H., Uetrecht C., Sinnige T., van Rosmalen M., Hausmann J., Heck A. J. R., Boelens R. & Rüdiger S. G. D. (2010). N-terminal domain of human Hsp90 triggers binding to the cochaperone p23. *PNAS*, Vol.108, No.2, pp. 580-585.

Kauppi B., Jakob C., Farnegardh M., Yang J., Ahola H., Alarcon M., Calles K., Engstrom O., Harlan J., Muchmore S., Ramqvist A.-K., Thorell S., Ohman L., Greer J., Gustafsson J.-A., Carlstedt-Duke J. & Carlquist M. (2003). The three-dimensional structures of antagonistic and agonistic forms of the glucocorticoid receptor ligand-binding domain: RU-486 induces a transconformation that leads to active antagonism. *J Biol Chem*, Vol.278, No.25, pp. 22748-22754.

Kim Y. S., Alarcon S. V., Lee S., Lee M. J., Giaccone G., Neckers L. & Trepel J. B. (2009). Update on Hsp90 inhibitors in clinical trial. *Curr Top Med Chem*, Vol.9, No.15, pp. 1479-1492.

Knoblauch R. & Garabedian M. J. (1999). Role for Hsp90-associated cochaperone p23 in estrogen receptor signal transduction. *Mol Cell Biol*, Vol.19, No.5, pp. 3748-3759.

Kosano H., Stensgard B., Charlesworth M. C., McMahon N. & Toft D. (1998). The assembly of progesterone receptor-Hsp90 complexes using purified proteins. *J Biol Chem*, Vol.273, No.49, pp. 32973-32979.

Lassle M., Blatch G. L., Kundra V., Takatori T. & Zetter B. R. (1997). Stress-inducible, murine protein mSTI1. Characterization of binding domains for heat shock proteins and in vitro phosphorylation by different kinases. *J Biol Chem*, Vol.272, No.3, pp. 1876-1884.

Li J., Richter K. & Buchner J. (2011). Mixed Hsp90-cochaperone complexes are important for the progression of the reaction cycle. *Nat Struct Mol Biol*, Vol.18, No.1, pp. 61-66.

Liou S.-T. & Wang C. (2005). Small glutamine-rich tetratricopeptide repeat-containing protein is composed of three structural units with distinct functions. *Arch Biochem Biophys*, Vol.435, No.2, pp. 253-263.

Louvion J.-F., Warth R. & Picard D. (1996). Two eukaryote-specific regions of Hsp82 are dispensable for its viability and signal transduction functions in yeast. *PNAS*, Vol.93, No.24, pp. 13937-13942.

MacLean M. & Picard D. (2003). Cdc37 goes beyond Hsp90 and kinases. *Cell Stress Chaperones*, Vol.8, No.2, pp. 114-119.

Magee J., Chang L., Stormo G. & Milbrandt J. (2006). Direct, androgen receptor-mediated regulation of the FKBP5 gene via a distal enhancer element. *Endocrinology*, Vol.147, No.1, pp. 590-598.

Marcu M. G., Chadli A., Bouhouche I., Catelli M. & Neckers L. M. (2000). The heat shock protein 90 antagonist novobiocin interacts with a previously unrecognized ATP-binding domain in the carboxyl terminus of the chaperone. *J Biol Chem*, Vol.275, No.47, pp. 37181-37186.

Matias P. M., Donner P., Coelho R., Thomaz M., Peixoto C., Macedo S., Otto N., Joschko S., Scholz P., Wegg A., Bäsler S., Schäfer M., Egner U. & Carrondo M. A. (2000a). Structural evidence for ligand specificity in the binding domain of the human androgen receptor: implications for pathogenic gene mutations. *J. Biol. Chem.*, Vol.275, No.34, pp. 26164-26171.

Matias P. M., Donner P., Coelho R., Thomaz M., Peixoto C., Macedo S., Otto N., Joschko S., Scholz P., Wegg A., Bäsler S., Schäfer M., Egner U. & Carrondo M. A. (2000b). Structural evidence for ligand specificity in the binding domain of the human androgen receptor: implications for pathogenic gene mutations. *J Biol Chem*, Vol.275, No.34, pp. 26164-26171.

McLaughlin S. H., Smith H. W. & Jackson S. E. (2002). Stimulation of the weak ATPase activity of human Hsp90 by a client protein. *J Mol Biol*, Vol.315, No.4, pp. 787-798.

McLaughlin S. H., Sobott F., Yao Z.-p., Zhang W., Nielsen P. R., Grossmann J. G. n., Laue E. D., Robinson C. V. & Jackson S. E. (2006). The co-chaperone p23 arrests the Hsp90 ATPase cycle to trap client proteins. *J Mol Biol*, Vol.356, No.3, pp. 746-758.

Mok D., Allan R. K., Carrello A., Wangoo K., Walkinshaw M. D. & Ratajczak T. (2006). The chaperone function of cyclophilin 40 maps to a cleft between the prolyl isomerase and tetratricopeptide repeat domains. *FEBS Lett*, Vol.580, No.11, pp. 2761-2768.

Moore T. W., Mayne C. G. & Katzenellenbogen J. A. (2010). Minireview: Not picking pockets: nuclear receptor alternate-site modulators (NRAMs). *Mol Endocrinol*, Vol.24, No.4, pp. 683-695.

Mousses S., Wagner U., Chen Y., Kim J. W., Bubendorf L., Bittner M., Pretlow T., Elkahloun A. G., Trepel J. B., Kallioniemi O.-P. & (2001). Failure of hormone therapy in prostate cancer involves systematic restoration of androgen responsive genes and activation of rapamycin sensitive signaling. *Oncogene*, Vol.20, No.46, pp. 6718-6723.

Murata S., Minami Y., Minami M., Chiba T. & Tanaka K. (2001). CHIP is a chaperone-dependent E3 ligase that ubiquitylates unfolded protein. *EMBO Rep*, Vol.2, No.12, pp. 1133–1138.

Nair S. C., A.Rimerman R., Toran E. J., Chen S., Prapapanich V., Butts R. N. & Smith D. F. (1997). Molecular cloning of human FKBP51 and comparisons of immunophilin interactions with Hsp90 and progesterone receptor. *Mol Cell Biol*, Vol.17, No.2, pp. 594-603.

Nathan D. F. & Lindquist S. (1995). Mutational analysis of Hsp90 function: interactions with a steroid receptor and a protein kinase. *Mol Cell Biol*, Vol.15, No.7, pp. 3917-3925.

Nelson G. M., Huffman H. & Smith D. F. (2003). Comparison of the carboxy-terminal DP-repeat region in the co-chaperones Hop and Hip. *Cell Stress Chaperones*, Vol.8, No.2, pp. 125-133.

Nelson G. M., Prapapanich V., Carrigan P. E., Roberts P. J., Riggs D. L. & Smith D. F. (2004). The heat shock protein 70 cochaperone Hip enhances functional maturation of glucocorticoid receptor. *Mol Endocrinol*, Vol.18, No.7, pp. 1620-1630.

Ni L., Yang C.-S., Gioeli D., Frierson H., Toft D. O. & Paschal B. M. (2010). FKBP51 promotes assembly of the Hsp90 chaperone complex and regulates androgen receptor signaling in prostate cancer cells. *Mol Cell Biol*, Vol.30, No.5, pp. 1243-1253.

Odunuga O. O., Hornby J. A., Bies C., Zimmermann R., Pugh D. J. & Blatch G. L. (2003). Tetratricopeptide repeat motif-mediated Hsc70-mSTI1 interaction. *J Biol Chem*, Vol.278, No.9, pp. 6896-6904.

Onuoha S. C., Coulstock E. T., Grossmann J. G. & Jackson S. E. (2008). Structural studies on the co-chaperone Hop and its complexes with Hsp90. *J Mol Biol*, Vol.379, No.4, pp. 732-744.

Pearl L. H. (2005). Hsp90 and Cdc37 - a chaperone cancer conspiracy. *Curr Opin Genet Dev*, Vol.15, No.1, pp. 55-61.

Periyasamy S., Hinds T., Jr., Shemshedini L., Shou W. & Sanchez E. R. (2010). FKBP51 and Cyp40 are positive regulators of androgen-dependent prostate cancer cell growth and the targets of FK506 and cyclosporin A. *Oncogene*, Vol.29, No.11, pp. 1691-1701.

Picard D. (2006). Chaperoning steroid hormone action. *Trends Endocrinol Metab*, Vol.17, No.6, pp. 229-236.

Picard D., Khursheed B., Garabedian M. J., Fortin M. G., Lindquist S. & Yamamoto K. R. (1990). Reduced levels of Hsp90 compromise steroid receptor action *in vivo*. *Nature*, Vol.348, 166-168.

Pirkl F. & Buchner J. (2001). Functional analysis of the Hsp90-associated human peptidyl prolyl cis/trans isomerases FKBP51, FKBP52 and CyP40. *J Mol Biol*, Vol.308, No.4, pp. 795-806.

Prapapanich V., Chen S., Nair S. C., Rimerman R. A. & Smith D. F. (1996a). Molecular cloning of human p48, a transient component of progesterone receptor complexes and an Hsp70-binding protein. *Mol Endocrinol*, Vol.10, No.4, pp. 420-431.

Prapapanich V., Chen S. & Smith D. F. (1998). Mutation of Hip's carboxy-terminal region inhibits a transitional stage of progesterone receptor assembly. *Mol Cell Biol*, Vol.18, No.2, pp. 944-952.

Prapapanich V., Chen S., Toran E. J., Rimerman R. A. & Smith D. F. (1996b). Mutational analysis of the hsp70-interacting protein Hip. *Mol Cell Biol*, Vol.16, No.11, pp. 6200-6207.

Pratt W. B. & Toft D. O. (1997). Steroid receptor interactions with heat shock protein and immunophilin chaperones. *Endocr Rev*, Vol.18, No.3, pp. 306-360.

Pratt W. B. & Toft D. O. (2003). Regulation of signaling protein function and trafficking by the Hsp90/Hsp70-based chaperone machinery. *Exp Biol Med*, Vol.228, No.2, pp. 111-133.

Prodromou C., Panaretou B., Chohan S., Siligardi G., O'Brien R., Ladbury J. E., Roe S. M., Piper P. W. & Pearl L. H. (2000). The ATPase cycle of Hsp90 drives a molecular 'clamp' via transient dimerization of the N-terminal domains. *EMBO J*, Vol.19, No.16, pp. 4383-4392.

Prodromou C., Siligardi G., O'Brien R., Woolfson D. N., Regan L., Panaretou B., Ladbury J. E., Piper P. W. & Pearl L. H. (1999). Regulation of Hsp90 ATPase activity by tetratricopeptide repeat (TPR)-domain co-chaperones. *EMBO J*, Vol.18, No.3, pp. 754-762.

Radanyi C., Chambraud B. & Baulieu E. E. (1994). The ability of the immunophilin FKBP59-HBI to interact with the 90-kDa heat shock protein is encoded by its tetratricopeptide repeat domain. *PNAS*, Vol.91, 11197-11201.

Ramsey A. J. & Chinkers M. (2002). Identification of potential physiological activators of protein phosphatase 5. *Biochemistry*, Vol.41, No.17, pp. 5625-5632.

Ramsey A. J., Russell L. C. & Chinkers M. (2009). C-terminal sequences of Hsp70 and Hsp90 as non-specific anchors for tetratricopeptide repeat (TPR) proteins. *Biochem J*, Vol.423, No.3, pp. 411-419.

Rao J., Lee P., Benzeno S., Cardozo C., Albertus J., Robins D. M. & Caplan A. J. (2001). Functional interaction of human Cdc37 with the androgen receptor but not with the glucocorticoid receptor. *J Biol Chem*, Vol.276, No.8, pp. 5814-5820.

Ratajczak T. (2001). Protein coregulators that mediate estrogen receptor function. *Reprod Fertil Dev*, Vol.13, 221-229.

Ratajczak T. & Carrello A. (1996). Cyclophilin 40 (CyP-40), mapping of its Hsp90 binding domain and evidence that FKBP52 competes with CyP-40 for Hsp90 binding. *J Biol Chem*, Vol.271, No.6, pp. 2961-2965.

Ratajczak T., Carrello A., Mark P. J., Warner B. J., Simpson R. J., Moritz R. L. & House A. K. (1993). The cyclophilin component of the unactivated estrogen receptor contains a tetratricopeptide repeat domain and shares identity with p59 (FKBP59). *J Biol Chem*, Vol.268, No.18, pp. 13187-13192.

Ratajczak T., Hlaing J., Brockway M. J. & Hahnel R. (1990). Isolation of untransformed bovine estrogen receptor without molybdate stabilization. *J Steroid Biochem*, Vol.35, No.5, pp. 543-553.

Ratajczak T., Ward B. K. & Minchin R. F. (2003). Immunophilin chaperones in steroid receptor signalling. *Curr Top Med Chem*, Vol.3, No.12, pp. 1348-1357.

Reynolds P. D., Ruan Y., Smith D. F. & Scammell J. G. (1999). Glucocorticoid resistance in the squirrel monkey is associated with overexpression of the immunophilin FKBP51. *J Clin Endocrinol Metab*, Vol.84, No.2, pp. 663-669.

Richter K., Walter S. & Buchner J. (2004). The co-chaperone Sba1 connects the ATPase reaction of Hsp90 to the progression of the chaperone cycle. *J Mol Biol*, Vol.342, No.5, pp. 1403-1413.

Riggs D. L., Cox M. B., Cheung-Flynn J., Prapapanich V., Carrigan P. E. & Smith D. F. (2004). Functional specificity of co-chaperone interactions with Hsp90 client proteins. *Crit Rev Biochem Mol Biol*, Vol.39, 279-295.

Riggs D. L., Cox M. B., Tardif H. L., Hessling M., Buchner J. & Smith D. F. (2007). Noncatalytic role of the FKBP52 peptidyl-prolyl isomerase domain in the regulation of steroid hormone signaling. *Mol Cell Biol*, Vol.27, No.24, pp. 8658-8669.

Riggs D. L., Roberts P. J., Chirillo S. C., Cheung-Flynn J., Prapapanich V., Ratajczak T., Gaber R., Picard D. & Smith D. F. (2003). The Hsp90-binding peptidylprolyl isomerase FKBP52 potentiates glucocorticoid signaling *in vivo*. *EMBO J*, Vol.22, No.5, pp. 1158-1167.

Roe S. M., Ali M. M. U., Meyer P., Vaughan C. K., Panaretou B., Piper P. W., Prodromou C. & Pearl L. H. (2004). The mechanism of Hsp90 regulation by the protein kinase-specific cochaperone p50cdc37. *Cell*, Vol.116, No.1, pp. 87-98.

Russell L. C., Whitt S. R., Chen M.-S. & Chinkers M. (1999). Identification of conserved residues required for the binding of a tetratricopeptide repeat domain to heat shock protein 90. *J Biol Chem*, Vol.274, No.29, pp. 20060-20063.

Scammell J. G., Denny W. B., Valentine D. L. & Smith D. F. (2001). Overexpression of the FK506-binding immunophilin FKBP51 is the common cause of glucocorticoid resistance in three New World primates. *Gen Comp Endocrinol*, Vol.124, No.2, pp. 152-165.

Schaufele F., Carbonell X., Guerbadot M., Borngraeber S., Chapman M. S., Ma A. A. K., Miner J. N. & Diamond M. I. (2005). The structural basis of androgen receptor activation: intramolecular and intermolecular amino-carboxy interactions. *PNAS*, Vol.102, No.28, pp. 9802-9807.

Scheufler C., Brinker A., Bourenkov G., Pegoraro S., Moroder L., Bartunik H., Hartl F. U. & Moarefi I. (2000). Structure of TPR domain-peptide complexes: critical elements in the assembly of the Hsp70-Hsp90 multichaperone machine. *Cell*, Vol.101, No.2, pp. 199-210.

Siligardi G., Hu B., Panaretou B., Piper P. W., Pearl L. H. & Prodromou C. (2004). Co-chaperone regulation of conformational switching in the Hsp90 ATPase cycle. *J Biol Chem*, Vol.279, No.50, pp. 51989-51998.

Silverstein A. M., Galigniana M. D., Chen M.-S., Owens-Grillo J. K., Chinkers M. & Pratt W. B. (1997). Protein phosphatase 5 is a major component of glucocorticoid receptor·Hsp90 complexes with properties of an FK506-binding immunophilin. *J Biol Chem*, Vol.272, No.26, pp. 16224-16230.

Silverstein A. M., Galigniana M. D., Kanelakis K. C., Radanyi C., Renoir J.-M. & Pratt W. B. (1999). Different regions of the immunophilin FKBP52 determine its association with the glucocorticoid receptor, Hsp90, and cytoplasmic dynein. *J Biol Chem*, Vol.274, No.52, pp. 36980-36986.

Sinars C. R., Cheung-Flynn J., Rimerman R. A., Scammell J. G., Smith D. F. & Clardy J. (2003). Structure of the large FK506-binding protein FKBP51, an Hsp90-binding protein and a component of steroid receptor complexes. *PNAS*, Vol.100, No.3, pp. 868-873.

Skinner J., Sinclair C., Romeo C., Armstrong D., Charbonneau H. & Rossie S. (1997). Purification of a fatty acid-stimulated protein-serine/threonine phosphatase from bovine brain and its identification as a homolog of protein phosphatase 5. *J Biol Chem*, Vol.272, No.36, pp. 22464-22471.

Smith D. F. (1993). Dynamics of heat shock protein 90-progesterone receptor binding and the disactivation loop model for steroid receptor complexes. *Mol Endocrinol*, Vol.7, No.11, pp. 1418-1429.

Smith D. F. (2004). Tetratricopeptide repeat cochaperones in steroid receptor complexes. *Cell Stress Chaperones*, Vol.9, No.2, pp. 109-121.

Smith D. F. & Toft D. O. (2008). Minireview. The intersection of steroid receptors with molecular chaperones: observations and questions. *Mol Endocrinol*, Vol.22, No.10, pp. 2229-2240.

Sullivan W. P. & Toft D. O. (1993). Mutational analysis of hsp90 binding to the progesterone receptor. *J Biol Chem*, Vol.268, No.27, pp. 20373-20379.

Tai P. K., Albers M. W., Chang H., Faber L. E. & Schreiber S. L. (1992). Association of a 59-kilodalton immunophilin with the glucocorticoid receptor complex. *Science*, Vol.256, No.5061, pp. 1315-1318.

Takayama S., Bimston D. N., Matsuzawa S.-i., Freeman B. C., Aime-Sempe C., Xie Z., Morimoto R. I. & Reed J. C. (1997). BAG-1 modulates the chaperone activity of Hsp70/Hsc70. *EMBO J*, Vol.16, No.16, pp. 4887-4896.

Taylor P., Dornan J., Carrello A., Minchin R. F., Ratajczak T. & Walkinshaw M. D. (2001). Two structures of cyclophilin 40: folding and fidelity in the TPR domains. *Structure*, Vol.9, No.5, pp. 431-438.

Tomlins S. A., Mehra R., Rhodes D. R., Cao X., Wang L., Dhanasekaran S. M., Kalyana-Sundaram S., Wei J. T., Rubin M. A., Pienta K. J., Shah R. B. & Chinnaiyan A. M. (2007). Integrative molecular concept modeling of prostate cancer progression. *Nat Genet*, Vol.39, No.1, pp. 41-51.

Tranguch S., Cheung-Flynn J., Daikoku T., Prapapanich V., Cox M. B., Xie H., Wang H., Das S. K., Smith D. F. & Dey S. K. (2005). Cochaperone immunophilin FKBP52 is critical to uterine receptivity for embryo implantation. *PNAS*, Vol.102, No.40, pp. 14326-14331.

Vaughan C. K., Gohlke U., Sobott F., Good V. M., Ali M. M. U., Prodromou C., Robinson C. V., Saibil H. R. & Pearl L. H. (2006). Structure of an Hsp90-Cdc37-Cdk4 complex. *Mol Cell*, Vol.23, No.5, pp. 697-707.

Vaughan C. K., Mollapour M., Smith J. R., Truman A., Hu B., Good V. M., Panaretou B., Neckers L., Clarke P. A., Workman P., Piper P. W., Prodromou C. & Pearl L. H. (2008). Hsp90-dependent activation of protein kinases is regulated by chaperone-targeted dephosphorylation of Cdc37. *Mol Cell*, Vol.31, No.6, pp. 886-895.

Wang Z., Chen W., Kono E., Dang T. & Garabedian M. J. (2007). Modulation of glucocorticoid receptor phosphorylation and transcriptional activity by a C-terminal-associated protein phosphatase. *Mol Endocrinol*, Vol.21, No.3, pp. 625-634.

Ward B. K., Allan R. K., Mok D., Temple S. E., Taylor P., Dornan J., Mark P. J., Shaw D. J., Kumar P., Walkinshaw M. D. & Ratajczak T. (2002). A structure-based mutational analysis of cyclophilin 40 identifies key residues in the core tetratricopeptide repeat domain that mediate binding to Hsp90. *J Biol Chem*, Vol.277, No.43, pp. 40799-40809.

Whitesell L. & Lindquist S. L. (2005). Hsp90 and the chaperoning of cancer. *Nat Rev Cancer*, Vol.5, No.10, pp. 761-772.

Wochnik G. M., Young J. C., Schmidt U., Holsboer F., Hartl F. U. & Rein T. (2004). Inhibition of GR-mediated transcription by p23 requires interaction with Hsp90. *FEBS Lett*, Vol.560, 35-38.

Wu B., Li P., Liu Y., Lou Z., Ding Y., Shu C., Ye S., Bartlam M., Shen B. & Rao Z. (2004). 3D structure of human FK506-binding protein 52: implications for the assembly of the glucocorticoid receptor/Hsp90/immunophilin heterocomplex. *PNAS*, Vol.101, No.22, pp. 8348-8353.

Xu M., Dittmar K. D., Giannoukos G., Pratt W. B. & Simons Jr S. S. (1998). Binding of hsp90 to the glucocorticoid receptor requires a specific 7-amino acid sequence at the amino terminus if the hormone-binding domain. *J Biol Chem*, Vol.273, No.22, pp. 13918-13924.

Yang J., Roe S. M., Cliff M. J., Williams M. A., Ladbury J. E., Cohen P. T. W. & Barford D. (2005). Molecular basis for TPR domain-mediated regulation of protein phosphatase 5. *EMBO J*, Vol.24, No.1, pp. 1-10.

Yang Z., Wolf I. M., Chen H., Periyasamy S., Chen Z., Yong W., Shi S., Zhao W., Xu J., Srivastava A., Sanchez E. R. & Shou W. (2006). FK506-binding protein 52 is essential to uterine reproductive physiology controlled by the progesterone receptor A isoform. *Mol Endocrinol*, Vol.20, No.11, pp. 2682-2694.

Young E. T., Saario J., Kacherovsky N., Chao A., Sloan J. S. & Dombek K. M. (1998). Characterization of a p53-related activation domain in Adr1p that is sufficient for ADR1-dependent gene expression. *J Biol Chem*, Vol.273, No.48, pp. 32080-32087.

Zhang M., Windheim M., Roe S. M., Peggie M., Cohen P., Prodromou C. & Pearl L. H. (2005). Chaperoned ubiquitylation--crystal structures of the CHIP U Box E3 ubiquitin ligase and a CHIP-Ubc13-Uev1a complex. *Mol Cell*, Vol.20, No.4, pp. 525-538.

6

The TPR Motif as a Protein Interaction Module – A Discussion of Structure and Function

Natalie Zeytuni and Raz Zarivach
*Department of Life Sciences, Ben-Gurion University of the Negev and the National Institute for Biotechnology in the Negev (NIBN)
Beer sheva,
Israel*

1. Introduction

Many biological functions involve the formation of protein-protein complexes. Indeed, protein–protein interactions are considered to be the center of all functional living cells. Proteins can interact with each other by various structural, chemical and physical means. Some of these interactions can be highly specific, accompanied by high affinity, while some proteins are more flexible and bind diverse proteins as ligands. A common group of proteins that participate in protein-protein interactions and serve as multi-protein complex mediators are the tetra-trico-peptide repeat (TPR) proteins. TPR-containing proteins were found to be involved in many diverse processes in eukaryotic cells, including synaptic vesicle fusion (Young et al. 2003), peroxisomal targeting and import (Brocard et al. 2006; Fransen et al. 2008) and mitochondrial and chloroplast import (Baker et al. 2007; Mirus et al. 2009). In addition, TRPs are required for many bacterial pathways, such as outer membrane assembly (Gatsos et al. 2008), bio-mineralization of iron oxides in magnetotactic bacteria (Zeytuni et al. 2011) and pathogenesis (Edqvist et al. 2006; Tiwari et al. 2009). In addition, mutations in TPR-containing proteins have been linked to a variety of human diseases, such as chronic granulomatous disease and Leber's congenital amaurosis (D'Andrea & Regan 2003).

To date, more than 5,000 TPR-containing proteins were identified in different organisms by bioinformatics tools. Of these, more than 100 structures have been determined and deposited in the Protein Data Bank. These structures demonstrate the tendency of TPR motifs to exist as an independent fold or as a segment of a fold within a protein. The available structures allow the study of a protein-protein interaction platform at atomic resolution. In general, TPR-containing proteins can serve as a study case for protein interactions as they display great binding variety.

In this chapter, we describe the basic TPR sequence and structure, as well as several examples of diverse TPR binding properties. These include:

1. Various amino acid sequences which give rise to the binding of diverse ligands that participate in different biological processes.
2. Variety of structural features and conformations that serve as TPR ligands
3. The role of TPR protein curvature angle in ligand specificity.

4. Protein surface electrostatic potential distribution and the contribution of such distribution to the diverse binding properties of TPRs.
5. Multiple binding pockets which allows TPR-containing proteins to serve as mediators in multi-protein complexes.
6. The broad distribution of homo-oligomerization states in TPR proteins.
7. Ligand computational docking.
8. TPR proteins and biotechnology.

2. TPR sequence and basic structure

The TPR represents a structural motif consisting of 34 amino acids, sharing a degenerate consensus sequence defined by a pattern of small and large hydrophobic amino acids. In this consensus sequence, there are no completely invariant residual positions. The consensus pattern of conserved residues involves positions 4, 7, 8, 11, 20, 24, 27 and 32, in reference to the single motif N-terminal (Fig. 1A). Residues type is highly conserved only at positions 8 (Gly or Ala), 20 (Ala) and 27 (Ala). The rest of the consensus positions displays a preference for small, large or aromatic amino acids rather than for a specific residue. In addition, important structural characteristic conservation can be found at the turns located between two helical segments which contain helix-breaking residues (D'Andrea & Regan 2003). Nowadays, TPR consensus sequences can be identified by most general sequence analysis programs, such as the Simple Modular Architecture Research Tool (SMART) or the PROSITE dictionary of protein sites and motif patterns. At the same time TPRpred is a specially designated tool that uses the profile representation of the known repeats to detect TPR motifs and other patterns of protein repeats.

The canonical unit of the TPR motif adopts a basic helix-turn-helix fold (Fig. 1B). Adjacent TPR units packed in parallel create a series of repeating anti-parallel α-helices that give rise to an overall super-helix structure. This super-helical twist is affected by the type of residue found between adjacent TPR motifs. The unique super-helix fold forms a pair of concave and convex curved surfaces (Fig. 1C). Concave and convex surfaces display some extent of flexibility, as well as amino acid variety, which permits the binding of diverse ligands, usually via the concave surface.

In today's era of complete genome sequencing, one can predict TPR-containing protein dispersion throughout different organisms. Indeed, TPR proteins were found to be common in all forms of life, namely bacteria, archaea and eukaryotes. In nature, TPR motifs can be found in tandem arrays of 3-16 sequential motifs within a given protein (Fig. 1D). Moreover, Kajander et al. (2007) designed and determined the structure of a non-natural recombinant TPR-containing protein incorporating 20 sequential motifs. By using an increasing number of repeat-containing proteins, these authors found a positive correlation between the number of repeats and protein thermo-stability.

3. Ligand binding diversity

TPR-containing proteins bind diverse ligands in different binding pockets. These ligands usually do not share sequence or secondary structure similarities. Despite the lack of a set of defined rules, binding is usually highly specific, with TPR-containing proteins able to identify their ligand within the dense cellular environment. To obtain such diverse binding,

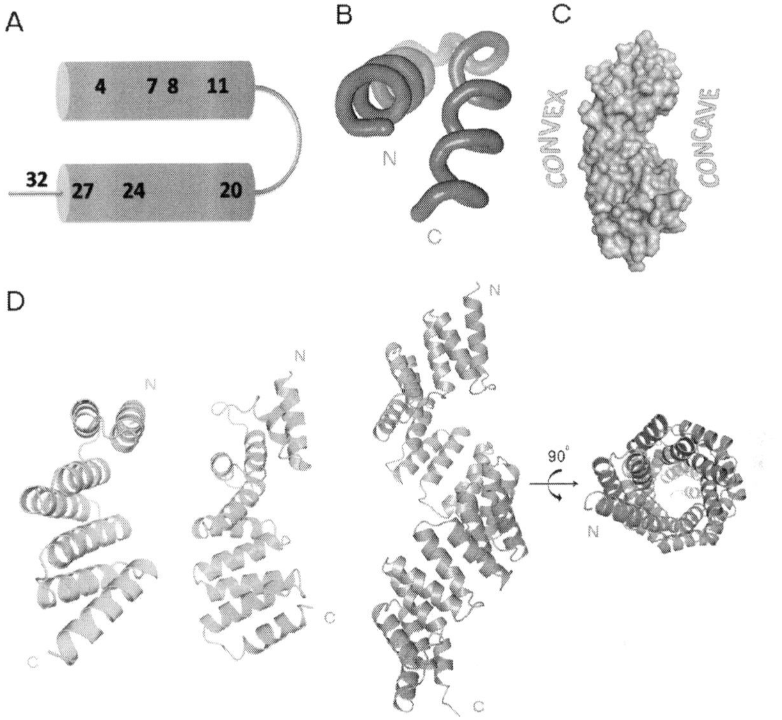

Fig. 1. TPR motif sequence and structure. (A) Schematic representation of the secondary structure arraignment of the 34 amino acids in a TPR motif. Numbers represent the conserved positions of amino acids within the motif. (B) The basic helix-turn-helix fold of a TPR motif canonical unit. (C) Surface representation of a TPR-containing protein displaying concave and convex surfaces. (D) Representative TPR-containing protein structures. From left, the TPR domain of Hop, containing three sequential motifs, MamA, containing five sequential TPR motifs, and the super-helix forming O-linked *N*-acetylglucosamine transferase TPR domain, containing 11 sequential TPR motifs, are shown in side and top view, in blue, green and orange respectively. Images B, C and D were generated by Pymol software (www.pymol.org).

several TPR-distinct folds serve as an interaction platform. This platform can display different surface residues in each binding surface, which later interacts in a specific manner with the ligand. Additionally, residue type influences the electrostatic nature of the binding surface, with, for example, arginine and lysine contributing positive charges, whereas glutamic acid and aspartic acid contribute negative charges. In addition, residues of different hydrophobicity and size can support hydrophobic interactions between the TPR protein and its ligand and therefore enhance protein-ligand specificity. Overall, the available TPR protein structures showing interaction with their ligands show that binding specificity cannot be attributed to a single force and that it is usually achieved by a combination of factors, such as residue type, hydrophobic pockets, charge and electrostatics.

3.1 Ligand sequence diversity

In this section, we describe two relatively simple TPR ligand interactions to demonstrate the multiple interaction types involved in binding and the correlation between TPR binding surfaces to the amino acid sequence of their ligands.

The first released structures of TPR domains bound to their ligand peptides were the two domains of the Hop protein (Scheufler et al. 2000). Hop is an adaptor protein which mediates the association of the molecular chaperones, Hsp70 and Hsp90. The TPR1 domain of Hop specifically recognizes the C-terminal heptapeptide of Hsp70, whereas the TPR2A domain binds the C-terminal pentapeptide of Hsp90 (Figs. 2A&B). Both C-terminal peptides share EEVD sequence ends and were found to bind in an extended conformation. An extended conformation allows the display of a maximal surface to the TPR domain and facilitates specific recognition of short amino acid stretches with sufficient affinity. Examination of solved crystal structures revealed that all electrostatic contacts between TPR domains and both peptides involve the EEVD sequence motif. These interactions include three classes of hydrogen bonding interaction, namely sequence-independent interactions with the peptide backbone, specific interactions with peptide side chains and the carboxylate of the C -terminal residue interaction with the TPR (Figs. 2C&D). The three strong hydrogen bonds formed between the peptide C-terminal carboxylate and the TPR1 and TPR2A domains conserved residues, Lys 8 (Lys 229), Asn 12 (Asn 233) and Asn 43 (Asn 264) respectively, allows the formation of a two carboxylate clamp. This clamp ensures the proper docking of the peptide ends to the TPR domains. Additional peptide residues located at the N-terminal, relative to EEVD motif, are exclusively engaged in hydrophobic and van der Waal's interactions. These contacts were found to be critical for peptide binding with a physiologically relevant high affinity. Other TPR domains which are known to bind Hsp70/Hsp90 proteins also contains identical residues which form the carboxylate clamp, suggesting that these TPR domains bind to the C-terminal carboxylate via a similar network of electrostatic interactions. The Hop and Hsp70/Hsp90 case studies presented here also demonstrate the importance of sequence conservation between TPR domains and their ligands involved in similar cellular functions.

Other example for the importance of TPR-ligand sequence conservation can be found in the peroxin 5 (PEX5) protein from the protozoan parasite, *Trypanasomas brucei*, the agent of human African trypanosomiasis (sleeping sickness). PEX5 is a cytosolic receptor which promotes cargo translocation across the glycosomic membrane, with the glycosome being a peroxisome-like organelle which hosts the metabolic reactions of the parasite. Two domains comprise PEX5, with the C-terminal domain consisting of 7 TPR motifs. The PEX5 C-terminal domain binds either the type 1 (PTS1) or type 2 (PTS2) peroxisomal targeting signal. The more common PTS1 sequence is a C-terminal tripeptide with the SKL amino acid sequence, or variants thereof, such as AKL or SHL. Additional residues upstream to the SKL sequence have also been implicated in binding to PEX5. The crystal structure of the C-terminal domain of PEX5 with five different PST1 fragments revealed that the protein does not fold as a sequential TPR protein with classic super-helical fold, but rather as two distinct sub-domains (Sampathkumar et al. 2008). The N-terminal sub-domain comprises TPR motifs 1-3 while the C-terminal sub-domain comprises TPR motifs 5-7. The fourth TPR motif that serves to interconnect the two sub-domains is only partial ordered and cannot be seen in the electron density resulted from the X-ray determination (Fig. 3A). The disordered nature of the fourth TPR motif implies its flexible tendencies and involvement in ligand-induced conformational changes that can promote cargo translocation. The PTS1 peptide is bound within the cavity found between the two sub-domains and interacts with residues from both, although the major binding contribution is attributed to the C-terminal sub-domain. The five PEX5 structures bound to five different peptides containing PTS1 sequences indicate that the ligand recognition mechanism involves three critical factors. The first is recognition of the C-terminal

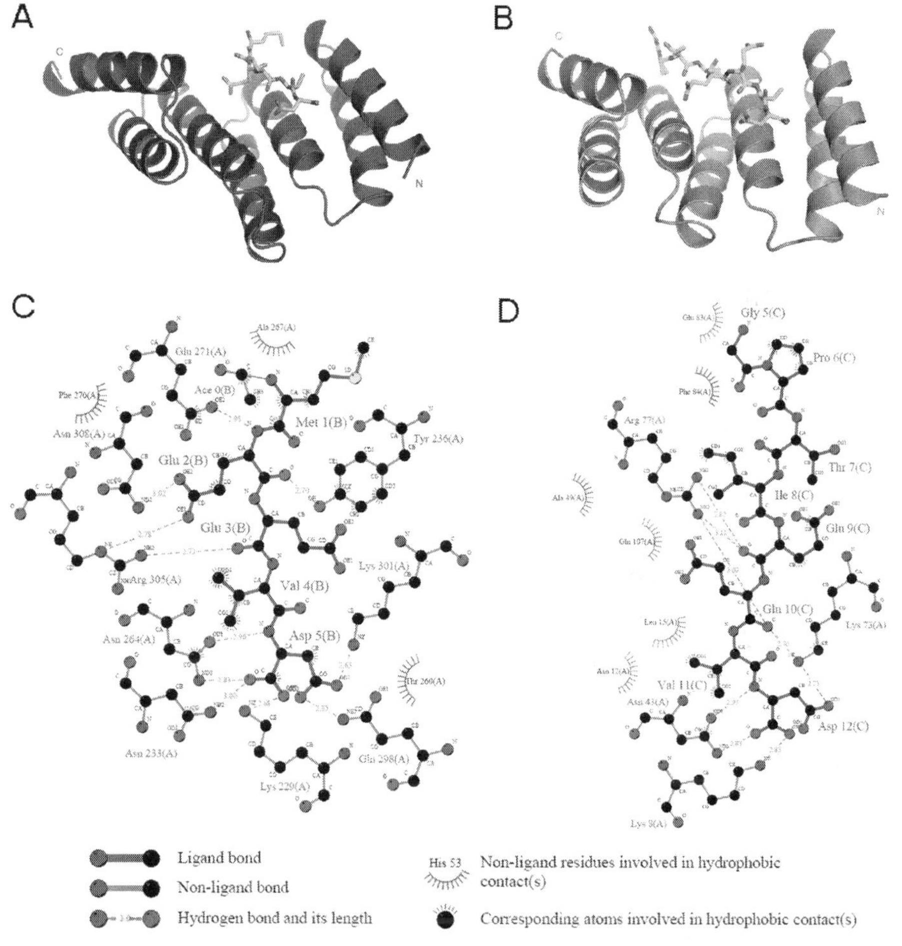

Fig. 2. TPR domains of the Hop protein and their interacting Hsp70/Hsp90 partner peptides. (A) The TPR2A domain of the Hop protein in complex with a Hsp90-derived peptide. (B) The TPR1 domain of the Hop protein in complex with a Hsp70-derived peptide. (C) Schematic representation of the TPR2A-hsp90 peptide interaction. (D) Schematic representation of the TPR1-Hsp70 peptide interaction. Images A and B were generated using Pymol software (www.pymol.org). Images C and D were generated using LigPlot software (Wallace et al. 1995).

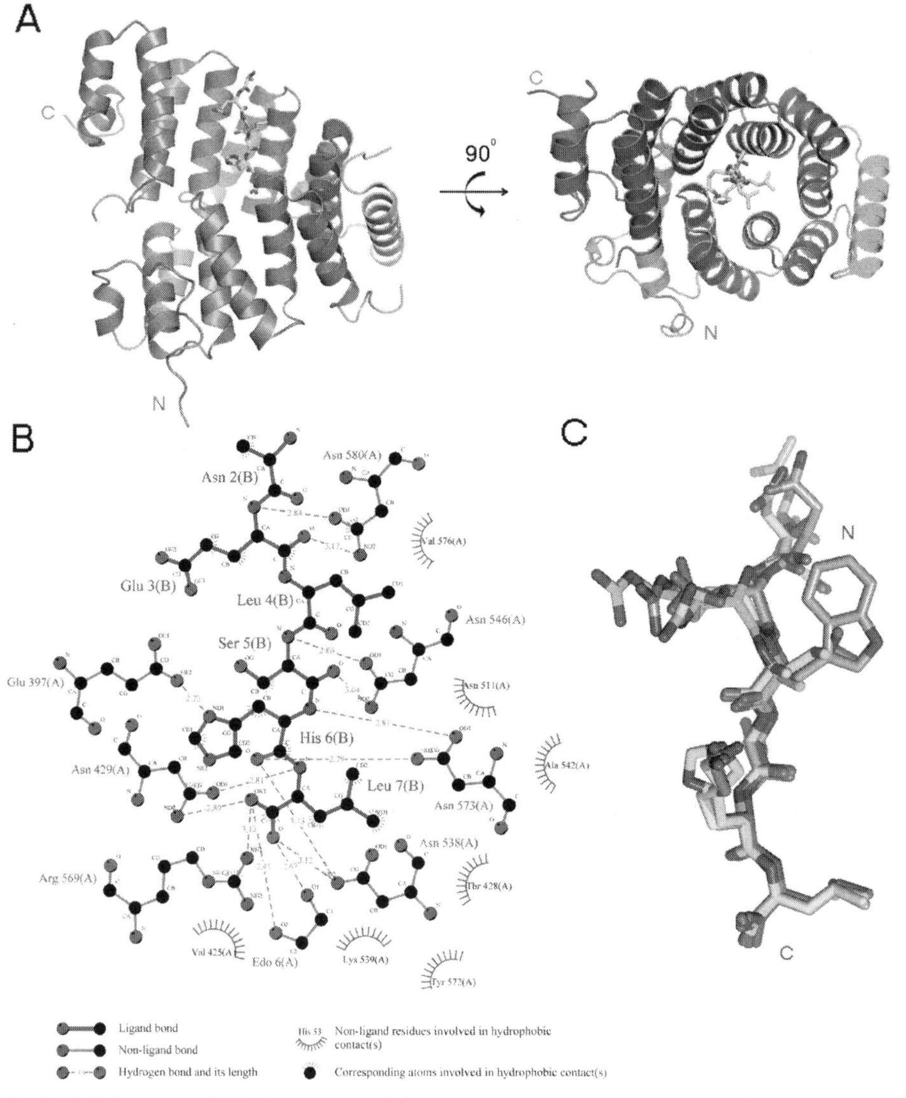

Fig. 3. TPR domains of PEX5 protein and their interacting PST1 partner peptides. (A) TPR domains of PEX5 in complex with a PST1 peptide (^{N}FNELSHL^{C}). PEX5 is represented as a cartoon, whereas the N-terminal domain, the fourth TPR motif and C- terminal domain are colored in blue, green and orange, respectively. The PST1 peptide is represented as sticks in yellow. (B) Schematic representation of the PEX5-PST1 peptide interactions. (C) Superposition of five PST1 peptides bound to PEX5, in green, blue, pink, yellow and light pink. Images A and C were generated using Pymol software (www.pymol.org). Image B was generated using LigPlot software (Wallace et al. 1995).

PST1 carboxylate in a similar manner to the Hop and Hsp70/Hsp90 complex by two PEX5 Asn residues and a single Arg residue, the second is hydrophobic embedding of the PTS1 C-terminal residue side chain and the third is multiple PTS1 backbone interactions with PEX5 Asn side chains (Fig. 3B). Overall, PTS1 peptides are bound in a similar manner despite substantial differences in amino acid composition. In addition, the spatial positions of the five Asn residues and a single Arg residue of PEX5 involved in backbone binding of PTS1 peptides are similar, emphasizing their significant function in diverse ligand binding (Fig. 3C).

3.2 Ligand secondary structure and length diversity

Secondary structure of a TPR-bound ligand varies between the coiled extended conformation to an α-helix or both. As mentioned previously, an extended conformation maximizes the ligand surface to the TPR domain and facilitates specific recognition of short amino acid stretches with sufficient affinity. Two examples of extended conformation peptide binding are Hop bound to Hsp70/Hsp90 peptides (Fig. 4A) and PEX5 bound to PTS1 peptides. Three TPR-containing proteins bound to their ligands display binding of long peptides that adopt both helical and extended conformations. The first structure of such binding is PscG-PscE in complex with the PscF peptide (Quinaud et al. 2007). PscG-PscE–PscF proteins are members of the bacterial Type III secretion system and play a specific role in formation of the needle that transports virulence effectors into target cell cytoplasm. PscG displays a three TPR motif fold with an additional C-terminal helix, even though its TPR fold could not have been predicted from its sequence. PscG also interacts with PscE through a hydrophobic platform formed by the N-terminal TPR motif of PscG. PscF is composed of two sub-domains, an extended coil (13 amino acids-long) and a C-terminal helix (17 amino acids-long). PscG and PscE fold into a cupped-hand form, whereas the amphipathic C-terminal helix of PscF is bound to the concave surface of PscG. The N-terminal of PscF is bound to the PscG convex surface (Fig. 4B).

The second structure displaying binding of a long peptide which adopts both helical and extended conformations is the TPR-containing protein, APC6, in complex with CDC26 (Wang et al. 2009). APC6 and CDC26 are both members of the multi-subunit anaphase-promoting complex (APC), an essential cell-cycle regulator. The crystallized APC6 TPR domain contains eight full TPR motifs and an additional C-terminal helix. APC6 adopts a solenoid-like structure, wrapping around the entire length of N-terminal region of CDC26 (26 amino acids). The bound CDC26 forms a rod-like structure with the first 12 amino acids adopting an extended conformation and the last 14 amino acids forming a helix (Fig. 4C). Interestingly, as the CDC26 peptide interacts with APC6, an additional non-TPR C-terminal helix with geometry that mimics two helices in a TPR motif appears. This inter-molecular TPR mimic continues the sinuous form of the overall structure and packs against the eight TPR motifs of APC6 to form a four-helix bundle.

A third structure displaying binding of two helices connected by a loop to a TPR-containing protein is Fis1 in complex with Caf4 (Zhang et al. 2007). Both Fis1 and Caf4 participate in the formation of yeast mitochondrial fission complex that controls the shape and physiology of the mitochondria. The Fis1 protein core is composed of two TPR motifs with two additional helices at each motif end and another N-terminal helix arm packed against the hydrophobic groove formed by the protein core. The Caf4 peptide adopts a U-fold with two helices formed at each end connected by a loop. The unique U-fold of Caf4 allows large scale interactions between the peptide and both the Fis1 core at the concave and convex surfaces (Fig. 4D).

Although these are only a few examples of TPR binding modes reflecting secondary structure conformations and peptides lengths, they describe well the diverse nature of ligand types that theTPR platform can bind.

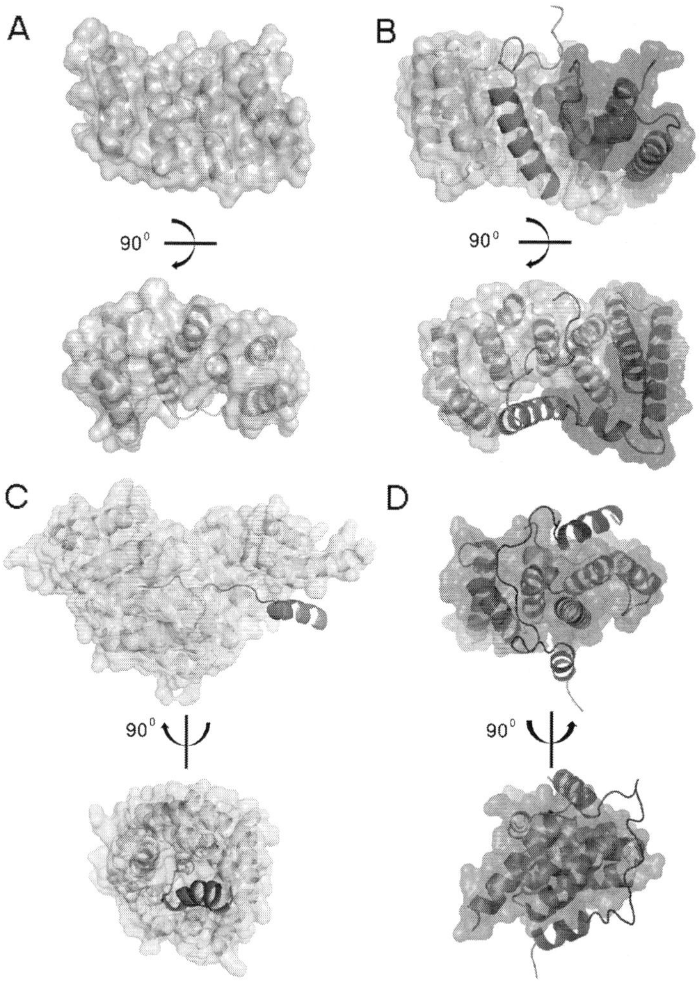

Fig. 4. TPR ligand secondary structure and length diversity. (A) TPR2A domain of Hop is shown in surface representation with secondary structure indicated in light pink, bound to Hsp90, in green. (B) A surface representation of a PscG-PscE dimer shown with secondary structure indicated in light blue and purple, respectively. PscF, in pink, is bound to the PscG-PscE dimer. (C) APC6 is shown in surface representation with secondary structure indicated in light green, bound to CDC26, in red. (D) Caf4 is shown in surface representation with secondary structure indicated in pink, bound to Fis1, in dark blue. Images were generated using Pymol software (www.pymol.org).

3.3 Curvature angle role and diversity

Major diversity can be seen in the overall shape and curvature angle displayed by TPR-containing proteins. There are many factors that determine the overall shape and curvatures angle of a protein. Among these are the number of repeating motifs, their arrangement as sequential motifs or separation into sub-domains, the presence of helix-breaking residues within turns between motifs, as well as residues type, protein function and others. Despite the apparent significance, no extensive study addressing the overall shape and curvature angle displayed by TPR-containing proteins has been reported. Indeed, articles presenting new TPR protein structures hardly refer to this property. The majority of TPR protein structures were determined by X-ray crystallography, a method that favors ordered proteins and can only tolerate a low extant of protein flexibly within the crystal. Since proteins and protein interactions are usually flexible and require the partners to be rigid, the use of X-ray crystallography to study protein interactions can be quite challenging. To overcome this challenge, structural biologists often use a relatively short peptide from the whole binding partner that participates directly in recognition and binding in their experiments. By using a small portion of the binding partner, the overall flexibility is reduced throughout the crystal, a trait that later improves crystallization and crystal quality. Due to method limitations, we usually obtain a close look at the detailed interaction and the forces participating in such interactions but cannot see the complete interaction or the spatial positions of the partners, nor detect remote stabilizing interactions. This incomplete observation of interactions between partners creates substantial difficulty in determining the exact role and importance of the curvature angle. Perhaps in the future, when new structure determination techniques emerge, a better understanding of the role of the curvature angle in TPR protein recognition and binding will be achieved.

4. TPR homo-oligomerization

TPR-containing proteins serve as mediators in the formation of multi-protein complexes. Within these complexes, TPR proteins or domains can be found as homo-oligomers, with oligomerization serving as a crucial factor for realizing proper protein function in the cell. TPR proteins displays a broad range of oligomerization states, such as monomers (Scheufler et al. 2000; Sampathkumar et al. 2008) and dimers (Lunelli et al. 2009; Z. Zhang et al. 2010). It is of note that a complex containing 24-26 TPR protein monomers has been described (Zeytuni et al. 2011). The protein surfaces involved in oligomer formation are diverse, as are the inter-molecule interaction types seen between monomers. In the following section, three interesting protein oligomerization forms will be presented.

The first TPR-containing protein displaying a dimer formation to be considered is invasion plasmid gene C (IpgC), a chaperone that binds two essential virulence factors of the pathogenic bacteria, *Shigella* (Lunelli et al. 2009). IpgC binds the invasion plasmid antigens, IpaB and IpaC which are responsible for epithelial cell invasion, membrane lysis of the phagocytic vacuole, contact hemolysis and macrophage cell death. The IpgC chaperone contains three TPR motifs, with two additional helices at the protein N-terminal and another helix at the protein C-terminal. The protein functional unit that allows efficient substrate binding is a dimer, with dimerization occuring in an asymmetric manner. Such asymmetric dimerization presents an interesting binding mode in which the first helix of one of the monomers binds on the convex surface presented by the other monomer (Fig. 5A). While the

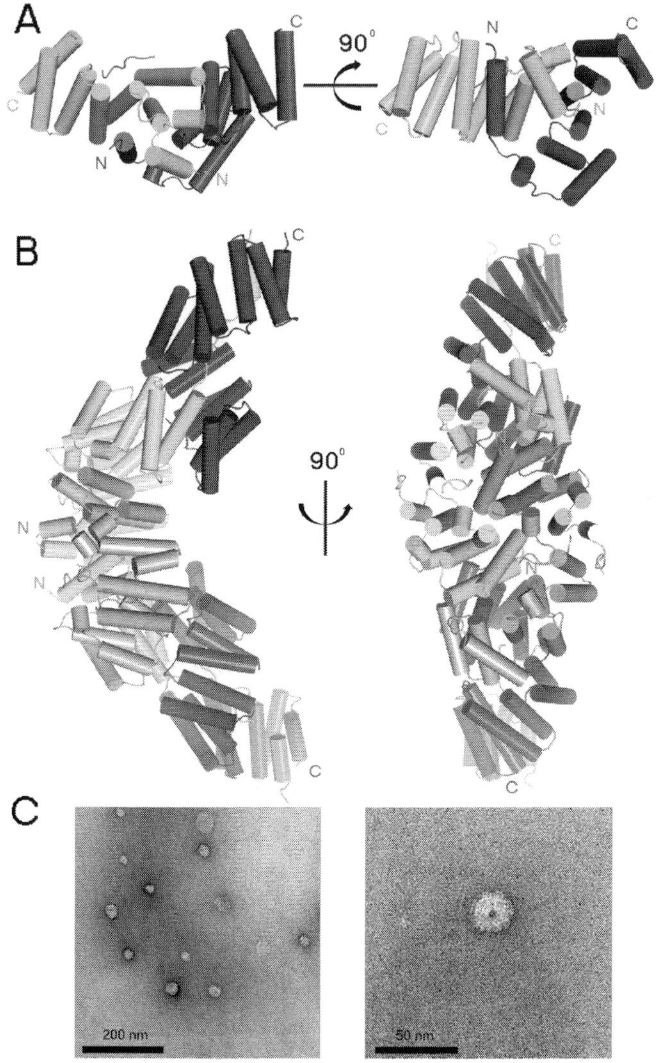

Fig. 5. Homo-oligomerization by -ontaining proteins. (A) Cylinder representation of a IpgC dimer, in green and purple, where one monomer is bound to an IpgB peptide, in pink. (B) The N-terminal domain of Cut9 promotes dimerization. Cylinder representation of the N-terminal domain of each monomer, shown in cyan and light pink. C-terminal domains are in blue and light purple. Hcn1 peptides are colored in green and yellow. (C) Transmission electron microscopy images of negatively stained MamA protein complexes. Images A, B were generated using Pymol software (www.pymol.org). The images shown 4C were taken by Zeytuni N. as described in Zeytuni et al. (2011).

first helix of one of the monomers displays inter-molecule binding, the first helix of the bound monomer appears to be found in a more packed, unbound conformation. The formation of a dimer interface at the TPR concave surface allows the binding of the IpaB peptide at the concave surface in an extended conformation. Moreover, the mode of IpgC dimerization demonstrates the significance of a convex surface as an additional protein interaction platform.

A second TPR-containing protein displaying dimer formation is *Schizosaccharomyces pombe* Cut9, the yeast homolog of human APC6. The Cut9 structure was determined while bound to Hcn1, the yeast homolog of human CDC26 in a similar binding conformation as APC6 in complex with CDC26 (Z. Zhang et al. 2010). Cut9 is composed of 14 TPR motifs forming a contiguous super-helix divided into two functionally and structurally distinct domains, namely an N-terminal domain comprising the first seven TPR motifs and a C-terminal domain comprising the last seven TPR motifs. The Cut9 subunits homo-dimerize through their N-terminal domains to generate a shallow 'V'-shaped molecule (Fig. 5B). The more globular N-terminal dimerization module forms the apex of the 'V'-shape, with narrower C-terminal TPR domains projecting away from the dimer interface. The majority of Cut9 interactions with Hcn1 involve the C-terminal TPR-containing domain. Dimerization of the N-terminal domains forms a tight interface in which the concave surface of each N-terminal domain encircles its dimer counterpart in an inter-lock clasp-like arrangement. In these interactions, the first two TPR motifs of a single monomer interact with residues lining the inner groove formed by the seven TPR motifs of the opposite Cut9 N-terminal domain.

An additional example involving dimerization is the TPR-containing protein, MamA, from *Magnetospirillium* magnetotactic bacteria species. MamA forms large homo-oligomeric complex of 24-26 monomers. This complex is presumed to serve as a wide platform for protein interaction during the iron-oxide bio-mineralization by the magnetotactic bacteria (Zeytuni et al. 2011). MamA contains five TPR motifs with an additional N-terminal putative TPR motif that was found to be responsible for oligomerization and complex formation. Through binding of the first helix of a single monomer to a binding surface displayed on a different monomer, a round-shaped complex of 14-20 nm in diameter with a central pore cavity is formed (Fig. 5C). The structural details of the monomer-monomer interaction remain unclear, since crystallization trials of MamA in complex with peptides of the first and/or second helices proved unsuccessful.

Overall, these three examples of homo-oligomrization by TPR-containing proteins describe additional TPR diversity and further establish the TPR motif as a broad platform for protein interactions.

5. Predicting TPR-ligand interactions

Today, available bioinformatics tools and the well-defined TPR profile allow us to identify TPR-containing proteins with great accuracy through analysis of amino acid sequence. However, these tools still cannot predict TPR-interacting partners and/or the region of interaction. As discussed in section 3.3, the majority of TPR-containing protein structures were determined by X-ray crystallography, a methodology in which the crystallization probability of protein complexes is significantly lower than the probability of crystallization of a single protein. Therefore, the majority of determined TPR proteins were crystallized in a non-complex form, not bound to their interacting partner, even if that partner was previously identified. In order to predict ligand binding, one should discriminate between two cases, the first involving an unidentified partner protein and the second involving an unknown binding region. In this section, we discuss these two cases and provide several examples of each.

5.1 Interacting protein prediction

A genomic approach for TPR-interacting protein identification was first presented by D'Andrea and Regan (2003). In their study, the authors generated a list of all TPR-containing proteins predicted from the *Saccharomyces cerevisiae* genome. Later, they used the generated list of 22 predicted TPR-containing proteins in protein-protein interaction databases to identify potential binding partners. Their search revealed about 80 potential interacting proteins, some of which are known to participate in multi-protein complex formation. Nevertheless, these authors could not exclude the possibility that these interacting proteins might not all interact directly with TPR domains within the multi-protein complexes.

Another approach uses information derived from the structures of unbound TPR-containing proteins. Certain properties of the binding partner can be deduced from a simple examination of a TPR protein structure, especially from its concave binding pocket, specifically its dimensions, residue composition and electrostatic potential. Although the concave surface serves as the common binding area, the convex surface can also participate in binding and should not be excluded from predictions.

Our first example is derived from the structure of the super-helical TPR domain of O-linked *N*-acetylglucosamine transferase (OGT). The TPR domain of OGT contains 11 motifs with an additional C-terminal helix and forms a homo-dimer through interactions at the convex surface. The inner surface of the elongated super-helix is highly conserved and contains an asparagine ladder. This asparagine ladder bears marked similarity to the array of conserved asparagines in the ARM-repeat importin α- and β-catenin proteins. In both ARM-repeat proteins, the asparagine side chains contribute to binding of the target peptide by forming hydrogen bonds with the peptide backbone. This structural similarity suggests that a similar binding mechanism for the OGT protein. In addition, the extensive surface generated by OGT is likely to represent several overlapping binding pockets which can accommodate multiple substrates (Fig. 6A). Furthermore, partner binding can rely on a mechanism similar to the mode of binding described earlier, in the case of CDC26 bound to APC6 (see Fig. 4C) (Wang et al. 2009).

Another example of TPR partner prediction is derived from the bacterial YrrB protein structure. YrrB is a *Bacillus subtilis* protein containing five TPR motifs with an additional C-terminal helix. The YrrB structure reveals a unique, highly negatively charged deep concave surface containing an aspartic acid array that can accommodate the binding of positively charged residues (Fig 6B). In order to discover new details on the role of the YrrB protein, functional gene cluster localization analysis was performed. Such analysis suggested that YrrB plays a role in mediating complex formation among RNA sulfuration components, RNA processing components and aminoacyl-tRNA synthetases. An opposite charge distribution to the YrrB protein was found in the TPR-containing protein, MamA, a protein that displays a highly positive concave binding surface, implying the electrostatic charge nature of ligand bindng (Fig. 6C). Although sharing a similar fold, MamA and YrrB demonstrate yet another variation among TPR-containing proteins, namely charge distribution. In general, binding partner identification is not straightforward and can be challenging, as the information obtained from the genetic approach can point to indirect interactions. Indeed, the structural approach can only provide clues for identifying the specific region of the partner involved in binding.

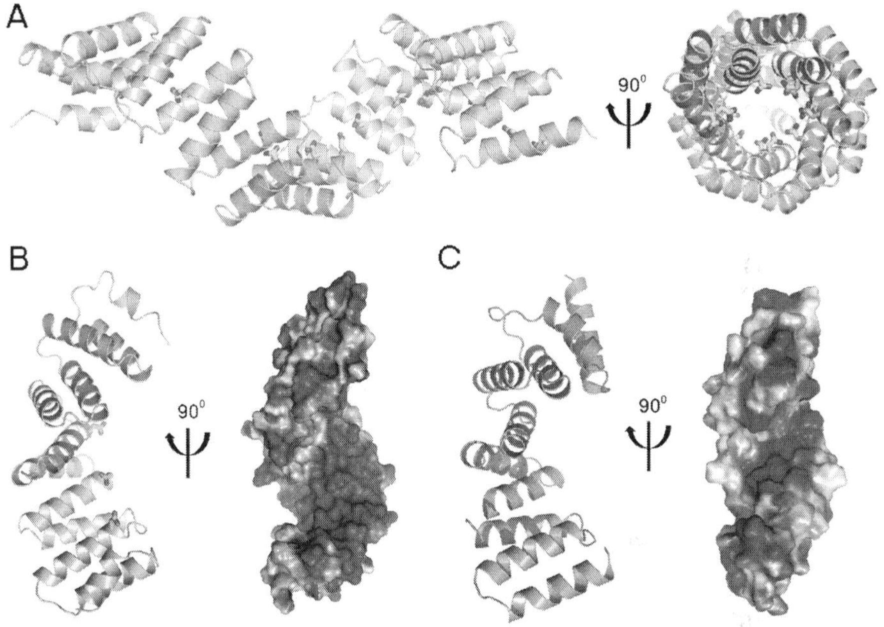

Fig. 6. Structural analysis of concave binding surface. (A) The TPR domain of O-linked *N*-acetylglucosamine transferase is presented as a grey cartoon, while the asparagine ladder within the inner surface of the super-helix is presented as yellow sticks. (B) *left* - YrrB in light orange cartoon. The aspartic acid ladder is presented as green sticks. *right* –Electrostatic surface potential representation, where blue and red represent positive and negative electrostatic potentials, respectively. The YrrB concave surface displays a highly negative potential distribution. (C) *Left* - MamA in light pink cartoon representation. *Right* – Electrostatic surface potential representation, where blue and red represent positive and negative potentials, respectively. The MamA concave surface displays a highly positive potential distribution. Images were generated using Pymol software (www.pymol.org). Surface electrostatic potential calculations were performed using the APBS plug-in (www.poissonboltzmann.org/apbs).

5.2 Interaction region prediction

Interaction region prediction requires at least two components, the TPR protein structure and the identified binding sequence of the partner. Over the past few years, several attempts to dock ligand peptides onto TPR-containing proteins have been made. These attempts mainly included manual docking of the ligand peptide onto the available TPR protein structure. The resulting models did not, however, consider side chain flexibility and conformational changes due to peptide binding and used limited energy minimization tools (Gatto et al. 2000; Kim et al. 2006). However, with the remarkable development of computational docking servers, more accurate models can be generated for TPR protein–peptide interactions. To date, however, no study has employed these advanced servers to demonstrate TPR-peptide docking, although successful docking have been recorded with other proteins involved in protein-protein interactions, such as PDZ domains and others (London et al. 2010; Raveh et al. 2010; Raveh et al. 2011). Overall, the prediction of interaction region is considered to be more accurate than binding partner prediction, since the former uses algorithms that consider chemical restrains, as well as energy minimization of the final model. In the near future, these tools might aid in overcoming challenges associated with crystallizing proteins with their binding partners and could provide important insight for molecular understanding of binding and recognition.

6. TPR design and biotechnology

The basic TPR fold resulting from its consensus sequence can be considered as a protein scaffold. Redesigning this stable basic scaffold by grafting functional residues involved in binding recognition and specificity enables the introduction of novel binding activities. TPR design includes three major steps. The first stage includes stable consensus scaffold generation by an alignment of natural TPR motifs (D'Andrea et al. 2003). Later, the minimal number of repeating motifs required for thermodynamic stability are determined (Kajander et al. 2007) and finally, functional residues are grafted onto the generated scaffold. An example for a successful designing process was the designed TPR module that binds the C-terminal peptide of Hsp90. This module has been designed by grafting Hsp90-binding residues from natural TPR proteins onto a consensus TPR scaffold to bind Hsp90 with greater affinity and specificity than natural co-chaperones. Introduction of this designed protein into breast cancer cells inhibited Hsp90 activity, presumably by out-competing the interaction of Hsp90 with its natural co-chaperones. Hsp90 inhibition resulted in misfolding and degradation of Hsp90-dependent proteins, such as HER2, and led to cancer cell death (Cortajarena et al. 2008).

TPR binding properties can also be used for specific identification of tagged proteins and can act as a functional substitute for antibodies in a wide range of applications (Jackrel et al. 2009). Conjugation of a peptide ligand sequence to the N or C termini of a desirable protein allows its identification by a TPR-containing protein that can bind the ligand peptide with sufficient affinity. This TPR-containing protein can be conjugated to a reporter protein, such as horseradish peroxidase, biotin or green fluorescent protein and can be later used in one or two-step western blot detection systems, replacing any requirement for antibodies. In addition, conjugation of a TPR-containing protein or domain directly to resin or beads can be used for affinity purification. Immobilizing a TPR-containing protein onto a resin permits the specific binding and enrichment of desirable proteins conjugated to a peptide ligand. Interaction dissociation at the elution step does not require extremely harsh pH conditions as may be needed with the use of antibodies.

7. Concluding remarks

The protein-protein interaction platform generated by TPR motifs can support the binding of diverse ligands. The elegant super-helical fold of TPR-containing proteins presents several binding surfaces that can promote the formation of multi-protein complexes. These binding surfaces can bind chemically distinct peptides in a variety of conformations with sufficient affinity. Therefore, it is not surprising that TPR-containing proteins are widespread across all kingdoms of life, where they participate in diverse cellular processes. From a molecular point of view, the diverse nature of interactions presented by TPR protein structures demonstrates multiple chemical forces involved in ligand binding and can promote the design of novel protein-protein interactions.

Overall, TPR-containing proteins hold great promise for protein engineering, therapeutics and biotechnology, as the basic TPR scaffold can be redesigned to modulate binding specificity and/or affinity towards desirable peptide ligands. As such, TPR-containing proteins can be inhibited by a designed ligand with higher affinity, serve as scaffolds to present proteins in nano-technological applications and more.

8. References

Baker, M. J., Frazier, A. E., Gulbis, J. M., and Ryan, M. T. (2007). Mitochondrial protein-import machinery: correlating structure with function. *Trends in Cell Biology* 17, 456-64.

Brocard, C., and Hartig, A. (2006). Peroxisome targeting signal 1: is it really a simple tripeptide? *Biochimica et Biophysica Acta* 1763, 1565-73.

Cortajarena, A. L., Yi, F., and Regan, Lynne (2008). Designed TPR modules as novel anticancer agents. *ACS Chemical Biology* 3, 161-6.

D'Andrea, L., and Regan, L (2003). TPR proteins: the versatile helix. *Trends in Biochemical Sciences* 28, 655-662.

Edqvist, P. J., Bröms, J. E., Betts, H. J., Forsberg, A., Pallen, M. J., and Francis, M. S. (2006). Tetratricopeptide repeats in the type III secretion chaperone, LcrH: their role in substrate binding and secretion. *Molecular Microbiology* 59, 31-44.

Fransen, M., Amery, L., Hartig, A., Brees, C., Rabijns, A., Mannaerts, G. P., and Veldhoven, P. P. Van (2008). Comparison of the PTS1- and Rab8b-binding properties of Pex5p and Pex5Rp/TRIP8b. *Biochimica et Biophysica Acta* 1783, 864-73.

Gatsos, X., Perry, A. J., Anwari, K., Dolezal, P., Wolynec, P. P., Likić, V. a, Purcell, A. W., Buchanan, S. K., and Lithgow, T. (2008). Protein secretion and outer membrane assembly in Alphaproteobacteria. *FEMS Microbiology Reviews* 32, 995-1009.

Gatto, G. J., Geisbrecht, B. V., Gould, S. J., and Berg, J. M. (2000). A proposed model for the PEX5-peroxisomal targeting signal-1 recognition complex. *PROTEINS: Structure, Function, and Genetics* 38, 241-6.

Jackrel, M. E., Valverde, R., and Regan, Lynne (2009). Redesign of a protein-peptide interaction: characterization and applications. *Protein Science* 18, 762-74.

Kajander, T., Cortajarena, A. L., Mochrie, S., and Regan, Lynne (2007). Structure and stability of designed TPR protein superhelices: unusual crystal packing and implications for natural TPR proteins. *Acta Crystallographica. Section D*, Biological crystallography 63, 800-11.

Kim, K., Oh, J., Han, D., Kim, E. E., Lee, B., and Kim, Y. (2006). Crystal structure of PilF: functional implication in the type 4 pilus biogenesis in Pseudomonas aeruginosa. *Biochemical and Biophysical Research Communications* 340, 1028-38.

London, N., Movshovitz-Attias, D., and Schueler-Furman, O. (2010). The structural basis of peptide-protein binding strategies. *Structure* (London, England: 1993) 18, 188-99..

Lunelli, M., Lokareddy, R. K., Zychlinsky, A., and Kolbe, M. (2009). IpaB-IpgC interaction defines binding motif for type III secretion translocator. *Proceedings of the National Academy of Sciences of the United States of America* 106, 9661-6.

Mirus, O., Bionda, T., Haeseler, A. von, and Schleiff, E. (2009). Evolutionarily evolved discriminators in the 3-TPR domain of the Toc64 family involved in protein translocation at the outer membrane of chloroplasts and mitochondria. *Journal of Molecular Modeling* 15, 971-82.

Quinaud, M., Plé, S., Job, V., Contreras-Martel, C., Simorre, J.-P., Attree, I., and Dessen, A. (2007). Structure of the heterotrimeric complex that regulates type III secretion needle formation. *Proceedings of the National Academy of Sciences of the United States of America* 104, 7803-8.

Raveh, B., London, N., Zimmerman, L., and Schueler-Furman, O. (2011). Rosetta FlexPepDock ab-initio: simultaneous folding, docking and refinement of peptides onto their receptors. *PloS One* 6, e18934.

Raveh, B., London, N., and Schueler-Furman, O. (2010). Sub-angstrom modeling of complexes between flexible peptides and globular proteins. *Proteins* 78, 2029-40.

Sampathkumar, P., Roach, C., Michels, P. a M., and Hol, W. G. J. (2008). Structural insights into the recognition of peroxisomal targeting signal 1 by Trypanosoma brucei peroxin 5. *Journal of Molecular Biology* 381, 867-80.

Scheufler, C., Brinker, a, Bourenkov, G., Pegoraro, S., Moroder, L., Bartunik, H., Hartl, F. U., and Moarefi, I. (2000). Structure of TPR domain-peptide complexes: critical elements in the assembly of the Hsp70-Hsp90 multichaperone machine. *Cell* 101, 199-210.

Tiwari, D., Singh, R. K., Goswami, K., Verma, S. K., Prakash, B., and Nandicoori, V. K. (2009). Key residues in Mycobacterium tuberculosis protein kinase G play a role in regulating kinase activity and survival in the host. *The Journal of Biological Chemistry* 284, 27467-79.

Wallace, A. C., Laskowski, R. A., and Thornton, J. M. (1995). LIGPLOT: a program to generate schematic diagrams o protein-ligand interactions. *Protein Engineering* 8, 127-134.

Wang, J., Dye, B. T., Rajashankar, K. R., Kurinov, I., and Schulman, B. a (2009). Insights into anaphase promoting complex TPR subdomain assembly from a CDC26-APC6 structure. *Nature Structural & Molecular Biology* 16, 987-9.

Young, J. C., Barral, J. M., and Ulrich Hartl, F. (2003). More than folding: localized functions of cytosolic chaperones. *Trends in Biochemical Sciences* 28, 541-547.

Zeytuni, N., Ozyamak, E., Ben-Harush, B., Davidov, G., Levin, M., Gat, Y., Moyal, T., Brik, A., Komeili, A., Zarivach, R., (2011). Self-recognition mechanism of MamA, a magnetosome-associated TPR-containing protein, promotes complex assembly. *Proceedings of the National Academy of Sciences of the United States of America* 108, 480-7.

Zhang, Y., and Chan, D. C. (2007). Structural basis for recruitment of mitochondrial fission complexes by Fis1. *Proceedings of the National Academy of Sciences of the United States of America* 104, 18526-30.

Zhang, Z., Kulkarni, K., Hanrahan, S. J., Thompson, A. J., and Barford, D. (2010). The APC/C subunit Cdc16/Cut9 is a contiguous tetratricopeptide repeat superhelix with a homo-dimer interface similar to Cdc27. *EMBO Journal* 29, 3733-44.

7

The Two DUF642 *At5g11420* and *At4g32460*-Encoded Proteins Interact *In Vitro* with the AtPME3 Catalytic Domain

Esther Zúñiga-Sánchez and Alicia Gamboa-de Buen
Universidad Nacional Autónoma de México
México

1. Introduction

The plant cell wall provides structural integrity to plant tissues and regulates cellular growth and form. The cell wall is a dynamic compartment that varies in composition and structure during plant development and in response to different environmental signals. During cell division, the cell plate is rapidly generated. The biogenesis of this new cell wall requires the delivery of vesicles containing newly synthesised material. Cell surface material that includes plasma membrane proteins and cell wall components can be also rapidly delivered to the forming cell plate (Dhonukshe et al., 2006). The three different layers that can compose the cell wall are the middle lamella, primary cell wall and secondary cell wall. The middle lamella, which is a pectinaceous interface, is deposited soon after mitosis to create a boundary between the two daughter nuclei and is important for the adhesion of neighbouring cells. The primary cell wall is deposited throughout cell growth and expansion. These two processes require a continuous synthesis and exportation of cell wall components that have to be reorganised in the cell wall network. The secondary cell wall is deposited when cell growth has ceased and is not present in all cell types.

1.1 Polysaccharide composition of the cell wall

The primary cell wall is composed of diverse polysaccharides (85-95%) and cell wall proteins with different functions (5-15%, CWP). Cellulose, hemicelluloses (e.g., xyloglucans) and pectins (e.g., homogalacturonans) are the main types of polysaccharides present in cell wall. Cellulose microfibrils confer rigidity to the cell wall and interact with hemicelluloses to provide structure to the network. These polysaccharide interactions could restrict access of enzymes to their substrates; however, this network can be modified during plant development by different proteins that interact with the network components or by enzymes that modify the polysaccharides (Harpster et al., 2002). The polysaccharides are not the only contributor to cell wall integrity during plant development. Recently, it was demonstrated that the presence of cellulose is essential to maintain the polar distribution of proteins at the plasma membrane. The polar distribution of PIN transporters for the phytohormone auxin is disrupted by a pharmacological interference with cellulose or by mechanical interference with the cell wall (Feraru et al., 2011). Pectins, which are a major component of primary cell wall, are a large group of complex polysaccharides that are synthesised in the Golgi and transported to the cell wall by secretory vesicles (Sterling et al., 2001). Methylesterification of homogalacturonan (HG) occurs in the plant Golgi apparatus, possibly by a S-adenosylmethionine (SAM) methyltransferase named Cotton Golgi-Related-3 (CGR3) (Held et al., 2011). HG is delivered to the cell wall in a highly methylesterified state, and the modulation of this state is a very important process in plant development. Highly esterified pectins are present in the proliferating zone of different tissues, whereas the cell walls of differentiating cells present abundant non-esterified pectins (Barany et al., 2010).

1.2 Protein composition of the cell wall

The cell wall composition is continuously modified by enzyme action during growth and development and in response to environmental conditions (Cassab, 1998). Proteins with enzyme activity and modulatory activity are present with different abundances in different cell types. Approximately 400 cell wall proteins that have been detected in cell wall proteomes have been classified into eight categories on the basis of predicted biochemical functions (Jamet et al., 2006). Members of seven of the eight groups have been previously defined as cell wall proteins involved in different aspects of cell wall dynamics. Many proteins have been detected in cell wall proteomes isolated from apoplastic fluids obtained from seedlings and rosette leaves (Charmont et al., 2005; Boudart et al., 2005), vegetative tissue that included etiolated hypocotyls and stem (Ishrad et al., 2008; Minic et al., 2007) and cell suspension cultures (Chivasa et al., 2002; Bayer et al., 2006, Bordereis et al., 2002). These proteins present a domain with an unknown function and are grouped together. The study of the function of the different families of this group of proteins will provide information about the dynamic processes of the cell wall.

1.2.1 Proteins acting on polysaccharides

Xyloglucan endotransglycosylase/hydrolase (XTH) is a family of glycosyl hydrolases that transglycosylate xyloglucan to allow expansive cell growth. These hydrolases are involved in cell growth, fruit ripening, and reserve mobilisation following germination in xyloglucan-storing seeds. In *Arabidopsis*, 33 genes have been identified that code for these hydrolases. Different temporal and spatial expression patterns for these *XTH* genes suggest that this family is involved in the change of cell wall properties related to every developmental stage. For example, *XHT5* is expressed in hypocotyls, root tips, and anther filaments, whereas *XHT24* is localised in vasculature tissue from the cotyledons, leaves, and petals. However, there is also an overlapping of the *XTH* gene expression pattern that suggests a combinatorial action of this enzyme group (Becnel et al., 2006).

Pectin modification is catalysed by a large family of pectin methylesterases (PMEs). In *Arabidopsis*, 66 genes have been suggested to potentially encode PMEs and are expressed differentially during organ and tissue development. A pro-domain is present in approximately 70% of the *Arabidopsis* PME family members (Micheli, 2001). It has been suggested that this domain has an inhibitory function during transportation to the cell wall by vesicles. The carboxylic fragment with the catalytic domain has been detected in cell wall proteomes, but the complete protein is required for secretion (Wolf et al., 2009). The interaction of PME with proteins that inhibit its activity, which are called pectin methyesterase inhibitors (PMEIs), contributes to the modulation of the degree of the methylesterified state of the pectin in the cell wall during different developmental processes (Pelloux et al., 2007). During pollen germination, the pollen tube wall presents highly methylesterified pectins in the tip region and weakly methylesterified pectins along the tube. It has been suggested that the activity of PMEs during pollen tube growth is highly regulated by PMEIs (Dardelle et al., 2010). Local relaxation of the transmitting tract cell wall also results from changes in the methylesterification of pectins that possibly facilitate the growth of the pollen tubes in the extracellular matrix of this female tissue (Lehner et al., 2010). An important role of pectin modifications in the regulation of cell wall mechanics in the apical meristem tissue has also been suggested (Peaucelle et al., 2011). The demethylesterification of pectin by PME activity results in random and contiguous patterns of free carboxylic residues. These contiguous patterns promote Ca^{++} binding, which generates a rigid cell wall. PMEs might also be involved in maintaining apoplastic Ca^{++} homeostasis. PME activity has been suggested to maintain apoplastic Ca^{++} homeostasis during heat shock. The resulting cell wall remodelling maintains the plasma membrane integrity to confer thermotolerance to the soybean (Wu et al., 2010). The random release of protons promotes pectin degradation by polygalacturonases, which are enzymes that also affect the pectin network. Polygalacturonases (PGs) promote pectin disassembly and might be responsible for various cell separation processes. PG activities are associated with seed germination, organ abscission, anther dehiscence, pollen grain maturation, fruit softening and decay, and pollen tube growth. In *Arabidopsis*, 69 genes encode PGs with different spatial and temporal patterns. For example, *At1g80170* is specifically expressed in the anther and pollen (González-Carranza et al., 2007).

Expansins are cell wall proteins that modify the mechanical properties of the cells to enable turgor-driven cell enlargement. Expansin genes are highly conserved in higher plants, and there are four different expansin families in plants. Multiple expansin genes are often expressed in association with developmental events such as root hair initiation or fruit growth. They are also involved in processes such as fruit ripening and abscission, although cell wall modification occurs without expansion. Expansins may also be involved in embryo growth and endosperm weakening during germination (Sampedro and Cosgrave, 2005). The localised expression of expansins is associated with the meristems and growth zones of the root and stems (Reinhardt et al., 1998).

1.2.2 Oxido-reductases

Peroxidases are implicated in many physiological phenomena that include cross-linking of cell wall components, defence against pathogens, and cell elongation. These enzymes have a great variety of substrates and can regulate growth by controlling the availability of elongation-promoting H_2O_2 in the cell wall (Passardi et al., 2004). In *Arabidopsis*, 73 genes have been reported to code for putative peroxidases (Valério et al., 2004), and AtPrx33 and AtPrx34 function is specifically related to root elongation (Passardi et al., 2006).

Germins are oligomeric enzymes with oxalate oxidase activity that are associated with the extracellular matrix. In *Arabidopsis*, this family contains 12 members that are expressed in almost every organ and developmental stage. *AtGer1* has been implicated in germination, whereas *AtGer2* is involved in seed maturation (Membré et al., 2000).

1.2.3 Proteases

Proteases cleave peptide bonds and are classified into four catalytic classes: Cys proteases, Ser carboxypeptidases, metalloproteases and Asp proteases. The *Arabidopsis* genome encodes 826 proteases that are classified into 60 families with high functional diversity. Plant proteases are key regulators of different biochemical processes that are related to meiosis, gametophyte survival, embryogenesis, seed coat formation, cuticle deposition, epidermal cell fate, stomata development, chloroplast biogenesis, and local and systemic defence responses (van der Hoorn, 2008). Some proteases have been detected in cell wall proteomes, especially in cell suspension cultures.

1.2.4 Proteins that have interacting domains with no enzymatic activity

LRR proteins are frequently implicated in protein-protein interactions and are localised in the different subcellular compartments (Kajava, 1998). The LRR superfamily includes polygalacturonase-inhibiting proteins (PGIPs) that are present in the cell wall and are involved in disease resistance as well as growth and development (Di et al., 2006). FLOR 1, a putative PGIP protein, has been detected in cell wall proteomes but is also localised intracellularly, as more than 70% of the PGIP in *Pisum sativum* was reported to be distributed in the cytoplasm (Acevedo et al., 2004; Hoffman & Turner, 1984).

Pectin methyl esterases inhibitors (PMEIs) are a diverse group of proteins that belong to the family of invertase inhibitors (INHs). PMEIs share with INHs a domain that is characterised by four conserved cysteine residues that can form two disulfide bonds (Juge, 2006). In *Arabidopsis*, there is an spatial patterning of cell wall PMEI at the pollen tip (Röckel et al., 2008).

Lectins are a diverse group of carbohydrate specific binding proteins that are involved in signal transduction (Lannoo et al., 2007). This group of proteins has interacting domains but does not show catalytic activity. The group presents with varying cellular localisation, which suggests a role in signal transduction between the different cellular compartments (Van Damme et al., 2004).

1.2.5 Proteins involved in signalling

In plants, there is a large subclass of receptor-like kinases that have extracellular LRRs in the receptor domain and are involved in signal transduction during development or defence (Clark et al., 1997). Arabinogalactan proteins (AGPs) are hydroxyproline-rich glycoproteins that are also involved in signalling. This family contributes to defensive, adhesive, nutrient and guidance function during pollen-pistil interactions (Cassab, 1998).

1.2.6 Proteins related to lipid metabolism

Lipases (LTPs) are hydrolytic enzymes with multifunctional properties. GDSL lipases are mainly involved in the regulation of plant development, morphogenesis, synthesis of secondary metabolites and defence responses (Ruppert et al., 2005).

1.2.7 Structural proteins

LRR-extensins were the only group of structural proteins detected in cell wall proteomes. This family may be involved in the local regulation of cell wall expansion. Eleven genes have been described in *Arabidopsis*; four of them are pollen specific (Baumberger et al., 2003).

1.2.8 Unknown proteins

Approximately 5 to 30% of the total proteins from different cell wall proteomes have been classified as hypothetical, expressed, putative, unknown or with a domain of unknown function (DUF), especially in cell suspension culture. A domain is considered to be a discrete portion of a protein that folds independently of the rest of the protein and possesses its own function. Eight DUF protein families (DUF26, DUF231, DUF246, DUF248, DUF288, DUF642, DUF1005, DUF1680) are represented by one (or more) member(s) of the cell wall proteomes.

DUF26 is a plant-specific protein family composed of 40 members in *Arabidopsis.* Some members include DUF26 receptor-like kinases (RLKs), which are also known as cysteine-rich RLK (CRKs). These proteins are involved in pathogen resistance and are transcriptionally induced by oxidative stress and pathogen attack (Wraczeck et al., 2010). *At5g43980* encodes a protein present in the apoplastic fluid from rosette leaves that has been described as a plasmodesmal protein (PDLP1) involved in cell-to-cell communication processes (Thomas et al., 2008). The other DUF26 protein, which was detected in the cell wall proteome from cell suspension cultures, has not yet been assigned a function.

DUF231 is present in the proteins of the *TRICHOME BIREFRINGENCE/TRICHOME BIREFRINGENCE-LIKE* (TBR/TBL) plant family with 46 members in *Arabidopsis*. The role of this family in cellulose biosynthesis has been recently described; *tbr* mutants presented decreased levels of crystalline secondary wall cellulose in trichomes and stems (Bischoff et al., 2010a). Loss of TBR also results in increased PME activity and reduced pectin esterification, which suggests that TBL/DUF231 proteins are "bridging" proteins that crosslink different cell wall networks (Bischoff et al., 2010b). *At5g06230* (TBL9) was found in a cell wall proteomic analysis of etiolated hypocotyls (Ishrad et al., 2008).

The domain unknown function 246 is considered to be a GDP-fucose o-fucosyltransferase domain in animals. This protein family has 16 members in *Arabidopsis,* and one of them, *At1g51630,* was detected in the proteome of cell suspension cultures.

DUF248 is a putative methyltransferase-related family of proteins with an ankyrin-like protein domain that is related to dehydration-responsive proteins. There are 29 proteins of this family in *Arabidopsis,* but only one, *At5g14430,* has been described in the cell wall proteome of cell suspension cultures (Bayer et al., 2006).

DUF288 is not a plant-specific family; this domain is also found in *Caenorhabditis elegans* proteins. In *Arabidopsis,* there are two members: *At2g41770* and *At3g57420*. *At3g57420* encode protein was purified from the apoplastic fluid of the cell wall proteome of rosette leaves (Boudart et al., 2005).

The DUF1005 domain has five integrants in *Arabidopsis* with two members that are similar to IMP dehydrogenase/GMP reductase from *Medicago trunc
ulata*. The integrant isolated from

the cell wall proteome of mature stems (*At4g29310*) does not have the other domain (Minic et al., 2007).

Two loci are described in *Arabidopsis* for the DUF1680 family, and one of them was purified from the cell wall proteome of mature stems.

The most important family of unknown proteins detected in cell wall proteomes is DUF642, which is a highly conserved plant-specific family that is present in angiosperms and gymnosperms (Albert et al., 2005, Vázquez-Lobo, personal communication). *Arabidopsis* has ten members. The *At3g08030*-encoded protein is present in all cell wall proteomes and is the only unknown protein that was also detected in a seed proteome from the *Arabidopsis* accession Cape Verde Island (Cvi) that has deeper seed dormancy (Chibani et al., 2006). *At2g41800* and *At1g80240*-encoded proteins were only found in cell suspension cultures (Bayer et al., 2006), whereas *At5g25460*-encoded protein was found in vegetative and cell wall suspension cultures. *At4g32460* and *At5g11420*-encoded proteins were both detected in apoplastic and vegetative tissues. The consistent presence of 6 members of this family in all cell wall proteomes suggest that the biochemical function of the DUF642 family is related to the regulation of the activity of cell-wall-modifying enzymes at different stages of plant development.

1.3 DUF642 family

The DUF642 protein family is highly conserved, is widespread in plants, and might be involved in important basic developmental processes. Members of this family have been observed in basal angiosperms such as *Amborella*, in both monocots and dicots and also in gymnosperm species. The relevance of the DUF642 family to plant evolution was discussed by Albert and collaborators (2005). The proteins encoded by the DUF642 gene family have a unique, highly conserved domain with no assigned function that shares similarity with the galactose-binding domain. The ten members of this family identified in *Arabidopsis* contain a signal peptide of 20 to 30 amino acids in the N-terminus region that could promote their localisation in the endomembrane system or in the cell wall. Three of the ten *Arabidopsis* genes (*At1g29980, At2g34510* and *At5g14150*) encode proteins have been described as glycosil-phosphatidyl-inositol anchored proteins (Figure 1) (Borner et al., 2003, Dunkley et al., 2006). The *At2g41800*-encoded protein has been detected in the *Arabidopsis* cell wall proteome. The proteins encoded by *At5g11420* and *At2g34510* contain a ATP/GTP binding site motif that has been described in many proteins involved in signal transduction processes.

Although a function has not yet been assigned for this family, it has been suggested that some members could be involved in different developmental processes. Organ-specific expression has been described for the flowers of two DUF642 members, *At3g08030* and *At5g11420* (Wellmer et al., 2004), and for the stems for a DUF642 *Medicago sativa* gene (Abrahams et al., 1995). *At4g32460, At5g14150* and *At2g41800* have been described as papillar cell-specific genes in flowers (Tung et al., 2005). Changes in DUF642 gene expression have been also detected under specific environmental conditions. Saline stress promotes the expression of *At2g41810* (Kreps et al., 2002), and an RNA increase in the three DUF642 *Arabidopsis* homologs (*At3g08030, At5g25460* and *At4g32460*) was described during the priming and germination of *Brassica oleracea* seeds (Soeda et al., 2005).

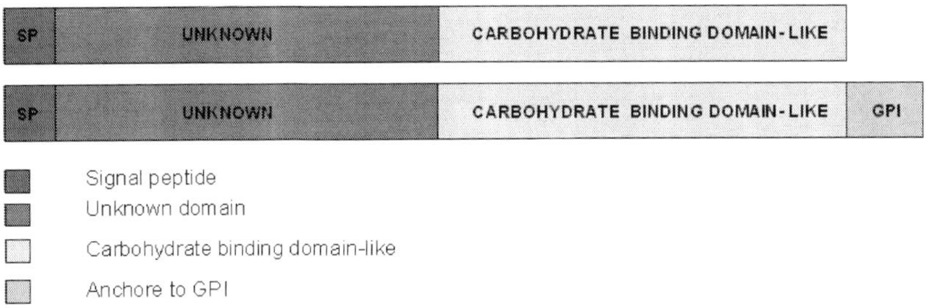

Fig. 1. DUF642 proteins have a basic structure divided into two subdomains and a signal peptide. N-terminus subdomain has not function or putative function assigned while C-terminus subdomain has homology with a carbohydrate binding domain. Some DUF642 proteins present in their C-terminus a GPI anchored motive.

We characterised the plant-specific DUF642 protein family using different approaches. We determined mRNA expression in different plant tissues, characterised sequence features and detected the potential interaction of proteins with two members of this family in *Arabidopsis* (*At5g11420* and *At4g32460*-encoded proteins). The proteins identified by LC/MS/MS analysis were the leucine-rich repeat protein FLOR1 (FLR1), a vegetative storage protein (VSP1), and a ubiquitous pectin methylesterase isoform (PME3) isolated from *Arabidopsis* flowers and leaves. Based on the structural characteristics of the DUF642 family of proteins and the associated affinity chromatography analyses, we propose that these proteins could interact specifically with other cellular components via their DUF642 domain and are therefore potentially involved in developmental plant processes. Our results provide a starting point for defining the function of the DUF642 family in plant development.

2. Materials and methods

2.1 Plant material and sample collection

Arabidopsis thaliana from the Columbia (Col) ecotype plants were grown on MS plates (1X Murashige and Skoog basal salt mixture, 0.05% MES, 1% sucrose as carbon source and 0.8% agar) in a REVCO growth chamber under a long photoperiod (16-h light 8-h darkness) at 20°C. Fifteen-day-old seedlings were transferred to pots containing Metro-Mix 200 (Scotts Company) soil and grown under the same controlled conditions.

2.2 Reverse transcriptase–polymerase chain reaction (RT-PCR)

Arabidopsis samples from different tissues were collected from 15-day-old seedlings and flowering plants, immediately frozen in liquid nitrogen and stored at -80°C until analysis. Total RNA from different tissues was isolated using TRIZOL according to the supplier's instructions (INVITROGEN™). cDNA templates for the amplification by PCR were prepared using SuperScript II reverse transcriptase (INVITROGEN™) according to the manufacturer's instructions. Based on the sequence of each gene member of the DUF642 family of *Arabidopsis*, the following primers were synthesised:

At2g41800: F 5'tcctcctcctatctctctgc 3' and R 5'aaacggttctcttcctgc 3';
At2g41810: F 5'atgggccaaaaaaacac 3' and R 5'atgtctctcgttctctctc 3';
At3g08030: F 5'ggttcccaaagccattattc 3' and R 5'acaatctcgtcaatgacagg3';
At5g25460: F 5'cttccttcttttcatcgcc 3' and R 5'acgagaaatcatcgctcc 3';
At5g11420: F 5'ccatgggcttcagtgacgggatg 3' and R 5'agatctgagtgtcttttcccgc 3';
At4g32460: F 5'gtgatagtgcttcttctccttcac 3' and R 5' agcgacgaatctcaatgac 3';
At1g80240: F 5'aaaagcagcactcctcttag 3' and R 5' atcattggtccctcacaac 3';
At1g29980: F 5'ccgagcaacaatagatgc 3' and R 5'actgtagaacgcaactctgg 3';
At2g34510: F 5'ttggtctctccattgtggc 3' and R 5'ccttaacgtcatcaatcacagg 3';
At5g14150: F 5´ttgcgcctcttcagattttt3'and R 5'cttctcaccagagccagtcc 3'.

Polymerase chain reaction (PCR) was performed under the following conditions: 94°C 5 min; 35 cycles of 94°C 30 sec, 60-62°C 30 sec, 72°C 1 min 30 sec, 72°C 5 min.

2.3 Sequence analysis and database search

The 10 DUF642 protein sequences of *Arabidopsis* were obtained from GenBank (NP_973938: *At1g29980;* NP_178141: *At1g80240;* AAC02768: *At2g41800;* AAC02767, NP_181712: *At2g41810;* AAC26689: *At2g34510;* AAO00904: *At3g08030;* ABF19001: *At4g32460;* NP_196919: *At5g14150;* AAN31807: *At5g11420* and AAP37805: *At5g25460*). A multiple sequence alignment, using only the DUF642 protein domain, was performed using ClustalW from the Bio Edit Sequence Alignment Editor. The possible secondary structure of the proteins coded for by *At5g11420* and *At4g32460* was compared on-line using the Draw an HCA (Hydrophobic Cluster Analysis) program (http://ca.expasy.org/tools/) as described in Gaboriaud et al. (1987).

2.4 Recombinant 5xHis-tagged DUF642 proteins and the resin-bound DUF642 protein affinity column

The entire open reading frame of the DUF642 genes *At5g11420* and *At4g32460*, without the signal-peptide-coding region, was amplified using PCR. The primers used for the *At5g11420* were MET11420 (5'ccatgggcttcagtgacgggatg3'), which includes an in-frame ATG, and primer 11420FIN2 (5'agatctagtgtcttttcccgca3'). For the amplification of the carboxyl-terminus truncated protein, the *At5g11420* (Δ11420) forward primer MET11420 and the reverse primer 11420FIN3 (5´agatctcggcttacgagcactgag3´) were used. *At4g32460* was amplified using the following primers: MET32460 (5'ccatgggcttcaatgatggactactacc3') and 32460FIN2 (5'agatctgcgtaaaacgtactgtaga3'). The amplified regions of these genes were cloned into the pQE60 vector using the NcoI and BglII restriction sites. A negative control was performed using the empty pQE60 vector. Protein expression and purification were performed following the supplier's instructions, and the recombinant proteins with the histidine tail were detected using western blot analysis with a Ni-NTA conjugate (QIAGEN). The three recombinant proteins were eluted as a single band and were identified to have the histidine tail. No protein was detected when the empty vector was used. The elution process was the only step omitted when the column was prepared for each recombinant protein.

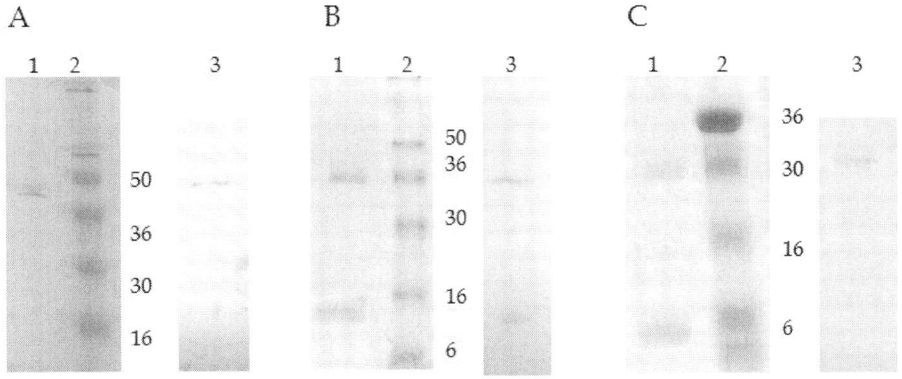

Fig. 2. Recombinant 5xHis-tagged DUF642 proteins.
A) Purification of the 32460 recombinant protein. 12% PAGE Gels were stained with Coomassie Blue. The column was eluted with 250 mM Imidazole (Lane 1). Western Blot of the eluted fraction (NiNta beads with phosphatase alkaline secondary antibody). The band of approximately 40 kDa corresponds to the calculated molecular weight for this protein (Lane 3).
B) Purification of the 11420 recombinant protein. 12% PAGE Gels were stained with Coomassie Blue. The column was eluted with 250 mM Imidazole (Lane 1). Western Blot of the eluted fraction (NiNta beads with phosphatase alkaline second antibody). The band of approximately 40 kDa corresponds to the calculated molecular weight for this protein (Lane 3).
C) Purification of the Δ11420 recombinant protein. 12% PAGE Gels were stained with Coomassie Blue. The column was eluted with 250 mM Imidazole (Lane 1). Western Blot of the eluted fraction (NiNta beads with phosphatase alkaline second antibody). The band of approximately 32 kDa corresponds to the calculated molecular weight for this protein (Lane 3).

2.5 Affinity chromatography of flower or leaf protein extracts

Frozen flowers or leaves from *Arabidopsis* plants (10-20 g) were ground with a mortar and pestle and placed in two 40 ml tubes with 14 ml of extraction buffer (50 mM Tris-HCl pH 7.5, 3 mM $MgCl_2$, 1 mM PMSF). The crude homogenate was centrifuged at 15,000 x*g* for 30 min, and in the case of DUF642 affinity columns, the supernatant was loaded onto a previously equilibrated DEAE-Sephacel column (2x 10 cm) with extraction buffer at 4°C. The resulting fraction was then used for affinity chromatography. The affinity column was prepared beforehand as described above and equilibrated with extraction buffer. The protein extracts from the different tissues were mixed for 1 h with the prepared resin at 25°C using gentle agitation in a ratio of 10 ml of extract/0.2 ml of agarose. The column was washed with 50 mM Tris-HCl pH 7.5, 5 mM $MgCl_2$ (50 vol) buffer to remove unbound proteins. Bound proteins were eluted with the same buffer containing different NaCl concentrations (100 to 1000 mM). These fractions were precipitated with cold acetone. Agarose and the empty vector column were used as negatives controls, and no bound proteins were detected (Gamboa et al., 2001).

The fractions obtained in the affinity chromatography assays were analysed on denaturing 12% SDS-PAGE gels and stained with silver. Bands of interest were extracted from the gels and sent to the Proteomics Platform of the Eastern Genomics Center, Quebec, Canada, where the in-gel digest and mass spectrometry experiments were performed. Tryptic digestion was performed according to Shevchenko et al. (1996) and Havlis et al. (2003). Peptide samples were separated by online reversed-phase (RP) nanoscale capillary liquid chromatography (nano/LC) and analysed by electrospray mass spectrometry (ES/MS/MS).

Database searching. All MS/MS samples were analysed using Mascot (Matrix Science, London, UK; version 2.2.0)

Criteria for protein identification. Scaffold (version Scaffold-01_07_00, proteome Software Inc. Pórtland Oregon, OR) was used to validate MS/MS-based peptide and protein identifications. Peptide identifications were accepted if they could be established at greater than 95.0% probability as specified by the Peptide Prophet algorithm (Keller et al., 2002). Protein identifications were accepted if they could be established at greater than 95.0% probability and contained at least 2 identified peptides. Protein probabilities were assigned by the Protein Prophet algorithm.

Only one protein was identified for the protein bands derived from the two chromatography steps, DEAE-Sephacel and affinity chromatography (11420 and 32460 column affinity protocols).

3. Results and discussion

3.1 Gene structure of the DUF642 family in *Arabidopsis thaliana*

The DUF642 domain was only present in the ten *Arabidopsis* members described before, and all members had the same gene structure, which consisted of three exons and two introns (Figure 3). The first intron encoded the signal peptide, and an alternative usage of the first exon was detected for *At1g29980* and *At3g08030*. The first intron was also included in the mRNA sequence for the *At3g08030* gene. The expression of two different mRNAs has been found in different tissues, which suggests a possibly different protein subcellular localisation.

3.2 DUF642 members are widely expressed in all *Arabidopsis thaliana* plant tissues

The RT-PCR expression analysis of the ten DUF642 genes in different tissues including seedlings, stems, cauline leaves, rosette leaves, flowers, inflorescences and roots is shown in Figure 4. The genes with broad expression patterns are *At1g80240, At5g11420, At5g25460* and *At2g41800,* whereas *At1g29980* and *At4g32460* were not detected in cauline leaves. *At2g41810* expression was restricted to inflorescence tissue. The *At2g41810*-encoded protein exhibits 81% identity and 89% similarity to the *At2g41800*-encoded protein. In the inflorescence tissue, the *At2g41800* transcript contained an additional region of 100 bp corresponding to the first intron, which suggests an alternative use of the first exon described for *At3g08030* and *At1g29980*. The gene with the most divergent sequence in the family, *At5g14150,* was also detected in the stem, flower, inflorescence, and root tissues and was detected at low levels in cauline leaves.

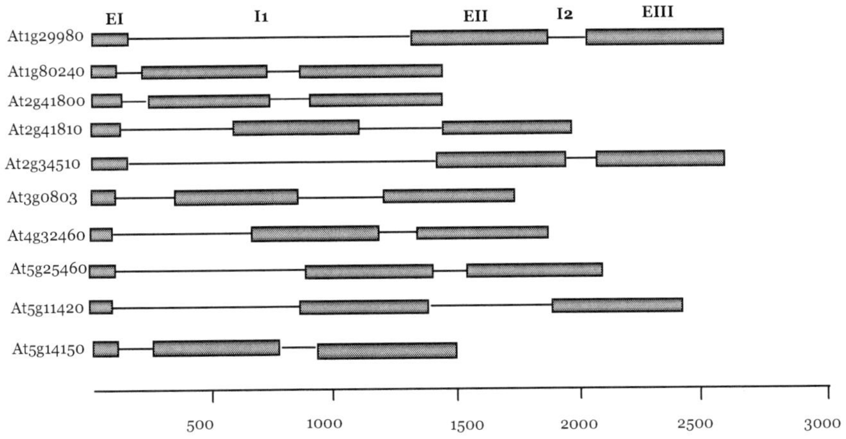

Fig. 3. Gene structure of DUF642 family in *Arabidopsis thaliana*.
EI: Exon 1, **I1**: Intron 1, **EII**: Exon 2, **I2**: Intron 2, **EIII**: Exon3

Our results are consistent with the microarray data described in the Gene Investigator Atlas (http://www.genevestigator.ethz.ch/), except for the *At2g41810* gene. We did not find *At2g41810* expression in the roots, but the Atlas indicated high expression. However, Kreps and collaborators (2002) demonstrated that the expression of this gene in the roots is induced by NaCl stress. These discrepancies in the results obtained in different studies could therefore be related to the different growth conditions used. Spatio-temporal expression analyses of this family will provide important information about its function. Cell-type-specific expression in the roots of the auxin-inducible DUF642 genes *At2g41800* and *At4g32460* was recently reported (Goda et al., 2004; Salazar-Iribe & Gamboa-deBuen, unpublished data).

Transcriptomic analyses suggest that the expression of this family of genes is also affected by different environmental conditions. The expression of genes that encode DUF642 proteins could be inhibited or stimulated by different pathogens. Indeed, invasion by necrotrophic pathogens or insect attack has been shown to significantly reduce the expression of *At5g11420*, *At5g25460*, *At4g32460* and *At1g29980* in plant tissues (Hu et al., 2008; Ehlting et al., 2008). Conversely, an increase of DUF642 gene expression in response to biotrophic organisms has been reported in *Arabidopsis* transcriptomic analyses of sink-heterologous structures, such as galls. Furthermore, the *At3g08030* and *At1g29980* genes have been found to be up-regulated in response to *Agrobacterium tumefaciens* and *Rhodococcus fascians* invasion (Depuydt et al., 2009, Lee et al., 2009). *At1g29980* has also been shown to be highly expressed in the giant cells induced by the root-knot nematode, *Meloidogyne incognita* (Barcalá et al., 2010), and the development of such sink structures is related to an increase in auxin (Grunewald et al., 2009). The study of the effect of nematode invasion on the gene expression of the DUF642 family will provide important functional insights.

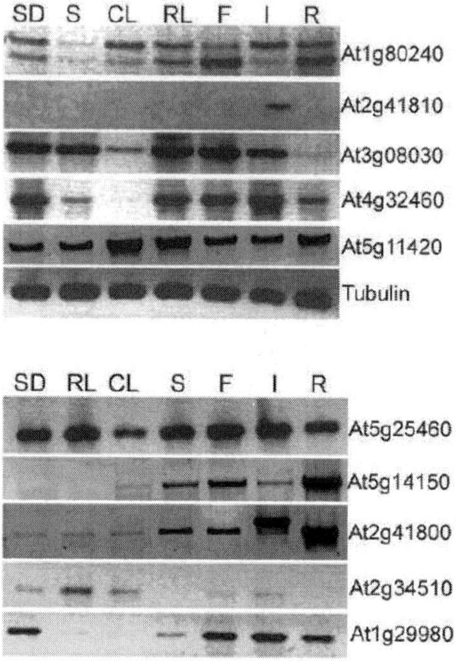

Fig. 4. RT-PCR expression of *Arabidopsis thaliana* DUF642 genes in various tissues. Seedlings (SD), rosette leaves (RL), cauline leaves (CL), stems (S), flowers (F), inflorescences (I), and roots (R). The expression of tubulin was analyzed simultaneously as an internal standard.

3.3 Comparison of the primary sequence of the ten *Arabidopsis thaliana* DUF642 family members

The DUF642 gene family encodes proteins with an estimated molecular mass ranging from 39 to 44 kDa. These proteins contain the DUF642 amino acid domain, preceded by a 20-30 amino acid signaling peptide on the amino terminus. This signaling peptide could be involved in the cell wall localisation of DUF642 proteins in several plant organs. Alignment analysis of the ten *Arabidopsis* members shows an extensive conservation of the DUF642 domain; the percentage of identical and similar amino acids varies from 30% to 85% and 43% to 92%, respectively (Figure 5A). About 30% of the amino acids distributed throughout the sequence of the DUF642 domain are hydrophobic. These residues are not identical, but they are similar among the different proteins. The comparison of the hypothetical secondary structure of *At5g11420* and *At4g32460*-encoded proteins shows that the hydrophobic clusters present are similar (Figure 5B). Four conserved cysteine residues are present in all of the sequences as previously described for the pectin methyl esterase inhibitors localised in the cell wall (Juge, 2006). Because no catalytic activity has yet been assigned to the DUF642 domain, this family could be involved in specific carbohydrate or protein interactions.

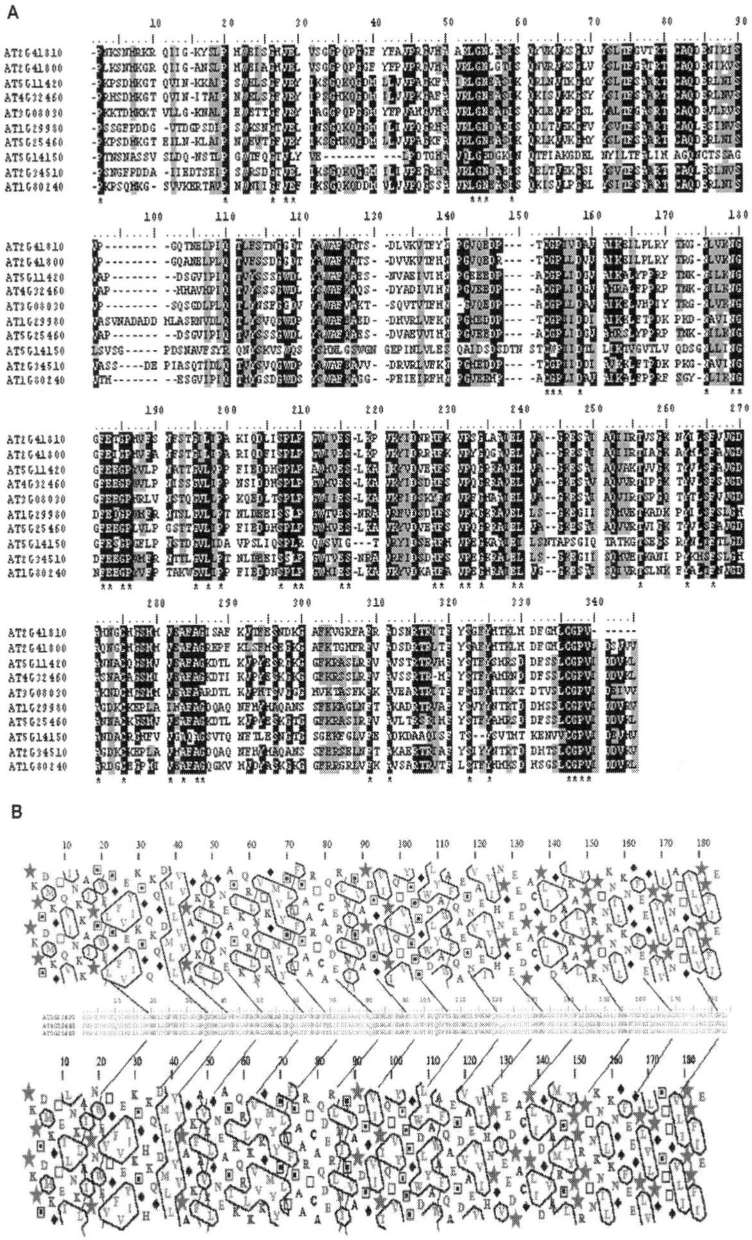

Fig. 5. DUF642 amino acid sequence and features.
(A) Clustal W alignment (BioEdit) of the DUF642 domain of the 10 *Arabidopsis* proteins is

shown. The N-terminal region (comprising the signal peptide) was eliminated for the alignment. Shading indicates conserved amino acid and dark shading indicates identities. (B). Secondary structure comparison of 11420, top, and 32460, bottom. The initial 180 amino acid sequences of the DUF642 domain of both proteins are compared using "Draw an HCA" online program (http://ca.expasy.org/tools/) (Gaboriaud et al., 1987). Amino acids forming putative hydrophobic clusters are grouped together. Compare similar patterns in both sequences. Star: P; dotted square: S; rhomb: G, and empty square: Y residues; other amino acids in standard abbreviation.

Most of the members of the DUF642 family have a broad expression pattern in different plant tissues. A putative redundancy of function in this family should be considered because of the high conservation of the DUF642 domain; however, it is important to describe the organ, cell type and specific stress-related expression patterns for each gene to determine the individual gene function (Wellmer et al., 2004).

3.4 DUF642 proteins have specific interactors in the flowers and leaves of *Arabidopsis thaliana*

Recombinant 32460 protein interacts *in vitro* with the LRR protein FLR1 (Q9LH52, *At3g12145*), with VSP1 (Q93VJ6, *At5g24780*) and with PME (Q9LUL7, *At3g14310*) in flowers, whereas in leaves, it interacts with the same PME (*At3g14310*) (Figure 6). The recombinant 32460 protein interacts *in vitro* with three proteins with sizes of 38 kDa, 37 kDa and 29 kDa from the flowers (Figure 6A). These proteins were identified as FLR1, PME, and VSP1, respectively (Figures 8A, B and C). It is important to note that FLR1 was not eluted by 500 mM NaCl, and VSP1 is only present in this fraction as determined in the interaction assay using the *At5g11420*-encoded protein. A 37 kDa band was purified in the three salt fractions from leaf extracts and was identified as the same PME isoform described for the flowers. A 29 kDa band was also eluted, and this protein was identified as a possible auxin-binding protein (Figure 6B). For all protein bands analysed, only a significant hit was assigned, as described in the material and methods.

The recombinant DUF642 11420-protein interacts *in vitro* with FLOR1 and VSP1 in flowers, but in leaves, it only interacts with PME (Figure 7). A high-purity protein fraction with two bands was obtained from the 11420-affinity column after the floral crude protein extracts were purified over several steps (Figure 7B). Different ionic strengths were used during elution; one 38 kDa band was eluted at 100 and 200 mM NaCl, whereas a 29 kDa band was obtained at 200 and 500 mM NaCl (see arrows in Figure 7B). The 38 kDa protein was identified as FLR1 (12% coverage) and the 29 kDa band as VSP1 (11% coverage), as described in the methods (Figures 8A and B). A Δ11420 protein without the carboxylic terminus that included the most divergent amino acid sequence was also used as a ligand. FLR1 was the only purified protein, which suggests that the carboxylic region is important for interaction with VSP1 (Figure 7C). *At5g11420* is expressed in all *Arabidopsis* tissues, and therefore, we were interested in the determination of the proteins in the leaves that interact with the *At5g11420*-encoded protein. The same procedure, using the affinity column with a leaf extract protein fraction, was used. In the first two fractions, two bands of 45 kDa and 32 kDa were detected. In the 500 mM NaCl fraction, three major bands of the following sizes were detected: 45 kDa, 32 kDa and 14 kDa (Figure 7D). The identified 32 and 14 kDa bands correspond to a PME (40% coverage, Q9LUL7, *At3g14310*). The PME 14 kDa band was also identified when the Δ11420 protein was used as the ligand (Figure 7E). The two lower-molecular-weight bands contained the carboxyl region that includes the catalytic domain of the PME, and therefore, it is possible that the differences in their electrophoretic mobility are the result of post-translational modifications.

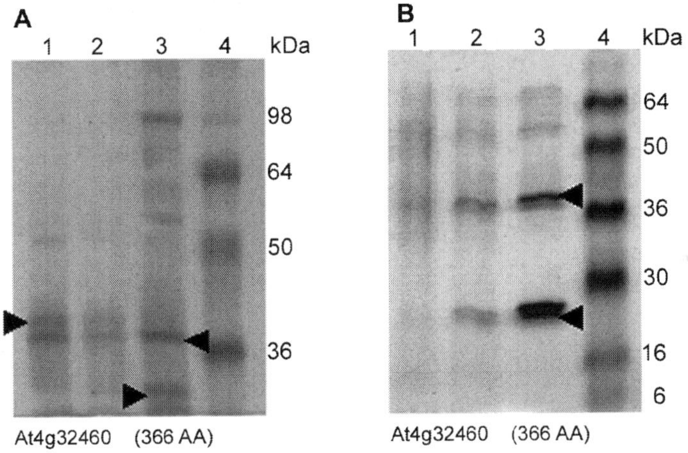

Fig. 6. 32460-protein *in vitro* interactors.
(A) Recombinant 32460 amino acid sequence. (B C) affinity chromatography assays of 32460 interactors from DEAE-Sephacel flow-through protein fraction from *Arabidopsis thaliana* flowers (B) and leaves(C). Silver staining of 12% SDS-PAGE gel showing: (1) NaCl 100 mM, (2) NaCl 200 mM, (3) NaCl 500 mM elution fractions, and (4) molecular weight reference.
(B) Flower interactors of the 32460 recombinant protein. In (1) and (2) two main protein bands are seen; with molecular masses of 38 and 37, corresponding to FLR1 (arrow in (1)) and PME (upper arrow in (3)) respectively. In (3) the two bands with molecular masses of 37 and 29 were identified as PME and VSP1 respectively (see arrows in (3)).
(C) Leaf interactors of the 32460 protein. Fraction (3) was highly enriched with two bands with molecular masses of approximately 37 and 29. The 37 kDa band was identified as the catalytic domain of a PME, while the 29 kDa band was identified as a possible auxin-binding protein.

The proteins that interacted *in vitro* with the DUF642 11420 and 32460 proteins, i.e., FLOR1 and AtPME3, were detected in the cell wall proteomes (Figures 7A, B and C). Similar expression patterns reflect a possible *in vivo* interaction. FLOR1 is an LRR protein related to polygalacturonase inhibitors (PGIPs) that are highly expressed in vascular and meristem tissues. An intracellular localisation of FLOR1 has been also reported (Acevedo et al., 2004). AtPME3 (*At3g14310*) is expressed in the vascular tissue of seedlings, leaves, stems and roots and is involved in adventitious root formation (Guénin et al., 2011). Recently, we demonstrated that *At4g32460* is also expressed in the meristems and in vascular tissue (Zúñiga-Sánchez & Gamboa-deBuen, unpublished data).

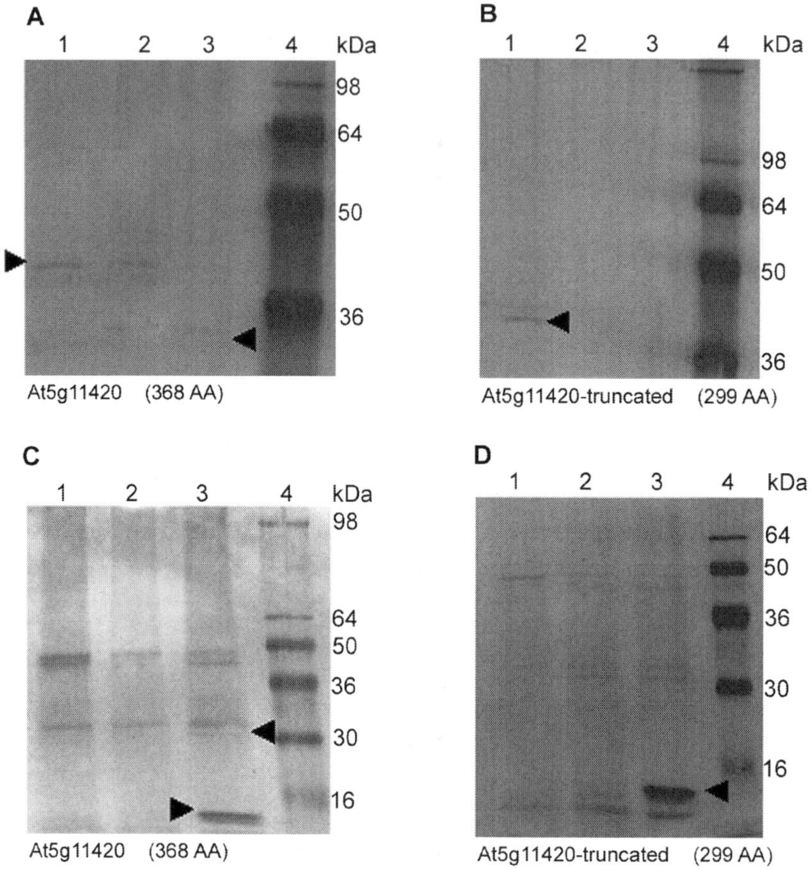

Fig. 7. 11420-protein *in vitro* interactors.
(ABCD) Affinity chromatography assays of 11420 interactors from DEAE-Sephacel flow-through protein fraction of *Arabidopsis thaliana* flowers (B,C) and leaves (D,E); Silver staining of 12% SDS-PAGE gels showing: (1) NaCl 100 mM, (2) NaCl 200 mM, (3) NaCl 500 mM elution fractions, and (4) molecular mass references.
(A) Flower interactors of the 11420 recombinant protein: Two protein bands are seen with molecular masses of 38 and 29 corresponding to FLR1 and VSP1 respectively (see arrows).
(B) Flower interactors of the 11420-truncated protein (Δ11420): In (1) only a 38kDa protein is detected, corresponding to FLOR1 (see arrow).
(C) Leaf interactors of the 11420 recombinant protein. In (1) and (2) two main protein bands are seen, corresponding to molecular masses of 37 and 32. In (3) Three proteins are detected with molecular masses of approximately 45, 32 and 14. The 32 kDa and 14 kDa bands were identified as the catalytic domain of a PME (see arrows).
(D) Leaf interactors of the 11420-truncated protein. The 14 kDa band shown in (3) (arrow) was identified as the same PME as in (C).

A

```
  1 MKLFVHLSIF FSILFITLPS SYSCTENDKN ALLQIKKALG NPPLLSSWNP RTDCCTGWTG
 61 VECTNRRVTG LSVTSGEVSG QISYQIGDLV DLRTLDFSYL PHLTGNIPRT ITKLKNLNTL
121 YLKHTSLSGP IPDYISELKS LTFLDLSFNQ FTGPIPGSLS QMPKLEAIQI NDNKLTGSIP
181 NSFGSFVGNV PNLYLSNNKL SGKIPESLSK YDFNAVDLSG NGFEGDAFMF FGRNKTTVRV
241 DLSRNMFNFD LVKVKFARSI VSLDLSQNHI YGKIPPALTK LHLEHFNVSD NHLCGKIPSG
```

B

```
  1 MKILSLSLLL LLAATVSHVQ SSASVPGLIE LLESNTIFGN EAELLEKEGL SINYPNCRSW
 61 HLGVETSNII NFDTVPANCK AYVEDYLITS KQYQYDSKTV NKEAYFYAKG LALKNDTVNV
121 WIFDLDDTLL SSIPYYAKYG YGTENTAPGA YWSWLESGES TPGLPETLHL YENLLELGIE
181 PIIISDRWKK LSEVTVENLK AVGVTKWKHL ILKPNGSKLT QVVYKSKVRN SLVKKGYNIV
241 GNIGDQWADL VEDTPGRVFK LPNPLYYVPS
```

C

```
  1 MAPSMKEIFS KDNFKKNKKL VLLSAAVALL FVAAVAGISA GASKANEKRT LSPSSHAVLR
 61 SSCSSTRYPE LCISAVVTAG GVELTSQKDV IEASVNLTIT AVEHNYFTVK KLIKKRKGLT
121 PREKTALHDC LETIDETLDE LHETVEDLHL YPTKKTLREH AGDLKTLISS AITNQETCLD
181 GFSHDDADKQ VRKALLKGQI HVEHMCSNAL AMIKNMTDTD IANFEQKAKI TSNNRKLKEE
241 NQETTVAVDI AGAGELDSEG WPTWLSAGDR RLLQGSGVKA DATVAADGSG TFKTVAAAVA
301 AAPENSNKRY VIHIKAGVYR ENVEVAKKKK NIMFMGDGRT RTIITGSRNV VDGSTTFHSA
361 TVAAVGERFL ARDITFQNTA GPSKHQAVAL RVGSDFSAFY NCDMLAYQDT LYVHSNRQFF
421 VKCLIAGTVD FIFGNAAVVL QDCDIHARRP NSGQKNMVTA QGRTDPNQNT GIVIQKCRIG
481 ATSDLQSVKG SFPTYLGRPW KEYSQTVIMQ SAISDVIRPE GWSEWTGTFA LNTLTYREYS
541 NTGAGAGTAN RVKWRGFKVI TAAAEAQKYT AGQFIGGGGW LSSTGFPFSL GL
```

Fig. 8. Sequences of protein bands identified by LC-MS/MS from the pull down essays using DUF642 proteins. Bands were excised from the gels and sent to the Proteomics Platform of the Eastern Genomics Center, Quebec, Canada for their identification. One protein with high hit was identified for each band sent. Peptides identified are shaded.
(A) FLR1 (Q9LH52, *At3g12145*) amino acid sequence showing all the peptides identified in different protein fractions.
(B) VSP1 (Q93VJ6, *At5g24780*) amino acid sequence showing all the peptides identified in different protein fractions.
(C) PME (Q9LUL7 *At3g14310*) amino acid sequence showing all the peptides identified in different protein fractions. Underlines show signal peptide (_ _ _), inhibitory domain (____) and catalytic region (___). Note that all the peptides identified for this protein match the catalytic domain.

Subcellular localisation is also an important criterion for putative *in vivo* protein interactions. Three bands of 37, 32 and 14 kDa were identified as fragments of the catalytic domain sequence from AtPME3 in leaf protein extracts. This electrophoretic pattern has been previously described in a purified citrus PME fraction. The enzymatic activity of the citrus PME fraction was not affected (Savary et al., 2002). However, this modification could be related to the subcellular localisation of AtPME3. The carboxylic 14 kDa fragment, which interacts with the *At5g11420*-encoded protein, was previously detected in the apoplastic fluid of rosette leaves (Boudart el al., 2005), whereas the complete AtPME3 catalytic domain that specifically interacts with 32460 protein was identified in the cell wall proteomes of different plant tissues (Feiz et al., 2006). FLOR1 was also detected in cell wall proteomes from different tissues.

The *in vitro* interactions of AtPME3 with the tested DUF642 proteins appear to be specific because no other PME was isolated with the affinity column. In particular, AtPME2 (*At1g53830*) shares a 90% sequence similarity to AtPME3, which is also present in the leaves. This result and the high similarity of the primary and secondary structures of both DUF642 proteins suggest that DUF642 proteins can interact with the same protein but with different isoforms that result from posttranslational modifications (Figures 6 and 7). A specific protein interaction of AtPME3 has been previously described. The cellulose-binding protein (CBP) secreted by the nematode *Heterodera schachtii* and that is involved in the infection process specifically interacts with AtPME3, and no interaction was detected with AtPME2 (Hewezi et al., 2008).

The interaction of PMEs with proteins is highly involved in cell wall remodelling. The interaction of PME with proteins that inhibit its activity contributes to the modulation of the methylesterified state of the pectin in the cell wall during different developmental processes (Pelloux et al., 2007). An important role of pectin modifications in the regulation of cell wall mechanics in the apical meristem tissue has been suggested (Peaucelle et al., 2011). In root tips, highly esterified pectins were found in the proliferating zone, and non-esterified pectins were abundant in the cell walls of differentiating cells (Barany et al., 2010). During pollen germination, the pollen tube wall presents highly methylesterified pectins in the tip region and weakly methylesterified pectins along the tube (Dardelle et al., 2010). It has been suggested that a local relaxation of the transmitting tract cell wall resulting from changes in the methylesterification of pectins could facilitate the growth of the pollen tubes in the extracellular matrix of this female tissue (Lehner et al., 2010).

4. Conclusions

The DUF642 domain contains a carbohydrate-binding module (CBM) that could be involved in cell wall polysaccharides. The presence of these modules has been described in enzymes from bacteria that hydrolyse hemicelluloses and pectins to degrade the plant cell wall (Kellet et al., 1990; Mc Kie et al., 2001). The function of these modules appears to be related to a precise targeting to polymers in specific regions of plant cell walls during developmental processes. Plant cell wall proteins can act as bridging proteins that target specific cell wall regions and crosslink different networks (Hervé et al., 2010). Additionally, 32460 and 11420 proteins interact *in vitro* with a PME and a LRR protein that are closely related to PGIPs. These two DUF642 proteins could be scaffold proteins that promote the complexation of PME and LRR proteins to prevent the targeting of non-esterified pectins by pectin-degrading enzymes such as polygalacturonases.

Our results suggest that FLOR1 and AtPME3 interact with the 11420 and 32460 DUF642 proteins, but the precise biochemical and biological functions remain to be determined.

5. Acknowledgments

This work was supported by PAPIIT grant IN220980 (Universidad Nacional Autónoma de México). EZS is a PhD student (Posgrado en Ciencias Biomédicas, UNAM) and received a CONACyT fellowship.

6. References

Abrahams, S., Hayes, C.M., & Watson, J. M. (1995). Expression pattern of three genes in the stem of Lucerne (*Medicago sativa*). *Plant Molecular Biology,* Vol. 27, No. 3, (February), pp 513-528, ISSN 0167-4412.

Acevedo, G.F., Gamboa, A., Páez-Valencia, J., Izaguirre-Sierra, M., & Alvarez-Buylla, R.E. (2004). FLOR1, a putative interaction partner of the floral homeotic protein AGAMOUS, is a plant-specific intracellular LRR. *Plant Science,* Vol.167, No. 2, (August), pp. 225-231, ISSN 0168-9452.

Albert, V.A., Soltis, D. E., Carlson, J.E., Farmerie, W.G., Kerr Wall, P., Ilut, D.C., Solow, T.M., Mueller, L.A., Landherr, L.L., Hu, Y.I., Buzgo, M., Kim, S., Yoo, M.J., Frohlich, M.W., Perl-Treves, R., Schlarbaum, S,E., Bliss, B.J., Zhang, X., Tanksley, S.D., Oppenheimer, D.G., Soltis, P.S., Ma, H., de Pamphilis, C.W., & Leebans-Mack, J.H. (2005). Floral gene resources from basal angiosperms for comparative genomics research. *BMC Plant Biology,* Vol. 5, No 5, (March), pp, ISSN 1471-2229.

Barcalá, M., García, A., Cabrera, J., Casson, S., Lindsey, K., Favery, B., García-Casado, G., Solano, R., Fenoll, C., & Escobar, C. (2010). Early transcriptomic events in microdissected *Arabidopsis* nematode-induced giant cells. *Plant Journal,* Vol. 61, No. 4, (February), pp. 698-712, ISSN 0960-7412.

Barany, I., Fadón, B., Risueño, M.C., & Testillano, P.S. (2010). Cell wall components and pectin esterification levels as markers of proliferation and differentiation events during pollen development and embryogenesis in *Capsicum annuum* L. *Journal of Experimental Botany,* Vol. 61, No. 4, (February), pp. 1159-1175, ISSN 0022-0957.

Baumberger, N., Doesseger, B., Guyot, R., Diet, A., Parsons, R.L., Clark, M.A., Simmons, M.P., Bedinger, P., Goff, S.A., Ringli, C., & Keller, B. (2003). Whole-genome comparison of leucine-rich repeat extensins in Arabidopsis and rice. A conserved family of cell wall proteins from vegetative and a reproductive clade. *Plant Physiology,* Vol. 131, No. 3, (March), pp. 1313-1326, ISSN 0032-0889.

Bayer, E.M., Bottrill, A.R., Walshaw, J., Vigouroux, M., Naldrett, M.J., Thomas, C.L., & Maule, A. (2006). Arabidopsis cell wall proteome defined using multidimensional protein identification technology. *Proteomics,* Vol. 6, No. 1, (January), pp. 301-311, ISSN 1615-9853.

Becnel, J., Natarajan, M., Kipp, A., & Braam, J. (2006). Developmental expression patterns of Arabidopsis XTH reported by transgenes and Genevestigator. *Plant Molecular Biology,* Vol. 61, No. 3, (June), pp. 451-467, ISSN 0167-4412.

Bischoff, V., Nita, S., Neumetzler, L., Schindelasch, D., Unrbain ,A., Eshed, R., Persson, S., Delmer, D., & Scheible, W.R. (2010a). TRICHOME BIREFRINGENCE and its homolog *At5g01360* encode plant-specific DUF231 proteins required for cellulose biosynthesis in Arabidopsis. *Plant Physiology,* Vol. 153, No. 2, (June), pp. 590-602, ISSN 0032-0889.

Bischoff, V., Selbig, J., & Scheible, W.R. (2010b). Involvement of TBL/DUF231 proteins into cell wall biology. *Plant Signaling & Behavior,* Vol. 5, No. 8, (August), pp. 1057-1059, ISSN 1559-2316.

Borderies, G., Jamet, E., Lafitte, C., Rossignol, M., Jauneau, A., Boudart, G., Monsarrat, B., Esquerré-Tugayé, M.T., Boudet, A., & Pont-Lezica, R. (2003). Proteomics of loosely bound cell wall proteins of *Arabidopsis thaliana* cell suspension cultures: a critical analysis. *Electrophoresis*, Vol. 24, No. 19-20, (October), pp. 3421-3432, ISSN 0173-0835.

Borner, G.H., Lilley, K.S., Stevens, T.J., & Dupree, P. (2003). Identification of glycosilphosphatidylinositol anchored proteins in Arabidopsis. A proteomic and genomic analysis. *Plant Physiology*, Vol. 132, No. 2, (June), pp. 568-577, ISSN 0032-0889.

Boudart, G., Jamet, E., Rossignol, M., Lafitte, C., Borderies, G., Jauneau, A., Esquerré-Tugayé, M.T., & Pont-Lezica, R. (2005). Cell wall proteins in apoplastic fluids of *Arabidopsis thaliana* rosettes: Identification by mass spectometrometry and bioinformatics. *Proteomics*, Vol. 5, No.1 , (January), pp. 212-221, ISSN 1615-9853.

Cassab, G.I. (1998). Plant cell wall proteins. *Annual Reviews of Plant Physiology and Plant Molecular Biology*, Vol. 49, pp. 281-309, ISSN 1040-2519.

Charmont, S., Jamet, E., Pont-Lezica, R., & Canut, H. (2005). Proteomic analysis of secreted proteins from *Arabidopsis thaliana* seedlings: improved recovery following removal of phenolic compounds. *Phytochemistry*, Vol. 66, No. 4, (February), pp. 453-461, ISSN 0031-9422.

Chibani, K., Ali-Rachedi, S., Job, C., Job, D., Jullien, M., & Grappin, P. (2006). Proteomic analysis of seed dormancy in Arabidopsis. *Plant Physiology*, Vol.142, No. 4, (December), pp. 1493-1510, ISSN 0032-0889.

Chivasa, S., Ndimba, B.K., Simon, W.J., Robertson, D., Yu, X.L., Knox, J.P., Bolwell, P., & Slabas, A.R. (2002). Proteomic analysis of *Arabidopsis thaliana* cell wall. *Electrophoresis*, Vol. 23, No. 11, (Junio), pp. 1754-1765, ISSN 0173-0835.

Clark, S.E., Williams, R.W., & Meyerowitz, E.M. (1997). The *CLAVATA1* gene encodes a putative receptor kinase that controls shoot and meristem size in Arabidopsis. *Cell*, Vol. 89, No. 4, (May), pp. 575-585, ISSN 0092-8674.

Dardelle, F., Lehner, A., Ramdani, Y., Basrdor, M., Lerouge, P., Driouich, A., & Mollet, J.C. (2010). Biochemical and immunocytological characterizations of Arabidopsis pollen tube cell wall. *Plant Physiology*, Vol. 153, No. 4, (August), pp. 1563-1576, ISSN 0032-0889.

Depuydt, S., Trenkamp, S., Fernie, A.R., Elftieh, S., Renou, J.P., Vuylsteke, M., Holsters, M., & Vereecke, D. (2009) An integrated genomics approach to define niche establishment by *Rhodococcus fascians*. *Plant Physiology*, Vol. 149, No. 3, (March), pp. 1366-1386, ISSN 0032-0889.

Dhonukshe, P., Baluska, F., Schlicht, M., Hlavacka, A., Samaj, J., Friml, J., & Gadella, T.W.J (2006). Endocytosis of cell surface material mediates cell plate formation during plant cytokinesis. *Developmental Cell*, Vol. 10, No. 1 (January), pp. 137-150, ISSN 1534-5807.

Di, C., Zhang, M., Xu, S., Cheng, T., & An, L. (2006). Role of poly-galacturonase inhibiting protein in plant defense. *Critical Reviews of Microbiology*, Vol. 32, No. 2, (January), pp 91-100, ISSN 1040-841X.

Dunkley, T.P., Hester, S., Shadforth, I.P., Runions, J., Weimar, T., Hamton, S.L., Griffin, J.L., Bessant, C., Brandizzi, F., Hawes, C., Watson, R.B., Dupree, P., & Lilley, K.S. (2006). Mapping the Arabidopsis organelle proteome. *Proceedings of the National Academy of Science*, Vol. 103, No. 11, (April), pp. 1128-1134, ISSN 1091-6490.

Ehlting, J., Chowrira, S.G., Matthews, N., Eschliman, D.S., Arimura, G., & Bohlmann, J. (2008). Comparative transcriptome analysis of *Arabidopsis thaliana* infested by diamond back moth (*Plutella xylostella*) larvae reveals signatures of stress response, secondary metabolism, and signalling. *BMC Genetics*, Vol. 9, No. 9 (April), pp. 154-160, ISSN 1471-2156.

Feiz, L., Irshad, M., Pont-Lezica, R.F., Canut, H., & Jamet, E. (2006). Evaluation of cell wall preparations for proteomics: A new procedure for purifying cell walls from Arabidopsis hypocotyls. *Plant Methods*, Vol. 2, No. 10, (May), pp. 13, ISSN 1746-4811.

Feraru, E., Feraru, M.I., Kleine-Vehm, J., Martiniere, A., Mouille, J., Vanneste, S., Vernhettes, S., Runions, J., & Friml, J. (2011). PIN polarity maintenance by the cell wall in Arabidopsis. *Current Biology*, Vol. 21, No. 4, (February), pp. 338-343, ISSN 0960-9822.

Gaboriaud, C., Bissery, V., Benchetrit, T., & Mornon, J.P. (1987) Hydrophobic cluster analysis: an efficient new way to compare and analyse amino acid sequences. *FEBS Letters*, Vol. 224, No. 1, (November), pp. 149-155, ISSN 0014-5793.

Gamboa, A., Páez-Valencia, J., Acevedo, G.F., Vázquez-Moreno, L., & Alvarez-Buylla, R.E. (2001). Floral transcription factor AGAMOUS interacts in vitro with a leucine-rich repeat and an acid phosphatase complex. *Biochemical and Biophysics Research Communications*, Vol. 288, No. 4, (November), pp. 1018-1026, ISSN 0006-291X.

Goda, H., Sawa, S., Asami, T., Fujioka, S., Shimada, Y., & Yoshida, S. (2004). Comprehensive comparison of auxin-regulated and brassinosteroid regulated genes in Arabidopsis. *Plant Physiology*, Vol. 134, No. 4, (April), pp. 1555-1573, ISSN 0032-0889.

González-Carranza, Z.H., Elliot, K.A., & Roberts, J.A. (2007). Expression of polygalacturonases and evidence to support its role Turing cell separation processes in *Arabidopsis thaliana*. *Journal of Experimental Botany*, Vol 58, No. 13, (October), pp. 3719-3730, ISSN 0032-0889.

Grunewald, W., van Noorden, G., Van Isterdaal, G., Beeckman, T., Gheysen, G., & Mathesius, U. (2009). A manipulation of auxin transport in plant roots during Rhizobium simbiosis and nematode parasitism. *Plant Cell* Vol. 21, No. 9, (September), pp. 2553-2562, ISSN 1040-4651.

Guenin, S., Mareck, A., Rayon, C., Lamour, R., Assoumou, Ndong, Y., Domon, J.M., Senechal, F., Fourner, F., Jamet, E., Canut, H., Percoco, G., Moville, G., Rolland, A., Rusterucci, C., Guerineau, F., Van Wuytswinkel, O., Gillet, F., Driovich, A., Lerouge, P., Gutierrez, L, & Pelloux, J. (2011). Identification of pectin methylesterase 3 as a basic pectin methylesterase isoform involved in adventitious rooting in *Arabidopsis thaliana*. *New Phytologist*, Vol. 192, No. 1, (October), pp. 114-126, ISSN 0028-646X.

Harpster, M.H., Brummell, D.A., & Dunsmuir, P. (2002). Suppression of a ripening-related endo-1,4-β-glucanase in transgenic pepper fruit does not prevent depolymerisation of cell wall polysaccharides during ripening. *Plant Molecular Biology*, Vol. 50, No. 3, (October), pp. 345-355, ISSN 0167-4412.

Havlis, J., Thomas, H., Sebela, M., & Shevchenko, A. (2003). Fast-response proteomics by accelerated in-gel digestion of proteins. *Analytical Chemistry*, Vol. 75, No. 6, (March), pp. 1300-1306, ISSN 0003-2700.

Held, M.A., Be, E., Zemelis, S., Withers, S., Wilkerson, C., & Brandizzi, F. (2011). CRG3: A golgi-localized protein influencing homogalacturonan methylesterification. *Molecular Plant*, Vol. 4, No. 5, (September), pp. 832-844, ISSN 1674-2052.

Hervé, C., Rogowski, A., Blake, A.W., Marcus, S.E., Gilbert, H.J., & Knox, J.P. (2010). Carbohydrate-binding modules promote the enzymatic deconstruction of intact plant cell walls by targeting and proximity effects. *Proceedings of the National Academy of Science*, Vol. 107, No. 34, (August), pp 15293-15298, ISSN 1091-6490.

Hewezi, T., Howe, P., Maier, T,R., Hussey, R.S., Mitchum, M.G., Davis, E.L., & Baum, T.J. (2009). Cellulose binding protein from the parsitic nematode *Heterodera schachtii* interacts with Arabidopsis pectin methyl-esterase: cooperative cell-wall modification during parasitism. *Plant Cell*, Vol. 20, No. 11, (November), pp. 3080-3093. ISSN 1040-4651.

Hoffman, R.M., and Turner, J.G. (1984), Occurrence and specificity of an endopolygalacturonase inhibitor in *Pisum sativum*. *Physiol. Plant Pathol.* Vol. 24, No. 1, (January), pp. 49-59, ISSN 0048-4059.

Hu, J., Barlet, X., Deslandes, L., Hirsch, J., Feng, X., Somssich, I., & Marco, Y. (2008). Transcriptional responses of *Arabidopsis thaliana* during wilt disease caused by soilborne phytopathogenic bacterium *Ralstonia solanacearum*. *PLoS ONE*, Vol. 3, No. 7, (July), e2589, ISSN 1932-6203.

Irshad, M., Canut, H., Borderies, G., Pont-Lezica, R., & Jamet, E. (2008). A new picture of cell wall protein dynamics in elongating cells of *Arabidopsis thaliana*; confirmed actors and new comers. *BMC Plant Biology*, Vol. 8, (September), pp. 94-103.

Jamet, E., Canut, H., Boudart, G., & Pont-Lezica, R.F. (2006). Cell wall proteins: a new insight through proteomics. *Trends in Plant Sciences*, Vol. 11, No. 1, (January), pp. 33-39. ISSN 1360-1385.

Juge, N (2006). Plant proteins inhibitors of cell wall degrading enzymes. *Trends in Plant Sciences*. Vol. 11, No. 7, (July), pp. 359-367. ISSN 1360-1385.

Kajava, A.V. (1998). Structural diversity of Leucine-rich repeat proteins. *Journal of Molecular Biology*, Vol. 277, No. 3, (April), pp. 519-527, ISSN 0022-2836.

Keller, A., Nesvizhskii, A.I., Kolker, E., & Aebersold, R. (2002). Empirical statistical model to estimate the accuracy of peptide identifications made by MS/MS and database search. *Analytical Chemistry*, Vol. 74, No. 20, (October), pp. 5383-5392, ISSN 0003-2700.

Kellet, L.E., Poole, D.M., Ferreira, L.M., Durrant, A.J., Hazlewood, G.P., & Gilbert, H.J. (1990). Xylanase B and arabinofuranosidase from *Pseudomonas fluorescens* subsp. cellulosa contain identical cellulose-binding domains and are encoded by adjacent genes. *Biochemical Journal*, Vol. 272, No. 2, (December), pp. 369-376. ISSN 0244-6021.

Kreps, J.A., Wu, Y., Chang, H-S., Zhu, T., Wang, X., & Harper, J.F. (2002). Transcriptome changes for Arabidopsis in response to salt, osmotic, and cold stress. *Plant Physiology*, Vol. 130, No. 4, (December), pp. 2129-2141, ISSN 0032-0889.

Lannoo, N., Vanderborre, G., Miersch, O., Smagghe, G., Wasternack, C., Peumans, W.J., & Van Damme, E.J. (2007). The jasmonate-induced exporession of the *Nicotiana tabacum* in leaf lectins. *Plant and Cell Physiology*, Vol. 48, No. 8, (August), pp. 1207-1218, ISSN 0032-0781.

Lee, C.W., Efetova, M., Engelmann, J.C., Kramell, R., Wastermack, C., Ludwig-Müller, J., Hedrich, R., & Deeken, R. (2009). *Agrobacterium tumefaciens* promotes tumor induction by modulating pathogen defense in *Arabidopsis thaliana*. *Plant Cell*, Vol. 21, No. 9, (September), pp. 2948-2962, ISSN 1040-4651.

Lehner, A., Dardelle, F., Soret-Morvan, O., Lerouge, P., Driouich, A., & Mollet, J.C. (2010). Pectins in the cell wall of *Arabidopsis thaliana* pollen tube and pistil. *Plant Signaling & Behavior,* Vol. 5, No. 10, (October), pp. 1282-1285.

Mc Kie , V.A., Vincken, J.O., Voragen, A.G., van der Broek, L.A., Stimson, E., & Gilbert H.J. (2001). A new family of rhamnogalacturonan lyases contains an enzyme that binds to cellulose. *Biochemical Journal,* Vol. 355, No. 1, (April), pp. 167-177. ISSN 0244-6021.

Membré, N., Bernier, F., Staiger, D., & Berna, A. (2000). *Arabidopsis thaliana* germin-like proteins: common and specific features point to a variety of functions. *Planta,* Vol. 211, No. 3, (August), pp.345-354, ISSN 0032-0935.

Micheli, F. (2001). Pectin methylesterases : cell wall enzymes with important roles in plant physiology. *Trends in Plant Science,* Vol. 6, No. 9, (September), pp. 414-418. ISSN 1360-1385.

Minic, Z., Jamet, E., Négroni, L., der Garabedian, A., Zivy, M., & Jouanin, L. (2007). A sub-proteome of *Arabidopsis thaliana* mature stems trapped on Concanavalin A is enriched in cell wall glycoside hydrolases. *Journal of Experimental Botany,* Vol. 58, No. 10, (May), pp. 2503-2512, ISSN 0022-0957.

Passardi, F., Penel, C., & Dunand, C. (2004). Performing the paradoxical: how plant peroxidises modify the cell wall. *Trends in Plant Science,* Vol. 9, No. 11, (November), pp. 534-540, ISSN 1360-1385.

Passardi, F., Tognolli, M., De Meyer, M., Penel, C., Dunand C., (2006). Two cell wall associated peroxidases from Arabidopsis influence root elongation. *Planta,* Vol. 223, No. 5, (April), pp. 965-974, ISSN 0032-0935.

Peaucelle, A., Braybrook, S.A., Le Guillou, L., Bron, E., Kuhlemeier, C., & Höfte, H. (2011). Pectin induced changes in cell wall mechanics underlie organ initiation in Arabidopsis. *Current Biology,* Vol. 21, No. 20, (October), 1720-1726, ISSN 0960-9822.

Pelloux, J., Rustérucci, C., & Mellerowicz, E.J. (2007). New insights into pectin methylesterase structure and function. *Trends in Plant Science,* Vol. 12, No. 6, (June), pp. 267-277, ISSN 1360-1385.

Reinhardt, D., Wittwer, F., Mandel, T., & Kuhlemeier, C. (1998). Localized upregulation of a new expansin gene predict the site of leaf formation in the tomato meristem. *Plant Cell,* Vol. 10, No. 9, (September), pp. 1427-143,. ISSN 1040-4651.

Röckel, N., Wolf, S., Kost, B., Rausch, T., & Greiner, S. (2008). Elaborate spatial patterning of cell wall PME and PMEI at the pollen tube tip involves PMEI endocytosis and reflects the distribution of esterified and de-esterified pectin. *Plant Journal,* Vol. 53, No. 1, (January), pp. 133-143, ISSN 0960-7412.

Ruppert, M., Woll, J., Giritch, A., Genady, E., Ma, X., & Stockigt, J. (2005). Functional expression of an ajmaline pathway-specific esterase from Rauvolfia in a plant-virus expression system. *Planta,* Vol. 222, No. 5, (November), pp. 888-898, ISSN 0032-0935.

Sampedro, J., & Cosgrave, D.J. (2005). The expansin superfamily. *Genome Biology,* Vol. 6, No. 12, (November), pp.242-250, ISSN 1474-7596.

Savary, B.J., Hotchkiss, A.T., & Cameron, R.G. (2002). Characterization of a salt independent pectin methylesterase purified from Valencia orange peel. *Journal of Agriculture and Food Chemistry,* Vol. 50, No. 12, (June), pp. 3553-3558, ISSN 0021-8561.

Shevchenko, M., Wilm, O., Vorm, O., & Mann, M. (1996). Mass spectrometric sequencing of proteins from silver-stained polyacrylamide gels. *Analytical Chemistry*, Vol. 68, No. 5 (March), pp. 850-858, ISSN 0003-2700.

Sterling, J.D., Quigley, H.F., Oerallana, A., & Mohnen, D. (2001). The catalytic site of the pectin biosynthetic enzyme alpha-1,4-galacturonosyltransferases is located in the lumen of the Golgi. *Plant Physiology*, Vol. 127, No. 1, (September), pp. 360-371, ISSN 0032-0889.

Thomas, C.L., Bayer, E.M., Ritzenthaler, C., Fernandez-Calvino, L., & Maule, A.J. (2008). Specific targeting of a plasmodesmal protein affecting cell-to-cell communication. *PLoS Biology*, Vol. 6, No.1, (January), e7, ISSN 1932-6203.

Tung, C.W., Dwyer, K.G., Nasrallah, M.E., & Nasrallah, J.B. (2005). Genome-wide identification of genes expressed in Arabidopsis pistils specifically along the path of pollen tube growth. *Plant Physiology*, Vol. 138, No. 2, (June), pp. 977–998, ISSN 0032-0889.

Van Damme, E.J., Lanoo, N., Fouquaert, E., & Peumans, W.J. (2004). The identification of inducible cytoplasmic/nuclear carbohydrate-binding proteins urges to develop novel concept about the role of plant lectins. *Glycoconjugate Journal*, Vol. 20, No. 7-8, pp. 449-460. ISSN 0282-0080.

van der Hoorn, R.A.L. (2008). Plant proteases: from phenotypes to molecular mechanisms. *Annual Review of Plant Biology*, Vol. 59, No. 191, (June), pp. 191-223, ISSN 1543-5008.

Wellmer, F., Riechemann, J.L., Alves-Ferreira, M., & Meyerowitz, E.M. (2004). Genome wide analysis of spatial gene expression in Arabidopsis flowers. *Plant Cell*, Vol. 16, No. 5 (May), pp. 1314- 1326, ISSN 1040-4651.

Wolf, S., Mouille, G., & Pelloux, J. (2009). Homogalacturonan methyl-esterification and plant development. *Molecular Plant* Vol. 5, No. 2 (September), pp. 851-860, ISSN 1674-2052.

Wraczeck, M., Brosché, M., Salojärvi, J., Kangasjärvi, S., Idänheimo, N., Mersmann, S., Robatzek, S., Karpinski, S., Karspinska, B., & Kangasjärvi, J. (2010). Transcriptional regulation of CRK/DUF26 group of receptor-like protein kinases by ozone and plant hormones in Arabidopsis. *BMC Plant Biology*, Vol. 10, No. 95, (May), pp 95-101, ISSN 1471-2229-

Wu, H.C., Hsu, S.F., Luo, D.L., Chen, S.J., Huang, W.D., Lur, H.S., & Jinn, T.L. (2010). Recovery of heat shock-triggered released apoplastic Ca2+ accompanied by pectin methylesterase activity is required for thermotolerance in soybean seedlings. *Journal of Experimental Botany*, Vol. 61, No. 10, (June), pp. 2843-2852, ISSN 0022-0957.

8

Protein-Protein Interactions and Disease

Aditya Rao, Gopalakrishnan Bulusu,
Rajgopal Srinivasan and Thomas Joseph
Life Sciences Division, TCS Innovation Labs, Tata Consultancy Services, Hyderabad
India

1. Introduction

Protein-protein interactions (PPI), in which, two or more proteins associate with each other by various means, are key to understanding all biological processes that occur within as well as between cells. In effect, biological processes are essentially interactions between multiple proteins (Zhang et al., 2011) with PPI networks controlling the flow of information both within and between biological processes.

Disruptions in PPI networks have been shown to result in diseases. This includes monogenic diseases such as hemophilia where a particular biochemical pathway is disrupted, as well as more complex diseases such as cancer, which involve several signaling pathways (Sam et al., 2007). Conversely, disruption of a set of PPI can lead to a particular disease or, in the case where the set is shared among several networks, to several diseases. While there is a wealth of protein-disease associations in the published literature that have been incorporated in PPI repositories, the challenge is to link such PPI to human disease (Ideker & Sharan, 2008).

In this chapter we discuss several examples of diseases that are caused by disruptions of PPI networks. Our goal is to illustrate through examples how the role of PPI in disease can be studied using a variety of computational tools and data sources. While we discuss tools and data sources that are of general interest, we also discuss methods for studying specific diseases and methods aimed at large scale analysis of PPI data to identify classes of diseases. In each case we provide specific examples from the literature and a brief discussion of the tools used.

2. PPI and disease example

Let us consider cerebral malaria as an example to understand how an analysis of PPI could be used to elucidate the molecular basis of disease. Here, a wide range of experimental and predicted human–*Plasmodium* (host-parasite), human-human (host-host) and *Plasmodium-Plasmodium* (parasite-parasite) PPI are combined and analyzed in the context of key events and processes of cerebral malaria, a dangerous infectious disease (Rao et al., 2010).

Cerebral malaria is a severe form of malarial infection, characterized by cerebral complications, such as neuronal damage and coma (Moxon et al., 2009). The disease is characterized by processes such as sequestration of infected red blood cells to cerebral capillaries and venules, systemic inflammation, hemostasis dysfunction and neuronal damage (van der Heyde et al., 2006; Wilson et al., 2008). PPI datasets from different sources were first obtained, summarized in Table 1. Since each dataset uses a different nomenclature system for the human and parasite proteins, a crucial step was to normalize all datasets using common gene names. This enabled creation of a unified host-parasite PPI dataset.

No	Source	Reference
1	Davis dataset	Davis et al., 2007
2	Dyer dataset	Dyer et al., 2007
3	Krishnadev dataset	Krishnadev & Srinivasan, 2008
4	Vignali dataset	Vignali et al., 2008
5	Literature PPI data	In-house manual curation

Table 1. Protein-Protein interaction datasets used in the cerebral malaria example.

An automated literature retrieval module was developed using Entrez Programming Utilities (Sayers et al., 2010) to retrieve the list of full-text articles relevant to the malarial parasite. This article set was pruned using the Medical Subject Headings (MeSH) controlled vocabulary for articles relevant to cerebral malaria. The resultant set was augmented by articles retrieved from the Google Scholar database using appropriate disease-specific query terms such as systemic inflammation, hemostasis dysfunction etc. This article corpus had two main uses:

- For extracting biochemical and signaling events of relevance in cerebral malaria.
- Identifying pairs of interacting proteins within the host, within the parasite and between host and parasite.

Gene Ontology (GO) cellular component annotations from PlasmoDB (Aurrecoechea et al., 2009), a comprehensive *Plasmodium* resource, were used to prune the unified PPI dataset using the approach of Mahdavi & Lin (2007). In the case of PPI involving parasite proteins, only those proteins that were annotated to be present on the parasite surface or were reported to be released during the relevant stage of the parasite were considered (Lyon et al., 1986). For the human protein annotations, tissue-specific annotations from UniProt (Hubbard et al., 2009) were used in the pruning process.

The resultant PPI subset was then analyzed by mapping the PPI to key events that influence the processes of the disease, as identified from the key review articles. The analysis showed the potential significance of apolipoproteins and heat-shock proteins on efficient *Plasmodium falciparum* erythrocyte membrane protein 1 (PfEMP1) presentation, role of the merozoite surface protein (MSP-1) in platelet activation, the role of albumin in astrocyte dysfunction and the effect of parasite proteins in transforming growth factor (TGF)-β regulation. The linking of these PPI to molecular events associated with the disease pathogenesis provides a basis for further experiments to determine the molecular basis of this fatal disease.

3. Tools of the trade

From the example, it is clear that the underpinnings for mapping PPI to disease are: (a) access to various repositories of PPI and (b) ability to filter these PPI in the context of disease and (c) using different tools for visualizing and analyzing the PPI in the context of diseases. Let us consider each of these in detail.

3.1 PPI repositories

There are a host of repositories that house experimental and predicted PPI data. The cerebral malaria example above considered malaria-specific PPI datasets. However, generic datasets such as BIND, DIP, HPRD, MINT, MIPS and STRING usually have the necessary PPI coverage required for a variety of disease studies.

The Biomolecular Interaction Network Database (BIND), a constituent database of the Biomolecular Object Network Databank, makes available a comprehensive collection of information for specific molecules such as proteins and small molecules (Bader et al., 2003). BIND has been one of the major sources of curated biomolecular interactions, especially PPI. The Database of Interacting Proteins (DIP) contains experimentally determined PPI (Salwinski et al., 2004). It has been created using both manual curation and computational approaches. The Human Protein Reference Database (HPRD) provides a platform to visually depict and integrate information, which are manually curated, pertaining to domain architecture, post-translational modifications, PPI networks and disease association for each protein of the human proteome (Prasad et al., 2009).

The Molecular INTeraction database (MINT) contains experimentally verified PPI that have been manually curated from the scientific literature (Ceol et al., 2010). The Mammalian Protein-Protein Interaction (MIPS) database is a collection of manually curated high-quality PPI data collected from the scientific literature by expert curators (Pagel et al., 2005). STRING is a database of known and predicted protein interactions (Szklarczyk et al., 2011). The interactions include direct (physical) as well as indirect (functional) associations.

Composite PPI resources are also available that integrate PPI data from some of these databases into a single resource. APID (Agile Protein Interaction DataAnalyzer), for instance, is one such resource that integrates experimentally validated PPI from databases such as BIND, DIP, HPRD and MINT, amongst others (Prieto et al., 2006). Protein Interaction Network Analysis (PINA) platform is another example of a composite PPI resource that integrates interactions from MINT, DIP, HPRD and MIPS, amongst others (Wu et al., 2009).

3.2 Integration and filtering of PPI

Databases and tools such as Reactome, GO, MeSH and the Entrez Programming Utilities are crucial for filtering the large number of PPI to obtain a PPI network relevant to a specific disease.

Reactome is a database of biological pathways from various organisms, especially humans (Matthews et al., 2009). This is manually curated by experts. It contains various entities such as proteins, chemicals, localization data, etc. The information in Reactome is cross-referenced to various standard bioinformatics databases such as Entrez Gene, UniProt, Ensembl, etc. The GO project attempts to standardize the description of gene and gene products across species and databases (The Gene Ontology Consortium, 2000). It consists of three ontologies that describe genes and gene products in relation to biological processes, molecular functions and cellular components.

MeSH (http://www.nlm.nih.gov/mesh/) is a controlled vocabulary thesaurus maintained by the National Library of Medicine. It is made up of terms naming descriptors in a hierarchical structure that permits searching at various levels of specificity. It currently has 16 major tree headings including "Diseases" and "Chemicals and Drugs". MeSH terms are used in various methods and tools to filter articles/abstracts and other data. Online Mendelian Inheritance in Man (OMIM) is a database of known Mendelian disorders and their related genes (Hamosh et al., 2002). Currently, there are around 12,000 genes described in this database. OMIM provides information on genotype-phenotype relationships in human Mendelian diseases.

Specific tools are available to access some of these databases, such as the Entrez Programming Utilities (Sayers et al., 2010). These are a set of server-side programs enabling a stable interface to utilize the Entrez query and database system at the National Center for Biotechnology Information. There are currently 38 databases in the Entrez system with a wide variety of information on nucleotide and protein structure and sequences, 3D-molecular structures, disease information and biomedical literature etc.

3.3 Visualization tools

Important tools that could aid in mapping PPI to disease include:

- Cytoscape (Shannon et al., 2003) for visualizing PPI datasets with nodes representing biological entities and edges representing the relationships between these entities
- Cell Circuits (Mak et al., 2007) for comparison of hand-curated pathway models to hypothetical models derived from large-scale 'omic' data.

Cytoscape plugins such as APID2NET (Hernandez-Toro et al., 2007) and PRINCIPLE (Gottlieb et al., 2011) are also very pertinent. APID2NET retrieves PPI data from the APID server for further analysis within the Cytoscape environment. PRINCIPLE, discussed later in this chapter, is built specifically for exploring PPI-disease associations. Given any disease as a query term, it provides a list of top-ranking genes associated with this disease and a Cytoscape visualization of the sub-networks formed by these genes and their direct interacting neighbors.

IPA (Ingenuity Systems, www.ingenuity.com) is an example of a commercially available platform that enables visualization of dynamically constructed pathway and network models.

4. PPI from literature

What happens when the PPI repositories do not have adequate coverage of the organism or specific protein-set under study? One possibility is that although such repositories do not have these PPI, the PPI have actually been reported in literature. One just needs to go look for them!

Let us consider an example of using text-mining to extract such PPI. In the cerebral malaria study, a basic text-mining approach has been used (Rao et. al., 2010). The article corpus was first checked for article-level co-occurrence of pairs of proteins. Full-text articles, wherever available, are automatically downloaded from the respective journal websites as Portable Document Format (PDF) files and converted to text format using the XPDF conversion utility (The FooLabs, http://www.foolabs.com/xpdf). All parasite and host proteins that occur in the full-text of each article were identified using a dictionary lookup approach, with PlasmoDB and UniProt/Ensembl being used to create the parasite and human protein dictionaries respectively. Only those articles that had at least one protein pair (host-parasite, host-host or parasite-parasite) were considered for further analysis.

Özgür et al. (2008) propose a more detailed approach based on integrating automatic text mining and network analysis methods to extract known disease genes and to predict unknown disease genes. They started by collecting an initial set of seed genes known to be related to a disease from curated databases such as OMIM. A disease specific gene network was created using advanced natural language processing techniques that capture both gene names as well as the semantic associations between them.

5. PPI Networks and SNPs

Genome-wide measurement technologies such as microarrays have provided an opportunity to identify genes that are mutated or differentially expressed. In particular, SNP-arrays have been very useful in such studies and have resulted in identification of several genes that are associated with disease-risk or poor prognosis (Karinen et al., 2011). Such genes typically affect cellular functions by altering signaling in regulatory PPI networks.

The mainstay of this approach is the fact that genes related to the same disease are also known to have protein products that physically interact (Navlakha & Kingsford, 2010). However, that by itself is only one crucial component. The other important component is that a genetic disease is associated with a linkage interval on the chromosome if SNPs in the interval are correlated with an increased susceptibility to the disease. These linkage intervals define a potential disease-causing gene set. The computational approach boils down to using both these sources of information—PPI networks and linkage intervals to predict relationships between genes and diseases.

Let us look at a method called CANGES to identify the genetic basis of disease (Karinen et al., 2011). The strength of the method lies in its ability to cohesively integrate many different pieces of information to arrive at testable hypotheses. Genome wide association studies have identified many variations that are possibly linked to one or more diseases. How does one go about prioritizing these variations to get to a set of genes that cause the disease? Clearly, one needs to bring in other known information to help arrive at a decision. The CANGES method combines pathway data, PPI data and genetic variation data with analytical tools to rapidly evaluate the disease causing potential of variations and thus focus attention on one or a few genes. Using this method, a set of 158 SNPs in the p53 gene were identified that plays a central role in cancers. These SNPs are likely to have pathogenic consequences. The same method has also been used, in conjunction with clinical patient data, to identify genes associated with glioblastoma multiforme. It is clear that in the future we will see many more such methods which bring together PPI with several other pieces of information and analytical tools to identify disease genes and gene networks.

Several computational methods can be used to identify causal genes central to gene-disease relationships from large PPI networks. The methods include network neighbors and neighborhood methods, unsupervised graph partitioning and Markov clustering, semi-supervised graph partitioning, random walks, network flow methods and several of their variants (Navlakha & Kingsford, 2010). Navlakha & Kingsford tested these on two large PPI networks: (a) one derived from the Human Protein Reference Database, consisting of 8776 proteins and 35,820 PPI and (b) the other derived from Online Predicted Human Interaction Database containing 9842 proteins and 73130 PPI. Annotations from OMIM were used to associate diseases with genes and linkage intervals. They observed that the performance of most methods showed a significant correlation with neighborhood homophily. Based on this, they suggest that homophily could be used to assess the quality of network-based predictions of disease-protein relationships. They also observed that the individual methods capture different kinds of structure in the network and these unique abilities can be used together in a consensus method to enhance prediction quality.

6. Structural significance of PPI

One important disease class in which a study of PPI could shed light is cancer. Let us look at a study that analyzed cancer proteins in human PPI networks (Kar et al., 2009). This study is important from a methodology perspective as it uses structural properties of the proteins present in the PPI network. Integrating three-dimensional protein structural information into PPI networks revealed important aspects about cancer-related proteins. Analysis of the structural properties of cancer-related interface proteins showed that the interfaces are, on an average, smaller in size, more planar, less tightly packed and more hydrophilic than those of non-cancer proteins. For instance, in a breast cancer network used in the study, there was significant accuracy in discriminating cancer-protein interfaces from the non-cancer interfaces. Thus, there seems to be a clear distinction between the interfaces.

In addition, they observed that cancer-related proteins tend to interact with their partners via multi-interface hubs, which comprise 56% of cancer-related proteins. Cancer protein networks are therefore more enriched in multi-interface proteins. Cancer proteins, in general, are longer and have larger surface areas. Thus, to participate in many PPI at the same time, these tend to be multi-interface hubs, with distinct interfaces interacting with different proteins.

The processes involved in obtaining relevant PPI with regard to a disease are shown in Figure 1.

7. PPI common across diseases

The hitherto discussed examples link diseases with their possible proteomic underpinnings. Research is also underway that focuses on bridging the gap between PPI and their association to different diseases. The goal is to bring out underlying PPI that are common amongst different sets of diseases. Diseases with overlapping clinical phenotypes are caused by mutations in functionally related genes. Since PPI are the strongest manifestation of a functional relationship between disease genes, applying a network model is an effective approach for revealing the associations among diseases (Zhang et al., 2011).

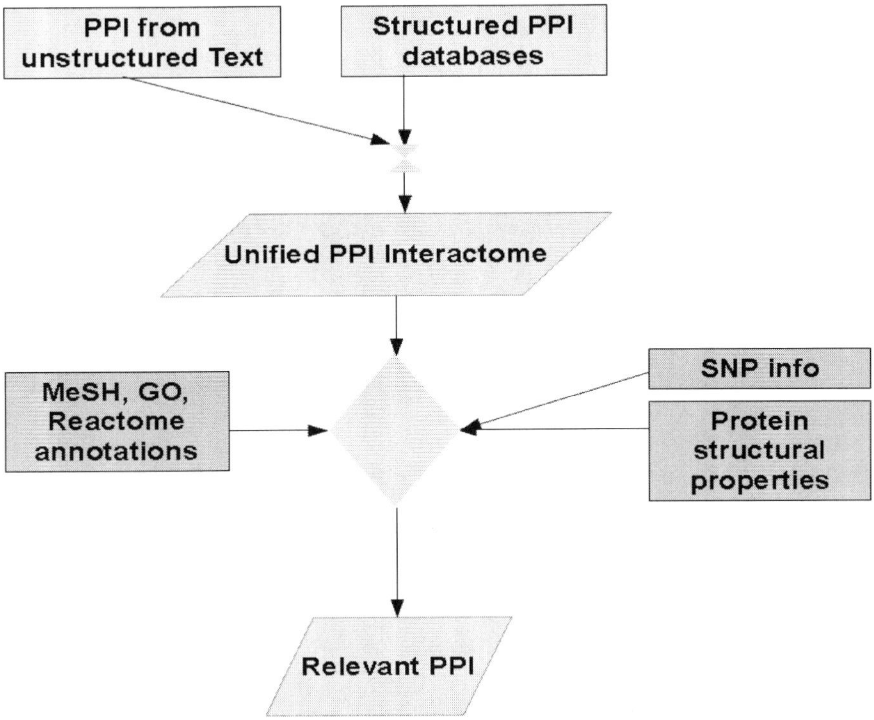

Fig. 1. Processes involved in obtaining relevant PPI with regard to a disease.

7.1 Background

Traditionally, diseases are defined as 'similar' mainly by their clinical appearance, with no correlation to underlying molecular processes. Conceptually, each monogenic disease has a collection of specific phenotypic features. This is true for about 2000 human single gene diseases with a defined genetic phenotype. Syndromes are defined in medicine as a set of phenotypes which, occurring together, serve to define a trait or disease. However, phenotypes very often overlap in the case of many syndromes. Recognition of this overlap brought about the concept of 'syndrome families' taking into account the common features shared between diseases (Sam et al., 2007).

The clustering of syndromes into these families in combination with genetic insights has led to the discovery that what were often thought as two different disorders were really variable expressions of the same disorder. On the other hand, it has long been known that mutations at different loci can lead to the same genetic disease. It has also been hypothesized that this genetic heterogeneity has its roots at the PPI level, suggesting that other genes associated with the phenotype also have some functional role. Therefore, it is plausible that functional properties of shared molecular networks reflect phenotypic overlap of diseases. Thus, PPI networks provide unique opportunities for exploring disease pathways (Sam et al., 2007).

Let us continue with cancer as a disease theme. Sam et al. (2007) highlight an example that links Fanconi's Anemia and cancer. Fanconi's Anemia is a hereditary DNA-repair deficiency disease characterized by defects in a set of DNA repair proteins, leading to, among others, hypersensitivity to DNA damaging agents. This disorder is caused by a mutation in any one of the genes in Fanconi's Anemia complementation group. Symptoms of the disease include anemia, several congenital malformations, etc. Importantly, patients suffering from it exhibit a strong predisposition to different cancers. In the study, this link was substantiated with 14 potential PPI common between Fanconi's Anemia and colorectal neoplasms.

7.2 PPI and common phenotypes

Let us consider another example where a PPI network has been systematically combined with disease-protein relationship data derived from mining GO annotations with phenotypic context (Sam et al., 2007). PPI associated with pairs of diseases were identified and the statistical significance of the occurrence of interactions in the protein interaction knowledgebase calculated. This study demonstrates that the associations between diseases are directly correlated to their underlying PPI networks. A subset of PhenoGO (Lussier et al., 2006; Sam et al., 2009) restricted to human diseases was examined to study the relationships between diseases according to the following criteria. Two basic types of relationships were considered, which determine whether two diseases share PPI networks: a) an identity relationship where common proteins are shared by two diseases, and b) direct interactions between protein A of one disease and protein B of another. A total of 10 pairs of diseases were identified that are significantly correlated due to their shared proteins and PPI. These pairs were analyzed based on mentions in literature, and their correlations were confirmed.

Xeroderma pigmentosum and Cockayne syndrome provide an example of how two diseases are correlated through their PPI networks. Xeroderma pigmentosum is a disorder causing susceptibility of the skin to ultraviolet radiation as a result of deficiencies in one of the XPA-XPG complementation group genes involved in nucleotide excision repair (NER). Cockayne syndrome results from deficiencies in transcription-coupled repair genes, like ERCC6 and ERCC8, leading to a number of conditions including abnormal sensitivity to sunlight (Sam et al., 2007; Spivak et al., 2004). There were 27 direct PPI and 5 common proteins shared amongst these two diseases. Majority of the proteins in the common networks between the two diseases are related to DNA repair processes - Global Genomic NER and Transcription-coupled NER. While the Global Genomic NER repairs lesions from non-transcribed regions of genome independent to transcription, the Transcription-coupled NER repairs UV induced damage in the transcribed strands of active genes. Both the diseases are seen to be associated with these processes, suggesting defects in the DNA damage repair processes are the cause of the diseases.

7.3 Of PRINCE and PRINCIPLEs

PRINCIPLE is very relevant tool specifically built for finding out common diseases based on PPI. It is a Cytoscape plugin implementation of the PRINCE algorithm (Vanunu, et al.,

2010). Given a query disease, it provides a list of top ranking genes associated with it and an additional visualization of the sub-networks formed by these top ranking genes and their direct interacting neighbors. The underlying logic is that genes causing similar diseases often lie close to one another in a PPI network (Oti & Brunner, 2007; Oti, et al., 2006). Given a disease as the query term, PRINCE (a) identifies a set of phenotypically similar diseases, (b) retrieves the known causal genes of these diseases based on their similarity to the query and (c) propagates the scores of the prior set of genes over a human PPI network to provide association scores for all genes. It uses a comprehensive set of weighted PPI compiled from disparate sources (Vanunu, et al., 2010), disease-disease similarity measures (van Driel, et al., 2006), and on the disease-gene associations present in OMIM.

7.4 Human disease network – The holy grail!

Zhang et al. (2011) constructed an expanded Human Disease Network by combining disease-gene information with PPI information. Work such as this is very important, since a network model to represent relationships between diseases is very useful in looking at relationships amongst diseases on a large scale. Analysis of the network's topological features and functional properties showed that the network was hierarchical. Most diseases in the network were connected to only a few diseases, while a small set of diseases were linked to many different diseases. Also, diseases in a specific disease category tended to cluster together, and genes associated with the same disease were functionally related. While this might intuitively sound obvious, it establishes a molecular basis for disease-disease associations.

The limitation of the network is that only known and available disease phenotypic data has been incorporated. However, as more data is made available in databases and in literature, this network provides an ideal template to analyze relationships amongst diseases from a PPI perspective.

8. The road ahead

This is a new field and there are many more approaches than what has been brought out in this chapter. For instance, Bandyopadhyay et al. (2006) use a network analysis of gene expression and PPI data to identify active pathways related to HIV pathogenesis. A functional analysis of the detected sub-networks provides useful insights into various stages of the HIV replication cycle. Chen et al. (2006) developed a framework to mine disease-related proteins from OMIM and PPI data. They demonstrate the power of their method by applying it to Alzheimer's disease. The key to their method is a scoring function that ranks proteins according to their relevance to a particular disease pathway.

Methods to arrive at high-precision predictions that are translatable to effective steps in disease prevention, diagnosis and prognosis should be the goal of PPI studies. The generated leads should be tested experimentally to determine their relevance.

9. References

Aurrecoechea C, Brestelli J, Brunk BP, Dommer J, Fischer S, Gajria B, Gao X, Gingle A, Grant G, Harb OS, Heiges M, Innamorato F, Iodice J, Kissinger JC, Kraemer E, Li W,

Miller JA, Nayak V, Pennington C, Pinney DF, Roos DS, Ross C, Stoeckert CJ Jr, Treatman C & Wang H. (2009). PlasmoDB: a functional genomic database for malaria parasites. *Nucleic Acids Research* D539-543.

Bader GD, Betel D & Hogue CW. (2003). BIND: the Biomolecular Interaction Network Database. *Nucleic Acids Research* 31:248-250.

Bandyopadhyay S, Kelley R & Ideker T. (2006). Discovering regulated networks during IV-1 latency and reactivation. *Pacific Symposium on Biocomputing* 354-366.

Ceol A, ChatrAryamontri A, Licata L, Peluso D, Briganti L, Perfetto L, Castagnoli L & Cesareni G. (2010). MINT, the molecular interaction database: 2009 update. *Nucleic Acids Research* (Database issue):D532-D539.

Chen JY, Shen C & Sivachenko AY. (2006). Mining Alzheimer disease relevant proteins from integrated protein interactome data. *Pacific Symposium on Biocomputing* 367-378.

Davis FP, Barkan DT, Eswar N, McKerrow JH & Sali A. (2007). Host pathogen protein interactions predicted by comparative modeling. *Protein Science* 16:2585-2596.

Dyer MD, Murali TM & Sobral BW. (2007). Computational prediction of host pathogen protein-protein interactions. *Bioinformatics* 23:59-66.

Gottlieb A., Magger O., Berman I., Ruppin E. & Sharan R. (2011). PRINCIPLE: A tool for associating genes with diseases via network propagation. *Bioinformatics.doi: 10.1093/bioinformatics/btr584*.

Hernandez-Toro J, Prieto C & De las Rivas J. (2007). APID2NET: unified interactome graphic analyzer. *Bioinformatics* 23:2495-2497.

Hamosh A, Scott AF, Amberger J, Bocchini C, Valle D & McKusick VA. (2002). Online Mendelian Inheritance in Man (OMIM), a knowledgebase of human genes and genetic disorders. *Nucleic Acids Research* 30:52-55.

Hubbard TJ, Aken BL, Ayling S, Ballester B, Beal K, Bragin E, Brent S, Chen Y, Clapham P, Clarke L, Coates G, Fairley S, Fitzgerald S, Fernandez-Banet J, Gordon L, Graf S, Haider S, Hammond M, Holland R, Howe K, Jenkinson A, Johnson N, Kahari A, Keefe D, Keenan S, Kinsella R, Kokocinski F, Kulesha E, Lawson D, Longden I, Megy K, Meidl P, Overduin B, Parker A, Pritchard B, Rios D, Schuster M, Slater G, Smedley D, Spooner W, Spudich G, Trevanion S, Vilella A, Vogel J, White S, Wilder S, Zadissa A, Birney E, Cunningham F, Curwen V, Durbin R, Fernandez-Suarez XM, Herrero J, Kasprzyk A, Proctor G, Smith J, Searle S & Flicek P. (2009). Ensembl 2009. *Nucleic Acids Research* 37:D690-697.

Ideker T & Sharan R. (2008). Protein networks in disease. *Genome Research* 18: 644-652.

Kann M, Ofran Y, Punta M & Radivojac P. (2006). Protein Interactions and disease. *Pacific Symposium on Biocomputing* 11:351-353.

Kar G, Gursoy A & Keskin O. (2009). Human cancer protein-protein interaction network: a structural perspective. *PLoS Computational Biology* 5:e1000601.

Karinen S, Heikkinen T, Nevanlinna H & Hautaniemi S. (2011). Data integration workflow for search of disease driving genes and genetic variants. *PLoS One* 6:e18636.

Krishnadev O & Srinivasan N. (2008). A data integration approach to predict host- pathogen protein-protein interactions: application to recognize protein interactions between human and a malarial parasite. *In Silico Biology* 8:235-250.

Lussier Y, Borlawsky T, Rappaport D, Liu Y, Friedman C. (2006). PhenoGO: assigning phenotypic context to gene ontology annotations with natural language processing. *Pacific Symposium on Biocomputing* 64-75.

Lyon JA, Haynes JD, Diggs CL, Chulay JD & Pratt-Rossiter JM. (1986). *Plasmodium falciparum* antigens synthesized by schizonts and stabilized at the merozoite surface by antibodies when schizonts mature in the presence of growth inhibitory immune serum. *Journal of Immunology* 136:2252-2258.

Mahdavi MA & Lin YH. (2007). False positive reduction in protein-protein interaction predictions using gene ontology annotations. *BMC Bioinformatics* 8:262.

Mak HC, Daly M, Gruebel B & Ideker T. (2007). CellCircuits: a database of protein network models. *Nucleic Acids Research* 35:D538-545.

Matthews L, Gopinath G, Gillespie M, Caudy M, Croft D, de Bono B, Garapati P, Hemish J, Hermjakob H, Jassal B, Kanapin A, Lewis S, Mahajan S, May B, Schmidt E, Vastrik I, Wu G, Birney E, Stein L & D'Eustachio P. (2009). Reactome knowledgebase of human biological pathways and processes. *Nucleic Acids Research* (Database issue): D619-622.

Moxon CA, Heyderman RS & Wassmer SC. (2009). Dysregulation of coagulation in cerebral malaria. *Molecular and Biochemical Parasitology* 166:99-108.

Navlakha S & Kingsford C. (2010). The power of protein interaction networks for associating genes with diseases. *Bioinformatics* 26:1057-1063.

Oti M & Brunner HG. (2007). The modular nature of genetic diseases. *Clinical Genetics* 71(1):1-11.

Oti M, Snel B, Huynen MA & Brunner HG. (2006). Predicting disease genes using protein-protein interactions. *Journal of Medical Genetics* 43(8):691-698.

Özgür A, Vu T, Erkan G & Radev DR. (2008). Identifying gene-disease associations using centrality on a literature mined gene-interaction network. *Bioinformatics* 24, i277-i1285.

Pagel P, Kovac S, Oesterheld M, Brauner B, Dunger-Kaltenbach I, Frishman G, Montrone C, Mark P, Stümpflen V, Mewes HW, Ruepp A, Frishman D (2005). The MIPS mammalian protein-protein interaction database. *Bioinformatics* 21(6):832-834.

Prasad TSK, Goel R, Kandasamy K, Keerthikumar S, Kumar S, Mathivanan S, Telikicherla D, Raju R, Shafreen B, Venugopal A, Balakrishnan L, Marimuthu A, Banerjee S, Somanathan DS, Sebastian A, Rani S, Ray S, Harrys Kishore CJ, Kanth S, Ahmed M, Kashyap MK, Mohmood R, Ramachandra YL, Krishna V, Rahiman BA, Mohan S, Ranganathan P, Ramabadran S, Chaerkady R & Pandey A. (2009). Human Protein Reference Database - 2009 Update. *Nucleic Acids Research* 37: D767-72.

Prieto C.& De Las Rivas J. (2006). APID: Agile Protein Interaction DataAnalyzer.Nucl. Acids Research 34:W298-W302.

Rao A, Kumar MK, Joseph T & Bulusu G. (2010). Cerebral malaria: Insights from host-parasite protein protein interactions. *Malaria Journal* 9:155.

Salwinski L, Miller CS, Smith AJ, Pettit FK, Bowie JU & Eisenberg D. (2004). TheDatabase of Interacting Proteins: 2004 update. *Nucleic Acids Research* 32:D449-451.

Sam L, Liu Y, Li J, Friedman C & Lussier YA. (2007). Discovery of protein interaction networks shared by diseases. *Pacific Symposium on Biocomputing* 76-87.

Sam LT, Mendonça EA, Li J, Blake J, Friedman C & Lussier YA. (2009). PhenoGO: an integrated resource for the multiscale mining of clinical and biological data. *BMC Bioinformatics* 10:S8.

Sayers EW, Barrett T, Benson DA, Bolton E, Bryant SH, Canese K, Chetvernin V, Church DM, Dicuccio M, Federhen S, Feolo M, Geer LY, Helmberg W, Kapustin Y, Landsman D, Lipman DJ, Lu Z, Madden TL, Madej T, Maglott DR, Marchler-Bauer A, Miller V,

Mizrachi I, Ostell J, Panchenko A, Pruitt KD, Schuler GD, Sequeira E, Sherry ST, Shumway M, Sirotkin K, Slotta D, Souvorov A, Starchenko G, Tatusova TA, Wagner L, Wang Y, John Wilbur W, Yaschenko E & Ye J. (2010). Database resources of the National Center for Biotechnology Information. *Nucleic Acids Research* 38:D5-16.

Shannon P, Markiel A, Ozier O, Baliga NS, Wang JT, Ramage D, Amin N, Schwikowski B & Ideker T. (2003). Cytoscape: a software environment for integrated models of biomolecular interaction networks. *Genome Research* 13:2498-2504.

Spivak G. (2004). The many faces of Cockayne syndrome. *Proceedings of the ational Academy of Sciences of the United States of America* 101(43):15273-15274.

Szklarczyk D, Franceschini A, Kuhn M, Simonovic M, Roth A, Minguez P, Doerks T, Stark M, Muller J, Bork P, Jensen LJ, von Mering C. (2011). The STRING database in 2011: functional interaction networks of proteins, globally integrated and scored. *Nucleic Acids Research* 39(Database issue):D561-568.

The Gene Ontology Consortium. (2000). Gene ontology: tool for the unification of biology. *Nature Genetics* 25(1):25-29.

The UniProt Consortium. (2009). The Universal Protein Resource (UniProt). *Nucleic Acids Research* 37:D169-D174.

van der Heyde HC, Nolan J, Combes V, Gramaglia I & Grau GE. (2006). A unified hypothesis for the genesis of cerebral malaria: sequestration, inflammation and hemostasis leading to microcirculatory dysfunction. *Trends in Parasitology* 22:503–508.

van Driel MA, Bruggeman J, Vriend G, Brunner HG & Leunissen JA. (2006). A text-mining analysis of the human phenome. *European Journal of Human Genetics* 14(5):535-542.

Vanunu, O., Magger, O., Ruppin, E., Shlomi, T. & Sharan R. Associating genes and protein complexes with disease via network propagation. (2010). *PLoS Computational Biology* 6, e1000641.

Vignali M, McKinlay A, LaCount DJ, Chettier R, Bell R, Sahasrabudhe S, Hughes RE & Fields S. (2008). Interaction of an atypical *Plasmodium falciparum* ETRAMP with human apolipoproteins. *Malaria Journal* 7:211.

Wilson NO, Huang MB, Anderson W, Bond V, Powell M, Thompson WE, Armah HB, Adjei AA, Gyasi R, Tettey Y & Stiles JK. (2008). Soluble factors from *Plasmodium falciparum*-infected erythrocytes induce apoptosis in human brain vascular endothelial and neuroglia cells. *Molecular and Biochemical Parasitology* 162:172–176.

Wu J, Vallenius T, Ovaska K, Westermarck J, Mäkelä TP, Hautaniemi S. (2009). Integrated network analysis platform for protein-protein interactions *Nature Methods* 6(1):75-77.

Zhang X, Zhang R, Jiang Y, Sun P, Tang G, Wang X, Lv H & Li X. (2011). The expanded human disease network combining protein-protein interaction information. *European Journal of Human Genetics* 19:783-788.

9

AApeptides as a New Class of Peptidomimetics to Regulate Protein-Protein Interactions

Youhong Niu[1,*], Yaogang Hu[1,*], Rongsheng E. Wang[1,*], Xiaolong Li[2,*], Haifan Wu[1], Jiandong Chen[2,**] and Jianfeng Cai[1,**]
[1]*Department of Chemistry, University of South Florida, Tampa, FL*
[2]*Department of Molecular Oncology, H. Lee Moffitt Cancer Center & Research Institute, Tampa, FL*
USA

1. Introduction

In human physiology, proteins at all times are synthesized, processed, degraded and post-translationally modified to varying degrees and at different rates, for their participations in a wide variety of activities to maintain normal functions of the body (Murray et al. 2007). In the successful proceedings of all the biological events, signals are consistently received and sent via physical contacts between proteins (Murray et al. 2007). The communication between proteins or the alternatively called non-covalent protein-protein interactions are thereby considered as important as proteins' own functions (Murray et al. 2007). Disruptions of these signaling pathways by either mutational changes or deregulation of one of the protein partners would result in a series of diseases (Murray et al. 2007). On the other hand, therapeutic approaches based on chemical agents could potentially inhibit protein-protein interactions, thereby restoring the balance of signaling pathways, and leading to the cure of diseases (Murray et al. 2007). However, it is quite challenging to develop chemical agents which can target protein-protein interactions (Arkin et al. 2004; Whitty et al. 2006). Unlike the traditional medicinal chemistry approach, in which small molecule inhibitors are developed to target the hydrophobic pocket of enzymes / kinases, chemical agents are now required to bind to large surfaces of proteins that are usually amphiphilic and flexible (Murray et al. 2007). Yet a number of successful stories have been reported (Murray et al. 2007). Taking the p53/MDM2 system as an example, the p53/MDM2 has been a model system for the inhibition of protein-protein interactions, and has been reported to be the targets of a wide variety of inhibitors (Oren 1999; Balint et al. 2001; McLure et al. 2004; Brooks et al. 2006).

The tumor suppressor protein p53 is a transcription factor that executes multiple anticancer functions. Through its binding to DNA, p53 can initiate the expression of several important proteins, which are responsible for DNA repair, induction of growth arrest to hold the cell

* These authors contributed to this work equally
** Corresponding author

cycle at the G1/S regulation point, as well as the initiation of apoptosis (Lowe et al. 1993; Pellegata et al. 1996; Liu et al. 2001). However, at the normal state, the p53 activity is down-regulated by the murine double minute 2 protein (MDM2) which binds to the α-helical transactivation domain near the N-terminus of p53 (Momand et al. 1992; Oliner et al. 1992). Cocrystal structure studies revealed that three hydrophobic side chains from Phe19, Trp23, and Leu26 of p53 make direct contacts with MDM2 and account for the primary interactions (Kussie et al. 1996). The binding of MDM2 not only inhibits p53 DNA-binding activity but also induces the proteosomal degradation of p53 (Haupt et al. 1997; Kubbutat et al. 1997) . In the event of stress, p53 protein is phosphorylated, which leads to a much reduced affinity between p53 and MDM2, and thereby reactivating p53 (Jimenez et al. 1999). Nonetheless, MDM2 is constantly over-expressed in tumor cells, which significantly blocks the activation of p53 pathway even during stress conditions, thereby leading to the uncontrolled tumor cell proliferation (Momand et al. 1998). The overproduction of MDM2 makes tumors less susceptible to programmed cell death and apoptosis as a result of chemotherapy and other cancer therapy. Hence, the disruption of MDM2/p53 interaction in tumor cells should stabilize p53, preventing it from degradations, and initiating a cascade of p53 pathways to eventually sensitize the tumor cells to death (Murray et al. 2007). To date, targeting the MDM2/p53 interaction has become an emerging therapeutic approach in anticancer treatment. Numerous efforts have been taken for the development of inhibitors such as natural products (De Vincenzo et al. 1995; Stoll et al. 2001; Duncan et al. 2003; Tsukamoto et al. 2006), small molecules (Zhao et al. 2002; Galatin et al. 2004; Vassilev et al. 2004; Ding et al. 2005; Grasberger et al. 2005; Hardcastle et al. 2006), and oligomers (Alluri et al. 2003; Hara et al. 2006; Robinson 2008; Hayashi et al. 2009; Michel et al. 2009; Bautista et al. 2010). Compared to small molecule approach, oligomers are easily programmable and are readily synthesized by solid phase synthesis. It is also believed that the larger size of oligomers relative to small molecules may bring them additional advantages to contact more protein surface area, which will lead to enhanced binding affinity (Murray et al. 2007).

However, oligomers made of natural peptides are subjected to biodegradations and are also immunogenic in vivo, which limit their practical applications, but on the other hand underscore the need for unnatural peptides (Patch et al. 2002). Peptidomimetics are a class of non-natural peptide mimics using the artificial backbones to mimic peptides' primary and secondary structures (Wu et al. 2008). Compared to traditional peptides, peptidomimetics have great proteolytic and metabolic stability and are believed to be less immunogenic, also with an enhanced bioavailability (Patch et al. 2002; Goodman et al. 2007; Wu et al. 2008). The development of peptidomimetics to disrupt MDM2/p53 has led to a diverse set of oligomers such as β-peptides (Seebach et al. 1996; Cheng et al. 2001; Kritzer et al. 2005), γ- and δ-peptides (Arndt et al. 2004; Trabocchi et al. 2005; Kumbhani et al. 2006), α/β-peptides (Horne et al. 2008; Horne et al. 2009), azapeptides (Graybill et al. 1992; Lee et al. 2002), α-aminoxy-peptides (Li et al. 2008), sugar-based peptides (Risseeuw et al. 2007; Tuwalska et al. 2008), peptoids (Simon et al. 1992), oligoureas (Boeijen et al. 2001; Violette et al. 2005), polyamides (Dervan 1986), and phenylene ethynylenes (Nelson et al. 1997), etc. Nonetheless, the development of peptidomimetics is far less straightforward, with the major limit lying in the availability of framework (Goodman et al. 2007). The search for peptidomimeitcs of a variety of backbones remains crucial in the research of peptide mimics, which would result in different classes of oligomers with diverse structures and functions (Goodman et al. 2007;

Horne et al. 2008; Gellman 2009). The development of new peptidomimetics would also facilitate the identification of novel therapeutic agents and help the understanding of protein folding and functions by using peptidomimetic probes, all of which are important to the progress in modern chemical biology research (Goodman et al. 2007; Horne et al. 2008; Gellman 2009).

2. Development of AApeptides

In the attempt to search for new peptide mimics for drug discovery and protein mimicry, we recently described a novel class of peptidomimetics termed "AApeptides", which is derived from N-acylated-N-aminoethyl amino acids that has been previously used as the building block for PNA (Winssinger et al. 2004; Dragulescu-Andrasi et al. 2006; Debaene et al. 2007). Compared to natural peptides, the repeating unit of the AApeptide is structurally similar to two adjacent residues of α-peptide, in which there are two side chains, one from the regular α-amino acid side chain, while the other one from a carboxylic acid residue appended to the tertiary amide nitrogen. Depending on the relative position of α-amino acid side chain, there are two types of AApeptides. The one with α-amino acid side chain at the α position is called α-AApeptide (Hu et al. 2011), while the other one with side chain at the γ position is called γ-AApeptide (Niu et al. 2011) (Figure 1).

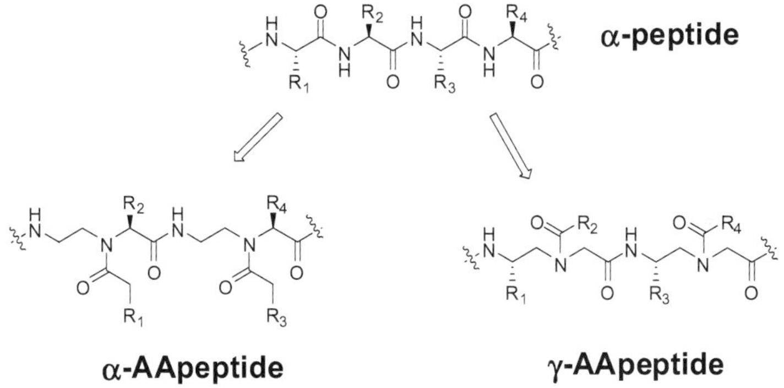

Fig. 1. Structures of a α-peptide and the corresponding AApeptides.

Both type of AApeptides project the same number of functional groups as conventional peptides with backbones of the same length. In addition, all the nitrogen atoms of AApeptides are involved in either secondary or tertiary amide bonds, in a way similar to natural α-peptide. Taken together, such AApeptides are designed to mimic the distance relationships and relative positions of amino acid side chains of natural peptides, so that they can reserve some functions of conventional peptides. It is also noteworthy that, even though AApeptides can mimic the structure as well as some activities of natural peptides, they are still different in backbone and should possess distinct hydrogen bonding properties and conformational flexibilities. The backbone of AApeptide is more flexible, with involved tertiary amide bonds potentially in cis/trans conformations, suggesting that the direct inter-conversion of sequences between AApeptides and natural peptides may not result in the same activity and functions.

3. Development of AApeptides for inhibition of p53/MDM2 interaction

For proof of concept, we demonstrated the facile synthesis and potential bioactivities of AApeptides by developing AApeptide based inhibitors of the p53/MDM2 model system, which has been a testing ground for freshly developed peptidomimetics of novel frameworks.

3.1 Design of AApeptide sequences

Previous reports indicated that synthetic agents displaying hydrophobic side chains of Phe19, Trp23, and Leu26, and in the orientation mimicking the array of these amino acids in p53 should compete with p53 in occupying the MDM2 cleft (Murray et al. 2007). Based on these findings, we designed four α-AApeptides and three γ-AApeptides to mimic the binding surface of p53 (Figure 2 and 3).

α-AA1
α-AA2
α-AA3
α-AA4

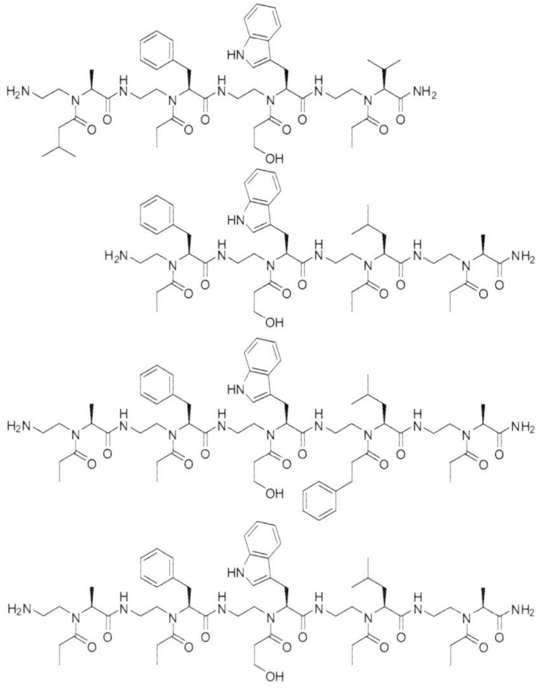

Fig. 2. α-AApeptide sequences designed for inhibition of p53 / MDM2 interaction. Figure is adapted from (Hu et al. 2011).

These AApeptides bear either some or all of the functional side chains of the three amino acids (Phe19, Trp23, and Leu26), which are designed to be the amino acid side chains at either α (for α-AApeptides) or γ positions (for γ-AApeptides). The other functional groups were randomly chosen, with most of them appended to the nitrogens of AApeptides through the formation of tertiary amide bonds.

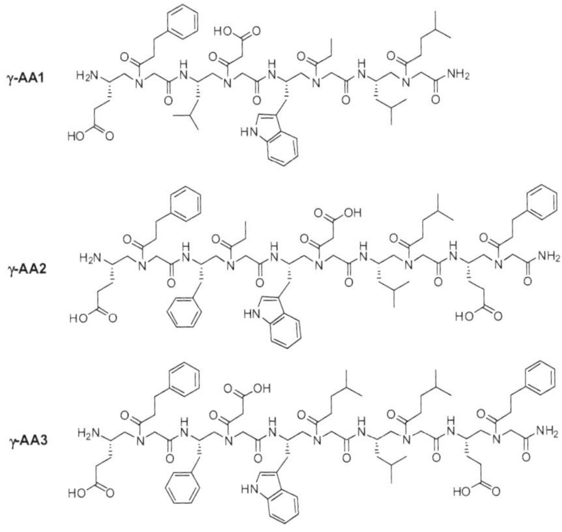

Fig. 3. γ-AApeptide sequences designed for p53/MDM2 disruption. Figure is adapted from (Niu et al. 2011).

3.2 Synthesis of AApeptides

In our initial attempt, we tried to synthesize AApeptides on solid phase resins through a direct sub-monomer strategy, in which the functional groups were introduced to the sequence step by step (Figure 4). Unfortunately, the presence of multiple secondary amines in the peptide backbone led to a constant over-alkylation during the reductive amination step. As a result, we only obtained a mixture of unidentified products after several coupling

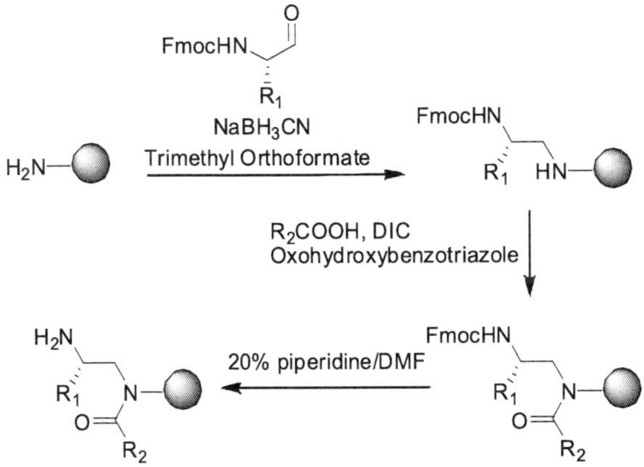

Fig. 4. Initial unsuccessful attempt to synthesize AApeptides on solid phase. Figure is adapted from (Niu et al. 2011).

cycles, as observed on HPLC after the cleavage of products from solid phase. We then carried out an alternative "monomer building block" strategy, in which building block was first synthesized in solution phase, and then assembled following the same procedure of standard solid phase synthesis of conventional peptides. In this route, AApeptide building blocks are readily prepared using commercially available agents at low cost.

For the synthesis of α-AApeptide building block (Figure 5), the carboxylic acid of amino acid was first protected to form the amino acid benzyl ester (**A**) or the amino acid tert-butyl ester (**B**). The resulting amino acid esters were then reacted with Fmoc-amino ethyl aldehyde by reductive amination to form secondary amines **2**, which were subsequently acylated with functional groups R^1. The coupling products **3** were finally deprotected with hydrogenlysis to remove the benzyl protecting group, or with trifluoroacetic acid to remove the tert-butyl protecting group. For the synthesis of γ-AApeptide building block (Figure 6), glycine benzyl ester was reacted with Fmoc-amino aldehydes through reductive amination, and the resulting intermediates **2** was subsequently acylated with carboxylic acid ended functional groups to form the coupled intermediate **3**. After a hydrogenation step, the desired γ-AApeptide building blocks **4** were finally obtained. For both α and γ AApeptides, a diverse set of conjugation conditions for the preparation of intermediates **3** were investigated. It was found that the use of activation agents such as HBTU/HOBt, DIC/HOBt, or PyBOP can provide the desired products in poor yields, and only when intermediate **2** was conjugated with a few types of carboxylic acids. After many trials, the coupling with oxohydroxybezotriazole / DIC emerged as the most efficient and can catalyze the successful conjugations of most carboxylic acids. It is also noteworthy that the derivatization of AApeptides is virtually limitless, since there are countless carboxylic acids available for acylation of nitrogen atom in the backbone. This specific feature allows the rapid generation of AApeptide library, which literally should have much more diversity than those libraries based on regular peptides, thereby expanding the versatility of oligomer libraries for potential applications in high-throughput screening based drug discovery and chemical biology research. With the prepared building blocks in hand, the solid phase synthesis of AApeptides was carried out on resins in a simple and highly efficient way. The sequences were finally obtained over 80% yield in crude and were purified by HPLC to achieve purities over 95%. Their identities were further confirmed by MALDI-mass spectrometry.

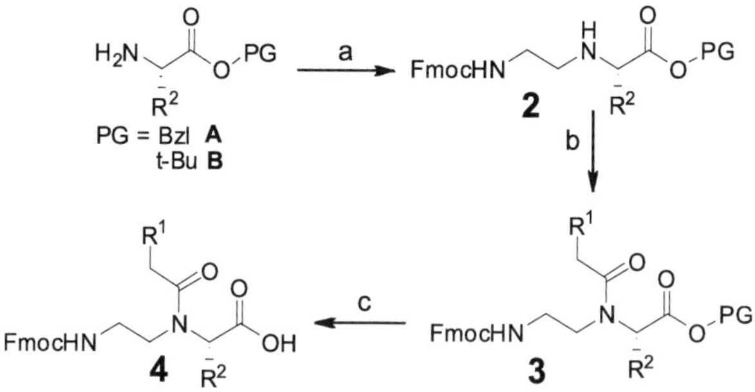

Fig. 5. Synthesis scheme of α-AApeptide building block. a) Fmoc-amino ethyl aldehyde, $NaBH_3CN$, overnight. b) R_1CH_2COOH, DhBtOH/DIC, overnight. c) Pd/C, H_2 for A; 50% TFA/CH_2Cl_2 for B. Figure is adapted from (Hu et al. 2011).

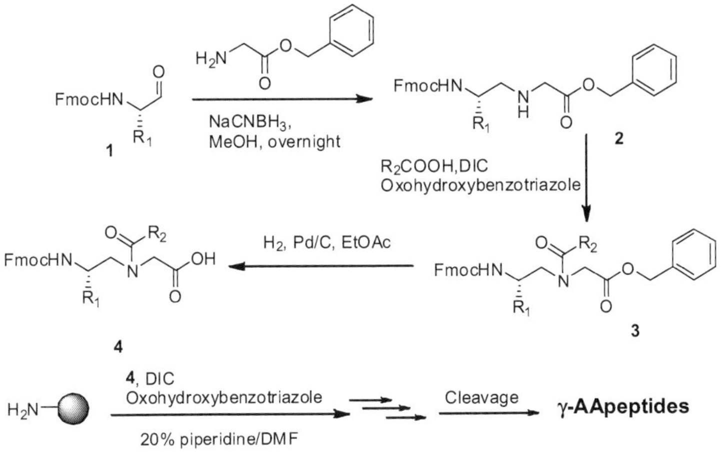

Fig. 6. Synthesis scheme of γ-AApeptide building block. Figure is adapted from (Niu et al. 2011).

3.3 ELISA assay of AApeptides for inhibition of p53/MDM2 interaction

These AApeptides were then tested by the ELISA assay for their inhibition of p53/MDM2 protein-protein interaction. Generally the ELISA plate was coated with p53, and then incubated with the mixture of MDM2 protein and AApeptide for one hour. The p53 bound MDM2 protein was then detected by MDM2 antibody and a secondary antibody conjugated with the horseradish peroxidase, which later on reacted with the TMB peroxidase substrate to manifest a yellowish color after acid quenching. The color intensity was monitored by absorbance at 450nm, the extent of which is directly proportional to the amount of bound MDM2 protein, and also reversely correlated to the inhibiting efficiency of AApeptide. The readings for each sample were then plotted against the concentration of α-AApeptide (Figure 7) or γ-AApeptide (Figure 8). Almost all AApeptides inhibit the p53-MDM2 binding when they are administrated at high concentrations.

Based on the plots of each peptide in Figure 7 and 8, the related IC_{50} values were calculated and summarized in Table 1. The published IC_{50} value of the wide type p53-drived peptide (Garcia-Echeverria et al. 2000) was also included for comparison. For α-AApeptide, **α-AA4** is the most prominent one, with an IC_{50} of 38μM, which is comparable to the previously reported β-peptides and peptoids (Knight et al. 2002; Kritzer et al. 2004; Hara et al. 2006), and is only 4-5 fold less potent than the reported p53-drived wild type peptide (Garcia-Echeverria et al. 2000). Consistent to previous reports (Kussie et al. 1996), the preliminary structure and activity relationship (SAR) study here suggests that the inclusion of functionalities of the three key amino acids "Phe, Typ, and Leu" are important to maintain the strong binding affinity, which are present in all sequences but **α-AA1**. Compared to **α-AA2**, the change of Leu to Val in **α-AA1** decreases the binding affinity to at least 10-fold. Further, that observation of **α-AA4**'s much higher activity than **α-AA2** indicates that better activities are possessed by longer sequences. The longer sequence may possibly have a better stabilized backbone conformation.

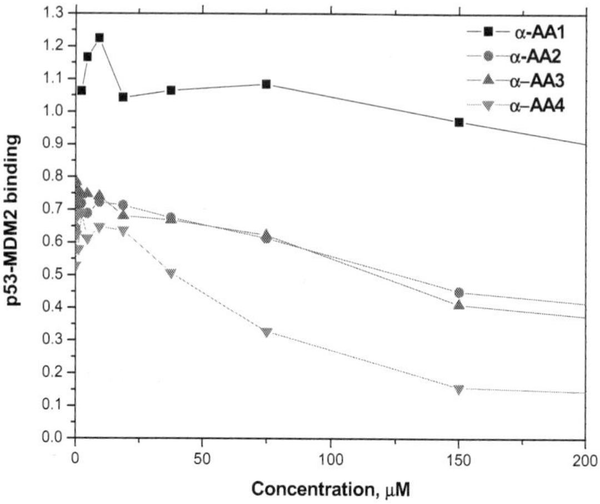

Fig. 7. Plots of ELISA assay for the inhibition of p53-MDM2 interaction by α-AApeptides Figure is adapted from (Hu et al. 2011).

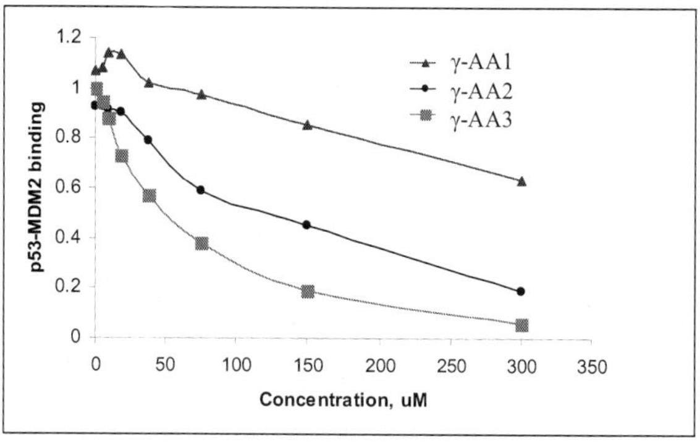

Fig. 8. Plots of ELISA assay for the inhibition of p53-MDM2 interaction by γ-AApeptides. Figure is adapted from (Niu et al. 2011).

AApeptides	IC_{50} (μM)
α-AA1	>1000
α-AA2	120 ± 10
α-AA3	120 ± 16
α-AA4	38 ± 8
γ-AA1	> 400
γ-AA2	120 ± 15
γ-AA3	50 ± 8
p53-derived peptide (Ac-QETFSDLWKLLP)	8.7

Table 1. ELISA results of AApeptides for the disruption of p53/MDM2.

Finally, since **α-AA4** differs from **α-AA3** in only one residue, the side chains not involved in the recognition of MDM2 may also play a substantial role in the binding event. In this case, the Phe side chain of **α-AA3** may be either in the hydrophobic binding cleft, or near the binding domain, clashing with the residues of MDM2 and thereby raising up their binding energy. For γ-AApeptides, **γ-AA3** turns out to be the most effective inhibitor, with an IC_{50} value of 50μM, which is comparable to the most active α-AApeptide and is also only a few fold less active than p53-derived peptide (Garcia-Echeverria et al. 2000). Compared to the others, replacement of Phe by Leu in **γ-AA1** peptide results in a significant loss of activity. Whereas **γ-AA2** and **γ-AA3** both have the required "Phe, Typ, and Leu", the slight difference of their side chain functionalities between Phe and Leu results in more than two fold difference in activity, suggesting that even the side chains are also involved in the binding pocket. This observation is similar to the situation of α-AA peptides.

3.4 Computer modelling for bioactive AApeptides

The ELISA results were further confirmed by preliminary computer modeling studies (Figure 9 and 10), which shows that the side chains of Phe, Trp and Leu in the energy-minimized structures of both **α-AA4** and **γ-AA3** are able to overlap perfectly with those residues in the helical domain of natural peptide p53, indicating that **α-AA4** and **γ-AA3** should be able to mimic the recognition of p53 to MDM2 very well. Compared to **γ-AA3**, **α-AA4** appears to prefer an extended conformation when interacting with MDM2.

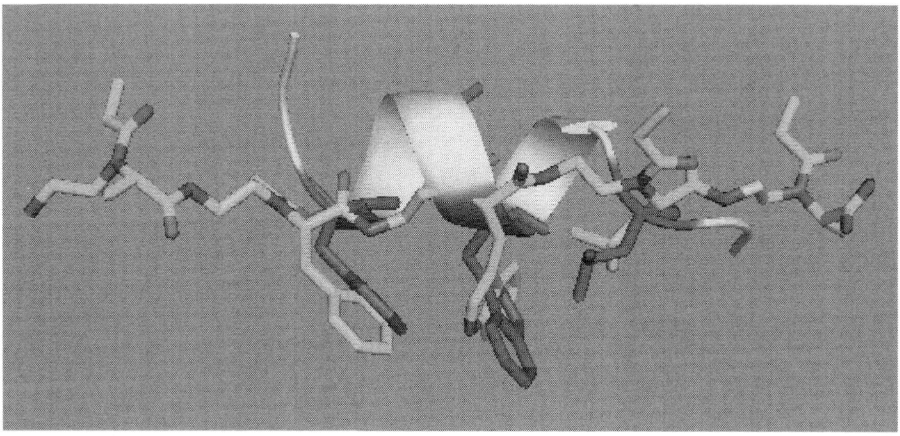

Fig. 9. Energy minimized (MM2) structures of **α-AA4** (green colored) and amino acids 17-29 of p53 helical domain (yellow colored). **α-AA4** is shown as sticks, and the three critical residues (Phe19, Trp23, and Leu 26) in p53 responsible for binding to MDM2 are also presented in sticks and colored in red. Figure is adapted from (Hu et al. 2011).

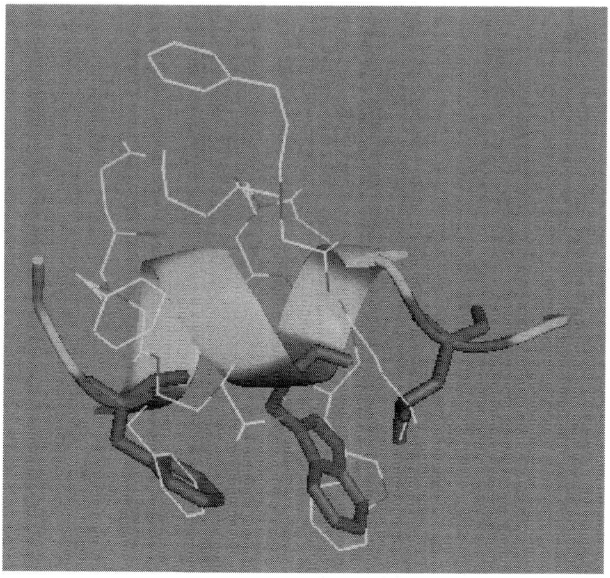

Fig. 10. Energy minimized (MM2) structures of **γ-AA3** (blue colored) is superimposed with the amino acids 17-29 of p53 helical domain (green colored). Three critical residues (Phe19, Trp23, and Leu 26) in p53 responsible for binding to MDM2 are presented in as sticks and colored in red. **γ-AA3** is shown as the wire frame presentation. Figure is adapted from (Niu et al. 2011).

3.5 Summary

Taken together, these results demonstrated AApeptides as a novel class of peptidomimetics. It is also noteworthy that both classes of AApeptides bear excellent selectivity, with different sequences giving different activities, instead of a random interaction with proteins. For example, **α-AA4** is the strongest inhibitor among all the demonstrated α-AApeptides, while **α-AA1** is a poor inhibitor, and **α-AA2**, **α-AA3** are weak inhibitors. Similarly, **γ-AA3** appears to be the best inhibitor, while **γ-AA1** turns out to be the worst.

Detailed structure-activity relationship studies for AApeptides with various lengths and distribution of functional groups along the backbone are currently ongoing, which should provide valuable information for rational design of AApeptide library for drug discovery and chemical biology research. Generally, much more potent AApeptide derivatives are expected with the stabilization of secondary structure, introduction of halogen atoms, and computer modeling-aided design (Hara et al. 2006; Murray et al. 2007; Michel et al. 2009).

4. Stability of AApeptides

One significant advantage of peptidomimetics is the superior resistance to proteolysis, owing to their unnatural backbones. To find out the stability of our AApeptides in this regard, representative sequences of α- and γ- AApeptides (**α-AA3**, **γ-AA3**) were incubated with proteases at the concentration of 0.1 mg/mL in 100 mM pH 7.8 ammonium bicarbonate buffer for 24 hours. **α-AA3** was mixed with chymotrypsin, trypsin, and pronase, respectively; and **γ-AA3** was mixed with chymotrypsin, thermolysin, and pronase, respectively. All the reaction mixtures were then analyzed by HPLC. The retention time and integrations of eluted peaks were compared with those of peaks representative of the starting materials.

As shown in figure 11 and figure 12, whereas conventional peptides are susceptible to proteolysis, especially by chymotrypsin and pronase, both **α-AA3** and **γ-AA3** are highly resistant to enzymatic hydrolysis within 24 hours. There are, however, a small shoulder observed for both types of AApeptides at 37°C with or even without incubation with proteases, which takes up less than 5% of the total volume and is presumably due to the isomerization of the syn/anti tertiary amide bonds in the peptide backbones.

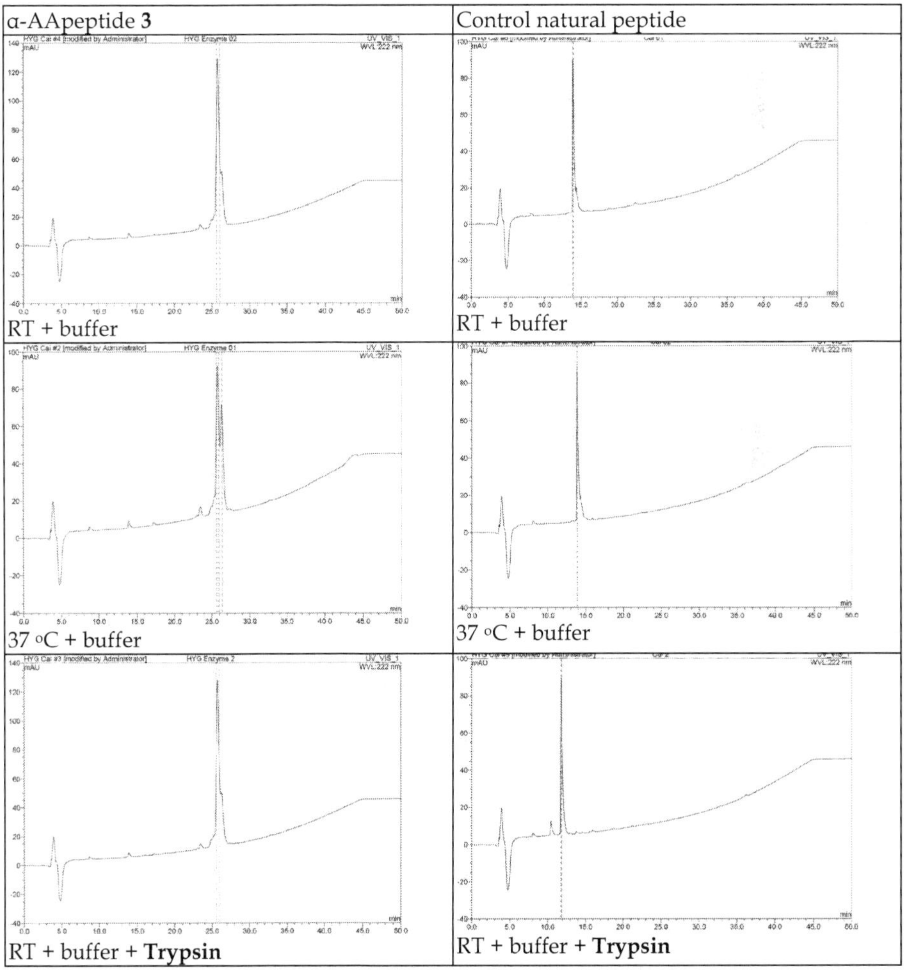

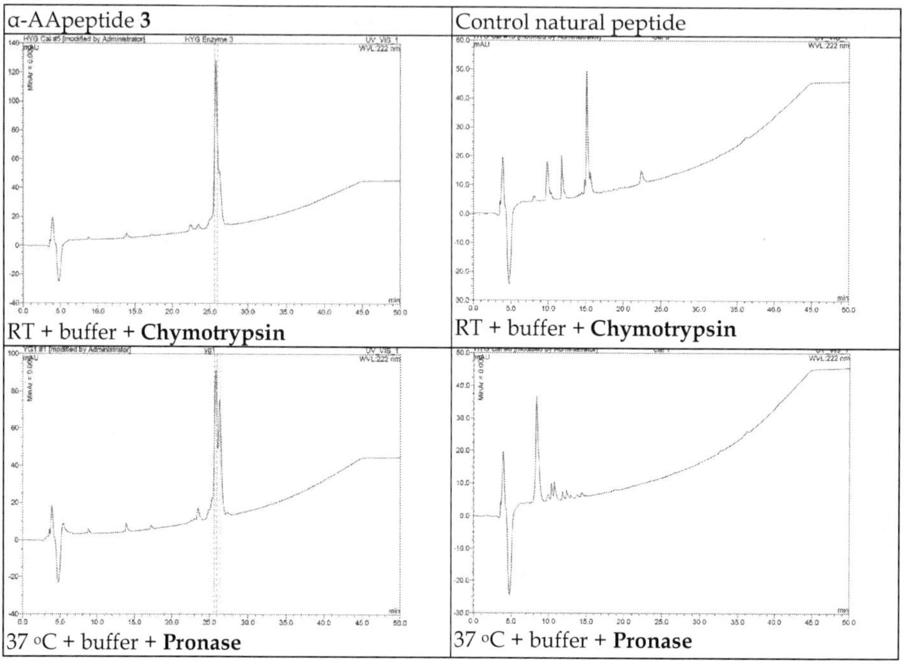

Fig. 11. Analytical HPLC spectra of **α-AA3** and control α-peptide after their incubations with different proteases. Figure is adapted from (Hu et al. 2011).

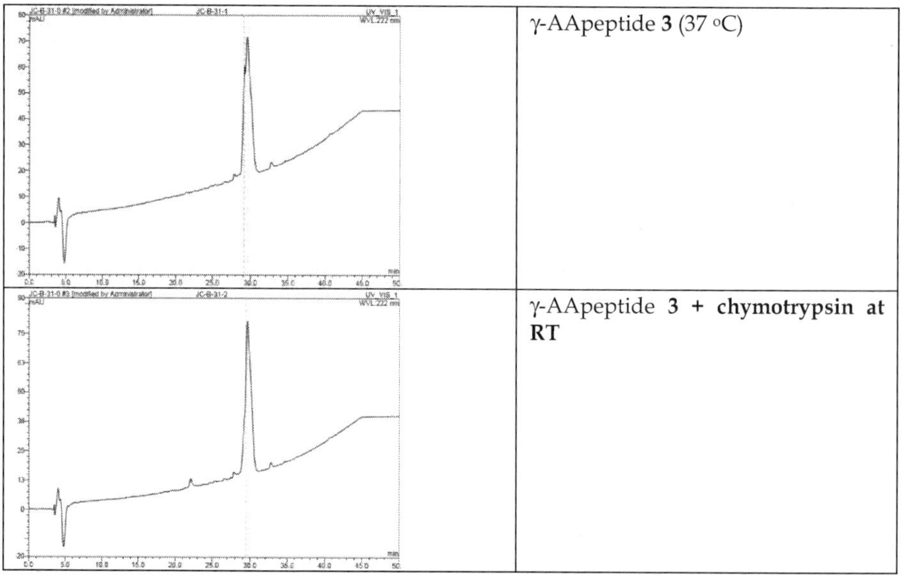

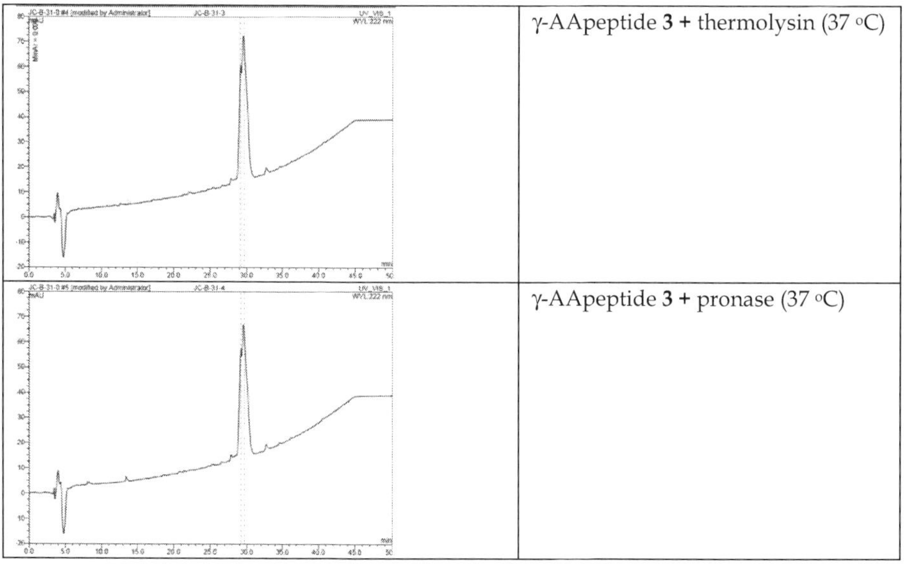

Fig. 12. Incubation of **γ-AA3** with different proteases. Figure is adapted from (Niu et al. 2011).

5. Conclusions

In conclusion, we have developed a new class of peptidomimetics - the AApeptides that can have amino acid side chains at either α-, or γ-position. This family of peptides can be readily synthesized on solid phase by standard monomer building block approach, using α or γ-substituted N-acylated-N-Fmoc-amino ethyl amino acid building blocks. Given the availability of countless types of carboxylic acids, the AApeptides are amenable to potential derivatizations with a wide variety of side chains in a simple and straightforward manner, indicating its promising applications in library based drug screening. The preliminary results show that AApeptides possess significant bioactivities including the mimicry of p53 to successfully inhibit the p53/MDM2 protein-protein interaction, the selectivity in binding MDM2 protein, and the excellent stability towards enzymatic degradations. Hence, it is conceivable that a continuing development of sequence-specific AApeptides would enrich the current types of functional peptidomimetics, and expand the applications of peptide mimics in biomedical research including the modulation of protein-protein interactions. Future work will involve the systematic studies using X-ray crystallography, Circular Dichroism (CD), and 2D-NMR to understand the structure requirements of AApeptides to adopt predicted conformations, which will help the design of functional AApeptides. More specifically, the optimizations of AApeptide sequences to achieve a better inhibition of p53/MDM2 interaction as well as other carbohydrates/proteins/nucleic acids interactions are also urgent and are currently under investigation.

6. Acknowledgement

This work is supported by USF start-up fund (Cai) and NIH CA 118210 (Chen).

7. References

Alluri, P. G., M. M. Reddy, et al. (2003). "Isolation of protein ligands from large peptoid libraries." *J Am Chem Soc* 125(46): 13995-4004.

Arkin, M. R. and J. A. Wells (2004). "Small-molecule inhibitors of protein-protein interactions: progressing towards the dream." *Nat Rev Drug Discov* 3(4): 301-17.

Arndt, H. D., B. Ziemer, et al. (2004). "Folding propensity of cyclohexylether-delta-peptides." *Org Lett* 6(19): 3269-72.

Balint, E. E. and K. H. Vousden (2001). "Activation and activities of the p53 tumour suppressor protein." *Br J Cancer* 85(12): 1813-23.

Bautista, A. D., J. S. Appelbaum, et al. (2010). "Bridged beta(3)-peptide inhibitors of p53-hDM2 complexation: correlation between affinity and cell permeability." *J Am Chem Soc* 132(9): 2904-6.

Boeijen, A., J. van Ameijde, et al. (2001). "Solid-phase synthesis of oligourea peptidomimetics employing the Fmoc protection strategy." *Journal of Organic Chemistry* 66(25): 8454-8462.

Brooks, C. L. and W. Gu (2006). "p53 ubiquitination: Mdm2 and beyond." *Mol Cell* 21(3): 307-15.

Cheng, R. P., S. H. Gellman, et al. (2001). "beta-Peptides: from structure to function." *Chem Rev* 101(10): 3219-32.

De Vincenzo, R., G. Scambia, et al. (1995). "Effect of synthetic and naturally occurring chalcones on ovarian cancer cell growth: structure-activity relationships." *Anticancer Drug Des* 10(6): 481-90.

Debaene, F., J. A. Da Silva, et al. (2007). "Expanding the scope of PNA-encoded libraries: divergent synthesis of libraries targeting cysteine, serine and metallo-proteases as well as tyrosine phosphatases." *Tetrahedron* 63(28): 6577-6586.

Dervan, P. B. (1986). "Design of sequence-specific DNA-binding molecules." *Science* 232(4749): 464-71.

Ding, K., Y. Lu, et al. (2005). "Structure-based design of potent non-peptide MDM2 inhibitors." *J Am Chem Soc* 127(29): 10130-1.

Dragulescu-Andrasi, A., S. Rapireddy, et al. (2006). "A simple gamma-backbone modification preorganizes peptide nucleic acid into a helical structure." *J Am Chem Soc* 128(31): 10258-10267.

Duncan, S. J., M. A. Cooper, et al. (2003). "Binding of an inhibitor of the p53/MDM2 interaction to MDM2." *Chem Commun (Camb)*(3): 316-7.

Galatin, P. S. and D. J. Abraham (2004). "A nonpeptidic sulfonamide inhibits the p53-mdm2 interaction and activates p53-dependent transcription in mdm2-overexpressing cells." *J Med Chem* 47(17): 4163-5.

Garcia-Echeverria, C., P. Chene, et al. (2000). "Discovery of Potent Antagonists of the Interaction between Human Double Minute 2 and Tumor Suppressor p53." *J Med Chem* 43(17): 3205-3208.

Gellman, S. (2009). "Structure and Function in Peptidic Foldamers." *Biopolymers* 92(4): 293-293.

Goodman, C. M., S. Choi, et al. (2007). "Foldamers as versatile frameworks for the design and evolution of function." *Nat Chem Biol* 3(5): 252-62.

Grasberger, B. L., T. Lu, et al. (2005). "Discovery and cocrystal structure of benzodiazepinedione HDM2 antagonists that activate p53 in cells." *J Med Chem* 48(4): 909-12.

Graybill, T. L., M. J. Ross, et al. (1992). "Synthesis and Evaluation of Azapeptide-Derived Inhibitors of Serine and Cysteine Proteases." *Bioorganic & Medicinal Chemistry Letters* 2(11): 1375-1380.

Hara, T., S. R. Durell, et al. (2006). "Probing the structural requirements of peptoids that inhibit HDM2-p53 interactions." *J Am Chem Soc* 128(6): 1995-2004.

Hardcastle, I. R., S. U. Ahmed, et al. (2006). "Small-molecule inhibitors of the MDM2-p53 protein-protein interaction based on an isoindolinone scaffold." *J Med Chem* 49(21): 6209-21.

Haupt, Y., R. Maya, et al. (1997). "Mdm2 promotes the rapid degradation of p53." *Nature* 387(6630): 296-9.

Hayashi, R., D. Wang, et al. (2009). "N-acylpolyamine inhibitors of HDM2 and HDMX binding to p53." *Bioorg Med Chem* 17(23): 7884-93.

Horne, W. S. and S. H. Gellman (2008). "Foldamers with heterogeneous backbones." *Acc Chem Res* 41(10): 1399-408.

Horne, W. S., L. M. Johnson, et al. (2009). "Structural and biological mimicry of protein surface recognition by alpha/beta-peptide foldamers." *Proc Natl Acad Sci U S A* 106(35): 14751-6.

Hu, Y., X. Li, et al. (2011). "Design and synthesis of AApeptides: a new class of peptide mimics." *Bioorg Med Chem Lett* 21(5): 1469-71.

Jimenez, G. S., S. H. Khan, et al. (1999). "p53 regulation by post-translational modification and nuclear retention in response to diverse stresses." *Oncogene* 18(53): 7656-65.

Knight, S. M., N. Umezawa, et al. (2002). "A fluorescence polarization assay for the identification of inhibitors of the p53-DM2 protein-protein interaction." *Anal Biochem* 300(2): 230-6.

Kritzer, J. A., J. D. Lear, et al. (2004). "Helical beta-peptide inhibitors of the p53-hDM2 interaction." *J Am Chem Soc* 126(31): 9468-9.

Kritzer, J. A., O. M. Stephens, et al. (2005). "beta-Peptides as inhibitors of protein-protein interactions." *Bioorg Med Chem* 13(1): 11-6.

Kubbutat, M. H., S. N. Jones, et al. (1997). "Regulation of p53 stability by Mdm2." *Nature* 387(6630): 299-303.

Kumbhani, D. J., G. V. Sharma, et al. (2006). "Fascicular conduction disturbances after coronary artery bypass surgery: a review with a meta-analysis of their long-term significance." *J Card Surg* 21(4): 428-34.

Kussie, P. H., S. Gorina, et al. (1996). "Structure of the MDM2 oncoprotein bound to the p53 tumor suppressor transactivation domain." *Science* 274(5289): 948-53.

Lee, H. J., J. W. Song, et al. (2002). "A theoretical study of conformational properties of N-methyl azapeptide derivatives." *Journal of the American Chemical Society* 124(40): 11881-11893.

Li, X., Y. D. Wu, et al. (2008). "Alpha-aminoxy acids: new possibilities from foldamers to anion receptors and channels." *Acc Chem Res* 41(10): 1428-38.

Liu, Y. and M. Kulesz-Martin (2001). "p53 protein at the hub of cellular DNA damage response pathways through sequence-specific and non-sequence-specific DNA binding." *Carcinogenesis* 22(6): 851-60.

Lowe, S. W., H. E. Ruley, et al. (1993). "p53-dependent apoptosis modulates the cytotoxicity of anticancer agents." *Cell* 74(6): 957-67.

McLure, K. G., M. Takagi, et al. (2004). "NAD+ modulates p53 DNA binding specificity and function." *Mol Cell Biol* 24(22): 9958-67.

Michel, J., E. A. Harker, et al. (2009). "In Silico Improvement of beta3-peptide inhibitors of p53 x hDM2 and p53 x hDMX." *J Am Chem Soc* 131(18): 6356-7.

Momand, J., D. Jung, et al. (1998). "The MDM2 gene amplification database." *Nucleic Acids Res* 26(15): 3453-9.

Momand, J., G. P. Zambetti, et al. (1992). "The mdm-2 oncogene product forms a complex with the p53 protein and inhibits p53-mediated transactivation." *Cell* 69(7): 1237-45.

Murray, J. K. and S. H. Gellman (2007). "Targeting protein-protein interactions: lessons from p53/MDM2." *Biopolymers* 88(5): 657-86.

Nelson, J. C., J. G. Saven, et al. (1997). "Solvophobically driven folding of nonbiological oligomers." *Science* 277(5333): 1793-6.

Niu, Y., Y. Hu, et al. (2011). "[gamma]-AApeptides: design, synthesis and evaluation." *New J Chem* 35(3): 542-545.

Oliner, J. D., K. W. Kinzler, et al. (1992). "Amplification of a gene encoding a p53-associated protein in human sarcomas." *Nature* 358(6381): 80-3.

Oren, M. (1999). "Regulation of the p53 tumor suppressor protein." *J Biol Chem* 274(51): 36031-4.

Patch, J. A. and A. E. Barron (2002). "Mimicry of bioactive peptides via non-natural, sequence-specific peptidomimetic oligomers." *Curr Opin Chem Biol* 6(6): 872-7.

Pellegata, N. S., R. J. Antoniono, et al. (1996). "DNA damage and p53-mediated cell cycle arrest: a reevaluation." *Proc Natl Acad Sci U S A* 93(26): 15209-14.

Risseeuw, M. D., J. Mazurek, et al. (2007). "Synthesis of alkylated sugar amino acids: conformationally restricted L-Xaa-L-Ser/Thr mimics." *Org Biomol Chem* 5(14): 2311-4.

Robinson, J. A. (2008). "Beta-hairpin peptidomimetics: design, structures and biological activities." *Acc Chem Res* 41(10): 1278-88.

Seebach, D., P. E. Ciceri, et al. (1996). "Probing the helical secondary structure of short-chain beta-peptides." *Helvetica Chimica Acta* 79(8): 2043-2066.

Simon, R. J., R. S. Kania, et al. (1992). "Peptoids: a modular approach to drug discovery." *Proc Natl Acad Sci U S A* 89(20): 9367-71.

Stoll, R., C. Renner, et al. (2001). "Chalcone derivatives antagonize interactions between the human oncoprotein MDM2 and p53." *Biochemistry* 40(2): 336-44.

Trabocchi, A., F. Guarna, et al. (2005). "gamma- and delta-amino acids: Synthetic strategies and relevant applications." *Current Organic Chemistry* 9(12): 1127-1153.

Tsukamoto, S., T. Yoshida, et al. (2006). "Hexylitaconic acid: a new inhibitor of p53-HDM2 interaction isolated from a marine-derived fungus, Arthrinium sp." *Bioorg Med Chem Lett* 16(1): 69-71.

Tuwalska, D., J. Sienkiewicz, et al. (2008). "Synthesis and conformational analysis of methyl 3-amino-2,3-dideoxyhexopyranosiduronic acids, new sugar amino acids, and their diglycotides." *Carbohydr Res* 343(7): 1142-52.

Vassilev, L. T., B. T. Vu, et al. (2004). "In vivo activation of the p53 pathway by small-molecule antagonists of MDM2." *Science* 303(5659): 844-8.

Violette, A., M. C. Petit, et al. (2005). "Oligourea foldamers as antimicrobial peptidomimetics." *Biopolymers* 80(4): 516-516.

Whitty, A. and G. Kumaravel (2006). "Between a rock and a hard place?" *Nat Chem Biol* 2(3): 112-8.

Winssinger, N., R. Damoiseaux, et al. (2004). "PNA-Encoded Protease Substrate Microarrays." *Chem Biol* 11(10): 1351-1360.

Wu, Y.-D. and S. Gellman (2008). "Peptidomimetics." *Accounts of Chemical Research* 41(10): 1231-1232.

Zhao, J., M. Wang, et al. (2002). "The initial evaluation of non-peptidic small-molecule HDM2 inhibitors based on p53-HDM2 complex structure." *Cancer Lett* 183(1): 69-77.

10

Protein Interactions in S-RNase-Based Gametophytic Self-Incompatibility

Thomas L. Sims
Department of Biological Sciences, Northern Illinois University
USA

1. Introduction

With well over 200,000 documented species (Mora et al., 2011) flowering plants (angiosperms) are among the most successful taxa on the planet. A major reason for the success of the angiosperms is self-incompatibility, a genetic and biochemical barrier to inbreeding that promotes outcrossing and diversity in populations. Plants exhibiting self-incompatibility have the ability to recognize (species-specific) pollen as being "self" or "non-self", with self (incompatible) pollen being rejected and non-self (compatible) pollen being accepted. S-RNase-based Gametophytic Self-Incompatibility (GSI) has been characterized in the Solanaceae, Rosacaeae, and Plantaginaceae (McClure et al., 2011; Meng et al. 2011; Chen et al., 2010; Sims & Robbins 2009), with the genetic, physiological and molecular basis of this form of GSI described in detail. To date, over a dozen different proteins have been identified that function in different parts of the GSI response; most of these have been tested for protein interactions with other GSI response pathway proteins. The two key recognition proteins: S-RNase (the style-expressed recognition component) and SLF (the pollen-expressed recognition component) interact with each other, and with other components of a putative SCF^{SLF} E3 ubiquitin ligase complex. Recently Kubo et al., 2010 demonstrated the existence of multiple SLF variant classes. Multiple S-RNase and SLF alleles are present in breeding populations (Richman et al., 1995, 1996, 2000), and it now seems probably that collaborative interaction of different SLF alleles and classes with different S-RNase alleles governs self/non-self recognition in GSI. In this review, I summarize the genetic basis of GSI, describe the different proteins identified that are thought to function in the GSI pathway, and describe what is known with regard to protein interactions underlying the function of self-incompatibility. Most of the work discussed here comes from studies in the Solanaceae and Plantaginaceae. Gametophytic self-incompatibility has also been studied extensively in the Rosaceae (e.g. Sassa et al., 2010). Work that demonstrates possible differences in the mechanism of GSI in Solanaceae/Plantaginaceae versus Rosaceae will be discussed as appropriate.

2. Genetic studies of gametophytic self-incompatibility

The first description of gametophytic self-incompatibility was by none other than Charles Darwin. As Darwin (1891) observed:

"....protected flowers with their own pollen placed on the stigma never yielded nearly a full complement of seed; whilst those left uncovered produced fine capsules, showing that pollen from other plants must have been brought to them, probably by moths. Plants growing vigorously and flowering in pots in the green-house, never yielded a single capsule; and this may be attributed, at least in chief part, to the exclusion of moths."

Since Darwin's observation, self-incompatibility systems in general, and GSI in particular, have interested both molecular and evolutionary biologists. As an example of self/non-self recognition, GSI presents interesting challenges in terms of molecular interactions, how recognition specificity is determined, and what types of sequences determine allelism. In terms of population genetics, evolutionary biologists have investigated questions of how GSI haplotypes are established and maintained over evolutionary time (Kohn, 2008).

2.1 The genetic basis of S-RNase-based gametophytic self-incompatibility

Early studies (de Nettancourt, 1977; Linskens 1975, Mather, 1943) demonstrated that self/non-self recognition was encoded by a single genetic locus, the S-locus, with pistil and pollen recognition components (termed "pistil-S" and "pollen-S", respectively). Both pistil-S and pollen-S have multiple alleles, such that a given S-locus recognition phenotype is now termed a S-locus haplotype. The S-locus ribonuclease (S-RNase) is pistil-S, and the S-locus F-box protein (SLF; SFB in Rosaceae) has been demonstrated to be pollen-S (Sijacic et al., 2004). During pollination, a match between S-RNase and SLF haplotypes results in pollen rejection (incompatibility). Lack of a match results in pollen acceptance (compatibility) and fertilization (see Figure 1). Recognition specificity in GSI is a cell-autonomous response, in that rejection or acceptance is specific to individual pollen tubes, and is not an "all or none" phenomenon. This can be demonstrated by the existence of "half-compatible" pollinations (Figure 1). In this case, pollen tubes expressing a SLF-specificity matching the S-RNase in the style are rejected, while other pollen tubes in the same style, with no haplotype match, grow normally and function for fertilization and seed set.

2.2 Tetraploidy results in self-compatibility due to competitive interaction

An intriguing aspect of GSI is that tetraploidy, in heterozygous individuals, leads to self-compatibility (Figure 1). In this case, heteroallelic, diploid pollen (e.g. S1-SLF/S2-SLF) is compatible on either a tetraploid style ($S_1S_1S_2S_2$) or a diploid style (S_1S_2). Haploid pollen (e.g. S_1 or S_2) remains incompatible on tetraploid styles (Figure 1). This phenomenon has been termed "competitive interaction" (de Nettancourt 1977). Competitive interaction is only observed in situations where the parent plant was heterozygous for S-locus haplotype. Tetraploids that are homozygous at the S-locus (homozygous plants can be obtained by bud-pollination), do not show competitive interaction, but remain self-incompatible. Competitive interaction is most likely the cause of GSI breakdown (compatibility) in induced pollen-part mutants (Golz et al., 1999, 2001). In mutants induced by radiation, Golz et al. (1999, 2001) showed that GSI breakdown was associated with partial duplications of S-haplotypes, in which the compatible pollen presumably phenocopied the heteroallelic condition found in tetraploids. Competitive interaction has been used as a test for the identity of pollen-S (Kubo et al., 2010; Sijacic et al., 2004), since transgenic plants, having diploid, heteroallelic pollen (two different SLF haplotypes) are self-compatible (Figure 1 and sections below).

3. Pistil-S and pollen-S

Although the genetic "identities" of pistil-S and pollen-S have been known for many years, the identification of specific proteins corresponding to these entities has been a more recent phenomenon. The S-locus ribonuclease (S-RNase) was initially cloned in 1986 (Anderson et al., 1986) with its identity as pistil-S confirmed eight years later (Lee et al., 1994, Murfett et al., 1994). Identification of SLF as pollen-S is far more recent. SLF was first identified by chromosomal walking (Entani et al., 2003; Lai et al., 2002; Wang et al., 2004) and subsequently confirmed as pollen-S using a competitive-interaction assay (Sijacic et al., 2004). As will be explained, the molecular nature of pollen-S appears to be far more complex than originally envisioned.

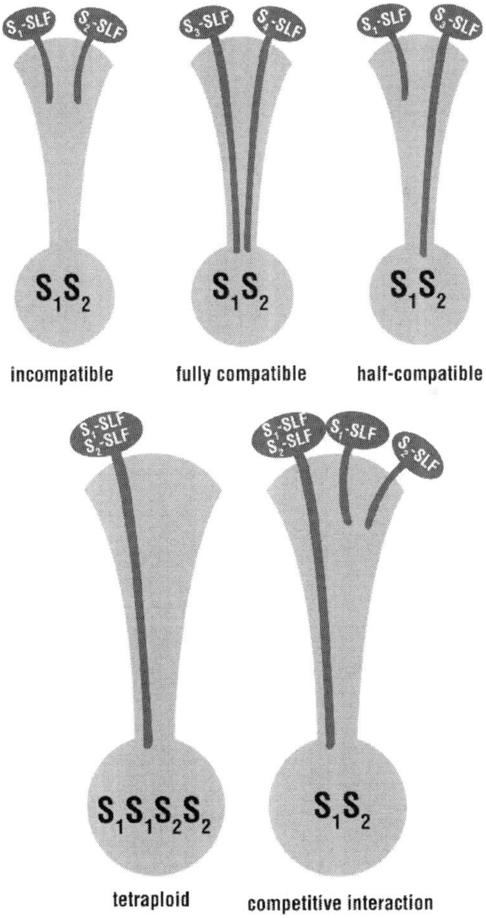

Fig. 1. **Genetic basis of gametophytic self-incompatibility.**

Figure 1 illustrates different types of pollinations with styles and pollen expressing different haplotypes at the S-locus. In an incompatible pollination (top left), a match of haplotypes between pollen and style results in incompatibility. No match of S-locus haplotypes (top, middle) results in full compatibility. A "half-compatible" cross results when half of the pollen carries a S-locus haplotype matching that of the style, but the other pollen is not matching. In this case, only the "matching" pollen tubes are rejected. The lower portion of the figure illustrates GSI breakdown in tetraploids (left figure) by competitive interaction. The figure at lower-right illustrates how competitive interaction can be used in transgenic assays to demonstrate that a particular gene (in this case, SLF) is pollen-S.

3.1 S-RNase is the pistil recognition component of GSI

The ability to selectively inhibit the growth of self pollen is determined in the style by a S-locus encoded ribonuclease known as the S-RNase. The S-RNase was first identified as a highly-expressed stylar protein that co-segregated with specific S-haplotypes (Anderson et al., 1986). The S-RNase gene is expressed at high levels late during development of the pistil (Clark et al., 1990), and encodes a secreted protein that accumulates to high levels in the transmitting tract of the style (Ai et al., 1990; Anderson et al., 1989). Comparative sequence analysis showed that S-RNase alleles have a high degree of sequence polymorphism, but that the polymorphism is not evenly distributed across the protein. Overall amino acid sequence identity can be less than 50% between allleles (Ioerger et al., 1991; Sims, 1993; Richman et al., 1995). Detailed sequence comparisons showed that S-RNase proteins have five highly conserved domains and two adjacent hypervariable domains, HVa and HVb (Ioerger et al., 1991; Sims, 1993). Although much of the sequence variability among S-RNase alleles is found in the two hypervariable regions, other portions of the protein are variable as well (Figure 2). The conserved domains C2 and C3 contain two histidine residues, His31 and His91, that along with Lys90 make up the catalytic site of the ribonuclease (Ida et al., 2001; Ishimizu et al., 1995). (Note that in different S-RNase alleles, the exact positions of these concerved amino acids vary by one or two positions).

Transgenic gain-of-function and loss-of-function experiments gave conclusive evidence that the S-RNase was the style-recognition component of GSI. Murfett et al. (1994) used a gain-of function approach, where the S_{A2}-RNase of *Nicotiana alata* (under control of a strong style-specific promoter) was transferred to a regenerable *N. lansgsdorfii* x *N. alata* hybrid. The transgenic plant remained compatible when pollinated with S_{C10} pollen from *Nicotiana alata,* but now showed the ability to reject S_{A2} pollen. Lee et al. (1994) used an antisense approach to down-regulate the *Petunia inflata* S_3-RNase in a S_2S_3 background. Plants with reduced levels of S_3-RNase were no longer capable of inhibiting S_3 pollen, although the transgenic plant showed otherwise normal GSI behavior. Lee et al. (1994) also used a gain-of-function approach, in which the S_3-RNase of *Petunia inflata* was transferred to a plant of the S_1S_2 genotype. Transgenic plants expressing the S_3-RNase at levels comparable to endogenous S-RNase had acquired the ability to reject S_3 pollen. These plants continued to reject S_1 and S_2 pollen, but set seed capsules when pollinated at an immature bud stage where the S-RNase is expressed at minimal levels (Clark et al., 1990; Lee et al., 1994). In these experiments, only the style recognition was altered. Pollen recognition specificity was not affected, confirming that a separate gene product from the S-RNase encoded the "pollen-S" component.

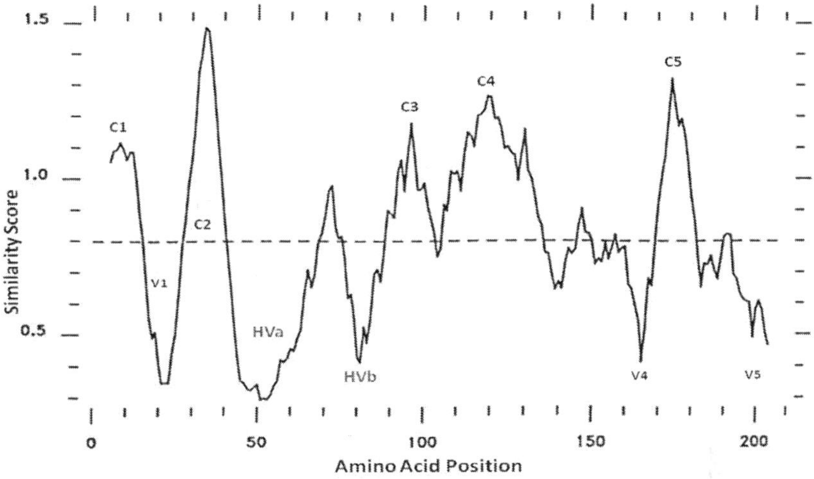

Fig. 2. **Graphical depiction of amino acid alignment among Solanaceae S-RNase alleles.** Amino acid sequences for eighteen S-RNase alleles were aligned using PlotSimilarity. The dotted line shows the average similarity score across the protein. Peaks above the line represent conserved regions (labeled C1 through C5). Valleys below the line represent more variable regions. Amino acids in hypervariable regions HVa and HVb were shown to be sufficient for determining S-RNase recognition specificity (after Sims, 1993).

Further work either analyzing spontaneous mutants (Royo et al., 1994) or using transgenic plants (Huang et al.,1994) demonstrated that ribonuclease activity of the S-RNase was required for pollen rejection. Royo et al. (1994) cloned and sequenced the S-RNase allele from a self-compatible S_cS_c accession of *Lycopersicon peruvianum* (now '*Solanum peruvianum*', http://solgenomics.net). The S_c allele sequence showed a change at amino acid 33 from the conserved histidine to asparagine. No other sequence changes were observed, and the authors concluded this change was correlated with both the loss of RNase activity in S_cS_c styles and with self-compatibility. In related work, Huang et al. (1994) used *in vitro* mutagenesis to construct a H93N variant of the *P. inflata* S_3-RNase, and introduced that construct in the S_1S_2 background. Unlike the results obtained when the wild-type S_3-RNase was transferred (Lee et al., 1994), the H93N S_3-RNase was unable to reject S_3 pollen. Reinforcing the critical role of ribonuclease activity in S-RNase function were earlier experiments indicting that degradation of pollen-tube RNA was associated with self-incompatibility. In those experiments (McClure et al., 1990) pollen RNA was labeled *in vivo* by watering plants with a solution containing ^{32}P-orthophosphate, then used for compatible or incompatible pollinations. Pollen tube RNA was degraded following incompatible pollination, but was not degraded following compatible pollination.

3.1.1 S-RNase recognition specificity is determined by hypervariable domains

Experiments investigating the basis for allele specificity in the S-RNase protein have focused on the role of the hypervariable regions. In one approach, Zurek et al. (1997) constructed chimeric S-RNase genes having different combinations of S_{A2} and S_{C10} conserved and variable domains, then expressed the chimeric proteins in transgenic plants. Although the transgenic styles had ribonuclease activity levels equivalent to self-incompatible controls, none of the chimeric S-RNase constructs could reject S_{A2} or S_{C10} pollen. By contrast, Matton et al. (1997) took advantage of two S-RNase allles in *Solanum chacoense* that were closely related in sequence, to make more limited alterations. The S_{11}- and S_{13}-RNase alleles of *S. chacoense* differ by only 10 amino acids, three of which are found in HVa and one in HVb. Matton et al. (1997) used *in vitro* mutagenesis to change the four S_{11} residues in the HVa and HVb regions to those found in S_{13}, then expressed the altered allele transgenically in the $S_{12}S_{14}$ background. Pollination with S_{11} and S_{13} pollen demonstrated that changing only these residues changed the recognition specificity of the transferred S-RNase from S_{11} to S_{13}. In an extension of this experiment (Matton et al., 1999), changing only two residues in HVa plus the HVb residue resulted in a "dual-specificity" S-RNase that retained the ability to reject S_{11} pollen while acquiring the ability to also reject S_{13} pollen. These experiments demonstrated that, at least for these two alleles, amino acid sequences in the hypervariable regions determine allelic specificity. The protein crystal structure has been determined for the S_{F11} S-RNase of *Nicotiana alata* (Ida et al., 2001). The two hypervariable regions are located on the surface of the S_{F11} S-RNase and readily accessible to solvent (Ida et al., 2001). These regions include all four of the equivalent residues to those targeted in the mutagenesis experiments of Matton et al. (1997, 1999). Another potential basis for allele specificity might be variability in carbohydrate modification of S-RNases, which are glycoproteins (Woodward et al., 1989). This does not appear to be the case, however as as elimination of the glycosylation site has no effect on the ability of S-RNase to reject pollen (Karunanandaa et al.,1994).

3.1.2 Both self and non-self S-RNases are imported into pollen tubes in vivo

Immunolocalization experiments, either using traditional TEM (Luu et al., 2000) or fluorescently-tagged antibodies hybridized to paraffin-embedded sections (Goldraij et al., 2006) demonstrate that both incompatible and compatible S-RNases are imported into pollen tubes. The authors of these two studies reached different conclusions about the location of S-RNase inside pollen tubes following compatible or incompatible pollinations. Luu et al., (2000) working with self-incompatible potato (*S. chacoense*), fixed pollinated styles, 18 hr post-pollination, with 0.5% glutaraldehyde, followed by embedding, hybridization with anti-S_{11} antibody and 20 nm gold-labeled secondary antibody, and visualization via TEM. S_{11}-RNase was taken up into both compatible and incompatible pollen tubes, and labeling was seen primarily in pollen-tube cytoplasm, with little labeling in the pollen-tube vacuole. Goldraij et al. (2006), working with *Nicotiana*, hybridized anti-S-RNase antibodies along with anti-callose, anti-aleurain (marker for vacuolar lumen) and anti-vPPase (marker for vacuolar membrane) to fixed, paraffin-embedded sections, then visualized fluorescence using confocal microscopy. These authors concluded that S-RNase was initially sequestered in a vacuolar compartment in pollen in both compatible and early-stage (16 hr) incompatible pollen tubes, but that this compartment broke down at later stages (36 hr) of incompatible pollinations, releasing S-RNase into the pollen-tube cytoplasm.

3.1.3 The cytotoxic model for pollen rejection

Current models for pollen rejection in GSI propose a cytotoxic mechanism, where RNA degradation in incompatible pollen tubes reduces protein synthesis resulting in a slowing or cessation of pollen tube growth and failure of incompatible pollen tubes to reach the ovary. This model is based on observations outlined in the previous sections, but is not without its caveats. S-RNases are imported into pollen tubes, and at least in self-incompatible pollinations (Goldraij et al., 2006) are freely distributed in the pollen-tube cytoplasm. The ribonuclease activity of S-RNases is required for pollen tube rejection, and generalized RNA degradation is associated with self-incompatibility, but not with cross-pollination. The ability to reject incompatible pollen tubes is also dependent on a threshold level of S-RNase expression and accumulation in the style. Both developmental (Clark et al., 1990; Shivanna, 1969) and transgenic assays (Lee et al., 1994; Murfett et al., 1994, Zurek et al., 1997) show that styles expressing the S-RNase at low-to-moderate levels are incapable of rejecting otherwise incompatible pollen. S-RNases, like other T2 ribonucleases, show no obvious substrate specificity, at least *in vitro* (Singh et al., 1991).

The cytotoxic model, while attractive and consistent with the majority of current evidence, cannot, however, fully explain some other observations. Grafting experiments (Lush & Clarke, 1996) where upper regions of incompatible styles were grafted onto compatible styles, and pollinated, showed that incompatible pollen tubes could recover, growing out of the incompatible style into the compatible style. Also, Walles & Han (1998) using conventional TEM, observed intact polysomes in incompatible pollen tubes after pollination. Last, there is little correlation between overall ribonuclease activity found in different styles with the level of self-incompatibility (Clark et al., 1990; Singh et al., 1991; Zurek et al., 1997), although it should be assumed that non S-RNase ribonucleases likely contribute to overall style RNase levels.

3.2 SLF is the pollen-recognition component of GSI

Although the S-RNase was identified and cloned early on, it took an additional 18 years before the S-locus F-box protein (SLF) was conclusively identified as "pollen-S". Even today, what, functionally constitutes "pollen-S" is not fully understood; recent work suggests that different SLF variants may act collaboratively to recognize S-RNases. Additionally, several other proteins are involved and/or required for recognition (see sections below).

Even before SLF was identified and cloned, the majority of experimental evidence pointed to pollen-S as an inhibitor of S-RNase activity in compatible pollen. Tetraploidy is associated with the breakdown of self-incompatibility in those cases where the parental diploid plant was heterozygous for two different S-locus haplotypes, but not when the parent plant was homozygous (Chawla et al., 1997; de Nettancourt, 1977; Entani et al., 1999). In early studies, Brewbaker & Natarajan (1960) induced pollen-part mutants of *Petunia* using irradiation (pollen part mutants are self-compatible, fertile as pollen parents, and show normal GSI behaviour in the style). Pollen-part mutants were obtained only when the irradiated parent was heterozygous, and were associated with centric chromosome fragments. Golz et al (1999, 2001) revisited this work, carrying out mapping and cytological analyses of pollen-part mutants of *Nicotiana alata* induced by gamma radiation. In all cases, the pollen-part mutants were associated with apparent duplications of part or all of the S-locus, either as centric chromosome fragments or as translocations. Luu et al. (2000), and later Goldraij et al. (2006) showed that both compatible and incompatible S-RNase proteins were imported into pollen tubes. Together, these results discredited a model where pollen-S was a "gatekeeper" preventing import of non-self S-RNases and suggested that compatible pollinations result from the specific inhibition of imported S-RNase proteins. According to models based on the results just described, pollen-S was an inhibitor of all S-RNases, except its own cognate S-RNase. Thus, compatible pollinations resulted from pollen-S inhibiting the action of any non-self S-RNase, while incompatible pollinations resulted from the inability of pollen-S to inhibit a co-evolved S-RNase.

3.2.1 Predictions for pollen-S

Prior to the actual isolation of pollen-S, there was a relative consensus with regard to the properties expected of this protein. Genetic studies had indicated that there was little or no recombination between pistil-S and pollen-S (de Nettancourt 1977), so both genes were expected to be tightly linked. That linkage, together with the observation that S-RNase alleles had diverged prior to speciation in the Solanaceae (some S-RNase alleles are more similar across species than within species) resulted in the assumption that S-RNase sequences and pollen-S sequences should be co-evolved. That is, most researchers fully expected that the degree of polymorphism among pollen-S alleles should be roughly equivalent to the polymorphism observed among S-RNase alleles (Kao & McCubbin, 1996). Pollen-S was also thought to interact directly with the S-RNase, with that interaction resulting in the inhibition of the action of the S-RNase in compatible pollinations. The sections below will illustrate that these assumptions were only partially correct.

3.2.2 Genetic and physical mapping of the S-Locus

The first attempt at mapping the S-locus was carried out by Tanksley and Loaiza-Figueroa (1985) who mapped the S-locus to chromosome I of *Lycopersicon peruvianum* using enzyme-linkage. RFLP mapping in potato (Gebhardt et al., 1991) demonstrated that chromosome I of tomato and potato were homeologous, and that the S-locus mapped to the same location in potato as in tomato. The S-locus was physically mapped in *Petunia hybrida* by fluorescence in-situ hybridization (FISH), using T-DNA inserts linked to the S-locus (ten Hoopen et al., 1998). Those experiments showed that in *Petunia hybrida,* the S-locus was located in a sub-centromeric region of chromosome III. Mapping of linked RFLP markers demonstrated synteny of the S-locus across four species in the Solanaceae (*Lycopersicon peruvianum, Nicotiana alata, Petunia hybrida* and *Solanum tuberosum*). Entani et al. (1999) carried out similar FISH experiments, but used cDNAs and genomic clones of the S-RNase instead of linked T-DNA inserts. Like ten Hoopen et al. (1998), Entani et al. (1999) found that the S-RNase gene was found in a subcentromeric region of chromosome III of *Petunia hybrida.* Both Li et al. (2000) and McCubbin et al. (2000) used RNA differential display to identify pollen-expressed cDNAs linked to the S-locus. Although not realized at the time, both of these differential display experiments identified cDNAs that would later turn out to true pollen-S genes. Part of the failure to recognize that these linked cDNAs did, in fact, encode pollen-S was the high degree of sequence identity between cDNAs isolated from different haplotypes as compared to the polymorphism previously observed for S-RNase alleles.

3.2.3 Gene walking identified pollen-S

The large amount of repetitive DNA sequences flanking S-RNase genes (Coleman & Kao, 1992; Matton et al., 1995), together with the subcentromeric location of the S-locus (ten Hoopen et al., 1998; Entani et al., 1999) were originally thought to preclude a map-based cloning approach for isolation of pollen-S (Kao & McCubbin 1996). Indeed, some of the early efforts to clone pollen-S involved T-DNA tagging (Harbord et al., 2000) yeast two-hybrid screens (Sims & Ordanic, 2001) or other protein-interaction methods (Dowd et al., 2000). Although these approaches provided important information regarding S-RNase-based GSI, none of them resulted in the identification of pollen-S. The first indication that pollen-S might be cloned using a map-based approach came from the work of Ushijima et al. (2001) who constructed an ~200 kb cosmid contig around the S-locus of *Prunus dulcis* (almond). When these authors carried out Southern blots with cosmid clones spanning the contig, with genomic DNA of different S-locus haplotypes, they observed that a ~70 kb region in the center of the contig was highly polymorphic across haplotypes, whereas either end of the contig showed a high degree of sequence similarity across haplotypes. This "island of polymorphism" presumably resulted from the known lack of recombination at the S-locus and was taken as defining the physical limits of the S-locus in *Prunus dulcis*.

Lai et al. (2002) were the first to report the isolation of the S-locus F-box gene, which would turn out to be pollen-S. Screening of a BAC library from *Antirrhinum hispanicum* with the S_2-RNase identified a 63 kb BAC clone, which was then fully sequenced. Of several putative ORFs identified, most were retrotransposons, however, the 'gene-11' ORF, when used to screeen a cDNA library, identified a pollen-expressed F-box protein, termed AhSLF-S_2. AhSLF-S_2 was located 9 kb downstream of the S_2-RNase gene, and appeared to be allele-specific, making it a good candidate for pollen-S. Similarly, Ushijima et al. (2003) sequenced the 70 kb region of *Prunus dulcis*, and identified a pollen-expressed, haplotype-specific, F-box gene, which they termed SFB. Using S-locus-specific cDNAs previously generated, and starting with BACs known to contain the S-RNase gene, Wang et al. (2004) identified an 881 kb contig surrounding the S-locus in *Petunia inflata*. Sequencing and analysis of a 328 kb region of this contig identified several genes, one of which, PiSLF, was highly similar to the F-box genes isolated from *Antirrhinum* and *Prunus*. Two previously identified S-linked F-box genes *A113* and *A134* (McCubbin et al., 2000) mapped outside of the 881 kb region.

3.2.4 Competitive interaction showed that SLF was pollen-S

The identity of SLF as pollen-S was confirmed by taking advantage of the phenomenon of competitive interaction in heteroallelic pollen (see section 2.2 and Figure 1). Sijajic et al., transferred the S_2-allele of SLF (PiSLF$_2$, but see nomenclature change in section 5.x, below) into a S_1S_1 line of *Petunia inflata*. First generation transgenic plants segregated two types of pollen, haploid S_1 pollen and heteroallelic S_1(PiSLF$_2$) pollen.. Self-pollination of the the S_1S_1(PiSLF$_2$) plant produced large fruits, indicating that the trangenic plant, formerly self-incompatible, was now self-compatible. Similarly, when S_1S_1(PiSLF$_2$) pollen was used to pollinate a non-transformed S_1S_1 plant, fruit set showed that the S_1S_1(PiSLF$_2$) pollen behaved as compatible pollen. Conversely, pollination of S_1S_1(PiSLF$_2$) styles with pollen from a non-transformed S_1S_1line produced no seed capsules, demonstrating that the loss of self-incompatibility in the transgenic plant was confined to the pollen. Analysis of progeny resulting from self-pollination demonstrated that all of the progeny inherited the transgene. Similar results were reported by Qiao et al. (2004b) who transferred the Ah-SLF_2 gene of *Antirrhinum hispanicum* into $S_{3L}S_{3L}$ *Petunia hybrida.* This experiment was conducted with two variations. In the first variation, a clone containing both Ah-S_2-RNase and Ah-SLF_2 was transferred to the host plant. Transgenic plants expressing both the *A. hispanicum* S-RNase and SLF were self-compatible, with the change in compatibility again confined to the pollen. Analysis of progeny confirmed that all inherited both the S_2-RNase and Ah-SLF_2. The change to the self-compatible phenotype was dependent on expression of the transgenes. In two individuals, neither the S_2-RNase transgene nor the endogenous S_{3L}-RNase was expressed at detectable levels, most likely due to co-suppression. Both of these progeny were completely self-incompatible. In the second variation, the Ah-SLF_2 cDNA alone, under control of the pollen-specific LAT52 promoter, was introduced into the $S_{3L}S_{3L}$ line. As above, the transgenic plants were self-compatible on the pollen side, but displayed normal self-incompatibility behavior when used as the style parent. These conversions of self-incompatibility to compatibility following pollen-specific expression of the SLF transgene is a direct phenocopy of the competitive interaction effect seen in heteroallelic pollen in tetraploid heterozygotes. One remarkable aspect of the work reported by Qiao et al (2004b) is that the *Antirrhinum* SLF protein can apparently cause competitive interaction in a completely different species.

3.2.5 SLF proteins appear to have different evolutionary history than S-RNases

Although many of the key predictions for the properties of pollen-S are indeed found for the SLF proteins (pollen-expression, interaction with S-RNases, competitive interaction), a surprising and confusing finding was the distinct lack of polymorphism among SLF proteins, together with the existence of multiple SLF-related proteins, originally termed SLFL (SLF-like) proteins. As will be discussed below (see section 8), many of these SLFL proteins may turn out to be true SLFs. Another confusing finding was apparent differences in the functional characteristics of SLF proteins in Solanaceae and Plantaginaceae compared with the equivalent SFB proteins in Rosaceae.

The first SLF proteins identified (Ah-SLF_1, Ah-SLF_2, Ah-SLF_4, Ah-SLF_5 in *Antirrhinum*) share 97% to 99% amino acid sequence identity. By contrast the related *Antirrhinum* S-RNase proteins share only 38% to 53% amino acid identity by pairwise BLASTp. Similarly, if not quite so dramatically, SLF proteins from *Petunia inflata* share ~ 90% amino acid sequence identify, while the corresponding S-RNase proteins share only about 70% amino acid sequence identity. Phylogenetic comparisons (e.g. Newbigin et al., 2008) present an even more striking picture. S-RNase sequences appear to be an ancient lineage; in the Solanaceae, S-RNases from one species are often more similar to a S-RNase from a different species than to other S-RNases within the same species. By contrast SLF sequences from an individual species cluster together. In addition, while the variability across S-RNases is clustered in variable and hypervariable regions (Figure 2), the variability across SLF alleles appears to be uniformly distributed across the protein. Because recombination between style and pollen recognition specificities is rarely if ever observed (de Nettancourt, 1977) the traditional assumption has been that pistil-S and pollen-S (S-RNase and SLF) have co-evolved and share the same evolutionary history. The actual observations, however appear to contradict that notion. One potential solution to this dilemma is that pollen-S may actually be comprised of multiple SLF protein variants, not a single SLF (see section 8).

4. Interaction assays identified other pollen proteins with roles in GSI

S-RNase and SLF are the pistil and pollen recognition components of GSI, however several other proteins with presumed or demonstrated roles in GSI have been identified and studied. Some of these proteins were first identified by protein-interaction screens with S-RNase or SLF, in other cases, a presumed role in GSI has been demonstrated using protein interaction assays. Figure 3 summarizes the interactions of pollen-expressed proteins with the S-RNase, specifics of these interactions are discussed below.

4.1 SBP1

Sims and Ordanic (2001) identified PhSBP1 (S-ribonuclease binding protein) in a yeast two-hybrid screen of a pollen cDNA library from S_1S_1 *Petunia hybrida.* The bait protein used for the screen was the N-terminal half of the S_1-RNase, containing domains C1 to C3 (see Figure 2). In subsequent pairwise interaction assays, PhSBP1 interacted with the same N-terminal construct of the S_3-RNase, and with subdomains (C2-HVa-HVb-C3, HVa-HVb) of both S_1- and S_3-RNases.

SBP1 did not show interaction with the C-terminal regions of either S-RNase (C4-V4-C5-V5 in Figure 2) nor with an unrelated bait protein, P53. ScSBP1 was isolated from *Solanum chacoense* (O'Brien et al., 2004), using a yeast two-hybrid screen with the HVa+HVb regions of the S_{11}-Rnase and the S_{13}-RNase as bait. Both the S_{11} and S_{13} baits interacted with ScSBP1, however a full-length S-RNase with a single amino acid change (H144L) at one of the active-site histidines failed to interact with SBP1. Similarly Hua & Kao (2006) used a partial bait (HVa-HVb-C3) of the *Petunia inflata* S2-RNase to screen a two-hybrid library and isolated PiSBP1. Similar to other reports (O'Brien et al., 2004; Sims & Ordanic, 2001) PiSBP1 did not interact with full-length S-RNase, with non-specific controls, or importantly, with a non-S-locus ribonuclease. Hua & Kao (2006) further showed that SBP1 interacted with $PiSLF_2$ and $PiSLF_1$ in both two-hybrid and pull-down assays, as well as with Cullin-1 and PhUBC1 (Sims, unpublished), an E2 conjugation enzyme protein from *Petunia hybrida.* Lee et al. (2008) used C-terminal domains of the style-transmitting-tract proteins TTS and 120K to screen a pollen two-hybrid library from *Nicotiana alata,* and also pulled out NaSBP1 from this screen. All of these reports (Hua & Kao, 2006; Lee et al., 2008; O'Brien et al., 2004; Sims & Ordanic 2001) showed that SBP1 was not pollen-specific, but was expressed in all tissues examined. SBP1 is non-allelic as well, as SBP1 isolated from S_1S_1 and S_3S_3 lines of *Petunia hybrida* are 100% identical. The SBP1 protein has two identifiable protein domains, a coiled-coil domain in the center of the sequence and a C-terminal RING-HC domain. RING-HC domains are characteristic of E3 ubiquitin ligases (Freemont 2000), and SBP1 has E3 ubiquitin ligase activity *in vitro* (Hua and Kao, 2008).

4.2 SSK1

Huang et al. (2006) used the *Antirrhinum hispanicum* SLF protein Ah-SLF_2 to screen a pollen yeast-two-hybrid library, and identified a SKP1-like protein that they named SSK1. AhSSK1 interacted with both Ah-SLF_2 and Ah-SLF_5 but not with proteins identified as SLF paralogs (Zhou et al., 2003). AhSSK1 futher interacted with a Cullin-1 protein. Sequence and phylogenetic analyses showed that AhSSK1 was related to, but distinct from canonical SKP1 proteins. In particular, AhSSK1 differed at several internal residues and also has a 7-residue C-terminal "tail" that extends beyond the "WAFE" sequence that terminates most plant SKP1 proteins (Huang et al., 2006; Zhao et al, 2010). Zhao et al. (2010) showed that SSK1 almost certainly plays a critical role in self-incompatibility. Using AhSSK1 as a guide, they isolated PhSSK1 from *Petunia hybrida.* PhSSK1, similar to AhSSK1 interacts with both SLF and Cullin-1 from *Petunia.* To directly test the role of PhSSK1, Zhao et al. (2010) used a RNAi construct of PhSSK1 to down-regulate this gene in $S_{3L}S_{3L}$ *Petunia hybrida.* When transgenic plants showing reduced levels of PhSSK1 in pollen were used as the pollen parent in crosses to S_1S_1 or S_vS_v *P. hybrida,* no seed capsules were produced. Conversely, when these same lines were used as pollen parent to a line defective for S-RNase expression (S_oS_o) normal seed capsules were produced, suggesting that down-regulation of SSK1 specifically affected cross-compatibility in the self-incompatibility response.

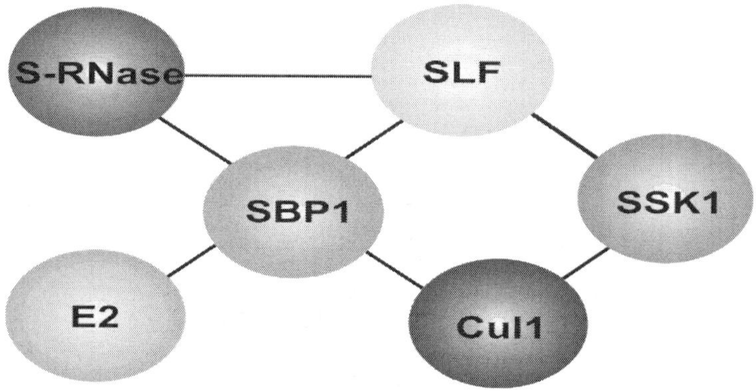

Fig. 3. **Protein interactions of pollen-expressed proteins in GSI.**
Lines between individual proteins indicate protein interactions that have been observed by yeast two-hybrid or pull-down assays. See the text for details.

5. Protein interactions between SLF and S-RNase proteins

S-locus F-box proteins were first cloned by chromosome-walking experiments to identify pollen-expressed proteins tightly linked to the S-RNase. Further studies examined the interaction between SLF proteins and S-RNases in detail. Qiao et al. (2004a) examined the interaction of the *Antirrhinum* Ah-SLF-S_2 protein with S-RNases using pull-down (His-tag), yeast two-hybrid, and co-immunoprecipitation assays. Pull-down assays demonstrated that the C-terminal portion of Ah-SLF-S_2 (lacking the F-box domain) interacted with S-RNase from style extracts of *Antirrhinum hispanicum.* The N-terminal F-box domain was incapable of interacting with S-RNase, and the full-length SLF protein could not be tested as it could not be expressed in *E. coli*. Similarly, the C-terminal portion of Ah-SLF-S_2 interacted with a full-length (lacking the signal peptide) S-RNase construct in yeast two hybrid assays; neither the F-box domain nor the full-length SLF protein showed interaction in the two-hybrid assays. Both of these assays also showed that Ah-SLF-S_2 interacted with different S-RNases, without any apparent allelic specificity. Co-immunoprecipitation assays where extracts from pollinated styles were immunoprecipitated using anti-Ah-SLF-S_2 antibody, then blotted with anti-S-RNase antibody showed that Ah-SLF-S_2 interacted with both S_2-and S_4-RNase *in vivo*. Qiao et al., (2004a) also tested interaction of Ah-SLF paralogs (Zhou et al., 2003) with S-RNase proteins. The SLF paralogs were identified as pollen-expressed SLF-like genes linked to the S-locus but distant from S-RNase or Ah-SLF-S_2. Similar SLF-like genes linked to the S-locus but outside of the core S-RNase-SLF contig had previously been identified in *Petunia inflata* (McCubbin et al., 2000; Wang et al., 2003). No interaction was observed between Ah-SLF-S_2 and any of the SLF paralogs. Recent data (see section 8) indicates the SLF-like genes (SLF paralogs) may, however be true SLF proteins, that recognize a subset of S-RNases rather than all S-RNases.

Hua and Kao (2006) also tested interactions between SLF and S-RNase allles in *Petunia inflata* using pull-down assays with His-tagged SLF and GST-tagged S-RNase constructs expressed in bacteria. These assays showed that both $PiSLF_1$ and $PiSLF_2$ interacted with the HVaHVbC3domain of the S_2-RNase, but that the non-self interactions ($PiSLF_1$:S_2-RNase) were far stronger than the self interactions ($PiSLF_2$:S_2-RNase) interactions. Similarly the reciprocal interactions of S_1- and S_2-RNase domains with His-tagged $PiSLF_1$ while showing some interaction in both cases, were far stronger for the non-self pairs than the self-pairs. Sims et al., (2010) used a fluorogenic substrate for β-galactosidase to quantify the strength of two-hybrid interactions between different domains of the S_1- and S_3-RNase of *Petunia hybrida* with SLF-S1 from *P. hybrida*. Similar to the results obtained by Hua & Kao (2006), both self and non-self S-RNases interacted with *P. hybrida* SLF-S1, but the interaction appeared stronger for the non-self interactions compared with the self-interactions.

One of the critical questions in GSI is that of how SLF and S-RNase alleles recognize each other as self versus non-self. This question has recently become even more complicated (see section 8) as it appears that proteins originally identified as SLF-like (SLFL), and not involved in GSI, may in fact be true SLF proteins. Chromosome-walking, differential display or degenerate PCR-cloning approaches (McCubbin et al., 2000; Wang et al., 2003,2004; Wheeler & Newbigin 2007; Zhou et al., 2003) in the Solanaceae and Plantaginaceae identified a number of SLF-like genes (SLFL) that were linked to the S-locus. These genes were thought not to be involved in self-incompatibility interactions specifically, since they did not show interaction with known S-RNases nor did they exhibit competitive interaction in transgenic assays (Hua et al., 2007; Meng et al., 2011). Hua et al., (2007) attempted to identify domains of SLF proteins that governed allelic specificity by iterative pairwise comparisons of SLF proteins with SLFL proteins. These comparisons identified three "SLF-specific" regions SR1, SR2 and SR3. Based on this identification, these authors then divided the SLF proteins into three domains: FD1, containing the F-box and SR1 (amino acids 1-110), FD2 (amino acids 111-259, including the SR2 region) and FD3 (amino acids 260-389, including SR3). Domain-swapping experiments, in which different chimeric proteins were tested for the ability to interact with the S_3-RNase in pull-down assays suggested that FD2 was the domain primarily responsible for SLF-S-RNase interactions. Testing chimeric constructs between $PiSLF_2$ and PiSLFLb-S_2 (a SLF-like protein in the same S_2 haplotype as $PiSLF_2$) failed to demonstrate functionality of the FD2 domain *in vivo* (Fields et al., 2010). That is, neither chimeric protein showed competitive interaction in transgenic assays. Given that most SLFL proteins now appear to be bona fide SLF variants (see section 8), the long-term significance of these assays is unclear. The different SLFL proteins used for sequence comparisons represent different classes of SLF variants (section 8) so that the "SLF-specific" domains identified may represent regions that are more similar within a particular SLF-variant class.

6. A role for ubiquitination in gametophytic self-incompatibility

The observed protein interactions described above, together with the properties of SLF, SBP1 and SSK1 all suggest that recognition of self versus non-self in gametophytic self-incompatibility involves ubiquitination pathways. Pollen-S (SLF), is an F-box protein, and F-box proteins are know the be the recognition component of SCF E3 ubiquitin ligases (Cardozo & Pagano, 2004; Hua et al., 2008; Sijacic et al., 2004). SBP1 (Hua & Kao 2006; Sims & Ordanic 2001; Sims et al., 2010) is a RING-HC protein. RING proteins are E3 ubiquitin ligases (Deshaies & Joazeiro 2009; Freemont 2000), and SBP1 has E3 ubiquitin ligase activity *in vitro* (Hua et al., 2007; Sims unpublished). AhSSK1 (Huang et al., 2006) and PhSSK1 (Zhao et al., 2010) are SKP1-like proteins (SKP1 is a scaffold component of SCF E3 ligases). Pollen extracts have been shown to ubiquitinate S-RNase proteins, albeit in an allele-independent manner (Hua & Kao 2006). Together, these results have lead to the proposal that a non-canonical SCF^{SLF}-like complex acts to recognize and ubiquitinate S-RNases, leading to the inhibition of S-RNase activity in compatible pollen tubes. This complex is proposed to differ from a canonical SCF complex, because neither SKP1 orthologues (Hua & Kao 2006; Huang et al., 2006; Zhao et al., 2010) or RBX1 (Hua & Kao 2006) interact with SLF or Cullin. Instead either (or both) SBP1 and SSK1 have been proposed to replace RBX1 and/or SKP1 in this complex (Sims 2007; Hua et al., 2008; Sims & Robbins 2009; Zhao et al., 2010). According to the simplest version of this model, recognition of non-self S-RNases by the SCF^{SLF} complex would lead to polyubiquitination and degradation of S-RNase by the 26S proteasome complex (Sims 2007; Hua et al., 2008). One prediction of the SCF^{SLF} ubiquitin ligase complex model is that down-regulation of SLF, SBP1 or SSK1 should render all pollen tubes incompatible, regardless of genotype. To date, down-regulation of SLF or SBP1 has not been reported. Down-regulation of PhSSK1 (Zhao et al., 2010) does, however, result in a switch from compatibility to incompatibility, in accordance with this model. Figure 4 summarizes the structure of the proposed SCF^{SLF} ubiquitin ligase complex.

7. Style-expressed proteins with roles in GSI

Several style-expressed proteins have been shown to either be required for pollen rejection, or to interact with S-RNase or SBP1 in different assays. Cruz-Garcia et al., (2005) immobilized the S_{c10}-RNase from *Nicotiana alata* on an Affi-Gel column, then tested the ability of different proteins from style extracts to bind to the immobilized S-RNase. NaTTS, Na120K and NaPELPIII stylar proteins all bound to the S_{c10}-RNase in a specific manner. All three of these are previously-characterized proteins that are secreted into the transmitting tract of the style and that interact with pollen tubes. Biochemical data suggests that these proteins and the S-RNase may form a complex that is taken up into pollen tubes. In an extension of these experiments, Lee et al., (2008) used the C-terminal domains of the TTS and the 120 K proteins in yeast two-hybrid screens of pollen cDNA libraries. In addition to interaction with SBP1 (see section 4.2) a putative cysteine protease, NaPCCP, interacted with both TTS and 120K. Figure 5 summarizes observed protein interactions of the style proteins.

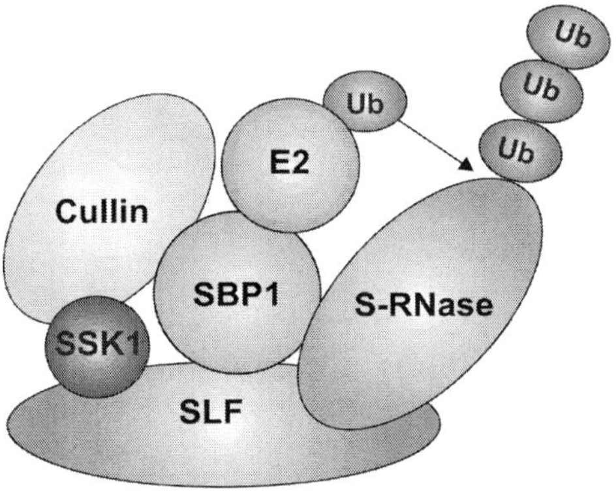

Fig. 4. **Proposed SCF^{SLF} ubiquitin ligase complex.**
Components of a proposed SCF^{SLF} complex are illustrated digrammatically. Specific contacts between components are based on protein-interaction assays summarized in Figure 3.

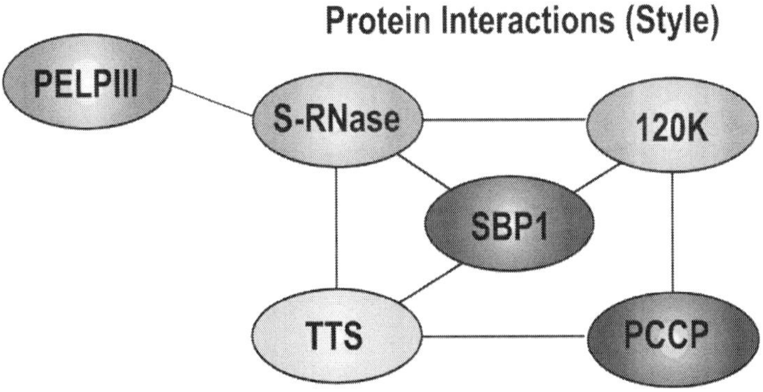

Fig. 5. **Interactions of the style proteins TTS, 120K and PELPIII with S-RNase and the pollen-expressed proteins SBP1 and PCCP.**

Two proteins, the 120k protein and a small asparagine-rich protein HT-B are required for the ability to reject pollen tubes in GSI (Hancock et al., 2005; McClure et al., 1999; O'Brien et al., 2002). Down-regulation of these genes by antisense (McClure et al., 1999) or RNAi (Hancock et al., 2005; O'Brien et al., 2002) resulted in an inability to block pollen tube growth, even though S-RNase proteins were expressed at normal levels. The 120K protein interacts with S-RNase and may be imported into pollen tubes in a complex with S-RNase and other style proteins (Cruz-Garcia et al., 2005). HT-B is also imported into pollen tubes (Goldraij et al., 2006), but whether in a complex with S-RNase and other style proteins, or separately, is not known.

8. Collaborative non-self recognition by SLF proteins in GSI

Although SLF proteins fulfill many of the expectations of pollen-S, different lines of evidence suggested that the nature of pollen-S may be more complicated than previously thought. First, as described, the evolutionary history of SLF proteins appeared far different than that of the linked S-RNase proteins, with the SLF proteins having evolved more recently and showing no evidence of co-evolution with S-RNase proteins as previously expected (Newbigin et al., 2008). Further, multiple SLF-like genes had been identified in *Petunia, Antirrhinum* and *Nicotiana* which appeared to be linked to the S-locus if not as close to the S-RNase as was SLF. It was unclear how the high degree of sequence identity among different SLF proteins could account for the ability to recognize and inhibit multiple different S-RNase alleles in populations.

Kubo et al. (2010) cloned additional SLF alleles from *Petunia hybrida*. [It should be noted here that *Petunia hybrida* is an "artificial" species created in the 19th century by crossing *Petunia axillaris* x *Petunia integrifolia*. *Petunia inflata* has been viewed either as synonymous with, a subspecies of, or very closely related to *Petunia integrifolia* (Stehmann et al., 2009). Thus S-locus haplotypes found in *Petunia hybrida* should be identical to those in either of the progenitor species.] When the SLF protein from *Petunia hybrida* S_7 haplotype was sequenced, it was found to be identical with the previously isolated SLF from *Petunia axillaris* S_{19}. What was striking, however was that the two S-RNases in these lines (S_7 versus S_{19}) were substantially different, having only 45% amino acid sequence identity. Reciprocal pollinations between S_7 and S_{19} confirmed that these two lines indeed had separate S-locus haplotypes. These results suggested that additional genes beyond SLF might constitute pollen-S. Further testing of SLF-S_7 showed that it could not cause competitive interaction in S_5S_7, S_7S_{11} or S_5S_{19} plants, but that it did show competitive interaction in S_7S_9 plants as well as in S_5S_{17} plants (Kubo et al., 2010 and supplemental material). Thus it appeared that individual SLF proteins could cause competitive interaction (i.e. act as pollen-S) with a limited subset of S-RNase alleles. Further analysis of proteins previously identified as SLF-like genes showed that these too, reacted with different subsets of S-locus haplotypes to cause competitive inhibition. Protein interaction assays demonstrated that there was direct correlation between SLF-S-RNase interaction and the ability to show competitive interaction in diploid heteroallelic pollen. Additional sequence comparisons demonstrated that SLF proteins could be grouped into at least six subclasses. Because Wheeler & Newbigin (2007) identified at least 10 different SLF-like genes in *Nicotiana*, and because not all of the previously-identified SLFL genes from *Petunia inflata* were included in the comparative sequence analysis of Kubo et al. (2010), it is possible that more than six SLF subclasses are present. As a result of this analysis, a new nomenclature for SLF proteins has been proposed. The original SLF isolates (e.g. PiSLF$_1$, PiSLF$_2$ etc.) have been renamed as the SLF1 variant class. Thus PiSLF$_1$ has been renamed S_1-SLF1, PiSLF$_2$ is now S_2-SLF1 and so on.

Genes previously identified as encoding SLFL proteins now comprise SLF2, SLF3, SLF4, SLF5, SLF6 and possibly additional SLF classes. The general nomenclature is thus $S_{haplotype\ ID}$-SLF(class).

These results led Kubo et al. (2010) to propose a "Collaborative Recognition" model for the interaction of SLF variants with different S-RNases. According to this model, different SLF variants can recognize separate but partially overlapping subsets of S-RNases. Thus S_7-SLF1 reacts with S_{17}-RNase and S_9-RNase, but not with S_{11}- or S_{19}-RNase, S_7-SLF2 reacts with S_9-, S_{11}- and S_{19}-RNases, but not with S_{17}-RNase. One important area of future research (see below) will be to further dissect the complexities of protein interactions between different SLF variants and S-RNase proteins.

9. Models for pollen recognition and rejection in GSI

At present, two different, but non-exclusive models have been proposed to explain the mechanism of pollen acceptance versus pollen-rejection in gametophytic self-incompatibility. Both models presume that incompatible pollen tubes are rejected via the cytotoxic action of the S-RNase, and that self-compatibility (or cross-compatibility) results from inhibiting S-RNase action, consistent with the presumed role of pollen-S as an inhibitor. Where the models differ is in the primary mechanism for S-RNase inhibition, as well as the "default" condition of pollination. One model proposes that pollen-S (SLF) acts via the SCF^{SLF} E3 ubquitin ligase complex to polyubiquitinate S-RNases, resulting in degradation by the 26 S proteasome complex. In this model, self-incompatibility (pollen rejection) would be the default pathway, unless the S-RNase is inhibited. The alternative model proposes that S-RNase imported into pollen tubes is sequestered in a vacuolar-like compartment. In this model, the default pathway is compatibility, unless SLF-S-RNase recognition leads to a breakdown of the compartment and release of the S-RNase.

9.1 The ubiquitination-degradation model

Much of the evidence for this model comes from protein-interaction assays, along with the known characteristics of the interacting proteins. Pollen-S (SLF), is an F-box protein, the recognition component of SCF E3 ubiquitin ligases (Cardozo & Pagano, 2004; Sijacic *et al.* 2004; Hua *et al.* 2008). SBP1 (Sims & Ordanic 2001; Hua & Kao 2006; Patel 2008, Sims *et al.*, 2010) is a RING-HC protein. RING proteins are E3 ubiquitin ligases (Freemont 2000, Deshaies & Joazeiro 2009), and SBP1 has E3 ubiquitin ligase activity *in vitro* (Hua *et al.* 2007; Sims unpublished). AhSSK1 (Huang *et al.* 2006) and PhSSK1 (Zhao *et al.* 2010) are SKP1-like proteins (SKP1 is a scaffold component of SCF E3 ligases). Pollen extracts have been shown to ubiquitinate S-RNase proteins, albeit in an allele-independent manner (Hua & Kao 2006). Together, these results have lead to the proposal that a non-canonical SCF^{SLF}-like complex acts to recognize and ubiquitinate S-RNases, leading to the inhibition of S-RNase activity in compatible pollen tubes. This complex is proposed to differ from a canonical SCF complex, because neither SKP1 orthologues (Hua & Kao 2006; Huang *et al.* 2006; Zhao et al 2010) or RBX1 (Hua & Kao 2006) interact with SLF or Cullin. Instead either (or both) SBP1 and SSK1 have been proposed to replace RBX1 and/or SKP1 in this complex (Sims 2007; Hua *et al.* 2008; Sims & Robbins 2009; Zhao *et al.* 2010). According to the simplest version of this model, recognition of non-self S-RNases by the SCF^{SLF} complex would lead to polyubiquitination and degradation of S-RNase by the 26S proteasome complex (Sims 2007; Hua et al., 2008). One prediction of the SCF^{SLF} ubiquitin ligase complex model is that down-regulation of SLF, SBP1 or SSK1 should render all pollen tubes incompatible, regardless of genotype. To date, down-regulation of SLF or SBP1 has not been reported. Down-regulation of PhSSK1 (Zhao *et al.* 2010) does, however, result in a switch from compatibility to incompatibility, in accordance with this model.

Although the ubiquitination-degradation model is attractive, several predictions of this model remain untested, and other predictions may (depending on interpretation) be contradicted by current evidence. The pattern of ubiquitination of the S-RNase *in vivo* is not known. Because K48-linked or K63-linked polyubiquitination, or monoubiquitination leads to different cellular outcomes for the tagged proteins, it will be important to determine what ubiquitination patterns occur in reponse to SLF:S-RNase interaction. Also, it is not clear whether large-scale degradation of S-RNase proteins occurs in compatible pollinations. The high level of secreted extracellular S-RNase that accumulates in the transmitting tract make it challenging to monitor the level of S-RNase proteins in pollinated styles. As stated earlier, the degradation model predicts that inactivation or down-regulation of SCF^{SLF} E3 ubiquitin ligase components should result in pollen rejection. This prediction appears to be sustained in the case of SSK1. Different SFB mutants characterized in the Rosaceae, however (Marchese et al., 2007; Sonneveld et al., 2005; Ushijima et al., 2004; Vilanova et al., 2006), all of which either truncate or delete the SFB protein, are self-compatible. Although these data (along with some other differences between Solanaceae/Plantaginaceae versus Rosaceae) have been interpreted as suggesting that GSI has a different mechanistic basis in these taxa, there is also a large degree of similarity in how GSI functions in Solanaceae/Plantaginaceae versus Rosaceae (e.g., S-RNase, F-box proteins) such that it may be premature to make a definitive judgement on that point (McClure et al., 2011). Figure 6 summarizes the basic ubiquitination-degradation model.

9.2 The sequestration model

Evidence for this model comes primarily from the work of Goldraij et al. (2006), who reported that S-RNase was sequestered in a vacuolar compartment in compatible pollinations. These authors fixed and paraffin-embedded pollinated styles of *Nicotiana alata,* then hybridized sections to antibodies for callose (pollen-tube cell wall marker), S_{c10}-RNase, 120K protein, HT-B, aleurain (vacuolar lumen marker) or vPPase (vacuolar membrane marker). They concluded that in a compatible pollination, S-RNase inside pollen tubes remained in a ribbon-like vacuole bounded by the 120K protein. HT-B levels in compatible pollinations were low or undetectable. In later stages of incompatible pollinations, conversely, S-RNase appeared to be released into the cytoplasm of pollen tubes, and HT-B levels remained higher than in compatible pollinations (Goldraij et al., 2006; McClure et al., 2011). This model is consistent with the results of RNAi down-regulation of HT-B and 120K, which prevent rejection of incompatible pollen. What this model only incompletely explains is the required S-RNase::SLF interaction leading to compatibility or incompatibility. Both genetic evidence and the protein-interaction data summarized in previous sections show that S-RNase and SLF must interact to determine self/non-self recognition. If S-RNase is sequestered in a vacuolar compartment, however and SLF is cytoplasmic, it is not clear how

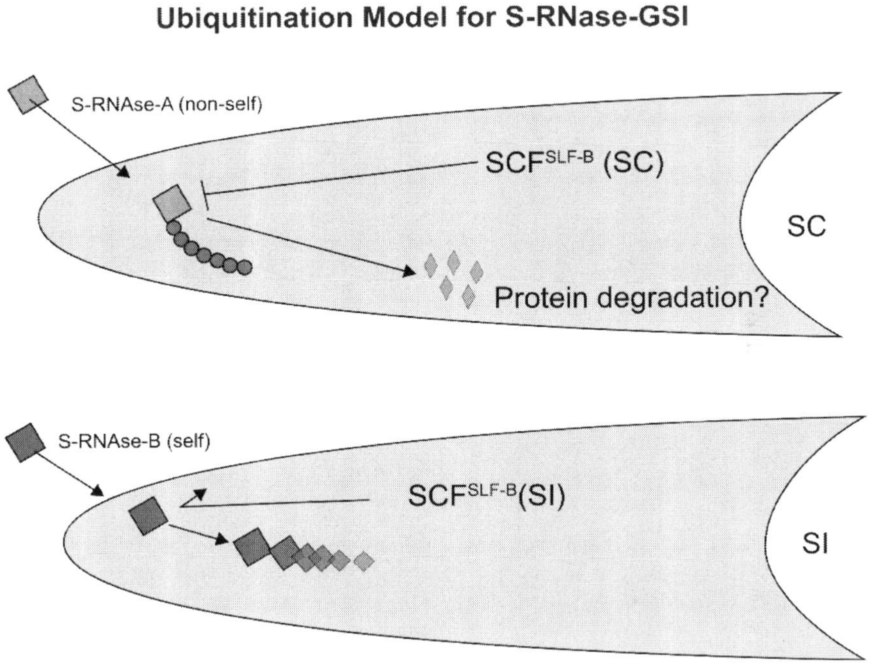

Fig. 6. **Ubiquitination-degradation model for gametophytic self-incompatibility.** According to this model, both non-self (S-RNase-A) and self (S-RNase-B) proteins are imported into pollen tubes (the mechanism of import is not defined, but probably does not involve a specific receptor). In compatible (non-self) pollen tubes a SCF^{SLF} E3 ubiquitin ligase complex targets ths S-RNase for polyubiquitination and degradation. In self-incompatible (self) pollen tubes, the SCF^{SLF} complex is incapable of targeting the S-RNase, which acts to degrade pollen tube RNA and inhibit protein synthesis and growth.

this interaction can take place. McClure et al. (2011) suggest that a small amount of S-Rnase may be able to escape the vacuolar compartment, possibly by retrograde transport, to interact with the SCF^{SLF} complex. In the case of an incompatible pollination, this interaction presumably leads to stabilization of HT-B, breakdown of the vacuolar compartment and release of the S-RNase. Figure 7 summarizes the essential aspects of the sequestration model.

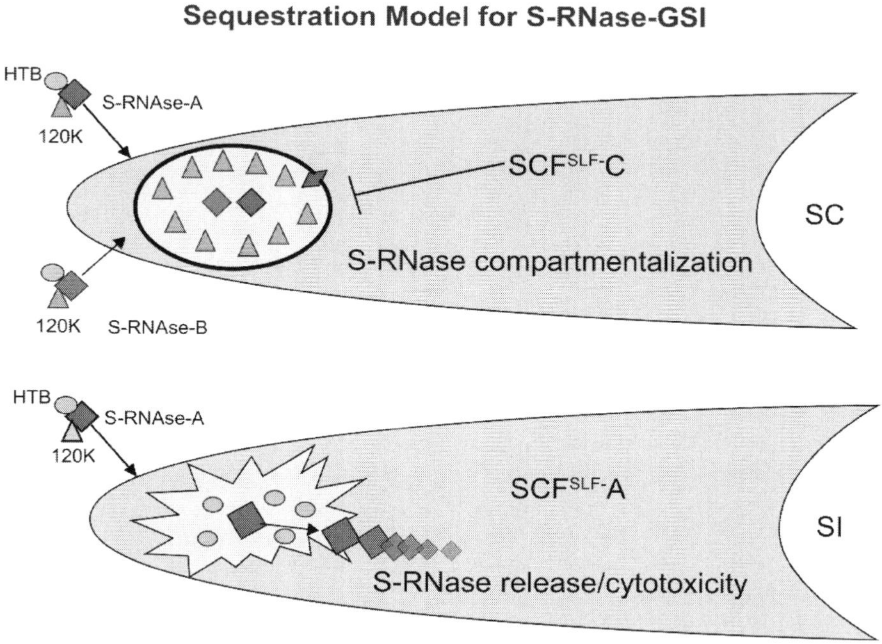

Fig. 7. **Sequestration model for gametophytic self-incompatibility.**
According to the sequestration model, S-RNases are imported into pollen tubes via an exocytotic mechanism, possibly in a complex with other style proteins (the complex shown with HT-B and 120K is speculative). S-RNase remains sequestered in a compatible pollination; in an incompatible pollination the vacuolar compartment breaks down releasing the S-RNase.

10. Current questions and future research

Although tremendous progress has been made in identifying genes and proteins involved in gametophytic self-incompatibility and in understanding much of the basic molecular biology of this phenomenon, many questions remain, and additional research is needed on nearly all aspects of GSI. In particular the collaborative recognition model raises the question of what is the exact nature of pollen-S? Do single SLF proteins interact one-on-one with individual S-RNases or can multiple SLFs interact simultaneously? Given the high degree of sequence identity with any specific class of SLF variants, what constitutes an allele? What protein interactions are required to make a determination of self versus non-self? If the sequestration model is correct, how do SLF and S-RNase even make contact? What is the specific role of ubiquitination in GSI interactions? Are S-RNases polyubiquitinated and degraded, or do different patterns of ubiquitination result in directing S-RNases to (or keeping them in) a membrane-bound compartment? What other proteins are needed for GSI interactions? Investigations not addressed in this review have suggested that proteins such as NaStEP (Busot et al., 2008) or Sli (Hosaka & Hanneman, 1998) may act as modifiers of the GSI response. What is the molecular basis for the quantitative, reversible, breakdown of GSI known as pseudo-self-compatibility, or PSC (Flaschenreim & Ascher 1979a, 1979b; Dana & Ascher 1985, 1986a, 1986b). What is the mechanism of uptake of S-RNases and other proteins into pollen tubes? Do Solanaceae/Plantaginaceae and Rosaceae really differ in fundamental mechanisms of GSI? More refined protein-interaction assays, suh as those using bi-molecular fluorescence complementation (Gehl et al., 2009), robust transgenic experiments, more complete information on genes and gene families involved in gametophytic self-incompatibility should all prove valuable in addressing these questions.

11. References

Ai Y, Singh A, Coleman CE, Ioerger TR, Kheyr-Pour A, Kao T-h. (1990) *Sexual Plant Reproduction,* 3, 130-138.

Anderson, M.A., Cornish, E.C., Mau, S.L., Williams, E.G., Hoggart, R., Atkinson, A., Bonig, I., Grego, B., Simpson, B., Roche, P.J., Haley, J.D., Penschow, J.D., Niall, H.D., Tregear, G.W., Coghlan, J.P., Crawford, R.J, & Clarke, A.E. (1986) *Nature,* 321, 38-44.

Anderson MA, McFadden GI, Bernatzky R, Atkinson A, Orpin T, Dedman H *et al.* (1989) The Plant Cell 1, 483-491.

Busot, G.Y., McClure, B., Ibarra-Sánchez, C.P., Jiménez-Durán, KI, Vázquez-Santana, S. & Cruz-Garcia, F. (2008) *Journal of Experimental Botany* 59, 3187-3201.

Cardozo T, Pagano M. (2004) *Nature Reviews Molecular Cell Biology,* 5, 739-751.

Chawla, B., Bernatzky, R., Liang, W. & Marcotrigiano, M. (1997) *Theoretical and Applied Genetics* 95, 992-996.

Chen, G., Zhang, B., Zhao, Z., Sui, Z., Zhang, H. & Xue, Y. (2010) *Journal of Experimental Botany* 61, 2027-2037.

Clark, KR, Okuley J, Collins PD, Sims TL. (1990) *The Plant Cell* 2, 815-826.

Coleman, C. & Kao, T.-H. (1992) *Plant Molecular Biology,* 18, 725-737.

Cruz-Garcia, F., Hancock, N., Kim, D. & McClure, B. (2005) *The Plant Journal,* 42, 295-304.

Dana MN, Ascher PD. (1985) *Journal of Heredity* 76, 468-470.

Dana MN, Ascher PD. (1986a) *Theoretical and Applied Genetics* 71, 573-577.

Dana MN, Ascher PD. (1986b) *Theoretical and Applied Genetics* 71, 578-584.

Darwin C (1891). *The Effects of Cross and Self Fertilisation in the Vegetable Kingdom.* (3rd ed). John Murray, London

de Nettancourt D. (1977) *Incompatibility in Angiosperms,* Springer-Verlag, ISBN 3-540-08112-7, Berlin

Deshaies RJ and Joazeiro CAP (2009) Annual Review of Biochemistry 78, 399-434.

Dodds PN, Ferguson C, Clarke AE and Newbigin E. (1999) *Sexual Plant Reproduction* 12, 76-87.

Dowd PE, McCubbin, AG., Wang, Verica, J.A., Tsukamoto, T., Ando, T. & Kao, T.-H., (2000) *Annals of Botany,* 85, 87-93.

Entani, T., Iwano, M., Shiba, H., Takayama, S., Fukui, K. & Isogai. A. (1999) *Theoretical and Applied Genetics* 99, 391-391.

Entani T, Iwano M, Shiba H, Che F-S, Isogai A Takayama S. (2003) *Genes to Cells* 8, 203-213.

Fields, A.M., Wang, N., Hua, Z. Meng, X. & Kao, T.-H., (2010) *Plant Molecular Biology* 74, 279-292.
Flaschenreim DR, Ascher PD. (1979a) *Theoretical and Applied Genetics* 54, 97-101
Flaschenreim DR, Ascher PD. (1979b) *Theoretical and Applied Genetics*, 55, 23-28
Freemont PS. (2000) *Current Biology* 10, 84-87.
Gebhardt, C., Ritter, E., Barone, A., Debener, T., Walkemeier, B., Schachtschabel, U., Kaufmann, H., Thompson, R.D., Bonierbale, M.W., Ganal, M.W., Tanksley, S.D. & Salamini, F. (1991) *Theoretical and Applied Genetics*, 83, 49-57.
Gehl C, Waadt R, Kudla J, Mendel RR and Hänsch (2009) *Molecular Plant* 2, 1051-1058.
Goldraij, A., Kondo, K., Lee, C.B., Hancock, C.N., Sivaguru, M., Vazquez-Santana, S., Kim, S., Phillips, T.E., Cruz-Garcia, F. & McClure, B. (2006) *Nature*, 439, 805-810.
Golz, J.F., Su, V., Clarke, A.E. & Newbigin, E. (1999) *Genetics* 152, 1123-1135.
Golz, J.F., Oh, H.Y., Su, V., Kusaba, M. & Newbigin, E. (2001) *PNAS* 98, 15372-15376.
Haglund K, Di Fiore PP, and Dikic I (2003) *Trends in Biochemical Sciences* 28, 598-604.
Hancock, N., Kent, L., & McClure, B. A. 92005) *The Plant Journal* 43, 716-723.
Harbord, R.M., Napoli, C.A. & Robbins, T.P. (2000) *Genetics* 154, 1323-1333.
Hosaka, K. & Hanneman, R.E. (1998) *Euphytica* 99, 191-197.
Hua Z, Kao T-H. (2006) *The Plant Cell* 18, 2531-2553.
Hua, Z., Meng, X. & Kao, T.-H. (2007) *The Plant Cell*, 19, 3593-3609.
Hua, Z. & Kao, T.-H. (2008) *Plant Journal* 54, 1094-1104.
Huang S, Lee H-S, Karunanandaa B, Kao T-H. (1994) *The Plant Cell* 6, 1021-1028.
Huang J, Zhao L, Yang Q, Xue Y. (2006) *The Plant Journal* 46, 780-793.
Ida K, Norioka S, Yamamoto M, Kumasaka T, Yamashita E, Newbigin E, Clarek AE,Sakiyama F, Sato M. (2001) *Journal of Molecular Biology* 314, 103-112.
Ioerger TR, Gohlke JR, Xu B, Kao T-h. (1991) *Sexual Plant Reproduction* 4, 81-87.
Ishimuzu, T., Miyagi, M., Norioka, S., Liui, Y.H., Clarke, A.e. & Sakiyama, F. (1995) *J. Biochem* 118, 1007-1013.
Kao, T.H., & McCubbin A.G. (1996) *PNAS* 93, 12059-12065.
Karunanandaa B, Huang S, Kao T-h. (1994) *The Plant Cell* 6, 1933-1940.
Kohn, J.R. (2008) What Geneologies of S-alleles Tell Us, In: *Self-Incompatibility in Flowering Plants, Evolution, Diversity, and Mechanisms*, Veronica E. Franklin-Tong, Editor, pp 103-121, Springer-Verlag, ISBN 978-3-540-68485-5, Berlin
Kubo, K.I., Entant, T., Takata, A., Wang, N., Fields, A.M., Hua, Z., Toyoda, M., Kawashima, S.I., Ando, T., Isogai, A., Kao, T.-H., & Takayama, S. (2010) *Science* 330, 796-799.
Lai Z, Ma W, Han B, Liang L, Zhang Y, Hong G, Xue Y. (2002) *Plant Molecular Biology* 50, 29-41.
Lee, C.B., Swatek, K.N. & McClure, B. (2008) *Journal of Biological Chemistry* 283, 26965-26973.
Lee, H.S., Huang, S. & Kao, T.-H., (1994) *Nature* 367, 560-563.
Li, J.H., Nass, N., Kusaba, M., Dodds, P.N., Treloar, N., Clarke, A.E. & Newbigin, E. (2000) *Theoretical and Applied Genetics*, 956-964.
Linskens, H.F. (1957) *Proceedings of the Royal Society of London, Series B.* 188,299-311.
Lush, W.M. & Clarke, A.E. (1996) *Sexual Plant Reproduction* 10, 27-35.
Luu DT, Qin KK Morse D, Cappadocia M. (2000) *Nature* 407, 649-651.
Marchese, A., Boskovic, R.I., Caruso, T., Raimondo, A., Cutuli, M. & Tobutt, K.R. (2007) *Journal of Experimental Botany* 58, 4347-4356.
Mather, K. (1943) *Journal of Genetics* 45, 215-235.

Matton DP Luu DT, Xike Q, Laublin G, O'Brien M, Maes O, et al. (1999) *The Plant Cell* 11, 2087-2098.

Matton DP, Maes O, Laublin G, Xike Q, Bertrand C, Morse D, Cappadocia M. 1997) *The Plant Cell* 9, 1757-1766.

McCubbin, A.G., Wang, X. & Kao, T.H. (2000) *Genome* 43, 619-627.

McClure, B.A., Gray, J.E., Anderson, M.A. & Clarke, A.e. (1990) *Nature* 347, 757-760.

McClure, B., Mou, B., Canevacsini, S. & Bernatzky, R. (1999) *PNAS* 96, 13548-13553.

McClure, B., Cruz-Garcia, F. & Romero, C. (2011) *Annals of Botany*, 108, 647-658

Meng, X., Sun, P. & Kao, T.-H. (2011) *Annals of Botany*, 108, 637-646.

Mora, C., Tittensor, D.P., Adl, S., Simpson, A.G.B. & Worm, B. (2011) *PLOS Biology*, 9, e1001127

Murfett, J., Atherton, T.L., Mou, B., Gasser, C.S. & McClure, B.A. (1994) *Nature* 367, 563-566.

Newbigin, E., Paape, T. & Kohn, J.R. (2008) *The Plant Cell*. 20, 2286-2292.

O'Brien, M., Kapfer, C., Major, G., Laurin, M., Bertrand, C., Kondo, K. Kowyama, Y. & Matton, D.P. (2002) *The Plant Journal* 32, 985-996.

O'Brien M, Major G, Chantha SC, Matton DP. (2004) *Sexual Plant Reproduction* 17, 81-87.

Patel, Avani (2008). Protein interactions between pistil and pollen components controlling gametophytic self-incompatibility. M.S. thesis, Northern Illinois University, United States -- Illinois. www.proquest.com

Qiao, H., Wang, H., Zhao, L., Zhou, J., Huang, J., Zhang, Y., & Xue, Y. (2004a) *The Plant Cell*, 16, 582-595.

Qiao H, Wang F, Zhao L, Zhou JL, Lai Z, Zhang YS. et al. (2004b) *The Plant Cell* 16, 2307-2322.

Richman, A.D., Kao, T.-H., Schaeffer, S.W. & Uyenoyama, M.K. (1995) *Heredity*, 75, 405-415.

Richman, A.D., Uyenoyama, M.K. & Kohn, J.R. (1996) *Heredity*, 76, 497-505.

Richman, A.R. & Kohn, J.R. (2000) *Plant Molecular Biology*, 42, 169-179.

Royo, J., Kunz, C., Kowyama, Y., Anderson, M., Clarke, A.e., & Newbigin, E. (1994) *PNAS* 91, 6511-6514.

Sassa, H., Kakui, H. & Minamkiawa, M. *Sexual Plant Reproduction* 23, 39-43.

Sijacic P, Wang X, Skirpan AL, Wang Y, Dowd PE, McCubbin AG, Huang S, Kao T-h. (2004) *Nature* 429, 302-305.

Sims TL (1993) Genetic Regulation of Self-Incompatibility. CRC Critical Reviews in Plant Sciences 12, 129-167.

Sims TL (2005) Pollen recognition and rejection in different self-incompatibility systems. Recent Research Developments in Plant Molecular Biology 2, 31-62.

Sims TL (2007) Mechanisms of S-RNase-based self-incompatibility. CAB Reviews, Perspectives in Agriculture, Veterinary Science, Nutrition and Natural Resources 2, No 058.

Sims TL and Ordanic M. (2001) *Plant Molecular Biology* 47, 771-783.

Sims, T.L. & Robbins, T.P. (2009) Gametophytic Self-Incompatibility in Petunia, In: *Petunia: Evolutionary, Developmental and Physiological Genetics, 2nd Edition,* Tom Gerats & Judith Strommer editors, pp 85-106, Springer, ISBN 978-0-387-84795-5, NY, USA.

Sims TL, Patel A and Shretsha P (2010) *Biochemical Society Transactions* vol 38, in press

Singh, A., Ai, Y. & Kao, T.-H. (1991) *Plant Physiology*, 96, 61-68.

Stehmann, J.R., Lorenz-Lemke, A.P., Freitas, L.B. & Semir J. (2009) The Genus Petunia, In: *Petunia: Evolutionary, Developmental and Physiological Genetics, 2nd Edition,* Tom Gerats & Judith Strommer editors, pp 1-28, Springer, ISBN 978-0-387-84795-5, NY
Sonneveld, T., Tobutt, K.R., Vaughan, S.P. & Robbins, T.P. (2005) *The Plant Cell* 17, 37-51.
Tanksley, S.D. & Loaiza-Figueroa, F. (1985) *PNAS* 82, 5093-5096
ten Hoopen, R., Harbord, R.M., Maes, T., Nanninga, N. & Robbins, ,T.P. (1998) *The Plant Journal* 16, 729-734.
Tsukamoto T, Ando T, Takahashi K, Omori T, Wataabe H, Kokubun H, Marchesi E and Kao T-H (2005) *Plant Molecular Biology* 57, 141-163.
Ushijima K, Sassa H, Tamura M, Kusaba M, Tao R, Gradziel TM et al. (2001) *Genetics* 158, 379-386.
Ushijima K, Yamane H, Watari A, Kakehi E, Ikeda K, Hauck NR et al. (2004) *The Plant Journal* 39, 573-586.
Ushijima K, Sassa H, Dandekar AM, Gradziel TM, Tao R, Hirano H. (2003) *The Plant Cell* 15, 771-781.
Vilanova, S., Badenes, M.L., Burgos, L., Martinez-Calvo, J., Llacer. G. and Romero, G. (2006) *Plant Physiology* 142, 629-641.
Walles, B. & Han S.P. (1998) *Physologia Plantarum* 103, 461-465.
Wang, Y., Wang, X., McCubbin, A.G. & Kao, T.-H., (2003) *Plant Molecular Biology* 53, 565-580.
Wang Y, Tsukamoto T, Yi K-W, Wang X, Huang A, McCubbin AG, Kao T-h. (2004) *Plant Molecular Biology* 54, 727-742.
Wheeler D. & Newbigin, E. (2007) *Genetics*, 177, 2171-2180.
Woodward, J.R., Bacic, A., Jahnen, W. & Clarke, A.E. (1989) *The Plant Cell*, 1, 511-514.
Zhao, L., Huang, J., Zhao, Z., Li, Q., Sims, T.L., & Xue, Y. (2010) *The Plant Journal*
Zhou J., Wang, F., MA, W., Zhang, Y., Han, B. & XUe, Y. (2003) *Sexual Plant Repoduction* 16, 165-177.
Zurek, D.M., Mou, B., Beecher, B. & McClure, B. (1997) *The Plant Journal* 11, 797-808

11

Direct Visualization of Single-Molecule DNA-Binding Proteins Along DNA to Understand DNA–Protein Interactions

Hiroaki Yokota
Institute for Integrated Cell-Material Sciences, Kyoto University
Japan

1. Introduction

Single-molecule fluorescence imaging has recently developed into a powerful method for studying biophysical and biochemical phenomena (Moerner, 2007), including DNA metabolism (Ha, 2004; Hilario & Kowalczykowski, 2010; Zlatanova & van Holde, 2006). While classical biochemical methods yield parameters that are ensemble averaged, single-molecule fluorescence imaging can be used to observe real-time behavior of individual biomolecules, allowing us to study their dynamic characteristics in great detail. Among the various biomolecules, protein molecules, which play central roles in many biological functions, are the prime targets for single-molecule imaging. Direct observations of single-molecule DNA-binding proteins acting on their DNA targets in real time have provided new insights into DNA metabolism. In this chapter, I focus primarily on recent advances in direct visualization of single-molecule DNA-binding proteins *in vitro*, especially on the key techniques employed for this visualization.

2. Fluorophores and fluorescence labeling methods

Direct visualization of protein dynamics by single-molecule fluorescence imaging requires labeling of target protein molecules with a fluorophore. To be used in single-molecule imaging, fluorophores must be (1) bright (have high extinction coefficients and high quantum yield), (2) photostable, and (3) relatively small so as not to perturb the functions of target protein molecules. With regard to photostability, fluorophores in single-molecule imaging tend to undergo photobleaching and blinking (repetitive fluorescence turning-on and -off) because of the high-power excitation. Agents that minimize such photophysical events have been reported (Aitken et al., 2008; Dave et al., 2009; Harada et al., 1990; Rasnik et al., 2006). In general, two classes of fluorophores are used for single-molecule fluorescence imaging: organic small-molecule fluorophores and quantum dots (Qdots) (Table 1). Despite the advantage of the labeling capability with genetic engineering, fluorescent proteins are not popularly applied to single-molecule fluorescence imaging because of their lower intensity and instability of their fluorescence emission. This section briefly describes fluorescence properties of dyes and Qdots and their fluorescence labeling methods.

2.1 Organic small-molecule fluorophores (dyes)

Organic small-molecule fluorophores or dyes with a molecular weight < 1 kDa are mainly used for covalent labeling of protein molecules. Frequently used dyes in single-molecule imaging are Cy3, Cy5, and Alexa derivatives. They emit sufficeint photons to be detected, but subsequently photobleach in approximately 1 minute, and often exhibit blinking (Table 1).

The most popular site-specific labeling method with dyes involves labeling the sulfhydryl group of the cysteine residues with a dye containing a maleimide group. The double bond of the maleimide group may undergo an alkylation reaction with the sulfhydryl group to form a stable thioester bond. The reaction is specific for thiols in the physiological pH range of 6.5–7.5. At pH 7.0, the reaction proceeds 1,000 times faster than its reaction with amines (Hermanson, 2008). Another popular labeling method, which is less site-specific, involves labeling the amine group of amino acids in protein molecules with a dye containing an *N*-hydroxysuccinimide ester (NHS). NHS ester reacts principally with the ε-amines of the lysine side chains and α-amines at the N-terminals (Hermanson, 2008).

2.2 Quantum dots (Qdots)

Qdots are inorganic semiconductor nanocrystals, typically composed of a cadmium selenide (CdSe) core and a zinc sulphide (ZnS) shell measuring 10–20 nm in diameter (Michalet et al., 2005). They are commonly used in single-molecule imaging owing to their resistance to photobleaching and extreme brightness (Table 1). They are characterized by broad absorption profiles, high extinction coefficients, and narrow and spectrally tunable emission profiles, depending on their sizes. Although they are resistant to photobleaching, they often exhibit blinking, which seems to be related to charging of the nanocrystal upon excitation (Table 1). However, recently, the problem of blinking was reported to be overcome by the use of a nanocrystalline CdZnSe core capped with a ZnSe semiconductor shell, in which the transition between ZnSe and CdSe is not abrupt, but radially graded (Wang et al., 2009).

Qdots are usually labeled to protein molecules via immunolabeling. The most popular method is based on an avidin-biotin interaction, an antigen-antibody reaction with the highest affinity. In this method, target protein molecules, which are conjugated with a biotin by chemical crosslinking or genetic engineering, can be labeled with avidin-coated Qdots that are commercially available.

2.3 Advances in site-specific labeling methods

Recent advances in tagging technologies based on genetic have engineering enabled site-specific targeting of fluorophores to protein molecules. One approach is to fuse the target protein molecule to a peptide or a protein recognition sequence, which then recruits a fluorophore coated with the proteins that have affinity for the specific peptide or the protein recognition sequence. For labeling of a Qdot used in single-molecule visualization, HA tag (Kad et al., 2010), Flag tag and HA tag (Gorman et al., 2010), and His tag (Dunn et al., 2011) have been used. Enzyme-mediated covalent protein labeling is also used, in which a recognition peptide is fused to the protein of interest and a natural or engineered enzyme ligates the small-molecule probe to the recognition peptide. This approach can confer highly specific and rapid labeling, with the benefit of a small directing peptide sequence. This approach involved the following pairs: Halo tag and dehalogenase, biotin ligase acceptor peptide and biotin ligase, and SNAP tag and O^6-alkylguanine-DNA alkyltransferase (Fernández-Suárez & Ting, 2008).

Fluorophore	λ_{em} (nm)	M. W. or size	ε ($M^{-1}cm^{-1}$)	Brightness	Resistance to photobleaching	Resistance to blinking
Dye						
Cy3 maleimide	570	791.0 Da	150,000 (550 nm)	++	++	++
Cy5 maleimide	670	817.0 Da	250,000 (649 nm)	++	+	+
Quantum dot						
Qdot655	~655	~15 to 20 nm	2,100,000 (532 nm)	+++	+++	+
Fluorescent protein						
EGFP	507	32.7 kDa	56,000 (484 nm)	+	+	+

Table 1. Size (or molecular weight) and fluorescence properties of commonly used dyes and a Qdot compared to those of a fluorescence protein.

3. Surface-coating methods

Minimization of the background noise arising from non-specific adsorption of the fluorescently labeled protein molecules on glass substrates is essential for their single-molecule fluorescence visualization. The non-specifically adsorbed protein molecules severely interfere with the visualization because the fluorescence associated with these molecules confounds the fluorescence signal from the target protein molecules. Poly (ethylene glycol) (PEG) and lipid are commonly used for reducing the non-specific protein adsorption on the glass substrate.

3.1 Poly (Ethylene Glycol) (PEG)

PEG is a biocompatible polymer that exhibits protein and cell resistance when immobilized onto metal (Prime & Whitesides, 1991), plastic (Ito et al., 2007), and glass surfaces (Cuvelier et al., 2003), which serves many applications especially in biosensors and medical devices. The feature of PEG comes from its high hydrophilicity and appreciable chain flexibility that induce an effective exclusion volume effect (Harris, 1992). To my knowledge, Ha et al. first used PEG coated glass substrates for single-molecule fluorescence imaging (Ha et al., 2002). The key factors affecting the suppression of nonspecific adsorption by PEG are its length and density (Harris, 1992), both of which are trade-offs. An increase in the chain length of PEG to construct a defined tethered chain layer results in a decrease in the density of the PEG chain due to the exclusion volume effect. PEG with a molecular weight of 5,000 appears to find a compromise between the trade-offs to maximize the suppression of non-specific adsorption of protein molecules on a glass surface (Heyes et al., 2004; Malmstena et al., 1998; McNamee et al., 2007; Pasche et al., 2003; Yang et al., 1999). A general approach to the immobilization of PEG onto a glass surface involves coupling of PEG to amine groups that have been conjugated on the surface by silanization (Figure 1a). Another approach involves coupling of PEG through poly (L-lysine) adsorption on the surface (Figure 1b). The inclusion of a small fraction of biotinylated PEG provides anchor points for tethering DNA (Section 4).

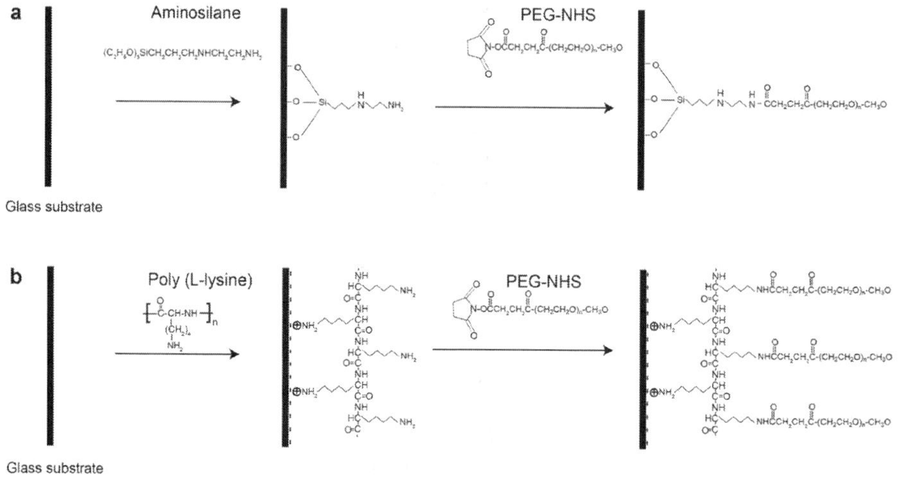

Fig. 1. PEG coating methods of glass substrates through either (a) silanization or (b) poly (L-lysine) adsorption.

3.1.1 PEGylation through aminosilanization

The standard PEG coating method on quartz slides and silicate coverslips for single-molecule fluorescence imaging is performed by silanization with *N*-2-(aminoethyl)-3-aminopropyl-triethoxysilane in methanol containing acetic acid for few hours (Figure 1 (a)) (Joo et al., 2006). The aminosilanization treatment yields a surface that is densely coated with exposed primary amines. A covalently attached PEG layer is formed on the surfaces by a PEGylation reaction with the amine reactive *N*-hydroxy-succinimidyl (NHS)-PEG (M.W. = 5,000 Da) dissolved commonly in freshly prepared 0.1 M sodium bicarbonate buffer (pH 8.3) for a couple of hours.

3.1.2 Improvement of non-specific adsorption suppression capability on silicate coverslips

I found that the standard PEG coating method mentioned above did not suppress non-specific adsorption as effectively on silicate coverslips as it did on quartz (Yokota et al., 2009). Therefore, an improved method for efficient PEG coating on silicates is required to reduce their non-specific adsorption. Then, I found that performing the PEGylation reaction in 50 mM 3-(*N*-morpholino)propanesulfonic acid (MOPS, pH 7.5) instead of the usual 0.1 M sodium bicarbonate (pH 8.3) reduced non-specific adsorption on silicate surfaces by up to an order of magnitude. Figure 2 shows the single-molecule fluorescence images of a cyanine dye-labeled protein (the *E. coli* helicase UvrD (Lohman et al., 2008) labeled with Cy5: Cy5-UvrD) non-specifically adsorbed on a silicate coverslip that was PEGylated in MOPS buffer as compared to one with no treatment and one coated following the standard method. The improvement is crucial since silicate coverslips are extensively used for single-molecule fluorescence microscopy with epi-illumination (Funatsu et al., 1995), highly inclined thin illumination (Tokunaga et al., 2008), and total-internal reflection fluorescence (Tokunaga et al., 1997) microscopies (sometimes combined with other techniques such as optical tweezers (Hohng et al., 2007; Lang et al., 2004; Zhou et al., 2011)). The improvement in non-specific adsorption using the new buffer condition for PEGylation was also observed on quartz, though to a lesser extent (Table 2) and with a Qdot-labeled protein molecule (UvrD labeled with Qdot655:Qdot655-UvrD) (Table 3). Figure 3 shows the single-molecule fluorescence images of a Qdot-labeled protein (Qdot-UvrD) non-specifically adsorbed on a silicate coverslip that was PEGylated in MOPS buffer as compared to one with no treatment. While PEGylation is required to reduce the non-specific adsorption of Qdot-labeled protein on both silicate and quartz surfaces, PEGylation in MOPS buffer reduced non-specific adsorption of Qdot-UvrD on glass substrates by a factor of approximately 2 as compared to PEG coating in sodium bicarbonate buffer (Table 3). The key issue determining the non-adsorption properties of PEG-coated glass surfaces is their coverage by the polymer (Harris, 1992), which is improved with the new method. PEG coating at pH 7.5 may be enhanced with respect to coating at higher pH owing to better stability of the NHS ester. Its lifetime is in the order of hours at physiological pH and decreases steeply at higher pH owing to increased hydrolysis (Hermanson, 2008).

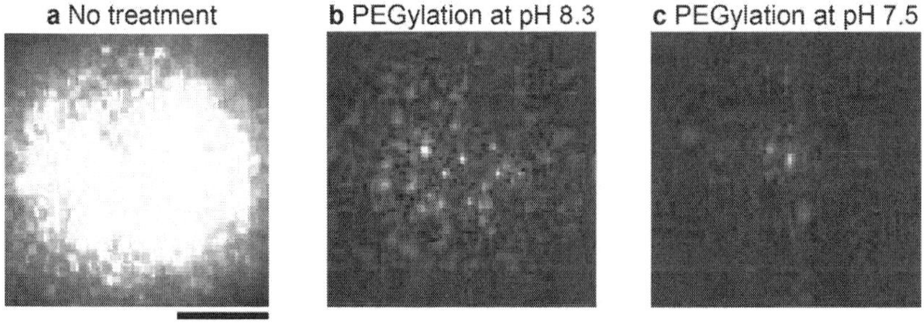

Fig. 2. Enhancement of protein-non-adsorption capability on silicate coated with PEG dissolved in 50 mM MOPS (pH 7.5). Fluorescence images of Cy5-UvrD non-specifically adsorbed on silicate with (a) no treatment, coated PEG dissolved in (b) 0.1 M sodium bicarbonate (pH 8.3) and (c) 50 mM MOPS (pH 7.5). The Cy5-UvrD concentration used was 2 nM. Scale bar, 10 μm.

	Silicate			Quartz		
	No treatment	PEG (pH 8.3)	PEG (pH 7.5)	No treatment	PEG (pH 8.3)	PEG (pH 7.5)
Mean number of Cy5-UvrD non-specifically adsorbed (/nM/1,000μm²)	1.6×10^3 $\pm 0.5 \times 10^3$ (21)	39 ±4 (9)	3.9 ±1.4 (30)	2.2×10^3 $\pm 0.3 \times 10^3$ (6)	1.4 ±0.2 (9)	0.78 ±0.20 (13)
Ratio to the number for PEG (pH 7.5)	4.1×10^2	10	1	2.8×10^3	1.8	1

The mean numbers of Cy5-UvrD non-specifically adsorbed per 1,000 μm² are normalized by the UvrD concentrations (nM) used for the experiments. Values represent mean ± standard deviation (*n*). (Reproduced from (Yokota et al., 2009) , Copyright (2009), with permission from The Chemical Society of Japan).

Table 2. Protein-non-adsorption capability on silicate coverslips and quartz slides (Cy5-UvrD).

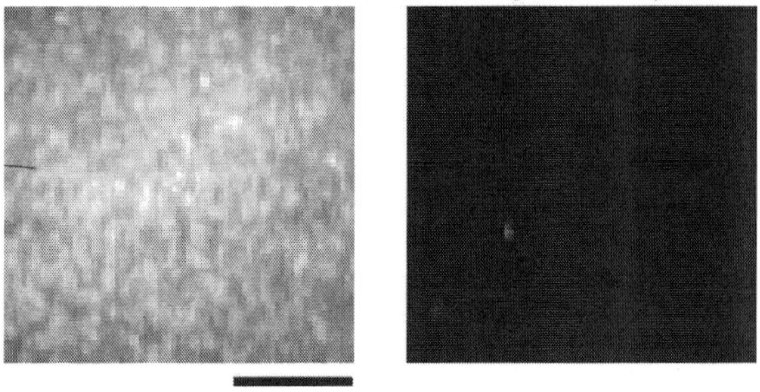

Fig. 3. Reduction of the non-specific adsorption of Qdot-labeled protein on glass substrates coated with PEG dissolved in 50 mM MOPS (pH 7.5). (b) Fluorescence images of Qdot-UvrD non-specifically adsorbed on silicate with no treatmemt (a) and coated with PEG dissolved in 50 mM MOPS (pH 7.5) (b). The Qdot-UvrD concentration used was 1 nM. Scale bar, 10 μm.

	Silicate			Quartz		
	No treatment	PEG (pH 8.3)	PEG (pH 7.5)	No treatment	PEG (pH 8.3)	PEG (pH 7.5)
Mean number of Qdot-UvrD non-specifically adsorbed (/nM/1,000 μm²)	6.8×10^3 $\pm 0.5 \times 10^3$ (8)	1.9 ±0.8 (23)	1.3 ±0.3 (22)	1.6×10^4 $\pm 0.3 \times 10^4$ (9)	1.7 ±0.2 (9)	0.51 ±0.20 (9)
Ratio to the number for PEG (pH 7.5)	5.2×10^3	1.5	1	3.1×10^4	3.3	1

The mean numbers of Qdot-UvrD non-specifically adsorbed per 1,000 μm² are normalized by the UvrD concentrations (nM) used for the experiments. Values represent mean ± standard deviation (*n*). (Reproduced from (Yokota et al., 2009) , Copyright (2009), with permission from The Chemical Society of Japan).

Table 3. Protein-non-adsorption capability on silicate coverslips and quartz slides (Qdot-UvrD).

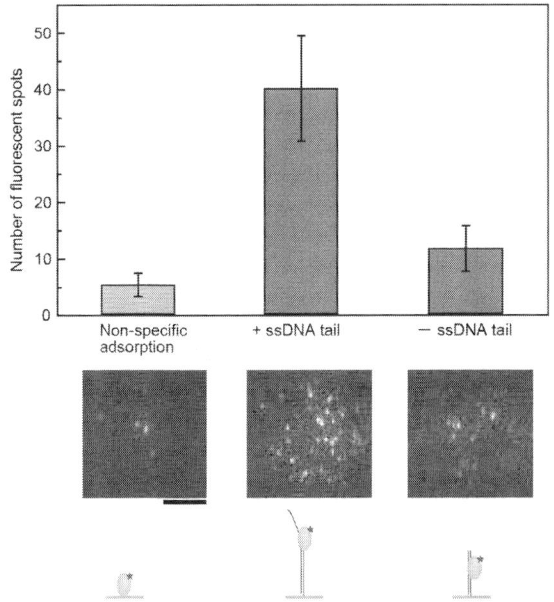

Fig. 4. Comparison of the numbers of Cy5-UvrD interacting with 18-bp dsDNA with or without an ssDNA tail immobilized on a PEG-coated silicate coverslip. Below are the single-molecule fluorescence images. The Cy5-UvrD concentration used was 2 nM. Scale bar, 10 µm. (Reproduced from (Yokota et al., 2009), Copyright (2009), with permission from The Chemical Society of Japan).

The results on the reduction of non-specific protein adsorption on glass surfaces indicate that our PEGylation method facilitates direct visualization of single-molecule interactions between fluorescently labeled protein molecules and DNA. Indeed, we could visualize how Cy5-UvrD interacts with a 18-bp dsDNA with or without an ssDNA tail immobilized on PEG-coated surfaces. The number of Cy5-UvrD fluorescent spots observed can be compared in Figure 4; this figure clearly shows that non-specific adsorption of Cy5-UvrD on the surface was effectively suppressed enough to conclude that Cy5-UvrD has higher affinity for the dsDNA with an ssDNA tail than for that without one; this finding is in agreement with that of a previous report (Maluf et al., 2003).

3.1.3 Single-molecule visualization of binding mode of helicase to DNA on silicate coverslips

Using microscopy, we could compare the number of UvrD molecules bound to 18-nt ssDNA, 4.7-kbp dsDNA, and 4.7-knt ssDNA immobilized on a PEG-coated silicate coverslip. Figure 5 shows a comparison of fluorescence intensity distributions of Cy5-UvrD bound to 18-nt ssDNA, 4.7-kbp dsDNA, or 4.7-knt ssDNA. In agreement with the results in Figure 4, Cy5-UvrD had low affinity for dsDNA, and thus, the fluorescence intensity distribution for 4.7-kbp dsDNA peaked at a fluorescence intensity that corresponds to that of single Cy5-UvrD, which was validated by the observation that most of the fluorescent spots photobleached in a single step. This was also the case with the experiment using 18-nt ssDNA. For the experiment using 4.7-knt ssDNA, the fluorescence intensity distribution shifted to a larger value, indicating that multiple Cy5-UvrD attached to the ssDNA.

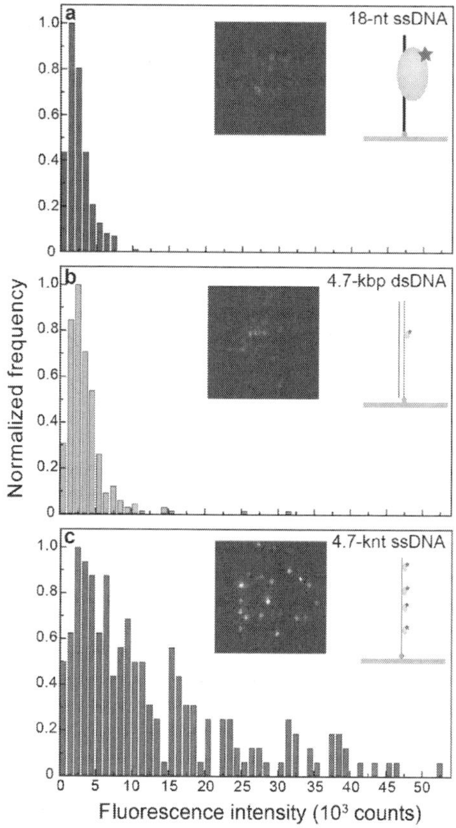

Fig. 5. Comparison of fluorescence intensity distributions of Cy5-UvrD bound to (a) 18-nt ssDNA, (b) 4.7-kbp dsDNA, or (c) 4.7-knt ssDNA immobilized on a PEG-coated silicate coverslip. The insets show the single-molecule fluorescence images. The Cy5-UvrD concentration used was 2 nM. Scale bar, 10 µm. (Reproduced from (Yokota et al., 2009) , Copyright (2009), with permission from The Chemical Society of Japan).

3.1.4 Stability of PEG-coated surfaces

The long-term stability of PEG has been examined in several studies. On examination of a PEG-coated coverslip surface for up to 1 month, the surface, immersed in 0.1 M sodium phosphate buffer (pH 7.4), has been found to degrade after 25 days (Branch et al., 2001). Another study examined the stability of PEG-modified silicon substrates that were incubated in PBS (37 °C, pH 7.4, 5% CO_2) for different periods of time up to 28 days. This study also concluded that PEG-modified surfaces retain their protein and cell repulsive nature for the period of investigation, i.e., 28 days (Sharma et al., 2004). With regard to PEG stability in the air, no degradation was reported after 75 days (Anderson et al., 2008). In contrast to these macroscopic experimental studies, PEG-coated surfaces used for single-molecule imaging imposes several challenges. The protein-non-adsorption capability of PEG-coated glass surfaces sharply deteriorated with time (incubation > 24 h) when incubated in a buffer (Yokota et al., 2009). Moreover, PEG-coated coverslips and quartz slides exhibited increasing non-specific adsorption with time when stored, exposed to the air, at room temperature (Yokota et al., 2009). PEG is known to degrade by oxidation (Han et al., 1997), and thus, to prevent the oxidative degradation, PEG-coated slides/coverslips can be stored in the dark under dry conditions at -20 °C until use (Joo et al., 2006).

3.1.5 PEGylation through poly(L-lysine) adsorption

PEGylation is also feasible through poly(L-lysine) adsorption on the glass surface (Figure 1b). The amino groups of the side chain of poly(L-lysine) are positively charged at pH values below 10, and are therefore easily attached to the negatively charged glass surface (Iler, 1979) through an electrostatic interaction at physiological pH. The protein non-adsorption capability is comparable to that of PEG-coated surfaces through aminosilanization (unpublished data). A similar strategy for PEGylation, which is not used for single-molecule fluorescence imaging, involves using a poly(L-lysine) grafted with PEG (PLL-g-PEG), a polycationic copolymer that is positively charged at neutral pH. The compound spontaneously adsorbs from aqueous solution onto negatively charged surfaces, resulting in the formation of stable polymeric monolayers and rendering the surfaces protein-resistant to a degree relative to the PEG surface density (Huang et al., 2001; Pasche et al., 2003).

3.2 Lipid bilayer

Artificial lipid membranes deposited on solid supports have proven to be useful for many types of biochemical studies (Chan & Boxer, 2007). The membranes provide a surface environment that is similar to that inside a living cell and can prevent any non-specific interactions of biomolecules on the surface. Lipid bilayers that are spontaneously formed on quartz surfaces have been used for single-molecule fluorescence imaging. The bilayers can be modified through the incorporation of lipids with various functional groups such as biotin or PEG. A mixture of 1,2-dioleoyl-*sn*-glycerophosphocholine dioleoylphosphatidylcholine (DOPC) with a small amount of 1,2-dioleoyl-*sn*-glycero-3-phosphoethanolamine-*N*-[methoxy(polyethylene glycol)-550] (mPEG 550-DOPE) was reported to help minimize non-specific binding of Qdot-tagged proteins to the lipid bilayer (Gorman et al., 2010). For experiments that require tethering DNA that is biotinylated at the end(s), avidin is attached either via non-specific interaction with the glass surface or via interaction with biotin head groups of the lipids present.

4. Platforms

A common feature among direct visualization of single-molecule DNA-binding protein molecules is the need for DNA to be tethered and extended. A pioneering platform published in 1993 employs dielectrophoresis between aluminium electrodes to extend λDNA (Fig. 6g) (Kabata et al., 1993). λDNA is a bacteriophage DNA that consists of 48,490 base pairs of double-stranded linear DNA with 12-nucleotide single-stranded segments at both 5′ ends (approximately 16.5 μm in total length). The platform can form "DNA belts" and has uncovered RNA polymerase dynamics along DNA, including sliding and jumping. Another pioneering platform published in 1998 employs optical tweezers, which was incorporated into a single-molecule fluorescence microscope, to extend λDNA (Fig. 6d) (Harada et al., 1999), which revealed strain-dependent DNA-binding kinetics of RNA polymerase. At that time, casein, a conventional blocking agent that is found in milk in large quantities, was used to block non-specific protein adsorption.

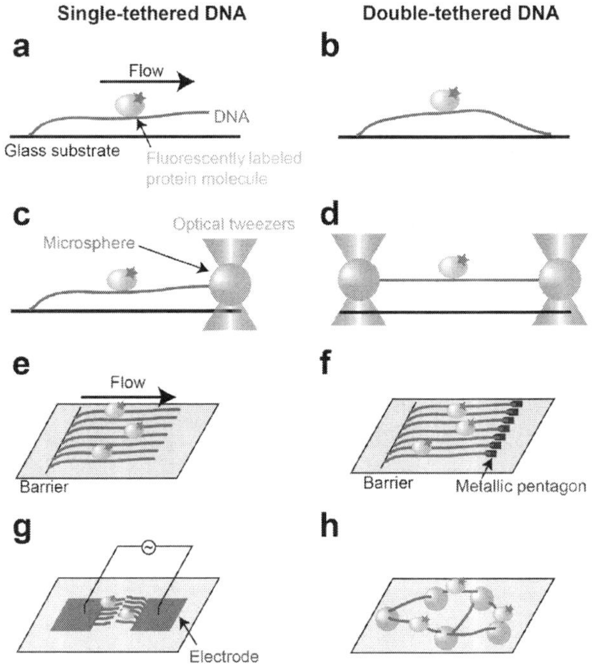

Fig. 6. Platforms used for direct visualization of single-molecule DNA-binding protein molecules along DNA. (a) DNA tethered at one end on a glass surface with buffer flow. (b) DNA tethered at both ends on a glass surface. (c) DNA tethered on a glass surface at one end and with optical tweezers at the other end. (d) DNA tethered with optical tweezers at both ends. (e) DNA curtains. (f) DNA racks. (g) DNA belts. (f) Tightropes.

To minimize non-specific protein adsorption, most of the visualization has been performed on glass surfaces modified by the coating methods discussed in the previous section. Many platforms capable of real-time visualization of single-molecule DNA-binding protein molecules along tethered DNA, most of which are used in combination with total-internal fluorescence microscopy (Funatsu et al., 1995; Tokunaga et al., 1997), have been developed to understand protein dynamics. Among many methods for DNA tethering by non-specific adsorption, or specific attachments mediated by a non-covalent interaction, the most popular DNA tethering method used for the platforms is tethering biotinylated λDNA on glass surfaces via an avidin-biotin interaction. The two single-stranded segments at both 5′ ends of λDNA, termed the *cos* site, are the sticky ends and can be easily biotinylated via ligation with corresponding the complimentary biotinylated oligos. The biotinylated λDNA can be tethered on glass surfaces at either one or both end(s) via an avidin-biotin interaction. In the following sections, I briefly review some of the modern platforms and present their target DNA protein molecules.

4.1 Platforms with continuous buffer flow

Tethering DNA at one end on the surface via an avidin-biotin interaction is the simplest method for tethering DNA. However, owing to the flexible nature of DNA, tethering alone cannot make the DNA extend sufficiently enough for use in single-molecule visualization with high spatial resolution. To stretch such DNA, some investigators use shear force generated by hydrodynamic flow (Fig. 6a) (Brewer & Bianco, 2008; Graneli et al., 2006; Greene & Mizuuchi, 2002; S. Kim et al., 2007; Lee et al., 2006; van Oijen et al., 2003). In this platform, buffer is continuously input into a flow cell made of sandwiched glass substrates. The glass substrates are coated with PEG (Blainey et al., 2006; Blainey et al., 2009; S. Kim et al., 2007) or a lipid bilayer (Graneli et al., 2006) to reduce non-specific adsorption of protein molecules. With this platform, dynamics of many single-molecule proteins have been uncovered. These include dynamics of single-protein polymers along DNA (Greene & Mizuuchi, 2002; Han & Mizuuchi, 2010; Tan et al., 2007) and one-dimensional diffusion of many DNA-binding protein molecules, for example, a base-excision DNA-repair protein along DNA (Blainey et al., 2006; Etson et al., 2010; Kochaniak et al., 2009; Komazin-Meredith et al., 2008; Tafvizi et al., 2011). A comparison of one-dimensional diffusion constants as a function of protein size with theoretical predictions indicates that DNA-binding proteins undergo rotation-coupled sliding along the DNA helix (Blainey et al., 2009).

4.2 Tethering DNA at multiple points on glass surfaces

As mentioned above, tethering DNA at one end on the surface alone is not sufficient for the single-molecule visualization. Several papers report methods of DNA tethering at many points via non-specific interaction or at both ends (Fig. 6b) on the surface via an avidin-biotin interaction. These include λDNA immobilized on a polystyrene-coated coverslip via a non-specific interaction (Kim & Larson, 2007) or on a lipid bilayer-deposited fused silica slide via an avidin-biotin interaction (Graneli et al., 2006), and T7 bacteriophage DNA immobilized on a silanized coverslip via an avidin-biotin interaction (Bonnet et al., 2008). Using these platforms, one-dimensional diffusion and transcription of single T7 RNA polymerases along DNA (Kim & Larson, 2007) and sliding and jumping of EcoRV restriction enzyme (Bonnet et al., 2008) along DNA were visualized.

4.3 DNA curtains

4.3.1 Single-tethered DNA curtains

To tether many DNA strands on the surface for a high-throughput single-molecule analysis, the Greene group has developed a new platform referred to as "DNA curtains" (Fig. 6e) (Fazio et al., 2008; Graneli et al., 2006; Visnapuu et al., 2008). This assay allows simultaneous study of up to hundreds of individual DNA strands anchored to a lipid bilayer and aligned with respect to one another within a single field-of-view with TIRFM. The DNA curtains were constructed in the following manner: (i) the surface of a glass slide was first mechanically etched to form lipid-diffusion barriers perpendicular to the direction of buffer flow. (ii) The flow cell was coated with a lipid bilayer by injecting lipid vesicles comprising DOPC, biotinylated lipids, and PEGylated lipids into the sample chamber (discussed in 3.2). (iii) After removing the excess vesicles by a buffer flush, neutravidin was injected into the flow cell. (iv) λDNA that was biotinylated at one end was injected into the flow cell. The PEGylated lipid enhanced protein non-specific adsorption suppression capability on the surface. The barriers formed on the flow cell could not be traversed by the lipids, and thus, buffer flow extended the lipid-tethered DNA in the flow direction and confined the DNA within the evanescent field generated by total-internal reflection. This platform has been used to study a broad range of DNA-binding proteins that function in homologous recombination (Prasad et al., 2006; Prasad et al., 2007) and mismatch repair (Gorman et al., 2007).

4.3.2 Double-tethered DNA curtains

The DNA curtains require buffer flow to extend DNA for single-molecule visualization of protein dynamics along the DNA. The hydrodynamic force exerted by the buffer flow can potentially influence the behavior of protein molecules; this influence is expected to be proportional to the hydrodynamic radius of the protein molecules under observation (Gorman & Greene, 2008; Tafvizi et al., 2008). To circumvent such an issue, the Greene group has developed double-tethered DNA curtains without the need of buffer flow during the visualization (Fig. 6f) (Gorman et al., 2010). The group used electron beam lithography to fabricate diffusion barriers with nanoscale dimensions, which allows for much more precise control over both the location and lateral distribution of the DNA within the curtains. These patterns, which the group termed "DNA racks," comprise linear barriers to lipid diffusion along with arrays of metallic pentagons, which are used for the scaffold of antibodies to anchor DNA at one end. The lipid-tethered DNA is first aligned along the linear barriers; subsequently, the DNA is anchored on the pentagons positioned at a defined distance downstream from the linear barriers. Once aligned and anchored, the "double-tethered" DNA curtains maintain their extended form. This sophisticated platform has been applied to biological systems such as nucleosomes and chromatin remodeling (Fig. 7) (Finkelstein et al., 2010; Gorman et al., 2010).

4.4 DNA tightropes

Another platform that did not require buffer flow was developed for single-molecule visualization of protein dynamics along extended DNA. In this platform, λDNA is suspended between poly (L-lysine)-coated microspheres non-specifically attached on a PEGylated surface to form "DNA tightropes" in a flow cell, allowing the DNA to maintain its extended form during the visualization (Fig. 6h) (Kad et al., 2010). Since DNA is negatively charged at physiological pH, DNA can attach to the positively charged poly (L-lysine) that is coated on the microspheres via an electrostatic interaction. Injection of high concentration of DNA (1.6 nM) forms many DNA tightropes, allowing a high-throughput single-molecule analysis by simultaneous observation of multiple DNA-binding protein

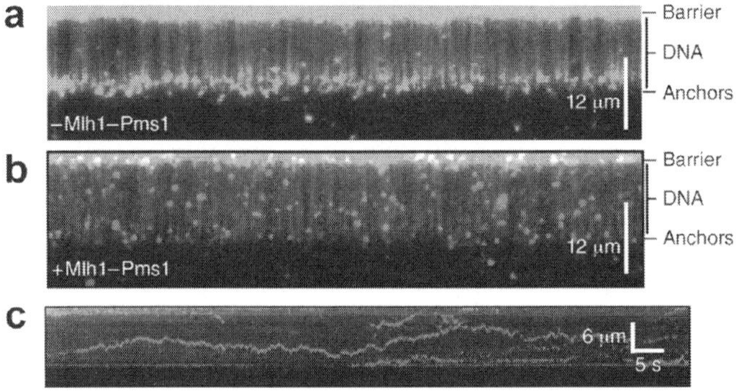

Fig. 7. Nanofabricated racks of DNA for visualizing one-dimensional diffusion of a postreplicative mismatch repair protein complex (Mlh1–Pms1). (a) YOYO1-stained λDNA curtains (green; 48,502 base pairs), (b) YOYO1-stained λDNA curtains with a Qdot-labeled protein complex (magenta), and (c) Kymogram illustrating the motion of the protein complex. (Modified from *Nature Structural & Molecular Biology* (Gorman et al., 2010) , Copyright (2010), with permission from Macmillan Publishers Ltd).

molecules. Since the DNA tightropes formed on the microspheres are off the glass surface, there is no interaction between the DNA and the glass surface, which may not interfere with the interactions with the protein molecules. Furthermore, the no buffer-flow condition enables the detection of various dynamic mechanisms of DNA repair proteins, including not only diffusion along DNA (Dunn et al., 2011) but also jumping from one DNA to another DNA (Fig. 8) (Kad et al., 2010).

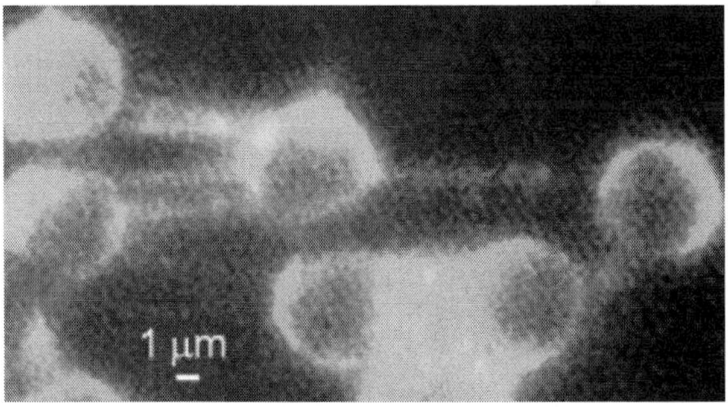

Fig. 8. Fluorescence image of Qdot-labeled single protein molecules (UvrA, red spots) bound to DNA tightropes (green). (Reproduced from (Kad et al., 2010), Copyright (2010), with permission from Elsevier).

4.4.1 Minimization of protein non-specific adsorption on microspheres by PEGylaton

One of the drawbacks of the DNA tightrope platform with poly (L-lysine)-coated microspheres is that many protein molecules are non-specifically adsorbed on the microspheres. Fluorescence from the Qdots that are labeled to the adsorbed protein molecules interferes with single-molecule fluorescence imaging. Kad et al. circumvented the interference by manually masking the bead portions on screen using an image analysis software before analysis (Kad et al., 2010). Here, I describe another tightrope platform developed to overcome the issue by PEGylation, in which the glass surface as well as microspheres are coated with biotinylated PEG. The excellent resistance of PEG to protein adsorption (discussed in 3.1) efficiently suppresses the non-specific adsorption of Qdot-labeled protein molecules on the microspheres. The biotin of the PEG allows the microspheres to be immobilized on the glass surface and biotinylated DNA to be tethered between the microspheres via an avidin-biotin interaction (Fig. 9). With this improved platform, dynamics of a DNA repair protein is currently being investigated.

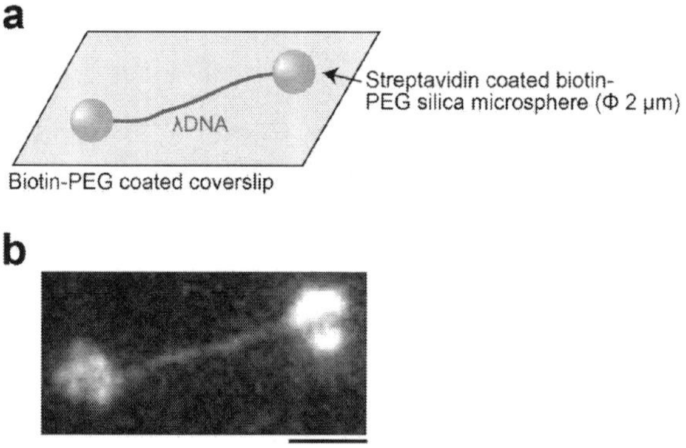

Fig. 9. A tightrope formed between streptavidin coated biotin-PEG silica microspheres immobilized on a biotin-PEG coated coverslip. (a) Schematic drawing of this tightrope platform. (b) Fluorescence image of YOYO1-labeled λDNA tethered between microspheres. Scale bar, 5 μm.

4.5 DNA extended by single-molecule DNA manipulation

Technical advances in combining single-molecule fluorescence visualization with single-molecule manipulation provide deeper insights into DNA–protein interactions. Some groups incorporated optical tweezers into a single-molecule fluorescence set-up (Fig. 6c,d) for studying DNA dynamics or DNA–protein interactions (Amitani et al., 2006; Comstock et al., 2011; Galletto et al., 2006; Harada et al., 1999; Hohng et al., 2007; Lang et al., 2004; van Mameren et al., 2009; Zhou et al., 2011). With such platforms, translocation (Amitani et al., 2006), association and dissociation (Harada et al., 1999; Zhou et al., 2011), filament assembly (Galletto et al., 2006), and disassembly (van Mameren et al., 2009) of protein molecules along DNA were visualized. Some of these studies investigated the effect of DNA tension on DNA–protein interactions. Recently, my colleagues and I have incorporated magnetic tweezers (Allemand et al., 2007), another apparatus for single-molecule DNA manipulation, into the single-molecule fluorescence set-up through collaboration with my research colleagues, which has allowed us to visualize single-molecule DNA helicase interacting with single dsDNA that is tethered by the magnetic tweezers.

5. Outlook

I have provided an overview recent of advances in direct visualization of single-molecule DNA-binding proteins, especially of key techniques used for the visualization. Fluorescence labeling methods as well as modern platforms employing biocompatible surface-coating methods described here have allowed the dynamics of single-molecule DNA-binding proteins to be elucidated. Frequently used fluorophores, dyes and Qdots have allowed visualization to a certain extent. However, the fluorophores have not satisfied all the requirements related to optimum size and photophysical properties, for example, dyes are small but short lived whereas Qdots are bright and resistant to photobleaching, but relatively large in size. Thus, it is highly desirable that new all-round fluorophores be synthesized or extracted from materials. One such promising fluorophore is the nitrogen-vacancy center, a crystal defect in diamonds (Aharonovich et al., 2011). The defect that emits near-infrared fluorescence exhibits neither blinking nor photobleaching, which is distinct from the common fluorophores.

Technical advances in combining single-molecule fluorescence visualization with single-molecule manipulation (discussed in 4.5) will provide deeper insights into DNA–protein interactions. Zero-mode waveguides (Levene et al., 2003), a nano-hole (<100 nm in diameter) array, is another promising microscopy for visualizing DNA–protein interactions. The waveguides attract increasing attention because compared to TIRFM, zero-mode waveguides make it possible to perform single-molecule fluorescence imaging under higher ligand concentration (μM range) conditions due to its extremely small excitation volume. And thus, a few biological processes were visualized at single-molecule resolution by zero-mode waveguides (Miyake et al., 2008; Sameshima et al., 2010; Uemura et al., 2010).

In a recent study (Kad et al., 2010), single-molecule interaction of two different DNA protein molecules along DNA was visualized by labeling them with different fluorophores, although most studies typically focused on the dynamics of single DNA binding-proteins. In the future, simultaneous visualization of different DNA-binding protein molecules may be possible by labeling the molecules with different fluorophores. This will help elucidate fundamental processes of DNA metabolism. In fact, biological processes in DNA metabolism involve various protein molecules as a form of macromolecular complexes. Moreover, the processes can be associated with chromatin for eukaryotes. To approach the protein dynamics in such an environment, a recent study investigated single-molecule interaction between a protein complex and chromatin (Gorman et al., 2010). Future developments in various fields, including platforms, fluorophores, surface coating, and fluorescence labelling, will provide fruitful information on complex DNA–protein interactions that cannot otherwise be explored.

6. Acknowledgments

I would like to thank the Harada group at Kyoto University and the Bensimon & Croquette group at École Normale Supérieure for supporting my research. The presented work performed in the groups was partly funded by JSPS and JST.

7. References

Aharonovich, I., Greentree, A. D. & Prawer, S. (2011). Diamond photonics. *Nature Photonics*, 5(7), 397-405.

Aitken, C. E., Marshall, R. A. & Puglisi, J. D. (2008). An oxygen scavenging system for improvement of dye stability in single-molecule fluorescence experiments. *Biophysical Journal,* 94(5), 1826-1835.

Allemand, J.-F., Bensimon, D., Charvin, G., Croquette, V., Lia, G., Lionnet, T., Neuman, K. C., Saleh, O. A. & Yokota, H. (2007). Studies of DNA-Protein Interactions at the single molecule level with magnetic tweezers. *Lecture Notes on Physics,* 711, 123-140.

Amitani, I., Baskin, R. J. & Kowalczykowski, S. C. (2006). Visualization of Rad54, a chromatin remodeling protein, translocating on single DNA molecules. *Molecular Cell,* 23(1), 143-148.

Anderson, A. S., Dattelbaum, A. M., Montano, G. A., Price, D. N., Schmidt, J. G., Martinez, J. S., Grace, W. K., Grace, K. M. & Swanson, B. I. (2008). Functional PEG-modified thin films for biological detection. *Langmuir,* 24(5), 2240-2247.

Blainey, P. C., van Oijen, A. M., Banerjee, A., Verdine, G. L. & Xie, X. S. (2006). A base-excision DNA-repair protein finds intrahelical lesion bases by fast sliding in contact with DNA. *Proceedings of the National Academy of Sciences of the United States of America,* 103(15), 5752-5757.

Blainey, P. C., Luo, G., Kou, S. C., Mangel, W. F., Verdine, G. L., Bagchi, B. & Xie, X. S. (2009). Nonspecifically bound proteins spin while diffusing along DNA. *Nature Structural & Molecular Biology,* 16(12), 1224-1229.

Bonnet, I., Biebricher, A., Porte, P. L., Loverdo, C., Benichou, O., Voituriez, R., Escude, C., Wende, W., Pingoud, A. & Desbiolles, P. (2008). Sliding and jumping of single EcoRV restriction enzymes on non-cognate DNA. *Nucleic Acids Research,* 36(12), 4118-4127.

Branch, D. W., Wheeler, B. C., Brewer, G. J. & Leckband, D. E. (2001). Long-term stability of grafted polyethylene glycol surfaces for use with microstamped substrates in neuronal cell culture. *Biomaterials,* 22(10), 1035-1047.

Brewer, L. R. & Bianco, P. R. (2008). Laminar flow cells for single-molecule studies of DNA-protein interactions. *Nature Methods,* 5(6), 517-525.

Chan, Y. H. & Boxer, S. G. (2007). Model membrane systems and their applications. *Current Opinion in Chemical Biology,* 11(6), 581-587.

Comstock, M. J., Ha, T. & Chemla, Y. R. (2011). Ultrahigh-resolution optical trap with single-fluorophore sensitivity. *Nature Methods,* 8(4), 335-340.

Cuvelier, D., Rossier, O., Bassereau, P. & Nassoy, P. (2003). Micropatterned "adherent/repellent" glass surfaces for studying the spreading kinetics of individual red blood cells onto protein-decorated substrates. *European Biophysics Journal,* 32(4), 342-354.

Dave, R., Terry, D. S., Munro, J. B. & Blanchard, S. C. (2009). Mitigating unwanted photophysical processes for improved single-molecule fluorescence imaging. *Biophysical Journal,* 96(6), 2371-2381.

Dunn, A. R., Kad, N. M., Nelson, S. R., Warshaw, D. M. & Wallace, S. S. (2011). Single Qdot-labeled glycosylase molecules use a wedge amino acid to probe for lesions while scanning along DNA. *Nucleic Acids Research,* 39(17), 7487-7498.

Etson, C. M., Hamdan, S. M., Richardson, C. C. & van Oijen, A. M. (2010). Thioredoxin suppresses microscopic hopping of T7 DNA polymerase on duplex DNA. *Proceedings of the National Academy of Sciences of the United States of America,* 107(5), 1900-1905.

Fazio, T., Visnapuu, M. L., Wind, S. & Greene, E. C. (2008). DNA curtains and nanoscale curtain rods: high-throughput tools for single molecule imaging. *Langmuir,* 24(18), 10524-10531.

Fernández-Suárez, M. & Ting, A. Y. (2008). Fluorescent probes for super-resolution imaging in living cells. *Nature Reviews Molecular Cell Biology,* 9(12), 929-943.

Finkelstein, I. J., Visnapuu, M. L. & Greene, E. C. (2010). Single-molecule imaging reveals mechanisms of protein disruption by a DNA translocase. *Nature,* 468(7326), 983-987.

Funatsu, T., Harada, Y., Tokunaga, M., Saito, K. & Yanagida, T. (1995). Imaging of single fluorescent molecules and individual ATP turnovers by single myosin molecules in aqueous solution. *Nature,* 374(6522), 555-559.

Galletto, R., Amitani, I., Baskin, R. J. & Kowalczykowski, S. C. (2006). Direct observation of individual RecA filaments assembling on single DNA molecules. *Nature,* 443(7113), 875-878.

Gorman, J., Chowdhury, A., Surtees, J. A., Shimada, J., Reichman, D. R., Alani, E. & Greene, E. C. (2007). Dynamic basis for one-dimensional DNA scanning by the mismatch repair complex Msh2-Msh6. *Molecular Cell,* 28(3), 359-370.

Gorman, J. & Greene, E. C. (2008). Visualizing one-dimensional diffusion of proteins along DNA. *Nature Structural & Molecular Biology,* 15(8), 768-774.

Gorman, J., Fazio, T., Wang, F., Wind, S. & Greene, E. C. (2010). Nanofabricated racks of aligned and anchored DNA substrates for single-molecule imaging. *Langmuir,* 26(2), 1372-1379.

Gorman, J., Plys, A. J., Visnapuu, M. L., Alani, E. & Greene, E. C. (2010). Visualizing one-dimensional diffusion of eukaryotic DNA repair factors along a chromatin lattice. *Nature Structural & Molecular Biology,* 17(8), 932-938.

Graneli, A., Yeykal, C. C., Prasad, T. K. & Greene, E. C. (2006). Organized arrays of individual DNA molecules tethered to supported lipid bilayers. *Langmuir,* 22(1), 292-299.

Greene, E. C. & Mizuuchi, K. (2002). Direct observation of single MuB polymers: evidence for a DNA-dependent conformational change for generating an active target complex. *Molecular Cell,* 9(5), 1079-1089.

Ha, T., Rasnik, I., Cheng, W., Babcock, H. P., Gauss, G. H., Lohman, T. M. & Chu, S. (2002). Initiation and re-initiation of DNA unwinding by the *Escherichia coli* Rep helicase. *Nature,* 419(6907), 638-641.

Ha, T. (2004). Structural dynamics and processing of nucleic acids revealed by single-molecule spectroscopy. *Biochemistry,* 43(14), 4055-4063.

Han, S., Kim, C. & Kwon, D. (1997). Thermal/oxidative degradation and stabilization of polyethylene glycol. *Polymer,* 38, 317-323.

Han, Y. W. & Mizuuchi, K. (2010). Phage Mu transposition immunity: protein pattern formation along DNA by a diffusion-ratchet mechanism. *Molecular Cell,* 39(1), 48-58.

Harada, Y., Sakurada, K., Aoki, T., Thomas, D. D. & Yanagida, T. (1990). Mechanochemical coupling in actomyosin energy transduction studied by *in vitro* movement assay. *Journal of Molecular Biology,* 216(1), 49-68.

Harada, Y., Funatsu, T., Murakami, K., Nonoyama, Y., Ishihama, A. & Yanagida, T. (1999). Single-molecule imaging of RNA polymerase-DNA interactions in real time. *Biophysical Journal,* 76(2), 709-715.

Harris, J. M. (1992). *Poly(Ethylene Glycol) Chemistry (Ethylene Glycol Chemistry : Biotechnical and Biomedical Applications).* New York: Plenum Press.

Hermanson, G. T. (2008). *Bioconjugate Techniques (2nd Edition)*: Academic Press.

Heyes, C. D., Kobitski, A. Y., Amirgoulova, E. V. & Nienhaus, G. U. (2004). Biocompatible surfaces for specific tethering of individual protein molecules. *Journal of Physical Chemistry B,* 108(35), 13387-13394.

Hilario, J. & Kowalczykowski, S. C. (2010). Visualizing protein-DNA interactions at the single-molecule level. *Current Opinion in Chemical Biology,* 14(1), 15-22.

Hohng, S., Zhou, R., Nahas, M. K., Yu, J., Schulten, K., Lilley, D. M. & Ha, T. (2007). Fluorescence-force spectroscopy maps two-dimensional reaction landscape of the holliday junction. *Science,* 318(5848), 279-283.

Huang, N.-P., Michel, R., Voros, J., Textor, M., Hofer, R., Rossi, A., Elbert, D. L., Hubbell, J. A. & Spencer, N. D. (2001). Poly(L-lysine)-g-poly(ethylene glycol) layers on metal oxide surfaces: surface-analytical characterization and resistance to serum and fibrinogen adsorption. *Langmuir,* 17(2), 489-498.

Iler, R. K. (1979). *The Chemistry of Silica.* New York: Wiley-Interscience.

Ito, Y., Hasuda, H., Sakuragi, M. & Tsuzuki, S. (2007). Surface modification of plastic, glass and titanium by photoimmobilization of polyethylene glycol for antibiofouling. *Acta Biomaterialia,* 3(6), 1024-1032.

Joo, C., McKinney, S. A., Nakamura, M., Rasnik, I., Myong, S. & Ha, T. (2006). Real-time observation of RecA filament dynamics with single monomer resolution. *Cell,* 126(3), 515-527.

Kabata, H., Kurosawa, O., Arai, I., Washizu, M., Margarson, S. A., Glass, R. E. & Shimamoto, N. (1993). Visualization of single molecules of RNA polymerase sliding along DNA. *Science,* 262(5139), 1561-1563.

Kad, N. M., Wang, H., Kennedy, G. G., Warshaw, D. M. & Van Houten, B. (2010). Collaborative dynamic DNA scanning by nucleotide excision repair proteins investigated by single-molecule imaging of quantum-dot-labeled proteins. *Molecular Cell,* 37(5), 702-713.

Kim, J. H. & Larson, R. G. (2007). Single-molecule analysis of 1D diffusion and transcription elongation of T7 RNA polymerase along individual stretched DNA molecules. *Nucleic Acids Research,* 35(11), 3848-3858.

Kim, S., Blainey, P. C., Schroeder, C. M. & Xie, X. S. (2007). Multiplexed single-molecule assay for enzymatic activity on flow-stretched DNA. *Nature Methods,* 4(5), 397-399.

Kochaniak, A. B., Habuchi, S., Loparo, J. J., Chang, D. J., Cimprich, K. A., Walter, J. C. & van Oijen, A. M. (2009). Proliferating cell nuclear antigen uses two distinct modes to move along DNA. *Journal of Biological Chemistry,* 284(26), 17700-17710.

Komazin-Meredith, G., Mirchev, R., Golan, D. E., van Oijen, A. M. & Coen, D. M. (2008). Hopping of a processivity factor on DNA revealed by single-molecule assays of diffusion. *Proceedings of the National Academy of Sciences of the United States of America,* 105(31), 10721-10726.

Lang, M. J., Fordyce, P. M., Engh, A. M., Neuman, K. C. & Block, S. M. (2004). Simultaneous, coincident optical trapping and single-molecule fluorescence. *Nature Methods,* 1(2), 133-139.

Lee, J. B., Hite, R. K., Hamdan, S. M., Xie, X. S., Richardson, C. C. & van Oijen, A. M. (2006). DNA primase acts as a molecular brake in DNA replication. *Nature,* 439(7076), 621-624.

Levene, M. J., Korlach, J., Turner, S. W., Foquet, M., Craighead, H. G. & Webb, W. W. (2003). Zero-mode waveguides for single-molecule analysis at high concentrations. *Science,* 299(5607), 682-686.

Lohman, T. M., Tomko, E. J. & Wu, C. G. (2008). Non-hexameric DNA helicases and translocases: mechanisms and regulation. *Nature Reviews Molecular Cell Biology,* 9(5), 391-401.

Malmstena, M., Emoto, K. & Alstine, J. M. V. (1998). Effect of chain density on inhibition of protein adsorption by poly(ethylene glycol) Based Coatings. *Journal of Colloid Interface Science,* 202(2), 507-517.

Maluf, N. K., Fischer, C. J. & Lohman, T. M. (2003). A dimer of *Escherichia coli* UvrD is the active form of the helicase in vitro. *Journal of Molecular Biology,* 325(5), 913-935.

McNamee, C. E., Yamamoto, S. & Higashitani, K. (2007). Preparation and characterization of pure and mixed monolayers of poly(ethylene glycol) brushes chemically adsorbed to silica surfaces. *Langmuir,* 23(8), 4389-4399.

Michalet, X., Pinaud, F. F., Bentolila, L. A., Tsay, J. M., Doose, S., Li, J. J., Sundaresan, G., Wu, A. M., Gambhir, S. S. & Weiss, S. (2005). Quantum dots for live cells, *in vivo* imaging, and diagnostics. *Science,* 307(5709), 538-544.

Miyake, T., Tanii, T., Sonobe, H., Akahori, R., Shimamoto, N., Ueno, T., Funatsu, T. & Ohdomari, I. (2008). Real-time imaging of single-molecule fluorescence with a zero-mode waveguide for the analysis of protein-protein interaction. *Analytical Chemistry,* 80(15), 6018-6022.

Moerner, W. E. (2007). Single-molecule Chemistry and Biology Special Feature: New directions in single-molecule imaging and analysis. *Proceedings of the National Academy of Sciences of the United States of America,* 104(31), 12596-12602.

Pasche, S., Paul, S. M. D., Vörös, J., Spencer, N. D. & Textor, M. (2003). Poly(L-lysine)-graft-poly(ethyleneglycol) assembled monolayers on niobium oxide surfaces: a quantitative study of the influence of polymer interfacial architecture on resistance to protein adsorption by ToF-SIMS and *in situ* OWLS. *Langmuir,* 19, 9216-9235.

Prasad, T. K., Yeykal, C. C. & Greene, E. C. (2006). Visualizing the assembly of human Rad51 filaments on double-stranded DNA. *Journal of Molecular Biology,* 363(3), 713-728.

Prasad, T. K., Robertson, R. B., Visnapuu, M. L., Chi, P., Sung, P. & Greene, E. C. (2007). A DNA-translocating Snf2 molecular motor: *Saccharomyces cerevisiae* Rdh54 displays processive translocation and extrudes DNA loops. *Journal of Molecular Biology,* 369(4), 940-953.

Prime, K. L. & Whitesides, G. M. (1991). Self-assembled organic monolayers: model systems for studying adsorption of proteins at surfaces. *Science,* 252(5010), 1164-1167.

Rasnik, I., McKinney, S. A. & Ha, T. (2006). Nonblinking and long-lasting single-molecule fluorescence imaging. *Nature Methods,* 3(11), 891-893.

Sameshima, T., Iizuka, R., Ueno, T., Wada, J., Aoki, M., Shimamoto, N., Ohdomari, I., Tanii, T. & Funatsu, T. (2010). Single-molecule study on the decay process of the football-shaped GroEL-GroES complex using zero-mode waveguides. *Journal of Biological Chemistry,* 285(30), 23159-23164.

Sharma, S., Johnson, R. W. & Desai, T. A. (2004). Evaluation of the stability of nonfouling ultrathin poly(ethylene glycol) films for silicon-based microdevices. *Langmuir,* 20(2), 348-356.

Tafvizi, A., Huang, F., Leith, J. S., Fersht, A. R., Mirny, L. A. & van Oijen, A. M. (2008). Tumor suppressor p53 slides on DNA with low friction and high stability. *Biophysical Journal,* 95(1), L01-03.

Tafvizi, A., Huang, F., Fersht, A. R., Mirny, L. A. & van Oijen, A. M. (2011). A single-molecule characterization of p53 search on DNA. *Proceedings of the National Academy of Sciences of the United States of America,* 108(2), 563-568.

Tan, X., Mizuuchi, M. & Mizuuchi, K. (2007). DNA transposition target immunity and the determinants of the MuB distribution patterns on DNA. *Proceedings of the National Academy of Sciences of the United States of America,* 104(35), 13925-13929.

Tokunaga, M., Kitamura, K., Saito, K., Iwane, A. H. & Yanagida, T. (1997). Single molecule imaging of fluorophores and enzymatic reactions achieved by objective-type total internal reflection fluorescence microscopy. *Biochemical and Biophysical Research Communications,* 235(1), 47-53.

Tokunaga, M., Imamoto, N. & Sakata-Sogawa, K. (2008). Highly inclined thin illumination enables clear single-molecule imaging in cells. *Nature Methods,* 5(2), 159-161.

Uemura, S., Aitken, C. E., Korlach, J., Flusberg, B. A., Turner, S. W. & Puglisi, J. D. (2010). Real-time tRNA transit on single translating ribosomes at codon resolution. *Nature,* 464(7291), 1012-1017.

van Mameren, J., Modesti, M., Kanaar, R., Wyman, C., Peterman, E. J. & Wuite, G. J. (2009). Counting RAD51 proteins disassembling from nucleoprotein filaments under tension. *Nature,* 457(7230), 745-748.

van Oijen, A. M., Blainey, P. C., Crampton, D. J., Richardson, C. C., Ellenberger, T. & Xie, X. S. (2003). Single-molecule kinetics of lambda exonuclease reveal base dependence and dynamic disorder. *Science,* 301(5637), 1235-1238.

Visnapuu, M. L., Duzdevich, D. & Greene, E. C. (2008). The importance of surfaces in single-molecule bioscience. *Molecular Biosystems,* 4(5), 394-403.

Wang, X., Ren, X., Kahen, K., Hahn, M. A., Rajeswaran, M., Maccagnano-Zacher, S., Silcox, J., Cragg, G. E., Efros, A. L. & Krauss, T. D. (2009). Non-blinking semiconductor nanocrystals. *Nature,* 459(7247), 686-689.

Yang, Z., Galloway, J. A. & Yu, H. (1999). Protein interactions with poly(ethylene glycol) self-assembled monolayers on glass substrates: diffusion and adsorption *Langmuir,* 15(24), 8405-8411.

Yokota, H., Han, Y. W., Allemand, J.-F., Xi, X. G., Bensimon, D., Croquette, V. & Harada, Y. (2009). Single-molecule visualization of binding modes of helicase to DNA on PEGylated surfaces. *Chemistry Letters,* 38, 308-309.

Zhou, R., Kozlov, A. G., Roy, R., Zhang, J., Korolev, S., Lohman, T. M. & Ha, T. (2011). SSB functions as a sliding platform that migrates on DNA via reptation. *Cell,* 146(2), 222-232.

Zlatanova, J. & van Holde, K. (2006). Single-molecule biology: what is it and how does it work? *Molecular Cell,* 24(3), 317-329.

12

Defining the Cellular Interactome of Disease-Linked Proteins in Neurodegeneration

Verena Arndt[1,*] and Ina Vorberg[1,2],
[1]*German Center for Neurodegenerative Diseases (DZNE), Bonn,*
[2]*Department of Neurology, Rheinische Friedrich-Wilhelms-University Bonn, Germany*

1. Introduction

Age-related neurodegenerative diseases are recognized as a major health issue worldwide. Due to our aging society, the number of people suffering from dementia is drastically increasing, creating serious challenges for society and the public health system. Our current paucity of effective treatments and total lack of cures, when coupled with this increasing prevalence, makes the exploration of novel strategies for therapeutic interventions of the utmost importance. Although many of the disease-causing proteins have been identified, the molecular mechanisms that underlie disease pathogenesis are still not fully understood. One common feature of neurodegenerative diseases is the accumulation of misfolded proteins into toxic oligomers and aggregates. Gaining extensive knowledge regarding the formation of these cytotoxic species, the cellular machineries that guarantee their persistence or clearance and the basis of their toxicity is essential for the development of both preventive and therapeutic interventions. A powerful approach to increase our knowledge on these disease processes is the use of proteomic tools to define the interaction networks of disease-related proteins. Significant technical progress has been made in the last decade that now allows high-throughput screening for protein-protein interactions on a proteome level. In this book chapter, we review the diverse proteomic approaches that have been used to define the interactomes of disease-linked proteins and the impact of these findings on the understanding of pathogenic processes.

1.1 Cellular mechanisms of neurodegeneration

The intracellular and extracellular aggregation of mutated and/or misfolded proteins into highly ordered β-sheet rich aggregates, termed amyloid, is a common hallmark of many neurodegenerative diseases such as Alzheimer's (AD), Parkinson's (PD) and Huntington's (HD) disease (Fig.1).

AD is the most common progressive neurodegenerative disease, affecting millions of people worldwide. AD patients show neuronal loss, the deposition of extracellular amyloid plaques consisting of amyloid β peptides (Aβ, ($Aβ_{40}$ and $Aβ_{42}$)) and intracellular neurofibrillary

---

* Corresponding Author

tangles consisting of hyperphosphorylated and cleaved Tau. Although various models have been proposed to explain the pathogenic processes underlying the disease, the exact mechanisms leading to AD are not fully understood (Mudher & Lovestone, 2002).

The amyloid cascade hypothesis proposes that aberrant cleavage of the amyloid precursor protein (APP) by two different proteases (β/γ-secretases) leads to the accumulation of aggregation-prone Aβ peptides that eventually cause disease through multiple mechanisms, including microglial infiltration, generation of reactive oxygen species and synaptic damage (Hardy & Selkoe, 2002). According to this model, early intracellular Aβ accumulation induces the aggregation of the microtubule associated protein Tau. In familial forms of AD, genetic mutations within the APP or γ-sectretase cause extensive formation of Aβ protofibrils, which leads to neurotoxicity.

The Tau or tangle hypothesis, however, claims that it is the disruption of microtubule binding and aggregation of Tau by phosphorylation or genetic mutations that initiates the disease cascade (Maccioni et al., 2010). The formation of neurofibrillary tangles subsequently leads to disintegration of the neuronal cytoskeleton, which causes the disruption of neuronal transport and cell death.

Although Tau and Aβ became the focus of extensive research, the exact disease mechanisms remain unclear. The complexity of the disease is reflected in the fact that many other susceptibility genes or proteins involved in development and progression of this disease have been identified in the last decade.

After AD, PD represents the second most prevalent neurodegenerative disease. It is characterized by the loss of dopaminergic neurons in the substantia nigra and formation of intracellular inclusions (Lewy bodies) consisting of α-synuclein, a presynaptic protein of unknown function (Martin, I. et al., 2011). Importantly, in idiopathic Parkinson's disease, α-synuclein inclusions first localize to defined brain areas and pathology appears to progress in a topographically predictable manner (Braak et al., 2004). The majority of cases are sporadic and age-related. However the identification of several mutated genes in the small number of early onset familial forms of PD implicate various proteins contributing to disease progression. So far, genetic mutations in α-synuclein (SNCA), parkin (PARK2), ubiquitin carboxyl-terminal esterase L1 (UCHL1), parkinson protein 7 (PARK7, DJ-1), PTEN-induced putative kinase 1 (PINK1), leucine-rich repeat kinase 2 (LRRK2), α-synuclein interacting protein (SNCAIP) and glucosidase, beta, acid (GBA) have been linked to familial forms of PD (Martin, I. et al., 2011).

Polyglutamine expansions within unrelated proteins are the underlying cause of nine different neurodegenerative pathologies, including HD, spinobulbar muscular atrophy and spinocerebellar ataxias (SCAs) (Hands & Wyttenbach, 2010). HD, a dominantly inherited neurodegenerative disease that is caused by an expansion of the polyglutamine tract in the Huntingtin (HTT) protein, is the most widely studied of these diseases. It manifests in movement disorder, psychological disturbances and cognitive dysfunction. Hallmarks of HD are cytoplasmic and nuclear inclusions that consist of N-terminal fragments of expanded HTT. Mutant HTT causes cellular dysfunction and neurodegeneration, probably through a combination of toxic gain-of-function and loss-of-function mechanisms. Many proteins have been found to localize to HTT inclusions and it has been postulated that mutant HTT interferes with the function of a diverse variety of cellular proteins, leading to toxic alterations of many pathways.

Although the disease-related proteins in AD, PD and HD have different identities, common cellular mechanisms underlie the formation of oligomeric complexes and amyloidogenic aggregates. Whether neurotoxicity is caused by a toxic loss-of-function of the disease-associated proteins or a toxic gain-of-function of the built up amyloids is still under debate. Gaining more insights into the cellular and molecular pathways leading to aggregation and neurotoxicity could, thus, help to identify general and specific targets for therapeutic intervention in a variety of neurodegenerative diseases.

1.2 Protein aggregation and neurotoxicity

Maintaining the functionality of the proteome in a highly dynamic cellular environment constantly exposed to physical, metabolic and environmental stresses is essential for cell survival. Protein homeostasis, or "proteostasis", is controlled by a highly interconnected network of different protein quality control pathways that balance protein folding, degradation and aggregation (Kettern et al., 2010). Molecular chaperones are central players in this network. According to the current concept of chaperone function, chaperones act as surveillance factors that scan the cell and recognize non-native proteins. Upon binding to a substrate protein, they prevent its aggregation by facilitating folding or disposal through the proteasomal (CAP - chaperone assisted proteasomal degradation) or autophagic (CMA - chaperone mediated autophagy, CASA - chaperone assisted selective autophagy) machineries. The association of regulatory cochaperones determines the function of the chaperone as a "folding" or "degradation" factor. Inefficient or unsuccessful folding of a substrate protein thereby enhances the probability that degradation-inducing cochaperones will associate and initiate ubiquitination of the client protein. The different protein quality control pathways do not operate independently from each other and their coordinated action allows the cell to adjust to different alterations that endanger the integrity of the proteome.

Aggregation of disease-linked proteins probably represents a second line of defense against cytotoxic effects of misfolded proteins. At this stage, aggregate formation is probably a cytoprotective mechanism by which the cell sequesters misfolded proteins or oligomers that cannot be degraded by the proteasome. It has been shown that aggregate formation is not an uncontrolled process, but can be defined as part of the cellular protein quality control (Tyedmers et al., 2010). Usually, protein aggregates can be cleared by macroautophagy (Martinez-Vicente & Cuervo, 2007). However, this second line of defense is also overtaxed in neurodegenerative diseases, as the enhanced production of misfolded proteins is cumulative. Thus, the continuous presence of large or numerous aggregates at later disease stages probably exerts a negative effect on many cellular functions and enhances toxicity. The inhibition of cell function may be due to either sterical hindrance of cellular processes, such as axonal transport, or to coaggregation of other proteins that are then depleted from the cell (Olzscha et al., 2011). Taken together, understanding the mechanisms that lead to aggregation of disease-linked proteins is crucial for the identification of processes that cause toxicity in neurodegeneration.

Sporadic neurodegenerative diseases are usually age-related and reflect the overtaxing of a large variety of cellular processes that normally control protein homeostasis (Douglas & Dillin, 2011). In contrast hereditary early-onset dementias are caused by a set of genetic mutations that lead to the constant production of misfolded and aggregation-prone proteins. Mutations causing familial neurodegenerative diseases have been shown to increase the aggregation tendency of several disease-associated proteins in different ways.

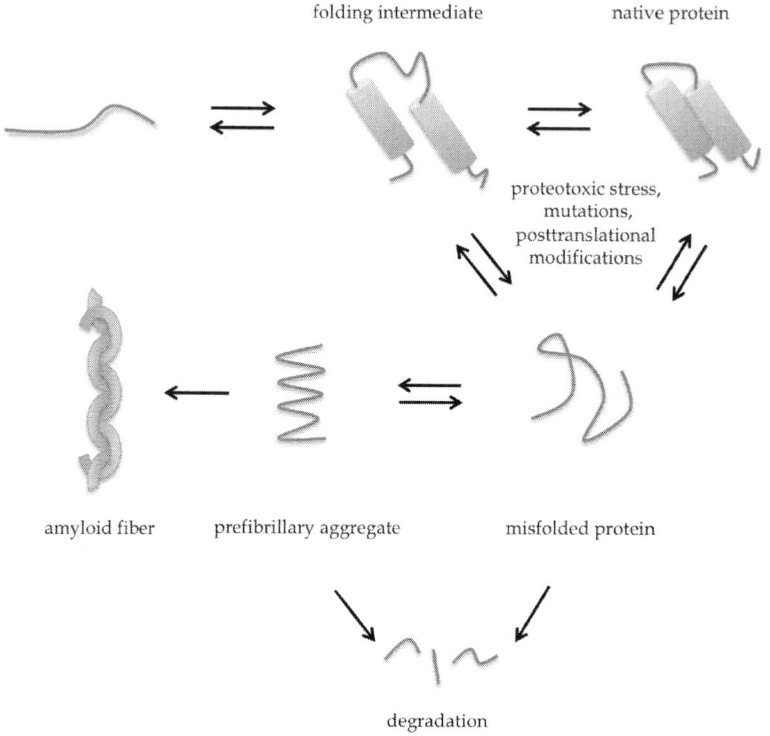

Fig. 1. Protein aggregation in neurodegenerative disease. To be functional, a protein has to fold into an appropriate three-dimensional structure. Aggregation of unfolded cellular proteins is usually prevented by the quality control machinery, which supports the folding process and ensures the removal of unfolded or misfolded proteins. Mutations that enhance the tendency of disease-linked proteins to aggregate, overexpression of aggregation-prone proteins or proteotoxic stress during aging may, however, overwhelm the cellular folding and degradation machineries. Once the quality control machinery is overwhelmed, protein aggregation may be a second line of defense to prevent cytotoxic effects of misfolded proteins or prefibrillary aggregates. Most of the disease-linked proteins in neurodegenerative diseases form highly ordered β-sheet-rich amyloid fibrils. The continued production of aggregation-prone proteins concomitant with decreased protein degradation leads to increased aggregation, likely leading to toxicity at later stages.

Often, mutations in the coding sequence of the disease-causing gene affect the propensity of the protein to misfold and, eventually, aggregate. In HD and several spinocerebellar ataxias (SCAs), a clear correlation between the lengths of the polyglutamine stretch in the disease-causing proteins and the tendency to aggregate has been observed (Martindale et al., 1998). Furthermore, three missense mutations in the α-synuclein gene that are associated with early-onset forms of PD (A30P, E46K and A53T) have been found to enhance the propensity of α-synuclein to aggregate (Li et al., 2001, Greenbaum et al., 2005). Several mutations in Tau have been linked to various neurodegenerative diseases; while Tau mutations affect different aspects of the protein's function, all result in increased aggregation (Wolfe, 2009).

Importantly, altered gene dosage is sometimes sufficient to cause disease. A genomic duplication or triplication involving the α-synuclein locus has been found to cause some forms of familial early-onset PD (Chartier-Harlin et al., 2004, Singleton et al., 2003). In addition, the higher risk for dementias with neuropathological features of AD in Down's syndrome has been attributed to the triplication of the APP gene in these patients (Rumble et al., 1989).

Mutations can, however, also alter posttranslational modifications of the disease-causing proteins. Proteolytic cleavage of disease-related proteins often precedes amyloid deposition and mutations in the coding region can also affect proteolytic processing of the disease-related protein. Generation of the neurotoxic Aβ peptide by β- and γ-secretase mediated sequential proteolysis of APP plays a central role in AD (O' Brien & Wong, 2011). Pathogenic APP mutations that cause early-onset familial AD are clustered around the α-, β-, γ-secretase cleavage sites and affect the ratio of $A\beta_{40/42}$, the latter having an increased propensity to form amyloid plaques (van Dam & De Deyn, 2006). The role of proteolytic cleavage of Tau in neurodegeneration is less well understood, but proteolysis of Tau can be clearly linked to aggregation and neurotoxicity (Wang et al., 2010). Proteolytic cleavage of several polyglutamine disease-linked proteins liberates toxic protein fragments that can form aggregates (Shao & Diamond, 2007). Finally, hyperphosphorylation of Tau and phosphorylation of α-synuclein and ataxin-1 has been shown to enhance their aggregation. Recently, other posttranslational modifications such as oxidation, sumoylation, ubiquitination or nitration have also been implicated in the aggregation of Tau, α-synuclein and polyglutamine-rich proteins (Beyer, 2006, Martin, L. et al., 2011, Pennuto et al., 2009). In addition, glycosylation affects the processing of APP (Georgopoulou et al., 2001).

2. Interactome mapping in diseases

In the last decades, significant progress has been made uncovering a large number of genetic mutations that are associated with a variety of genetically inherited disorders, summarized in the Online Mendelian Inheritance in Man database (OMIM) (Amberger et al., 2011). Clinical symptomatology, however, is less dependent on single mutations than on how whole organisms and systems are altered.

The pathology of one single gene mutation mapped to a specific disease is rarely caused by the malfunction of just one mutated gene product, but rather reflects perturbations of the whole interaction network in which the altered protein is embedded. This concept is in line with the finding that cellular proteins linked to the same disease exhibit a high tendency to interact with each other (Barabasi et al., 2011). In the interactome, the corresponding disease module, thus, consists of a group of proteins that function together in a cellular pathway or process whose breakdown results in a specific pathophenotype. To understand local network perturbations underlying the diseases phenotype, it is necessary to systematically explore the complex interaction network in which the disease-associated proteins are interconnected. Systems-based approaches to human diseases can, thus, lead to the identification of new disease genes and pathways. As drug discovery starts to concentrate on network-based targets rather than single gene targets, a deeper understanding of the disease module is crucial for the development of treatments (Morphy & Rankovic, 2007). Moreover the identification of disease modules can help to uncover subnetworks of interacting proteins that are shared between diseases with similar pathologies, such as neurodegeneration. A powerful approach to detect disease modules in neurodegenerative diseases is the interactome mapping of the different disease-linked proteins in combination with the generation of a complete map of the human interactome. The integration of the interactome of a diseases module with gene expression data or structural proteomic data will significantly advance our understanding of the pathophysiology of the disease in search for effective treatments.

2.1 The human interactome

To date, most attempts to map protein-protein interaction networks have been made using model organisms such as the yeast *Saccharomyces cerevisiae,* the nematode *Caenorhabditis elegans* and the fruitfly *Drosophila melanogaster*. This model organism usage is due to the better and earlier annotation of their genomes. In the last decade, several attempts to map the human interactome have been made, representing a first crucial step towards the understanding of cellular interconnectivity and the identification of networks that play a key role in human diseases (Vidal et al., 2011). In addition, the combination of high-throughput datasets and literature-based protein-protein interactions into databases helped to extend existing interactome maps and made information obtained from high-throughput screens more accessible (Tab. 1).

Database	**Description**	**Webadress**
BioGRID	*Biological General Repository for Interaction Datasets*	http://thebiogrid.org/
BIND	*Biomolecular Interaction Network Database:* component database of BOND (*Biomolecular Object Network Databank*) integrates a range of component databases including Genbank and BIND, the Biomolecular Interaction Network Database, resource for cross database searches	http://bond. unleashedinformatics. Com/
DIP	*Database of Interacting Proteins*	http://dip.doe-mbi.ucla.edu/dip/ Main.cgi
GWIDD	*Genome Wide Docking Database:* integrated resource for structural studies of protein-protein interactions on genome-wide scale	http://gwidd. bioinformatics.ku.edu/

Database	Description	Webadress
Human Protein Inter-action Database	interactions of HIV-1 proteins with host cell proteins, other HIV-1 proteins, or proteins from disease organisms associated with HIV / AIDS	http://www.ncbi.nlm.nih.gov/RefSeq/HIVInteractions/index.html
HPID	*Human Protein Interaction Database*	http://wilab.inha.ac.kr/hpid/
HPRD	*Human Protein Reference Database:* integrates information on domain architecture, post-translational modifications, interaction networks and disease association for each protein in the human proteome	http://www.hprd.org/
HUGE ppi	*Human Unidentified Gene-Encoded large proteins*	http://www.kazusa.or.jp/huge/ppi/
IBIS	*Inferred Biomolecular Interactions Server*	http://www.ncbi.nlm.nih.gov/Structure/ibis/ibis.cgi
IntAct	curated from published protein-protein interaction data	http://www.ebi.ac.uk/intact/
MINT	*Molecular Interaction Database*	http://mint.bio.uniroma2.it/mint/Welcome.do
MIPS	*Mammalian Protein-Protein Interaction Database*	http://mips.helmholtz-muenchen.de/proj/ppi/
NetPro	Comprehensive database of protein-protein and protein-small molecules interaction	http://www.molecularconnections.com/home/en/home/products/NetPro
POINeT	prediction of the human protein-protein interactome based on available orthologous interactome datasets	http://point.bioinformatics.tw/main.do?file=architecture.txt&selectMenu=environment.ini

Table 1. A selection of protein-protein interaction databases.

In the past years, improvement of techniques for high-throughput interaction screening has helped to generate high-quality protein interaction data for many organisms, including humans. In addition, an empirical framework has been proposed to evaluate the quality of data generated by high-throughput mapping approaches and to estimate the size of interactome networks (Venkatesan et al., 2009). Based on empirical sizing, the authors predict the human interactome to consist of ~130,000 interactions. Previous estimates of human interactome size range from 150,000 - 650,000 interactions (Venkatesan et al., 2009). To date ~23000 human protein-protein interactions have been reported. With the authors' estimation that 42% of these reported interactions represent true positives, only ~8% of the full interactome size has been identified so far (Venkatesan et al., 2009). Comprehensive mapping of the human interactome will require further development of complementary, systematic, unbiased and cost-effective high-throughput mapping approaches.

2.2 Experimental strategies for mapping protein-protein interactions

A large variety of experimental and computational methods can be used to identify protein-protein interactions. Two complementary methods are primarily applied for the large-scale mapping of protein-protein interactions. Mapping of binary interactions was first accomplished by a high-throughput adaption of the yeast two-hybrid (Y2H) system, originally developed by Fields and Song (Fields & Song, 1989). Mapping of indirect protein associations within protein complexes can be carried out by a combination of affinity purification and mass spectrometry (AP/MS) (Rigaut et al., 1999). Unfortunately both techniques have their limitations in terms of quality and coverage. Quality assessment of datasets from different interaction studies has shown that Y2H and AP/MS data provide the same quality but represent different subpopulations of the interactome, resulting in networks with different topologies and biological functions (Seebacher & Gavin, 2011). Binary maps from Y2H screens were enriched for transient signalling interactions and interactions between different protein complexes, whereas data generated from AP/MS experiments preferentially detected interaction within a protein complex. The combinatorial application of both techniques is, therefore, recommended to obtain more complete protein interaction maps. The lack of a complete interactome map can be overcome to a certain extent by using data-mining strategies to identify sub-networks from different incomplete interactome maps.

2.3 The yeast two-hybrid system (Y2H)

Originally developed by Field and Song (Field & Song, 1989) to detect the interactions of two proteins, Y2H has been further adapted to high-throughput screening (Koegl & Uetz, 2007). The Y2H is based on the finding that many transcription factors can be divided into a DNA-binding domain (*BD*) and an activation domain (*AD*) that maintain their functionality when separated and recombined. In the two-hybrid approach the BD (e.g. from the yeast Gal4 or the *E. coli* LexA protein) is fused to a protein of interest to generate the bait (Fig. 2A). The prey is constructed by the fusion of the *AD* (e.g. from the yeast Gal4 *or* the heterologous B42 peptide) with a set of open reading frames (ORFs). Bait and prey fusions are then coexpressed in yeast. When the bait and prey proteins interact, a functional transcription factor is reconstituted. Reconstitution of the transcription factor is detected by measuring the activity of a reporter gene. Several reporter genes have been used so far. In the "classic" two-hybrid, auxotrophic markers such as HIS3 and LEU2 allow selection by growth on selective media lacking histidine or leucine. Another commonly used reporter is the bacterial β-galactosidase. Recently, GFP has been successfully used as a reporter gene and many others are under investigation. Although Y2H has proved valuable in the confirmation and identification of many protein-protein interactions, there are experimental limitations of this system.

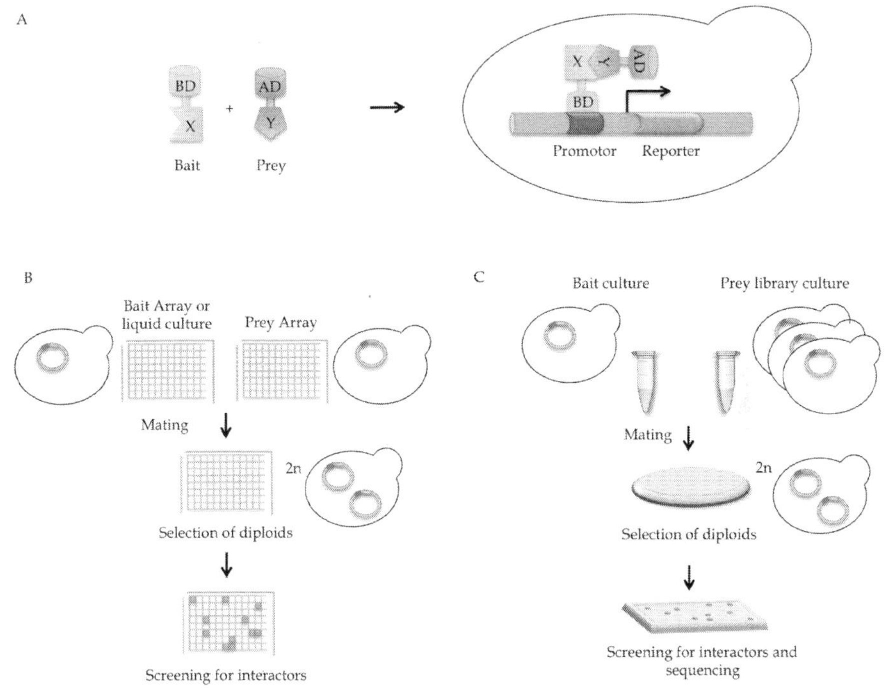

Fig. 2. The yeast two-hybrid system and high-throughput adaptions.
A Principle of the yeast two-hybrid system. To test for a direct interaction, proteins X and Y are coexpressed as fusions with the binding domain (*BD*), or bait, and the activation domain (*AD*), or prey, of a transcription factor (e.g. Gal4) in yeast. If protein X and Y interact directly a functional transcription factor is reconstituted which induces transcription of a reporter gene that allows detection of X-Y interaction (e.g. GFP) or selection for X-Y interaction (e.g. HIS3).
B Workflow of the matrix or array approach for high-throughput interaction screening. A prey array is generated by dispensing yeast clones that each express a different *AD* fusion in a multiwell plate. In an automated step, the prey array is then pinned on a multiwell plate containing yeast clones that express the bait (*BD* fusion). Prey and bait clones are allowed to mate and diploids are selected based on the expression of selection genes. If the bait and prey proteins interact directly, then the expression of a reporter gene that allows screening for interactors of the bait protein is induced.
C Principle of the exhaustive or random library screen for high-throughput screening. In this screen, a bait fusion is screened against a pooled prey library consisting of ORFs or ORF fragments. Diploids and positive interactors are selected based on growth of selection plates. In contrast to the array approach each positive clone has to be picked and sequenced after selection to identify the prey protein.

To obtain worthwhile Y2H results, the bait and prey fusion proteins must properly fold and not be hindered from proper interaction. Their interaction must be stable and not require posttranslational modification. Furthermore, the interaction in the Y2H takes place in the nucleus, though this restriction can be circumvented by the use of protein fragments that fold more efficiently and are able to translocate to the nucleus. *AD* and *BD* fusions alone can also sometimes auto-activate reporter gene expression. This is excluded by performing a self-activation assay of the bait and prey constructs. In this test, a control plasmid coding for an unrelated *BD* or *AD* fusion-protein is added along with the construct of interest. The auto-activation background of a HIS3 reporter can be reduced by the addition of the HIS3 inhibitor, 3-AT. The usage of the opposite fusion construct or a different protein fragment is also often effective in suppressing auto-activation. In addition, several alternative genetic screening techniques have been developed to allow the detection of interactions for transcription activators and membrane proteins (Auerbach et al., 2002).

2.3.1 Large-scale yeast two-hybrid screens

The most powerful application of the Y2H system, one that can generate comprehensive protein-protein interaction maps, is the unbiased screening of whole libraries. Two adaptions of the "classical" Y2H for high-throughput screening are the "matrix approach" or "array approach" and the "exhaustive library screening" or "random library screening" approach (Fig. 2B and 2C) (Koegl & Uetz, 2007).

In the array approach, a set of defined prey proteins is tested for interaction with a bait protein (Fig. 2B). Bait and prey constructs are individually transformed into isogenic reporter strains of opposite mating types. Yeast clones expressing a single *AD* fusion are dispensed into single wells of a multi-well plate, generating a matrix of *AD* fusions. The array of *AD* fusions is then spotted onto another multi-well plate containing one yeast clone expressing a single *BD* fusion for mating. In this way, all *BD* fusions are mated to *AD* arrays. A positive interaction is then detected by the ability of a diploid cell to activate the reporter gene (e.g. growth on selective media). As multiple assays can be performed with the same system under identical conditions, they can later be compared. To exclude false positives, experiments are usually performed in duplicates and only interactions found in both experiments are considered to be true. As whole genomes are available as ordered clone sets, each component of the array has a known identity and no sequencing is required after identifying a positive interaction. To accelerate the screening procedure, a pooling strategy can be applied. In the pooled array screen, *AD* fusions of known identity are tested as pools of *AD* fusions against the *BD* strains. This means that the identification of interaction partners from a positive pool requires retesting of all members. It is also possible to use pools of baits against an *AD* array.

A second approach to analyse the interactome of whole proteomes is the "exhaustive library screening" or "random library screening" approach (Fig. 2C). In this assay, a *BD* fusion is screened against a library of *AD* fusions of full Length ORFs or ORF fragments. For this approach, it is not necessary to know the sequence of the whole genome, as random prey libraries can be generated from randomly cut genomic DNA. In addition a large number of cDNA libraries is commercially available. Like the array screen, the bait and prey constructs are individually transferred into isogenic yeast strains of opposing mating types. However, in contrast to the matrix approach, the different *AD* fusions are not separated on an array.

Instead each *BD* fusion strain is mated with pooled *AD* fusion strains. After mating of the two strains, the diploid yeast cells are plated on selective media to screen for interactions. The identity of the *AD* fusion has to be determined by yeast colony PCR of positive colonies followed by sequencing.

In terms of quality assessment, rigorous evaluation and filtering of the raw data will enhance the quality and reliability of results. Routine testing of positive interactions is performed in duplicate and protein interactions that are not reproducible are discarded. From interactome studies, it is estimated that the coverage of array-based two-hybrid screens is only ~20% (Rajagopala et al., 2011). This high false negative rate is probably caused by the technical limitations of the system. Furthermore, different attempts to map the yeast interactome show very little overlap with each other and with annotated protein-protein interactions, probably due to the usage of different Y2H systems (Uetz et al., 2000, Ito et al., 2001). Recent attempts to generate a high quality dataset of the yeast interactome have revealed over 1000 new interactions (Yu et al., 2008). Further technical development and improvement is needed to increase the coverage of interactome screens. However, as the Y2H only covers a subset of interactions, a complementary AP/MS approach is necessary to generate a comprehensive map of the interactome (Yu et al., 2008).

2.4 Affinity purification coupled to mass spectrometry

Most proteins function in cellular processes as multi-subunit protein complexes. A well-established method to identify protein co-complexes is based on affinity purification followed by mass spectrometry (Fig.3) (Bauer & Kuster, 2003). The classical coimmunoprecipitation protocol is commonly used to detect whether two proteins interact in cellular systems, but can be also used to identify new interaction partners. In this approach, one protein is affinity captured along with its associated proteins by a specific antibody immobilized on Protein A or G sepharose. Experimental details like the affinity tag, lysis conditions, incubation time and washing conditions have a significant impact on the output and need to be optimized depending on the protein complex stability and localization. To preserve protein-protein interactions and native conformations, relatively mild conditions should be used during lysis. TritonX-100 or NP-40 are widely used in cell lysis buffers as non-ionic detergents and efficiently lyse membranes but are mild enough to preserve protein-protein interactions. Other variables that can influence the outcome of the affinity purification are salt concentrations, divalent cation concentrations and pH. The purified complex is rinsed several times to remove unspecifically bound proteins and complexes are eluted from the resin by low or high pH, high salt concentrations, by competition with a counter ligand or by adding Laemmli buffer to the beads. After purification, the isolated protein complex is usually separated on 1D or 2 D gels and protein lanes or spots are digested by trypsin and subsequently analyzed by mass spectrometry. It is also possible to precipitate eluted protein complexes and trypsinize the pellet. Comparison of the proteins detected in the sample with a negative control leads to the identification of specific interaction partners. Although low background binding is favoured, the stringency of washing steps needs to be adjusted to preserve more dynamic or transient interactions. Nonspecific binding can be reduced by adding low levels of detergent or by adjusting the salt concentration.

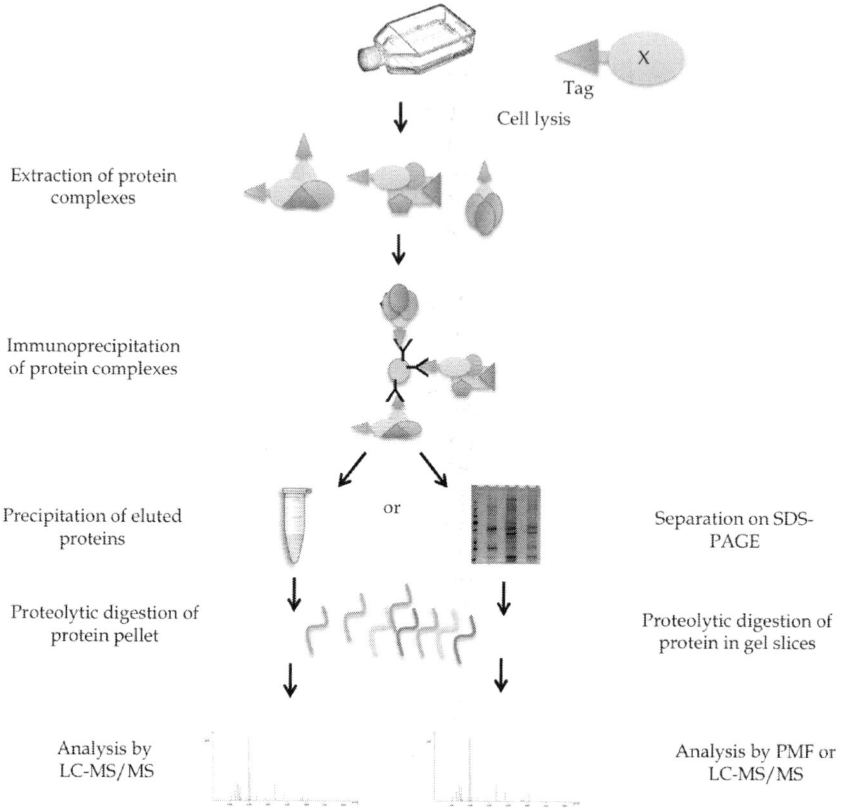

Fig. 3. Workflow of AP/MS for high throughput screening. Cells that express an epitope-tagged form of the protein of interest are lysed to extract protein complexes. The protein complexes are then isolated using a tag- specific antibody. The antibody-bound protein complex is immobilized on Protein A or Protein G sepharose and non-specific interactors are eliminated by several wash steps. As a control, the immobilized antibody is incubated with an untagged cell extract or a cell extract containing a tagged control protein (e.g. GFP). After purification, the protein complexes can be either separated on a gel or eluted and precipitated. The gel slices or the protein pellet are then digested to generate peptide fragments that can be analyzed by mass spectrometry to identify co-complexed proteins.

2.4.1 Large scale AP/MS screens

Proteome wide interaction studies need to meet certain criteria, such as reproducibility and low background binding. To circumvent the need of a specific antibody, the bait protein can be fused to an epitope tag that allows the pull down of different protein complexes using the same antibody under standardized conditions. A large variety of different epitopes has been used in the past and are commercially available (e.g. HA, FLAG, MYC). One disadvantage of this technique, however, is that the epitope-tagged proteins are overexpressed in the cell. Cytotoxicity induced by overexpression can be prevented by the usage of inducible expression systems. The tag might also interfere with protein folding and localization. Despite these limitations, coimmunoprecipitation via an epitope tag has been successfully used in proteome wide screens (Gavin et al., 2002, Ho et al., 2002).

Another coimmunoprecipitation protocol that is frequently used for high-throughput experiments is the tandem affinity purification (TAP) developed by Rigaut (Rigaut et al., 1999). The basic concept is similar to the coimmunoprecipitation of epitope tagged proteins. The main difference, however, is the use of two tags. One commonly used TAP tag consists of an IgG binding protein (Protein A) linked to a calmodulin-binding domain (CBD) via a TEV protease recognition site. Protein complexes are purified by the incubation of lysates with IgG sepharose beads that capture the Protein A tag. After washing the immobilized complex is eluted by TEV cleavage. Eluted protein can then be bound to a calmodulin sepharose column via their CBD domain. As the binding to calmodulin is calcium dependent, immobilized protein complexes can be eluted by addition of EDTA. The major advantage of this technique over one-step purification is increased specificity, reducing background binding to levels suitable for large-scale analysis (Gavin et al., 2002, Krogan et al., 2006). However, the long purification protocol preserves only stable interactions. For this reason, one-step purification with an epitope-tagged protein of interest is preferable if a broader range of interactions should be captured. Although the stringency of the purification protocol needs to be adjusted, the high sensitivity of mass spectrometry requires a minimum of non-specific binding to avoid laborious post-experimental filtering, especially in the case of large-scale experiments. An alternative approach to eliminate non-specific interactors without loosing low abundant or transiently interacting proteins is the use of quantitative mass spectrometry (Kaake et al., 2010).

The analysis of the isolated protein complex typically starts from separation of the complex components on 1D SDS-PAGE gels based on their molecular weight or 2D gels based on their charge and molecular weight followed by protease digestion. Usually, trypsin or Lys-C are employed to cleave the protein in peptide fragments because they generate peptides that have basic amino acids at their C-termini, which is favourable for detection and sequencing by mass spectrometry. In large-scale experiments, however, all purified proteins are precipitated and digested together to avoid time-consuming separation steps by gel electrophoresis and instead protein samples are fractionated by liquid chromatography (LC) after tryptic digest. Depending on the complexity of the sample, different approaches for protein identification by mass spectrometry can be used (Bauer and Kuster 2003). Complex peptide mixtures are analyzed by liquid chromatography in combination with tandem mass spectrometry (LC-MS/MS), whereas protein samples that have been separated on 1D SDS-PAGE or 2D gels and have therefore a lower complexity are analyzed by peptide mass fingerprinting (PMF).

PMF is an analytical method in which the absolute masses of peptides from a tryptic digest can be measured by matrix-assisted laser desorption/ionization time-of-flight mass spectrometry (MALDI-TOF). The measured masses of the sample are then looked up in a database that contains predicted peptide masses that have been generated by *in silico* digests for every protein. The results are statistically analyzed to find a significant overlap between the experimentally generated and predicted peptide masses. The identification method is based on the assumption that it is very unlikely that the exact same combination of peptide masses is found in more than one protein. Although this method can be applied to protein mixtures, the presence of different proteins complicates the analysis. Therefore, PMF is usually applied to identify proteins or protein mixtures isolated from 1 or 2D gels. Mixtures of more than five proteins require the additional use of tandem mass spectrometry for reliable protein identification.

Tandem mass spectrometry, uses a combination of sequence and mass information for the identification of proteins (MS/MS). In a first step, the masses of peptides from the tryptic digest are measured. Single peptides are then isolated from the mixture in the mass spectrometer and collided with inert gas molecules, which leads to fragmentation of the peptide backbone. Peptide fragments generated by this method differ in length by one amino acid. Measurement in the mass spectrometer allows determination of mass differences between two adjacent fragments that indicate a particular amino acid residue. The peptide mass, fragment mass and sequence is then compared against one or more databases to identify the proteins in the sample. Each peptide is individually identified and attributed to a protein in the mixture. For the analysis of complex samples, tandem mass spectrometry is generally combined with liquid chromatography to fractionate the peptide mixture before analysis by tandem mass spectrometry (LC-MS/MS).

Due to the sensitivity of the method and depending on the stringency of the purification protocol, post-experimental filtering is necessary to eliminate non-specific interactors. The ultimate goal of post-experimental analysis is the maximum reduction of false positives while maintaining the maximum coverage. Usually proteins that coimmunoprecipitate with a protein of interest are compared to proteins that are detected in a control sample to define unspecific interactions. Each of the experiments is commonly done at least in duplicate to identify high confidence interacting proteins. Whereas reproducibility seems to be a good indicator for true positive interactions, the quality assessment by comparisons with existing datasets depends on the selected gold standard. An ideal gold standard reference data set would be confirmed by other sources and should be generated by a comparable method to that being applied. Yu et al. (Yu et al., 2008) used an approach in which they tested several reference databases using various techniques to select a good gold standard.

2.5 Quality assessment of interactome mapping datasets

The identification of disease-modules and the interactome mapping of disease-linked proteins relies on the quality of the available reference datasets and the filtering of the experimentally generated interaction data. In early studies the precision of interactome maps was estimated by the integration of other biological attributes, such as gene ontology, or comparison with literature curated datasets. However, these methods suffer limitations in the estimation of data quality, as they need to be complete and unbiased. Recent efforts to establish an empirical framework for protein interaction maps will improve the estimation of accuracy and sensitivity for interaction maps generated by high-throughput interaction screens (Venkatesan et al., 2009). This empirical framework evaluates four different quality parameters of the currently used methods to estimate quality:

1. *Screening completeness*: the fraction of overall possible ORF pairs that are tested to generate the interaction map.
2. *Assay sensitivity*: the fraction of all interactions that could be identified by the applied experimental method.
3. *Sampling sensitivity*: the fraction of all detectable interactions that are identified in one trial of an assay under specific experimental conditions.
4. *Precision:* the fraction of known true positive pairs in the dataset.

Estimation of these parameters offers a quantitative idea of the coverage and accuracy of an interaction map and, when used in a standardized way, enables the comparison of different datasets.

3. Interactome mapping of disease-linked proteins in neurodegenerative diseases

Although great efforts have been made to uncover specific and shared pathways underlying neurodegeneration, many cellular mechanisms involved remain to be uncovered. As many disease-linked proteins function in multisubunit complexes and are functionally embedded in large cellular networks, gaining extensive knowledge of the architecture and composition of these complexes and networks is crucial to identify disease-linked pathways and to establish new disease markers. Based on this knowledge, diagnosis of early-onset dementias could be significantly improved and new therapeutic strategies could be developed. For age-related dementias, the discovery and characterization of shared pathways could pave the way towards common therapeutic interventions. To uncover the networks and complexes in which disease-linked proteins are embedded, interaction studies of many disease linked proteins, such as AD–associated (Chen et al., 2006, Krauthammer et al., 2004, Liu et al., 2006, Norstrom et al., 2010, Perreau et al., 2010, Soler-Lopez et a., 2010, Tamayev et al., 2009), PD-associated (Engelender et al., 1999, Meixner et al., 2011, Schnack et al., 2008, Suzuki, 2006, Woods et al., 2007, Zheng et al., 2008), HD-associated (Goehler et al., 2004, Kaltenbach et al., 2007) and ataxia-associated proteins (Kahle et al., 2011, Lim et al., 2006) as well as of prion protein (PrP) (Nieznanski, 2010) have been conducted. In addition, an interactome study that utilizes artificially designed amyloid-like fibrils has been performed and focused more on the general mechanisms underlying the toxic gain-of-function of β-sheet rich proteins (Olzscha et al., 2011).

3.1 Interactome mapping of disease- linked genes by large-scale Y2H

Several Y2H studies have been conducted to identify interaction partners of disease-related proteins in neurodegenerative diseases. In the following section, we highlight several studies that have contributed significant insight into disease networks by using new approaches or developing new ideas.

3.1.1 Generating the AD interaction network

An interesting approach to generate a complete interaction map for AD has been employed by Soler-Lopez et al. (Soler-Lopez et al., 2010). The authors elegantly combined different strategies to establish the most complete AD interaction network to date by exploring the interactomes of all known AD-linked proteins. The integration of their data with literature-based information provides new insights into the molecular interplay of different functional modules within the AD network and has helped to identify new candidate genes. Whereas most of the other interactome mapping approaches only focused on one prominent disease-linked protein, this study underlines the need to study the complete set of disease-associated genes in order to get the whole picture of a complex disorder like AD.

Although extensively studied, the cellular mechanisms that underlie the neurophathological changes associated with AD remain elusive. Despite the central role of APP and Tau, AD is a genetically complex disease and several genetic risk factors have been identified in the last decades. As it has been shown that the susceptibility/causative genes for many diseases are often interconnected within the same biological module, Soler-Lopez et al. applied different network biology strategies in their recently published study to build the most complete AD related interactome (Soler-Lopez et al., 2010).

The authors used available information from the Online Mendelian Inheritance in Man database to assemble a set of 12 well established AD related genes that they termed "seed". To be classified as a seed gene at least one mutation of the gene must be associated with AD in the OMIM Morbid Map. Interestingly, quantification of the degree of connectivity by computing the minimal path length between seed genes shows that the seed genes are connected by three links on average, whereas control sets (randomly picked genes and randomly picked disease causing genes from different disorders) were connected by more then four links on average. Starting from their defined AD seed genes, the authors developed a strategy to discover new disease genes. For this purpose they aimed to identify proteins that match at least one of the following criteria: 1.) the encoded proteins must directly interact with the seed genes, 2.) the gene must locate to a known genetic susceptibility locus and 3.) the gene must have altered expression in AD.

To define the interaction network of their selected seed proteins, they performed Y2H interaction screens. Baits were constructed for nine genes from the seed (three ORFs were not available) and yeast clones for the individual baits were transformed with an adult brain cDNA prey library. Each experiment was done in five replicates. 72 positive interactions resulted from this screen and were retested by pairwise cotransformant Y2H arrays. 32 high confidence interactors for the seed proteins could be validated.

In a second approach, the authors performed a pairwise candidate screen based on published genome wide association studies that identified four chromosomal regions (7q36, 10q24, 19q13.2, 20p) containing unknown AD susceptibility genes. They focused on 185 genes that are located in chromosomal regions linked to AD. Of those, 44 candidate genes known to be coexpressed with AD genes and suitable for Y2H screening were tested for their interaction with the 9 seed proteins. A systematic matrix-based Y2H was performed to identify interactors of the AD seed among the candidates extracted from the AD linked chromosomal regions. Two different technical approaches, a mating and a cotransformation screen, were used to perform the Y2H. With the identified interactions from the two Y2H, the authors generated a high-confidence interaction core (HC). The interactors identified in the library screen had to be validated to be included in the HC and the interactors from the pairwise candidate screen had to activate at least two reporter genes to be defined as high-confidence interaction. The final HC contained 8 seed genes (no HC interactors were identified for one seed gene) and 66 interactors, 27 from the library screen and a further 39 from the pairwise candidate screen. Interestingly, the different interactor set showed no overlap and only a low overlap with other studies, highlighting the importance of utilizing multiple approaches to increase coverage.

The HC interactions were validated by complementary strategies, such as GST pull down assays, coimmunoprecipitations and colocalization studies. Due to the high stringency applied for the identification of HC interactors, 87% of the interactors could be validated. The analysis of the HC identified four novel direct interactions between well-established AD related proteins (APP, A2M, APOE, PSEN1, PSEN2), implicating a possible link between plaque formation and inflammatory processes and providing insights into the regulation of APP cleavage. The assembled network showed an enrichment of 3 biological processes (oxidation reduction, regulation of apoptosis and negative regulation of cell motion), 5 molecular functions (protein binding, mono-oxygenase activity, oxygen binding, actin binding and integrin binding) and 6 cellular compartment terms (cytoplasm, pseudopodium, platelet alpha granule lumen, cytosol, cytoskeleton and internal side of plasma membrane) as well as altered expression of 17 out of 66 interactors, based on microarray data (Blalock et al., 2004). Furthermore, 6 out of 58 direct interactors were listed in the AlzGene database, whereas none of the non-interacting proteins were found in the database. Network analysis suggests a role for the programmed cell death 4 protein (PDCD4) in regulating neuronal death. PCDC4 was found to bind to the seed proteins PSEN2 and APOE and is located in a functionally homogenous network module enriched for translational elongation. As PCDC4 is upregulated in AD brain tissue, the authors suggest that the protein plays a role in Aβ toxicity. Another interesting member in the AD interactome network (AD-PIN) is ESCIT (evolutionary conserved signalling intermediate in Toll pathway), which links the redox signalling and immune response modules. Based on published data, the authors hypothesize that ESCIT may represent a molecular link between mitochondrial processes and AD lesions. Taken together, newly identified genes in the HC are likely related to AD onset or progression.

To generate a complete AD-PIN, the interaction partners of the HC were retrieved from literature-curated databases. The AD-PIN was then used to identify functional modules within the network that help to link processes potentially involved in AD and to identify the relationship between new candidate genes.

Taken together, this integrative approach combined stringent interaction screening and extensive validation by complementary strategies to build a comprehensive disease interaction map and highlight the importance of a network view of disease and the necessity of data integration from different sources when exploring disease interactomes.

3.1.2 The ataxia-ome

Disorders having a common clinical presentation likely also have common altered pathways. Examples of such diseases are familial spinocerebellar ataxias. A multitude of disease-associated mutations have been discovered, each leading to gain or loss of normal protein function. These mutations all inevitably lead to loss of Purkinje cells through an as yet unknown common pathway. An interesting study by Lim et al. used an Y2H approach to explore the interaction network of 54 proteins involved in 23 ataxias (Lim et al., 2006).

This study nicely showed that interactome mapping of different disease related genes can help to identify common pathways shared by diseases with similar presentation, such as neurodegenerative disorders.

Similar to the AD interactome study, the authors first defined a set of 23 ataxia-causing genes (11 recessive and 12 dominant, including the polyglutamine-mediated SCAs) whose mutations are linked to ataxias in humans or mice. Paralogs and an additional 31 directly interacting proteins were grouped together as ataxia-associated proteins. Bait and prey constructs for each of the genes were constructed to perform matrix-based mating type reciprocal screens against the human ORFeome. To minimize false positives, they used a stringent version of the Y2H system that expresses fusion proteins at low levels. Yeast clones containing single ataxia baits were mated with yeast clones of the opposite mating type containing 188 different ORF minilibrary pools (Rual et al., 2005). In a second screen, reciprocal interactions were tested between human ORFeome baits and ataxia preys to exclude effects caused by misfolding of the fusion protein and to include autoactivation baits. The overlap of the two screens comprises only 5.2% of the observed interactions, which is typical for reciprocal studies (Rual et al., 2005). To include a tissue specific library, the authors screened a human brain cDNA library using the same experimental setup. 29 interactions were overlapping in both screens, indicating that screening different types of libraries can enhance the coverage of the interactome map, as splice forms of a single gene are often spread across cDNA libraries. 83% of the interactions were validated experimentally by GST pull down assays from HEK293 cells with randomly picked interaction pairs. Analysis of the interaction sets found that 72% of those with compartment annotations colocalize together and 98% of the interactors share a GO branch, demonstrating the high quality of the Y2H generated interaction network.

A large interconnected network between the 36 ataxia genes and 541 preys was revealed for ataxia processes. In addition, 13 ataxia causing baits were linked directly or through common interacting proteins, indicating that the proteins in this interconnected network are functionally linked.

To build the ataxia network, Y2H data were integrated with interaction data from available databases. The network was then divided into a direct ataxia network, which contains first-order protein interactions, and the expanded ataxia network, which contains additional second- or third-order interactions. Further comparison of this ataxia network with networks generated from unrelated disease proteins showed that ataxia network nodes have a shorter path length, with a higher number of single hub interconnections. Surprisingly, 18 out of 23 ataxia causing proteins interacted directly or indirectly.

The ataxia interactome assembled by Lim et al. shows that for different ataxias, the similar pathophenotype is indeed caused by the alteration of shared pathways and processes. Main hubs in this ataxia network are involved in DNA repair and maintenance, transcription, RNA processing and protein quality control. Taken together, this kind of approach shows that genes and proteins that are involved in the same or related disorders are highly interconnected, often operating in the same pathways. In this way, interactome studies can prove valuable in identifying the shared pathways underlying related phenotypes.

3.2 Interactome mapping of disease-linked genes by AP/MS

Several AP/MS approaches have been successfully applied to identify interaction partners of disease-linked proteins in different neurodegenerative diseases. The following section summarizes two quantitative approaches that aim to identify distinct disease-relevant interaction networks.

3.2.1 QUICK screen to identify a PD-associated interaction network

Although PD represents a complex disorder for which several genetic risk factors have been identified, no complementary PD network has yet been generated, unlike for AD (Soler-Lopez et al., 2010). To explore the physiological function of the disease associated leucine-rich repeat kinase 2 (LRRK2), Meixner et al. used a new AP/MS protocol to identify the interaction network under stoichiometric constraints, called QUICK (quantitative immunoprecipitation combined with knockdown) (Meixner et al., 2011, Selbach & Mann, 2006). The QUICK screen aims to detect protein-protein interactions at endogenous levels and in their normal cellular environment by using a combination of SILAC, RNA interference, coimmunoprecititation and quantitative MS. For the QUICK approach, NIH3T3 cells were transduced with a lentiviral shRNA construct to knock down endogenous LRRK2. WT and LRRK2 knock down cells were then grown in SILAC medium containing either normal heavy isotope labeled lysine or arginine. For coimmunoprecipitation, an LRRK2 specific antibody was crosslinked to Protein G sepharose. Cells were lysed in buffer containing 1% NP-40 and phosphatase inhibitors. Equal numbers of heavy and light SILAC labeled cells were incubated with LRRK2 sepharose and then pooled. Purified LRRK2 complexes were eluted after several washing steps in Laemmli buffer. Separation of the proteins was achieved by SDS-PAGE gel electrophoresis, followed by tryptic digestion of gel slices. Interaction partners were identified by subsequent LC-MS/MS. Identified interaction partners were verified using the same approach, without prior SILAC labeling. Using this protocol, 36 proteins were identified as high confidence LRRK2 interactors. Bioinformatic analysis and integration of interaction data from different databases showed that these proteins are mainly members of the actin family and actin-associated proteins, pointing to an important role for LRRK2 in actin cytoskeleton based processes. Additional experiments demonstrated that LRRK2 binds to F-Actin *in vitro* and regulates its assembly. Knockdown of LRRK2 leads to morphological alterations and shortened neurite processes in primary neurons, further indicating a physiological role for LRRK2 in cytoskeletal organization. As the experimental strategy of the QUICK approach aims to identify interactors at physiological conditions (endogenous expression of the target protein) with specific antibodies, this method is well suited to explore the physiological function of other disease-linked proteins. Moreover, using knock-down or knock-out cells for control immunoprecipitations with a specific antibody proved as a suitable control. So far, this approach is not suited for large-scale applications involving several candidate proteins, but is of interest for generating interactome maps for one protein.

3.2.2 Interactome mapping of amyloid-like aggregates

Ordered proteinacious aggregates with high ß-sheet content, termed amyloid, are a characteristic of many neurodegenerative diseases. Whether this aggregation is cytoprotective or cytotoxic is still under debate. One hypothesis for aggregate toxicity is that these aggregates sequester cellular proteins, thereby leading to functional impairment (Chiti & Dobson, 2006).

To uncover the gain-of-function toxicity of amyloid-like fibrils in general, Olzscha et al. defined the interactome of artificial β-sheet proteins designed to form amyloids (West et al., 1999). The advantage of this method is that none of the artificial proteins have endogenous interaction partners, which allows for the identification of common amyloid interacting proteins independent of the identity of the aggregating protein. β4, 17 and 23, which differ in their β-sheet propensity, with β23 having the highest tendency to form β-sheets, were selected for interactome mapping. As a control, the authors used an α-helical protein with a similar amino acid composition (α-S284). Interaction analysis of the different β-sheet proteins was performed using SILAC followed by LC-MS/MS. Constructs coding for MYC-labelled proteins were transfected into HEK293 cells and proteins were labelled with light, medium or heavy arginine and lysine isotopes. Different experimental setups were chosen: (1) empty MYC tag vector, MYC α-S284, MYC β23; (2) MYC α-S284, MYC β4, MYC β17; (3) MYC β4, MYC β17, MYC β23 for quantitative MS. Lysates from heavy, medium and light labelled cells were mixed in a 1:1:1 ratio and amyloidogenic aggregates were isolated using anti-MYC coupled magnetic beads to avoid loss of the protein aggregates due to centrifugation. After purification the bound proteins were eluted and processed for LC-MS/MS analysis. Experiments were performed in triplicate. Proteins were defined as interactors if they were enriched relative to the α-S284 control or one of the β proteins with >95% confidence in two of the three repeats. Interactors were validated by coimmunoprecipitation and western blotting or immunofluorescence analysis. Bioinformatic analysis of the amyloid interactomes revealed that aggregates sequester a large amount of large, unstructured proteins that occupy essential hub position in housekeeping functions such as transcription and translation, chromatin regulation, vesicular transport, cell motility and morphology, as well as protein quality control. The authors hypothesize that amyloidogenic aggregation coaggregates a metastable subproteome, thereby causing perturbations in essential cellular networks leading to toxicity.

3.3 Interactome mapping by combinatorial approaches: The HD interactome

As Y2H and AP/MS approaches identify different groups of interactors, the combination of the two experimental strategies may prove valuable for the generation of comprehensive interaction maps of disease-linked proteins. In a study on HD, Kaltenbach et al. used these two approaches to generate a comprehensive HD interactome, which was then tested for modulators of polyglutamine toxicity. Importantly, the two approaches revealed distinct but biologically relevant interactions (Kaltenbach et al., 2007).

In this study the authors performed Y2H and AP/MS approaches in parallel to generate a comprehensive HTT interactome. Assuming that HTT and its interactors are functionally linked and regulate each other, the HTT interactome should be enriched for genetic modifiers of neurotoxicity. To test this hypothesis, the authors generated a *Drosophila* model of polyglutamine toxicity. For the AP/MS, different recombinant TAP-tagged HTT fragment baits were constructed (HTT 1-90: Q23, Q48, Q75; HTT 443-1100 and HTT-2758-3114) and purified from bacteria for pull down assays with mouse or human brain tissue and mouse muscle tissue. Purified complexes were then analyzed by MS. HTT fragments 1-90 were also used to probe HeLa, HEK293 and M17 neuroblastoma cell lysates. The interaction lists from the different pull downs were filtered by excluding proteins observed in a control pull down with the TAP tag alone and subjected to statistical analysis. The interaction lists were compared to a database of high-scoring peptides from 88 other unrelated pull downs. By using this approach, the authors generated a list of 145 mouse and human specific interactors with WT and expanded HTT fragments. In a complementary assay, a set of additional HTT interactors was identified by Y2H with HTT fragment baits (WT: 23Q and mutant: >45Q) against prey libraries from 17 different human tissue cDNA sources. Again, only baits generated from N-terminal fragments gave reproducible results. Filtering of the data was performed to generate a high confidence interaction dataset. Only interactions that were observed at least three times in the Y2H were integrated into the dataset; interactors were further compared to a database of Y2H interaction screens such that promiscuous interactors could be excluded. Finally, all positive prey constructs were retested by cotransformation into yeast strains expressing bait constructs. Previously published HTT interactors were included regardless of whether they matched the quality assessment criteria. A total of 104 interactors were identified by this Y2H approach.

The interactome was tested for possible genetic interaction in a fly model of polyglutamine toxicity. In this model, the N-terminal fragment of human HTT cDNA, including a 128Q expansion in exon 1, is expressed in the eye. The resulting neurodegenerative phenotype is visible by examining the retinal histology. Interactors that have orthologues in the fly and for which suitable fly stocks for overexpression or partial loss-of-function were available were selected from both high-confidence datasets. 60 interactors divided equally between interactors from the Y2H and AP/MS approach were tested as modifiers of polyglutamine mediated toxicity. 80% of the tested interactors altered the readout of more than one allele in a single background or a single allele in multiple backgrounds. The high confidence interaction dataset was further validated by coimmunoprecipitation studies using lysates from WT and HD mice. Data from this study strongly suggest that Y2H and AP/MS are of comparable quality in identifying biologically relevant interactions. Gene ontology analysis demonstrated that modifiers cluster into different groups, such as cytoskeletal organization, signal transduction, synaptic transmission, proteolysis and regulation of transcription and translation. Based on their experimentally generated interaction datasets, the authors built an interaction map of HD by integrating interaction data from different databases.

Interacting proteins of the HD network that were confirmed as modifiers in the fly model were shown to have diverse biological functions, such as synaptic transmission, cytoskeletal organization, signal transduction and transcription. In addition the authors revealed an unknown association between the HTT fragment and components of the vesicle secretion apparatus, suggesting that modulation of SNARE-mediated neurotransmitter secretion may be a physiological function of HTT. The involvement of HTT in these processes suggests a model in which mutant HTT interferes with different cellular pathways, leading to pathology.

4. Summary

Proteins do not operate alone, but instead function as a large interconnected network. Disease-related mutations, consequently, disrupt not only the function of individual proteins, but also the larger network in which these proteins function, only thereby leading to clinically relevant pathology. While many complex diseases are caused by an array of mutations in any of a multitude of genes, these all lead to a similar pathology. Focusing not just on individual proteins but, rather, on the network of proteins altered in diseases will prove essential to both identifying new candidate genes for this and related disease, as well as to provide new targets for the development of therapeutic interventions. A combination of different experimental and bioinformatic approaches has proved valuable in the network analysis of neurodegenerative diseases. Further studies are necessary to generate high quality, comprehensive datasets, which can be used to identify shared disease pathways.

5. Acknowledgements

We thank Devon Ryan, Daniele Bano, Donato A. Di Monte and Moritz Hettich for providing a critical reading of the manuscript and for helpful suggestions.

6. References

Amberger, J., Bocchini, C. & Hamosh, A. (2011). A new face and new challenges for Online Mendelian Inheritance in Man (OMIM®). *Human Mutation*, Vol. 32, No. 5, (May2011), pp.(564-567), ISSN 1059-7794

Auerbach, D., Thaminy, S., Hottiger, M.O. & Stagljar, I. (2002). The post-genomic era of interactive proteomics: facts and perspectives. *Proteomics*, Vol. 2, No. 6, (June 2002), pp. (611-623), ISSN 1615-9861

Barabási, A.L., Gulbahce, N. & Loscalzo, J. (2011). Network medicine: a network-based approach to human disease. *Nature Reviews Genetics,* Vol. 12, No. 1, (January 2011), pp. (56-68), ISSN 1471-0064

Bauer, A. & Kuster, B. (2003). Affinity purification-mass spectrometry. Powerful tools for the characterization of protein complexes. *European Journal of Biochemistry*, Vol. 270, No. 4, pp. (570-578), ISSN 1432-1033

Braak, H., Ghebremedhin, E., Rüb, U., Bratzke, H. & Del Tredici, K. (2004). Stages in the development of Parkinson's disease-related pathology. *Cell and Tissue Research,* Vol. 318, No. 1, (October 2004), pp. (121-134), ISSN 1432-0878

Beyer, K. (2006). Alpha-synuclein structure, posttranslational modification and alternative splicing as aggregation enhancers. *Acta Neuropathologica,* Vol. 112, No. 3, (September 2006), pp. (237-251), ISSN 1432-0533

Blalock, E.M., Geddes, J.W., Chen, K.C., Porter, N.M., Markesbery, W.R. & Landfield, P.W. (2004). Incipient Alzheimer's disease: Microarray correlation analyses reveal major transcriptional and tumor suppressor responses. *Proceedings of the National Academy of Sciences of the United States of America*, Vol. 101, No. 7, (February 2004), pp. (2173–2178), ISSN 1091-6490

Chartier-Harlin M.C., Kachergus J., Roumier C., Mouroux V., Douay X., Lincoln S., Levecque C., Larvor L., Andrieux J., Hulihan M., Waucquier N., Defebvre L., Amouyel P., Farrer M. & Destée A. (2004). Alpha-synuclein locus duplication as a cause of familial Parkinson's disease. *Lancet*, Vol. 364, No. 9440, (October 2004), pp. (1167-1169), ISSN 0140-6736

Chen, J.Y., Shen, C. & Sivachenko, A.Y. (2006). Mining Alzheimer disease relevant proteins from integrated protein interactome data. *Pacific Symposium on Biocomputing,* Vol.11, pp. (367-378), ISSN 1793-5091

Chiti, F. & Dobson, C.M. (2006). Protein misfolding, functional amyloid, and human disease. *Annual Review of Biochemistry*, Vol. 75, (2006), pp. (333-366), ISSN 0066-4154

Douglas, P.M. & Dillin, A. (2011). Protein homeostasis and aging in neurodegeneration. *The Journal of Cell Biology,* Vol. 190, No. 5, (September 2010), pp. (719-729), ISSN 1540-8140

Engelender, S., Kaminsky, Z., Guo, X., Sharp, A.H., Amaravi, R.K., Kleiderlein, J.J., Margolis, R.L., Troncoso J.C., Lanahan, A.A., Worley, P.F., Dawson, V.L., Dawson, T.M. & Ross, C.A. (1999). Synphilin-1 associates with alpha-synuclein and promotes the formation of cytosolic inclusions. *Nature Genetics,* Vol. 22, No. 1, (May 1999), pp. (110-114), ISSN 1546-1718

Fields, S. & Song, O. (1989). A novel genetic system to detect protein-protein interactions. *Nature,* Vol. 340, No. 6230, (July 1989), pp. (245-246), ISSN 0028-0836

Gavin, A.C., Bösche, M., Krause, R., Grandi, P., Marzioch, M., Bauer, A., Schultz, J., Rick, J.M., Michon, A.M., Cruciat, C.M., Remor, M., Höfert, C., Schelder, M., Brajenovic, M., Ruffner, H., Merino, A., Klein, K., Hudak, M., Dickson, D., Rudi, T., Gnau, V., Bauch, A., Bastuck, S., Huhse, B., Leutwein, C., Heurtier, M.A., Copley, R.R., Edelmann, A., Querfurth, E., Rybin, V., Drewes, G., Raida, M., Bouwmeester, T., Bork, P., Seraphin, B., Kuster, B., Neubauer, G. & Superti-Furga, G. (2002). Functional organization of the yeast proteome by systematic analysis of protein complexes. *Nature,* Vol. 415, No. 6868, (January 2002), pp. (141-147), ISSN 0028-0836

Georgopoulou, N., McLaughlin, M., McFarlane, I. & Breen, K.C. (2001). The role of post-translational modification in beta-amyloid precursor protein processing. *Biochemical Society Symposium,* Vol. 67, (2001), pp. (23-36), ISSN 0067-8694

Goehler, H., Lalowski, M., Stelzl, U., Waelter, S., Stroedicke, M., Worm, U., Droege, A., Lindenberg, K.S., Knoblich, M., Haenig, C., Herbst, M., Suopanki, J., Scherzinger, E., Abraham, C., Bauer, B., Hasenbank, R., Fritzsche, A., Ludewig, A.H., Büssow, K., Coleman, S.H., Gutekunst, C.A., Landwehrmeyer, B.G., Lehrach, H. & Wanker, E.E. (2004). A protein interaction network links GIT1, an enhancer of huntingtin aggregation, to Huntington's disease. *Molecular Cell,* Vol. 15, No. 6, (September 2004), pp. (853-865), ISSN 1097-2765

Greenbaum, E.A., Graves, C.L., Mishizen-Eberz, A.J., Lupoli, M.A., Lynchm D.R., Englander, S.W., Axelsen, P.H. & Giasson, B.I. (2005). The E46K mutation in alpha-synuclein increases amyloid fibril formation. *The Journal of Biological Chemistry,* Vol. 280, No. 9, (March 2005), pp. (7800-7807), ISSN 1083-351X

Hands, S.L. & Wyttenbach, A. (2010). Neurotoxic protein oligomerisation associated with polyglutamine diseases. *Acta Neuropathologica,* Vol. 120, No. 4, (October 2010), ISSN 1432-0533

Hardy, J. & Selkoe, D.J. (2002). The amyloid hypothesis of Alzheimer's disease: progress and problems on the road to therapeutics. *Science,* Vol. 297, No. 5580, (July 2002), pp. (353-356), ISSN 1095-9203

Ho, Y., Gruhler, A., Heilbut, A., Bader, G.D., Moore, L., Adams, S.L., Millar, A., Taylor, P., Bennett, K. & Boutilier, K., et al. (2002). Systematic identification of protein complexes in Saccharomyces cerevisiae by mass spectrometry. *Nature,* Vol. 415, No. 6868, pp. (180-183), ISSN 0028-0836

Ito, T., Chiba, T., Ozawa, R., Yoshida, M., Hattori & M., Sakaki, Y. (2001). A comprehensive two-hybrid analysis to explore the yeast protein interactome. *Proceedings of the National Academy of Sciences* of the United States of America, Vol. 98, No. 8, (April 2001), pp. 4569-4574, ISSN 1091-6490

Kaake, R.M., Wang, X. & Huang, L. (2010). Profiling of protein interaction networks of protein complexes using affinity purification and quantitative mass spectrometry. *Molecular & Cellular Proteomics,* Vol. 9, No. 8, (August 2010), pp. (1650-1665), ISSN 1535-9484

Kahle, J.J., Gulbahce, N., Shaw, C.A., Lim, J., Hill, D.E., Barabási, A.L. & Zoghbi HY. (2011). Comparison of an expanded ataxia interactome with patient medical records reveals a relationship between macular degeneration and ataxia. *Human Molecular Genetics,* Vo. 20, No. 3, (February 2011), pp. (510-527), ISSN 1460-2083

Kaltenbach, L.S., Romero, E., Becklin, R.R., Chettier, R., Bell, R., Phansalkar, A., Strand, A., Torcassi, C., Savage, J., Hurlburt, A., Cha, G.H., Ukani, L., Chepanoske, C.L., Zhen, Y., Sahasrabudhe, S., Olson, J., Kurschner, C., Ellerby, L.M., Peltier, J.M., Botas, J. & Hughes, R.E. (2007) Huntingtin interacting proteins are genetic modifiers of neurodegeneration. *PLoS Genetics,* Vol. 3, No. 5, (May 2007), pp. (e82), ISSN 1553-7390

Kettern, N., Dreiseidler, M., Tawo, R. & Höhfeld, J. (2010). Chaperone-assisted degradation: multiple paths to destruction. *Biological Chemistry,* Vol. 391, No. 5, (May 2010), pp. (481-489), ISSN 1437-4315

Koegl, M. & Uetz, P. (2007). Improving yeast two-hybrid screening systems. *Briefings in Functional Genomics and Proteomics,* Vol. 6, No.4, (December 2007), pp. (302-312), ISSN 2041-2647

Krauthammer, M., Kaufmann, C.A., Gilliam, T.C. & Rzhetsky, A. (2004). Molecular triangulation: bridging linkage and molecular-network information for identifying candidate genes in Alzheimer's disease. *Proceeding of the National Academy of Sciences of the United States of America,* Vol. 101, No. 42, (October 2004), pp. (15148-15153), ISSN 0027-8424

Krogan, N.J., Cagney, G., Yu, H., Zhong, G., Guo, X., Ignatchenko, A., Li, J., Pu, S., Datta, N., Tikuisis, A.P., Punna, T., Peregrin-Alvarez, J.M., Shales, M., Zhang, X., Davey, M., Robinson, M.D., Paccanaro, A., Bray, J.E., Sheung, A., Beattie, B., Richards, D.P., Canadien, V., Lalev, A., Mena, F., Wong, P., Starostine, A., Canete, M.M., Vlasblom, J., Wu, S., Orsi, C., Collins, S.R., Chandran, S., Haw, R., Rilstone, J.J., Gandi, K., et al. (2006). Global landscape of protein complexes in the yeast saccharomyces cerevisiae. *Nature,* Vol. 440, No. 7084 (March 2006), pp. (637-643), ISSN 0028-0836

Li, J., Uversky, V.N. & Fink, A.L. (2001). Effect of familial Parkinson's disease point mutations A30P and A53T on the structural properties, aggregation, and fibrillation of human alpha-synuclein. *Biochemistry,* Vol. 40, No. 38, (September 2001), pp.(11604-11613), ISSN 0006-2960

Lim, J., Hao, T., Shaw, C., Patel, A.J., Szabó, G., Rual, J.F., Fisk, C.J., Li, N., Smolyar, A., Hill, D.E., Barabási, A.L., Vidal, M. & Zoghbi, H.Y. (2006). A protein-protein interaction network for human inherited ataxias and disorders of Purkinje cell degeneration. *Cell,* Vol. 125, No. 4, (May 2006), pp. (801-814), ISSN 0092-8674

Liu, B., Jiang, T., Ma, S., Zhao, H., Li, J., Jiang, X. & Zhang, J. (2006). Exploring candidate genes for human brain diseases from a brain-specific gene network. *Biochemical* and *Biophysical* Research *Communications,* Vol. 349, No. 4, (November *2006),* pp. (1308-1314), ISSN 0006-291X

Maccioni, R.B., Farías, G., Morales, I. & Navarrete, L. (2010). The revitalized tau hypothesis on Alzheimer's disease. *Archives of Medical Research,* Vol. 41, No. 3, (April 2010), pp.(226-231), ISSN 0188-4409

Martin, I., Dawson, V.L. & Dawson, T.M. (2011). Recent advances in the genetics of Parkinson's disease. *Annual Review of Genomics and Human Genetics,* Vol. 22, No. 12, (September 2011), pp. (301-325), ISSN 1527-8204

Martin, L., Latypova, X. & Terro, F. (2011). Post-translational modifications of tau protein: implications for Alzheimer's disease. *Neurochemistry International,* Vol. 58, No. 4, (March 2011), pp. (458-471), ISSN 0197-0186

Martindale, D., Hackam, A., Wieczorek, A., Ellerby, L., Wellington, C., McCutcheon, K., Singaraja, R., Kazemi-Esfarjani, P., Devon, R., Kim, S.U., Bredesen, D.E., Tufarom F. & Hayden, M.R. (1998). Length of huntingtin and its polyglutamine tract influences localization and frequency of intracellular aggregates. *Nature Genetics,* Vol. 18, No.2, (February 1998), pp. (150-154), ISSN 1546-1718

Martinez-Vicente, M. & Cuervo, A.M. (2007). Autophagy and neurodegeneration: when the cleaning crew goes on strike. *Lancet Neurology,* Vol. 6, No. 4, (April 2007), pp. (352-361), ISSN 1474-4422

Meixner, A., Boldt, K., Van Troys, M., Askenazi, M., Gloeckner, C.J., Bauer, M., Marto, J.A., Ampe, C., Kinkl, N. & Ueffing, M. (2011). A QUICK screen for Lrrk2 interaction partners--leucine-rich repeat kinase 2 is involved in actin cytoskeleton dynamics. *Molecular & Cellular Proteomics,* Vol. 10, No. 1, (January 2011), pp. (M110.001172), ISSN 1535-9484

Morphy, R. & Rankovic, Z. (2007). Fragments, network biology and designing multiple ligands. *Drug Discovery Today,* Vol. 12, No. (3-4), (February 2007), pp. (156-160), ISSN 1359-6446

Mudher, A. & Lovestone, S. (2002). Alzheimer's disease-do tauist and Baptist finally shake hands? *Trends in Neurosciences,* Vol. 25, No. 1, (January 2002), pp. (22-26), ISSN 0166-2236

Nieznanski, K. (2010). Interactions of prion protein with intracellular proteins: so many partners and no consequences? *Cellular and Molecular Neurobiology,* Vol. 30, No. 5, (July 2010), pp. (653-666), ISSN 1573-6830

Norstrom, E.M., Zhang, C., Tanzi, R. & Sisodia, S.S. (2010). Identification of NEEP21 as a ß-amyloid precursor protein-interacting protein in vivo that modulates amyloidogenic processing in vitro. *Journal of Neuroscience,* Vol. 30, No. 46, (November 2010), pp. (15677-15685), ISSN 1529-2401

O'Brien, R.J. & Wong, P.C. (2011). Amyloid precursor protein processing and Alzheimer's disease. *Annual Review of Neuroscience,* Vol. 34, (July 2011), pp. (185-204), ISSN 0147-006X

Olzscha, H., Schermann, S.M., Woerner, A.C., Pinkert, S., Hecht, M.H., Tartaglia, G.G., Vendruscolo, M., Hayer-Hartl, M., Hartl, F.U. & Vabulas, R.M. (2011). Amyloid-like aggregates sequester numerous metastable proteins with essential cellular functions. *Cell*, Vol. 144, No. 7 (January 2011), pp. (67-78), ISSN 0092-8674

Pennuto, M., Palazzolo, I. & Poletti, A. (2009). Post-translational modifications of expanded polyglutamine proteins: impact on neurotoxicity. *Human Molecular Genetics*, Vol. 15, No. 18 (R1), (April 2009), pp. (R40-47), ISSN 1460-2083

Perreau, V.M., Orchard, S., Adlard, P.A., Bellingham, S.A., Cappai, R., Ciccotosto, G.D., Cowie, T.F., Crouch, P.J., Duce, J.A., Evin, G., Faux, N.G., Hill, A.F., Hung, Y.H., James, S.A., Li, Q.X., Mok, S.S., Tew, D.J., White, A.R., Bush, A.I., Hermjakob, H. & Masters, C.L. (2010). A domain level interaction network of amyloid precursor protein and Abeta of Alzheimer's disease. *Proteomics*, Vol. 10, No. 12, (June 2010), pp. (2377-2395), ISSN 1615-9861

Rajagopala, S.V. & Uetz, P. (2011). Analysis of protein-protein interactions using high-throughput yeast two-hybrid screens. *Methods in Molecular Biology*, Vol. 781, (2011), pp. (1-29), ISSN 1064-3745

Rigaut, G., Shevchenko, A., Rutz, B., Wilm, M., Mann, M. & Séraphin, B. (1999). A generic protein purification method for protein complex characterization and proteome exploration. *Nature Biotechnology*, Vol. 17, No. 10, (October 1999), pp. (1030-1032), ISSN 1087-0156

Rual, J.F., Venkatesan, K., Hao, T., Hirozane-Kishikawa, T., Dricot, A., Li, N., Berriz, G.F., Gibbons, F.D., Dreze, M., Ayivi-Guedehoussou, N., Klitgord, N., Simon, C., Boxem, M., Milstein, S., Rosenberg, J., Goldberg, D.S., Zhang, L.V., Wong, S.L., Franklin, G., Li, S., Albala, J.S., Lim, J., Fraughton, C., Llamosas, E., et al. (2005). Towards a proteome-scale map of the human protein-protein interaction network. *Nature*, Vol. 437, No. 7062, (October 2002), pp. (1173-1178), ISSN 0028-0836

Rumble, B., Retallack, R., Hilbich, C., Simms, G., Multhaup, G., Martins, R., Hockey, A., Montgomery, P., Beyreuther, K. & Masters, C.L. (1989). Amyloid A4 protein and its precursor in Down's syndrome and Alzheimer's disease. *The New England Journal of Medicine*, Vol 320, No. 22, (June 1989), pp. (1446-1452), ISSN 0028-4793

Schnack, C., Danzer, K.M., Hengerer, B. & Gillardon, F. (2008). Protein array analysis of oligomerization-induced changes in alpha-synuclein protein-protein interactions points to an interference with Cdc42 effector proteins. *Neuroscience*, Vol. 154, No. 4, (July 2008), pp. (1450-1457). ISSN 0306-4522

Seebacher, J., Gavin, A.C. (2011). SnapShot: Protein-protein interaction networks. *Cell*, Vol. 144, No. 6, (March 2011), pp. (1000), ISSN 0092-8674

Selbach, M. & Mann, M. (2006). Protein interaction screening by quantitative immunoprecipitation combined with knockdown (QUICK). *Nature Methods*. Vol. 3, No. 12, (December 2006), pp. (981-983), 1548-7091

Shao, J. & Diamond, M.I. (2007). Polyglutamine diseases: emerging concepts in pathogenesis and therapy. *Human Molecular Genetics*, Vol. 16, Spec No. 2, (October 2007), pp. (R115-123), ISSN 1460-2083

Singleton, A.B., Farrer, M., Johnson, J., Singleton, A., Hague, S., Kachergus, J., Hulihan, M., Peuralinna, T., Dutra, A., Nussbaum, R., Lincoln, S., Crawley, A., Hanson, M., Maraganore, D., Adler, C., Cookson, M.R., Muenter, M., Baptista, M., Miller, D., Blancato, J., Hardy, J. & Gwinn-Hardy, K. (2003). alpha-Synuclein locus triplication causes Parkinson's disease. *Science*, Vol. 302, No. 5646, (October 2003), pp. (841), ISSN 1095-9203

Soler-López, M., Zanzoni, A., Lluís, R., Stelzl, U. & Aloy, P. (2011). Interactome mapping suggests new mechanistic details underlying Alzheimer's disease. *Genome Research*, Vol. 21, No. 3, (March 2011), pp. (364-376), ISSN 1943-0264

Suzuki, H. (2006). Protein-protein interactions in the mammalian brain. *The Journal of Physiology*, Vol. 575, No. 2, (September 2006), pp. (373-377), ISSN 1469-7793

Tamayev, R., Zhou, D. & D'Adamio, L. (2009). The interactome of the amyloid beta precursor protein family members is shaped by phosphorylation of their intracellular domains. *Molecular Neurodegeneration*, Vol.14, No. 4 (July 2009), pp. (28), ISSN 1750-1326

Tyedmers, J., Mogk, A. & Bukau, B. (2010). Cellular strategies for controlling protein aggregation. *Nature Reviews Molecular Cell Biology*, Vol. 11, No. 11, (November 2010), pp. (777-788), ISSN 1471-0080

Uetz, P., Giot, L., Cagney, G., Mansfield, T.A., Judson, R.S., Knight, J.R., Lockshon, D., Narayan, V., Srinivasan, .M, Pochart, P., Qureshi-Emilim, A., Li, Y., Godwin, B., Conover, D., Kalbfleisch, T., Vijayadamodar, G., Yang, M., Johnston, M., Fields, S. & Rothberg, J.M. (2000). A comprehensive analysis of protein-protein interactions in Saccharomyces cerevisiae. *Nature*, Vol. 403, No. 6770, (February 2000), pp. (623-627), ISSN 0028-0836

Van Dam, D. & De Deyn, P. P. (2006). Drug discovery in dementia: the role of rodent models. *Nature Reviews Drug Discovery*, Vol. 5, No. 11, (November 2006), pp. (956-970), ISSN 147-41776

Venkatesan, K., Rual, J.F., Vazquez, A., Stelzl, U., Lemmens, I., Hirozane-Kishikawa, T., Hao, T., Zenkner, M., Xin, X., Goh, K.I., Yildirim, M.A., Simonis, N., Heinzmann, K., Gebreab, F., Sahalie, J.M., Cevik, S., Simon, C., de Smet, A.S., Dann, E., Smolyar, A., Vinayagam, A., Yu, H., Szeto, D., Borick, H., Dricot, A., Klitgord, N., Murray, R.R., Lin, C., Lalowski, M., Timm, J., Rau, K., Boone, C., Braun, P., Cusick, M.E., Roth, F.P., Hill, D.E., Tavernier, J., Wanker, E.E., Barabási, A.L. &Vidal, M. (2009). An empirical framework for binary interactome mapping. *Nature Methods*, Vol. 6, No. 1, (January 2009), pp. (83-90), ISSN 1548-7105

Vidal, M., Cusick, M.E. & Barabási, A.L. (2011). Interactome networks and human disease. *Cell*, Vol. 144, No. 6, (March 2011), pp. (986-998), ISSN 0092-8674

Wang, Y., Garg, S., Mandelkow, E.M. & Mandelkow, E. (2010). Proteolytic processing of tau. *Biochemical Society Transactions*, Vol. 38, No. 4, (August 2010), pp: (955-961), ISSN 1470-8752

West, M.W., Wang, W., Patterson, J., Mancias, J.D., Beasley, J.R. & Hecht, M.H. (1999). De novo amyloid proteins from designed combinatorial libraries. *Proceeding of the National Academy of Sciences of the United States of America*, Vol. 96, No. 20, (September 1999), pp. (11211-11216), ISSN 0027-8424

Wolfe, M.S. (2009). Tau mutations in neurodegenerative diseases. *The Journal of Biological Chemistry*, Vol. 284, No. 10, (March 2009), pp. (6021-6025), ISSN 1083-351X

Woods, W.S., Boettcher, J.M., Zhou, D.H., Kloepper, K.D., Hartman, K.L., Ladror, D.T., Qi, Z., Rienstra, C.M. & George, J.M. (2007). Conformation-specific binding of alpha-synuclein to novel protein partners detected by phage display and NMR spectroscopy. *The Journal of Biological Chemistry*, Vol. 282, No. 47, (November 2007), pp. (34555-34567), INSS 1083-351X

Yu, H., Braun, P., Yildirim, M.A., Lemmens, I., Venkatesan, K., Sahalie, J., Hirozane-Kishikawa, T., Gebreab, F., Li, N., Simonis, N., Hao, T., Rual, J.F., Dricot, A., Vazquez, A., Murray, R.R., Simon, C., Tardivo, L., Tam, S., Svrzikapa, N., Fan, C., de Smet, A.S., Motyl, A., Hudson, M.E., Park, J., Xin, X., Cusick, M.E., Moore, T., Boone, C., Snyder, M., Roth, F.P., Barabási, A.L., Tavernier, J., Hill, D.E. & Vidal, M. (2008). High-quality binary protein interaction map of the yeast interactome network. *Science*, Vol. 322, No. 5898, (October 2008) pp. (104-110), ISSN 1095-9203

Zheng, X.Y., Yang, M., Tan, J.Q., Pan, Q., Long, Z.G., Dai, H.P., Xia, K., Xia, J.H. & Zhang, Z.H. (2008). Screening of LRRK2 interactants by yeast 2-hybrid analysis. *Zhong Nan Da Xue Xue Bao Yi Xue Ban*, Vol. 33, No. 10, (October 2008), pp. (883-891), ISSN 1672-7347

13

Biochemical, Structural and Pathophysiological Aspects of Prorenin and (Pro)renin Receptor

A.H.M. Nurun Nabi[1] and Fumiaki Suzuki[2]
[1]*Department of Biochemistry and Molecular Biology, University of Dhaka, Dhaka, Bangladesh;*
[2]*Laboratory of Animal Biochemistry, Faculty of Applied Biological Sciences, Gifu University, 1-1 Yanagido, Gifu, Japan*

1. Introduction

Our knowledge of understanding the complex role of renin angiotensin system (RAS) or RA system in human physiology has been widened for more than 100 years and it is rapidly increasing day-by-day. Over the last two decades discoveries of angiotensin converting enzyme 2 (ACE 2), putative receptor for angiotensin (Allen et al., 2006; Aronsson et al. 1988; Bickerton & Buckley, 1961; Cooper et al., 1996; Deschepper et al., 1986; Dzau et al., 1986; Epstein et al., 1970), and (pro)renin receptor have laid the foundation of many new hypotheses in the context of their biochemical actions, physiological effects and activation of second messenger pathways. Thus, scientists have started to reconsider the complex biochemistry and physiology of RAS. The primary and main role of this system is to regulate homeostasis of body fluid that ultimately maintains the blood pressure (Kobori et al., 2006; Oparil & Haber, 1974a; Oparil & Haber; 1974b). This system catalyzes a liver product, angiotensinogen, to generate a small decapeptide, angiotensin-I (Ang-I). Angiotensin converting enzyme (ACE), thus, converts Ang-I into octapeptide, angiotensin-II (Ang-II). Ang-II acts directly within the central nervous system to increase blood pressure (Bickerton & Buckley, 1961). Injection of purified Ang-II peptide around the hypothalamus in rat brain stimulated thriving drinking response (Epstein et al., 1970). The physiological actions of the most potent hormone peptide are mediated via G-protein coupled angiotensin II type 1 (AT1) and angiotensin II type 2 (AT2) receptors. Ang-II facilitates vasoconstriction, cell proliferation, cell hypertrophy, anti-natriuresis, fibrosis, atherosclerosis using AT1 (Ito et al., 1995) while, via AT2 receptor, the peptide elicits vasodilation, anti-proliferation, anti-hypertrophy, anti-fibrosis, anti-thrombosis, anti-angiogenesis (Siragy & Carey, 1997; Goto et al., 1997; Gross et al., 2000). The classical renin angiotensin system with the generation of different peptides and their physiological effects has been presented in Figure 1.

The systemic or classical renin angiotensin system has usually been viewed as the blood-borne cascade whose ultimate product Ang-II plays the pivotal endocrine role. Plasma renin activity is the most accepted clinical marker of circulating RAS. However, circulating RAS remained unsuccessful to describe the autocrine and paracrine functions mediated by RAS within specific tissue sites particularly in heart, kidney, brain and vasculature. Transgenic

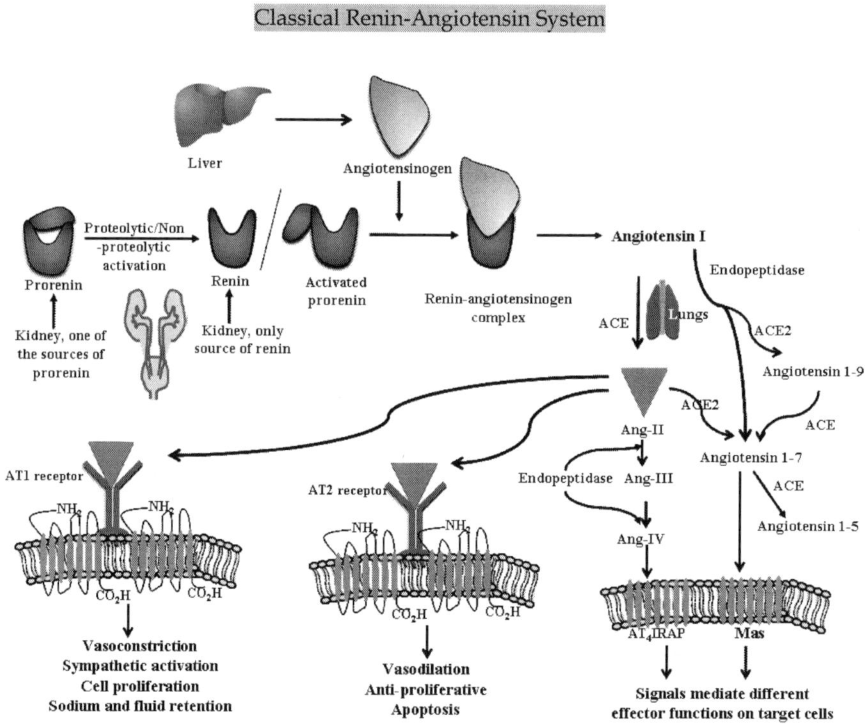

Fig. 1. Schematic diagram of the classical renin-angiotensin (RA) system shows angiotensin-II dependent pathway mediated different physiological effects via angiotensin type 1 (AT1) and type 2 (AT2) receptors. Renin, secreted from kidney, regulates the rate limiting step of this pathway by converting its liver originated macromolecule substrate angiotensinogen into a short peptide, angiotensin-I (Ang-I). Angiotensin-I, thus, is converted into angiotensin-II (Ang-II) by angiotensin converting enzyme. Other peptide products via stimulation of enzymes and receptor subtypes on target cells can also mediate physiological functions.

animals facilitate to demonstrate the existence of tissue RAS parallel to and independent of systemic RAS. Thus, local RA system has been ensured in intracellular compartments (de Mello, 1995; van Kesteren et al., 1997; Admiraal et al., 1999), interstitial fluids (de Lannoy et al., 1997, 1998), cardiac cells including fibroblasts, endothelial cells, myocytes, macrophages (de Lannoy et al., 1998; Hokimoto et al., 1996; Sun et al., 1994) as well as on the cell membrane (Danser et al., 1992; Neri Serneri et al., 1996). All the circulating components of renin angiotensin system i.e., renin, angiotensinogen, ACE, and Ang I and II though not produced but have been identified in cardiac tissue (Campbell et al., 1993; Danser et al., 1994). As a consequence, presence of local RAS in the heart could contribute to the pathogenesis of congestive heart failure, cardiac hypertrophy and remodeling, and reperfusion arrhythmias (Yusuf et al., 1991: Ruzicka et al., 1994; Schieffer et al., 1994; Van

Gilst et al., 1984). Direct action of Ang II within the central nervous system causes increased blood pressure. Also, presence of renin and endogenous production of angiotensin have established the existence of local RAS in the central nervous system (Bickerton and Buckley et al., 1961; Fischer-Ferraro et al., 1971; Ganten et al., 1971).

1.1 The main players associated with renin angiotensin system

The RA system is initiated by its rate-limiting enzyme renin (37 kDa with 340 amino acid residues) which catalyzes its only known substrate angiotensinogen. Renin is only secreted from kidney as preprorenin and levels of renin in the plasma of nephrectomized animals is not detectable. Professor Robert A. Tigerstedt and his student Per G. Bergman for the first time reported a "pressor" substance in the kidney extract more than 100 years ago, which caused increase in blood pressure in experimental animals and later, coined that substance as renin (Tigerstedt & Bergman, 1898). Renin, also known as aspartyl proteinase having an optimum pH of 5.5 to 7.5 instead of 2.0 to 3.4, has no known physiological effect other than the proteolysis of angiotensinogen (Murakami & Inagami, 1975; Inagami & Murakami, 1977; Matoba *et al.*, 1978; Figueiredo *et al.*, 1985: Dzau *et al.* 1979; Yokosawa *et al.*, 1978; Yokosawa, 1980; Hirose, 1982; Pickens *et al.*, 1965). The neutral pH is necessary to show its activity in plasma. The renin gene is also expressed in other tissues such as adrenal gland, gonads, placenta, pituitary, brain and hypothalamus (Hirose et al., 1978; Naruse et al., 1981, 1982; Pandey et al., 1984; Deschepper et al., 1986; Dzau et al., 1987; Paul et al., 1987; Suzuki et al., 1987; Tada et al., 1989). These extra renal renins have been thought to play a part in the tissue renin-angiotensin system proposed by several investigators (de Mello, 1995; van Kesteren et al., 1997; Admiraal et al., 1999; de Lannoy et al., 1997, 1998; Hokimoto et al., 1996; Sun et al., 1994; Danser et al., 1992, 1994; Neri Serneri et al., 1996; Campbell et al., 1993; Yusuf et al., 1991: Ruzicka et al., 1994; Schieffer et al., 1994; Van Gilst et al., 1984; Bickerton and Buckley et al., 1961; Fischer-Ferraro et al., 1971; Ganten et al., 1971).

Removal of the 23 amino acid residues from the C-terminus of preprorenin generates prorenin. Prorenin (45-47 kDa containing 406 amino acid residues), the pre-active form of renin, is predominantly synthesized by granular cells of the juxtaglomerular apparatus (JGA) in the terminal afferent arteriole (Schnermann & Briggs, 2008; Schweda et al., 2007) and principle cells of the collecting ducts (Prieto-Carrasquero et al., 2004; Rohrwasser et al., 1999; Kang et al., 2008). Prorenin is also synthesized in many other tissues like adrenal glands (Ganten et al., 1974, 1976; Ho and Vinson, 1998), zona glomerulosa (Doi et al., 1984; Deschepper, et al., 1986; Brecher et al., 1989), eye, Müller cells, mast cells (Krop et al., 2008), ovarian follicular fluid (Glorioso et al., 1986), and theca cells (Do et al., 1988), uterus (Derkx et al., 1987; Itskovitz et al., 1987), myometrium/decidual cells (Shaw et al., 1989), placenta (Lenz et al., 1991), chorionic cells, testis and leydig cells (Sealey et al., 1988). The submandibular gland in some mice strains produces a large amount of renin, which is a product of the Ren-2 renin gene distinct from the renal renin gene, Ren-1 (Cohen et al., 1972; Wilson et al., 1981; Holm et al., 1984) and this action is mediated by prorenin converting enzyme present in submandibular gland of the same mice strains (Kim et al., 1991). Prorenin, in the juxtaglomerular cells of the kidney, is converted to mature renin by the limited endoproteolysis after paired basic residues, Lys-Arg to remove the 43-amino acid residues containing prosegment sequence. The concentration of prorenin in human plasma is 10 times higher than that of mature renin though the physiological role of prorenin is still not clear and the relative concentration of prorenin to renin varies at different conditions. Thus, conversion of prorenin i.e., activation of prorenin to renin plays important role in the regulation of RA system. Certain proteases like trypsin or cathepsin were found to activate prorenin by cleaving the residue prosegment reversibly (Inagami et al., 1980; Shinagawa et al., 1990, 1994; Kikkawa et al., 1998; Jutras et al., 1999; Taugner et al., 1985; Wang et al., 1991; Jones et al., 1997). However, many tissues store prorenin but do not process it to active renin. Though extra-renal sources of prorenin are evident, kidney is the major source of plasma prorenin. Renin and prorenin have long been considered as the separate mediators of tissue and circulating systems (Sealy & Rubattu, 1989). *In vitro,* when prorenin is acidified at pH 3.3 or exposed to low temperature (< 4°C) or allowed to interact with antibodies designed from its prosegment sequences (protein-protein interaction), it mediates intrinsic catalytic activity without removal of the prosegment sequence from its N-terminus through a reversible change in conformation (Derkx et al., 1979, 1983, 1987a & b, 1992; Pitarresi et al., 1992, Suzuki et al., 2000, 2003).

Both renin and non-proteolytically activated prorenin catalyze angiotensinogen, a 6 kDa protein macromolecule found also in adipose tissues to generate a small decapeptide called angiotensin I. Both neonatal and adult rat cardiac cells express mRNA for angiotensinogen (Dostal et al., 1992; Malhotra et al., 1994; Zhang et al., 1995; Liang et al., 1998; Sadoshima et al., 1993), while van Kesteren and colleagues (1999) were unable to detect angiotensinogen in neonatal rat cardiac cells or in the conditioned medium of these cells using radioimmunoassay. Secreted angiotensinogen in the cultured medium of neuronal cells has been identified. Generated renin product, angiotensin I thus, further converted into angiotensin II by the action of ACE.

The (pro)renin receptor or (P)RR is now considering as another important regulatory component in renin-angiotensin system. However, ongoing research works have revealed its association both in angiotensin II-dependent and –independent pathways which also play pivotal role in the developmental processes.

2. (Pro)renin receptor, a new family member of renin angiotensin system

It's been almost a decade since the full length (pro)renin receptor or (P)RR was cloned (Nguyen et al., 2002). However, earlier the same group (Nguyen et al., 1996, 1998) reported high affinity binding of ^{125}I renin to primary and immortalized human mesangial cells (0.2 and 1.0 nM, respectively) in a time-dependent fashion that could attain saturable state. The (P)RR does not internalize the ligands inside the cells rather activates renin and prorenin after binding to generate angiotensin I and second messenger pathway by activating proteins involved in signaling. In the late nineties of the last century, the mannose 6-phosphate/insulin-like growth factor II (M6P/IGF2) receptor was found on rat cardiac myocytes (van Kesteren et al., 1997) and human endothelial cells (Admiraal et el., 1999; Saris et al., 2001) that could bind and internalize renin/prorenin (van Kesteren et al., 1997). However, such binding and internalization could not generate any angiotensin peptides intracellularly. Besides, existence of renin/prorenin receptor independent of mannose 6-phosphate such as renin binding protein (RnBP), renin/prorenin binding protein (ProBP) in rat tissues, vascular renin binding protein have also been confirmed (Takahashi et al., 1983;

Tada et al., 1992; Campbell et al., 1994; Sealy et al., 1996) that bind with different binding affinities to their ligands.

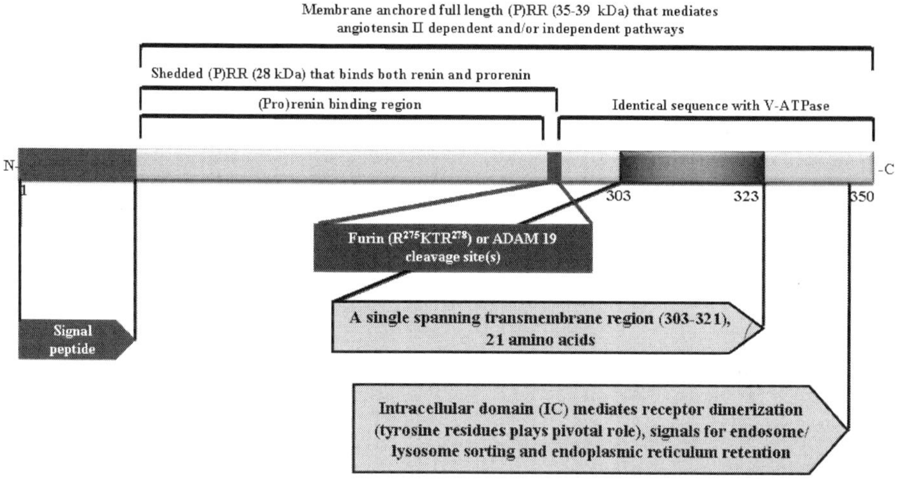

Fig. 2. Structure of (pro)renin receptor protein. The receptor protein is composed of three basic constituents with an N-terminal domain, which is the (pro)renin binding site, a single spanning transmembrane sequence that traverse through the plasma membrane and the intracellular cytoplasmic domain that recently has been identified as the important region required for the dimerization of (P)RR.

The (P)RR, expressed on the cell surface, is a 350 amino acid (39 kDa) containing protein with a single spanning transmembrane domain encoded from the X-chromosome. A short signal peptide is present at the N-terminus end of the unglycosylated large extracellular domain with ~310 residues and the transmembrane domain has putative 20-amino acid residues followed by a ~19-amino acid containing intracellular cytoplasmic (IC) domain (shown in Figure 2). Ubiquitous expression of (P)RR has been demonstrated with the highest amount of mRNA found in brain, heart and placenta while lower amount was expressed in liver, pancreas and kidney (Nguyen et al., 2002). It is reported that (P)RR expressed in VSMCs in human (P)RR transgenic rats can be recycled between intracellular compartment and cell membrane (Batenburg et al., 2007). The (P)RR is also localized on the membrane of stromal adipose cells (Achard et al., 2007), in the neurons of neonatal rats (Shan et al., 2008), on COS-7 cells (Nabi et al., 2007), in glomerular mesangial cells, the subendothelium of renal arteries, podocytes, and distal nephron cells (Nguyen et al., 2002, 1996, 1998) of human and rat kidneys; U937 monocytes (Feldt et al., 2008b) and also in intracellular compartments or on the surface of vascular smooth muscle cells (VSMCs) (Sakoda et al., 2007; Zhang et al, 2008), in endoplasmic reticulum (Schefe et al., 2006; Yoshikawa et al., 2011), golgi apparatus (Contrepas et al., 2009; Yoshikawa et al., 2011), cytosol (Contrepas et al., 2009; Cousin et al., 2009) and found in plasma (Cousin et al., 2009). Expression of (P)RR is also categorized in the subfornical organ (SFO), paraventricular nucleus, the supraoptic nucleus, the nucleus of the tractus solitarius (NTS), or the rostral ventrolateral medulla regions of brain that were believed to be involved in the central regulation of cardiovascular function and volume homeostasis (Contrepas et al., 2009), in the frontal lobe of human brain and pituitary (Takahashi et al., 2010). Retina is also a source of (P)RR (Satofuka et al., 2009). In particular, it is localized to pericytes in retinal vessels, endothelial cells, and, mostly in retinal ganglion cells and glia (Wilkison-Berka et al., 2010). Moreover, predominant expression of (P)RR using immunohistochemistry and *in situ* hybridization on the epical membrane of acid secreting cells in the collecting duct has been reported (Advani et al., 2009).

Full length rat and human recombinant (P)RR with transmembrane followed by cytoplasmic domains were expressed in baculovirus expression system, and identified in the cellular fraction (Nabi et al., 2006; Du et al., 2008; Kato et al., 2008). On the other hand, human (P)RR containing only the extracellular domain lacking transmembrane part was found secreted in the culture medium (Kato et al., 2009). Also, human (P)RR was successfully expressed in the Bombyx mori multiple nucleopolyhedrovirus (BmMNPV) and found in the silkworm larvae as well as in the fat body of silkworm larvae. ELISA and surface plasmon resonance technique in BIAcore assay system confirmed the renin/prorenin binding ability i.e., the functional bioactivity of (P)RR expressed and fractionated from silkworm and baculovirus expression system (Nabi et al., 2006; Du et al., 2008; Kato et al., 2008, 2009).

A protease called furin was found to be responsible for shedding of endogenous (P)RR in trans-golgi by cutting at positions $R^{275}KTR^{278}$ near the N-terminus of transmembrane sequence (Cousins et al., 2009). This soluble form of (P)RR [s(P)RR, 28 kDa] was detected in the conditioned cultured medium and also in human plasma using co-precipitation experiment with human renin. Another protease ADAM19 sheds intracellular (P)RR from golgi apparatus into the extracellular space (Yoshikawa et al., 2011). Moreover, constitutively secreted soluble form of (P)RR (~30 kDa) shedded from the cell surface was found in the cultured medium of human umbilical vein endothelial cells (HUVECs) (Biswas et al., 2011) that could also bind recombinant human prorenin with a nanomolar order, similar to what was reported for full length (P)RR on the cell surface (Nguyen et al., 2002; Batenburg et al., 2007; Nurun et al., 2007) or from the baculovirus expression system (Nabi et al., 2006). Non-proteolytic activation of prorenin occurred when it interacted with s(P)RR in the soluble phase and this was confirmed by Western blot analysis. Also, activated prorenin showed renin activity by generating Ang I from sheep angiotensinogen (Biswas et al., 2011). However, the enzymatic properties of renin after binding to (P)RR is yet to be determined. These phenomenons have been depicted in Figure 3 (vi).

C-terminal domain of (P)RR is identical to "M8-9," a truncated protein of 8.9 kDa that co-purified with a proton-ATPase of bovine chromaffin granule membranes (Ludwig et al., 1998). At the gene level (P)RR from human, rat, and mice showed 95% sequence homology, while at the amino acid level they showed 80% homology. Phylogenetic analyses also revealed that the sequences for (P)RR are not only conserved within the closely related species but also similar sequences are present in the remote species. The IC-domain of (P)RR mediates the signal transduction pathways and promyelocytic zinc finger (PLZF) protein has been identified as an associated protein that interacts with the IC-domain to down regulate expression of the receptor. (P)RR has also been reported to exist as a dimer *in vivo* (Schefe et al., 2006). Recent evidences suggest that short and relatively flexible loop of IC segment generates the driving force in the process of dimerization of (P)RR and tyrosine residues of IC contribute in dimerization dominantly (Zhang et al., 2011).

2.1 (Pro)renin receptor and its ligand: interaction of (pro)renin with (pro)renin receptor

Interaction of renin and prorenin with (pro)renin receptor instigates two pathways: one leads to generation of angiotensin II that ultimately contribute to the activation of local RA system via angiotensin II-dependent pathway as in case of classical circulating RA system and the other one leads to signal transduction mediated by angiotensin II-independent pathway outlined in Figure 3.

2.1.1 Binding mechanism and activation of renin angiotensin system

Binding of human renin to human (P)RR increases local angiotensin production as it is manifested by the increased (4/5-fold) substrate affinity of (P)RR-bound renin compare to free form of soluble renin (Nguyen et al., 2002). On the other hand, human renin bound to recombinant human/rat (P)RR and free form of soluble renin showed similar binding affinity for the substrate, sheep angiotensinogen at the micromolar order (Nabi et al, 2006; Nurun et al., 2007). However, kinetic data analyses revealed that prorenin preferentially binds to (P)RR and such binding initiates angiotensin I generation (Nabi et al., 2006; 2009b). Full length rat (P)RR expressed and isolated from the baculovirus expression system had almost 3 times higher binding affinity (K_D = 8.0 nM) for rat prorenin than that of mature rat renin (K_D = 20.0 nM) *in vitro* (Nabi et al., 2006). Receptor-bound rat prorenin also had similar affinity for the substrate (K_m = 3.3 μM) sheep angiotensinogen as it was for rat renin. On the other hand, receptor-bound renin showed higher molecular activity (10 nM·h) compared to free form of mature renin and receptor-bound activated prorenin (1.25 and 1.1 nM·h) (Nabi et al., 2006).

Ninety% of rat and fifty% of human prorenin (at 2.0 nM of initial concentration) bound to their respective (P)RR over expressed on the membrane of COS-7 cells and the K_Ds were estimated to be 0.89 and 1.8 nM, respectively. Receptor-bound rat and human prorenin showed 30% and 40% activity, respectively, in comparison with the activity of trypsinized prorenin molecules (Nurun et al., 2007). A similar binding and activation patterns of prorenin to human (pro)renin receptor expressed in VSMCs of transgenic rats (K_D = 6.0 nM) (Baternburg et al., 2007) and of rat prorenin by rat (P)RR expressed in cultured VSMCs were observed (Zhang et al., 2010). Differences in the K_D values of rat prorenin bound to the immobilized receptors on the synthetic surfaces and the membrane-anchored receptor could be due to the presence of some other associated proteins that might have stabilized the (P)RR on the membrane. Surface plasmon resonance technique in BIAcore assay system revealed almost four times higher binding affinity of human prorenin (1.2 nM) to the *in vitro* synthesized human recombinant (P)RR compared to that of human mature renin (4.4 nM) (Nabi et al., 2009b). The immobilized receptors bind recombinant human renin and prorenin with the dissociation constant (K_D) values of 1.2 and 4.4 nM, respectively. Also, the data obtained from the BIAcore kinetic study showed that association rate of prorenin to (P)RR is higher than that of mature renin (1.8×10^7 and 2.16×10^6 $M^{-1}.s^{-1}$, respectively) (Nabi et al., 2009b).

The binding mechanism of renin and prorenin to the (pro)renin receptor has also been proposed depending on the ground work led by Suzuki and colleagues who demonstrated the importance of "handle" ($I^{11P}FLKR^{15P}$) and "gate" ($T^{7P}FKR^{10P}$) region peptides designed from the prosegment sequence of prorenin in the non-proteolytic activation of prorenin via protein-protein interaction (Suzuki et al., 2003). Later, another peptide called "decoy" ($R^{10P}IFLKRMPSI^{19P}$ including the "handle" sequence) that mimics the N-terminus sequence of human prorenin prosegment showed its high binding affinity to the recombinant (P)RR and this affinity explains the probable reason for high binding affinity of prorenin for (P)RR. Decoy peptide has got binding affinity to (P)RR at the nanomolar order similar to that of prorenin (Nurun et al., 2007; Nabi et al., 2009a, 2009b). Even after 28 days of administration, fluorescent tagged handle region peptide (HRP) was recognized in the renal glomeruli and tubular lumen (Ichihara et al., 2006a; Kaneshiro et al., 2007). However, a signal of these fluorescent molecules is from the intact form of HRP or not is still arguable. This argument becomes even stronger from the findings of Leckie and Bottrill (2011). They synthesized part of prosegment sequence, RIFLKRMPSIR (it contains an additional arginine residue at the C-terminus of the decoy) and its scrambled sequence (SRRMIFPIKLR) to find out a novel binding sites in human umbilical vein endothelial cells using liquid chromatography coupled with tandem mass spectrometry (LC-MS/MS). Finally, they concluded that the binding of the human prorenin peptide $R^{10}IFLKRMPSIR^{20}$ to HUVEC proteins is not specific for amino acid sequence and probably involves a general peptide/protein uptake mechanism without detecting a specific prorenin prosegment binding sites (Leckie & Bottrill, 2011). Moreover, decoy peptide containing fluorescent component (carboxyfluorescein) either at N-terminus or C-terminus showed different binding affinity for (P)RR compared to that of wild type decoy *in vitro* (Nabi et al., 2010). Recombinant (P)RR coupled to CM5 sensor chips in BIAcore assay system (Nabi *et al*, 2009a, 2009b), immobilized on synthetic surfaces (Nabi *et al*, 2009b), (P)RR over expressed on COS-7 cells (Nurun *et al*, 2007) have revealed that decoy inhibits binding of renin/prorenin to (P)RR. The inhibitory constant (K_i) for the peptide was found at the nanomolar order. Also, subsequent *in vivo* studies have been carried out to show beneficial role of decoy peptide in ameliorating the end-stage organ damage related disorders by abolishing non-proteolytic activation of prorenin via inhibition of prorenin binding to (P)RR (Ichihara et al., 2004; 2006b & c; Kaneshiro et al., 2007; Satofuka et al., 2006, 2007, 2009; Wikinson-Berka et al., 2010).

Interestingly, decoy peptide has also been found to inhibit binding of renin to (P)RR and this action of decoy on renin is yet to be clarified. Based on these annotations and on the tertiary structure of renin as well as predicted tertiary structure of prorenin, the possibility of having a common site in both renin and prorenin through which these molecules can interact with the (P)RR other than the decoy peptide sequence was hypothesized. A new sequence ($S^{149}QGVLKEDVF^{158}$) that localizes in the flexible junctional region between the N- and C-domains of renin/prorenin termed as the "hinge" has recently been reported to have such pivotal role for renin/prorenin binding to (P)RR (Nabi et al., 2009b). The K_D for the binding of the "hinge" peptide to (P)RR was five times higher than that of the decoy and estimated to be 17 nmol/L. The "hinge" showed higher binding affinity to the (P)RR than that of another peptide ($A^{248}KKRLFDYVV^{257}$) from the C-domain of renin/prorenin molecule,. Like the decoy, "hinge" peptide also reduced the resonance signal of renin/prorenin binding to (P)RR as observed in BIAcore, and equilibrium state analysis revealed this paradigm as a competitive inhibition with the K_i of 37.1 and 30.7 nmol/L, respectively (Nabi et al., 2009b). Therefore, these data suggest that not only the decoy peptide but also the "hinge" region peptide together accounted for the higher binding affinity of prorenin and hence, prorenin molecule has at least two high affinity sites while renin has a single site for their binding to (P)RR. Considering the nanomolar binding affinities of renin/prorenin and handle region peptide, Duncan J Campbell in one of his review article suggested that the (pro)renin receptor may have at least two separate binding domains, one domain is for renin and the other one is for prorenin prosegment and/or HRP (Campbell, 2008). Though, prorenin has two regions to interact with (P)RR, but to confirm the existence of different binding sites within (P)RR for its ligands, three dimensional structure of (P)RR has to be elucidated.

Activation of renin angiotensin system or in other words, generation of Ang-I by (P)RR mediated non-proteolytically activated prorenin depends on the sources of prorenin. Human prorenin showed higher binding affinity to both human and rat (P)RR compared to that of rat prorenin (Biswas et al., 2010a). More interestingly, either bound to human or rat (P)RR, molecular activity of non-proteolytically activated human prorenin was 2-4 fold higher than that of rat prorenin (Biswas et al., 2010), which could be due to the slow activation rate through change in conformation of rat prorenin compared to that of human prorenin after protein-protein interaction. Contribution of prorenin prosegment in the non-proteolytic activation mechanism was reported earlier *in vitro* (Suzuki et al., 2000). Chimera of human renin and rat prosegment showed very slow activation like native rat prorenin compared to the chimera of rat renin and human prosegment. Thus, it could be proposed that the prosegment sequence of prorenin played a pivotal role for the activation of prorenin molecules. More concisely, species specific regions within the prorenin prosegment like "handle" (Nurun et al., 2007; Suzuki et al., 2003) and decoy peptides (Nurun et al., 2007; Nabi et al., 2009a, b) actually crucial for the interaction of prorenin with (P)RR and also, for the non-proteolytic activation mediated by protein-protein interaction. Activation of rat prorenin through change in conformation at acidic condition required long time, even days to month (Suzuki et al., 2000). However, (P)RR mediated activation of rat prorenin has been observed within hours using recombinant (P)RR on *in vitro* synthetic surface system (Nabi et al., 2006; Biswas et al., 2010a) or overexpressing (P)RR on COS-7 cells (Nurun et al., 2007) or on rat VSMCs (Batenburg et al., 2007; Zhang et al., 2008). This might be the result of quick conformational change of prosegment of rat prorenin exerted by the interaction of one protein (receptor) with the other (ligand).

Furthermore, while cosidering the binding mechanism of renin and prorenin to their receptor, (P)RR has not only been discussed from the ligand's point of view, rather primary structure of (P)RR has also got similar attention for explaining the possible mechanism of receptor's involvement in ligand binding. On the other hand, though three dimensional (3D) structure of renin (Dhanaraj et al., 1992) and a predicted 3D model of prorenin (Suzuki et al., 2003; Nabi et al., 2009b) are available but due to lack of 3D structure of the receptor, mechanism for interaction of (pro)renin can not be explained from the receptor's point of view. However, several anti-(P)RR antibodies designed from the middle part (^{107}DSVANSIHSLFSEET^{121} named as anti-107/121 antibody) and C-terminus [^{221}EIGKRYGEDSEQFRD^{235} and ^{237}SKILVDALQKFADD^{250}; close to the N-terminus of transmembrane region of the receptor, named as anti-221/235 and 237/250 antibodies, respectively] regions of (pro)renin receptor have been used in many studies (Nabi et al, 2009a, 2009b; Nabi et al., 2012). Depending on the flexibility of the anti-(P)RR antibody associated (pro)renin receptor, it would show its binding affinity towards the ligands. The calculated binding affinities of prorenin were $2.9x10^{-9}$, $1.2x10^{-9}$ and $1.74x10^{-9}$ nM, when (P)RR was immobilized or occupied by anti-107/121, anti-221/235 (Nabi et al, 2009a, 2009b) and 237/250 antibodies (Nabi et al., 2012), respectively. The recombinant (P)RR tagged with six histidine residues was synthesized in a cell free *in vitro* system using wheat germ lysate. It was hypothesized that the His tag sequence at the C-terminal end would retain the transmembrane characteristics of (P)RR *in vitro*. So, (P)RR occupied by the anti-His tag antibody would indicate its native binding pattern while interacting with the ligands. Study showed that the binding affinity of prorenin to anti-His tag antibody-bound (P)RR was 7.8 nM (Nabi et al., 2009a) and other studies using over expressed (P)RR on the cell surface showed comparable nanomolar order of binding affinity of prorenin to (P)RR (Nguyen et al., 2002; Nurun et al., 2007; Batenburg et al., 2007). Reports available so far indicate that binding region for prorenin within (P)RR resides possibly further upstream region of the amino acid residue at position 107, which could be more close to the N-terminal region(s) of the receptor.

2.1.2 Initiation of second messenger pathways

Binding of renin/prorenin to (P)RR initiates an intracellular signaling pathway that is independent of angiotensin II mediated pathway. Both renin and prorenin stimulated p42/p44 mitogen-activated protein kinase (MAPK) or ERK1/2 that leads to up-regulation of transforming growth factor-β1 release in mesangial cells, PAII, collagens, fibronectin (Huang *et al*, 2006; Huang *et al*, 2007; Sakoda *et al*, 2007) and cyclooxygenase 2 (Kaneshiro et al., 2006; Nguyen, 2006) as shown in Figure 3 (ii). Moreover, prorenin also activated p38 mitogen-activated protein kinase and simultaneously phosphorylate heat-shock protein-27 in cardiomyocytes (Sasris et al., 2006). Prorenin and renin induced activation of extracellular protein kinases (ERK) 1/2 in monocytes has also been reported (Feldt et al., 2008b). In the kidneys of diabetic mice, activation of all the three members of MAPK family including ERK, p38 and c-Jun NH_2-terminal kinase (JNK) was observed (Ichihara et al., 2006a), whereas another report (Sakoda et al., 2007) revealed activation of ERK not p38 and JNK upon activation of (P)RR via its ligand, prorenin. A protein called promyelocytic zinc finger (PLZF) has also been identified and this has been found to be associated with the cytoplasmic domain of the receptor (Schefe et al. 2006). Binding of prorenin to the receptor drives translocation of PLZF to the nucleus by stimulating P13K p85 pathway that ultimately generates short negative feedback loop to down regulate (P)RR expression [depicted in Figure 3 (iii)]. Furthermore, (P)RR is a component of the Wnt [wingless-type mouse mammary tumor virus (MMTV) integration site family] receptor complex. Wnt proteins are highly conserved secreted signaling molecules and regulators of multiple biological and pathological processes (Logan and Nusse, 2004). The signaling mechanism mediated by Wnt receptor in conjunction with (P)RR and H^+-VATPase has been explained in detail later in this chapter. The detail of the intracellular signaling pathway activated and mediated by the (pro)renin receptor has been depicted and categorically presented in Figure 3.

3. Pathophysiology of prorenin and (pro)renin receptor

Hepatocyte specific prorenin transgenic rat revealed direct pathophysiological action of prorenin. Prorenin is not activated in liver and less than 2% of the total circulating prorenin found to be active in plasma. Diabetic subjects with microalbuminuria had very high prorenin to renin ration. Before the onset of microalbuminuria levels of prorenin begins to increase, and in conjunction with the glycated haemoglobin, the prorenin levels in plasma could be used to predict the occurrence of later microalbuminuria (Deinum et al., 1999). The circulating prorenin is responsible for developing hypertrophy of cardiomyocytes, glomerulosclerosis and atherosclerosis of small to medium sized artery, indicating elevated prorenin itself causes cardiomyopathy, glomerulosclerosis and atherosclerosis. Use of angiotensin converting enzyme inhibitors and angiotensin-II type 1 receptor blockers play protective role in end-stage organ damage in patients with hypertension and diabetes by suppressing the circulating RA system. Yet, low amount of renin activity is still evident in the plasma of these under treatment diabetic and hypertensive subjects which could ultimately be attributed to the enhanced tissue RA system. Thus, reasons behind the direct involvement of prorenin in the pathology of hypertension, diabetes and heart failure remained unclear. Receptor- associated prorenin system (RAPS), a novel phenomenon, sheds light on this direct action of prorenin. (P)RR, the new member of the RA system, has set a new perception about the physiological functions, activation mechanism and pathophysiological roles of renin/prorenin by activating angiotensin II-dependent or -independent pathways [Figures 3 (i), (ii), (iii)]. It has its own intracellular signalling pathways. Non-proteolytic activation of prorenin after interacting with (P)RR hypothesized that this activation mechanism of prorenin plays a pivotal role in the regulation of tissue RA system and end-organ damage in diabetic animals. (P)RR mRNA and protein expression are up-regulated in the hearts and kidneys of rats with congestive heart failure (Hirose et al., 2009). Thus, (P)RR in the heart can act as a capturing molecule for renin/prorenin which ultimately explain the presence of local renin-angiotensin system in heart, which can't synthesize renin. In diabetes, enhanced activity of oxidative stress and AT1 receptor are associated with up-regulation of (P)RR and this could be suppressed using AT1 receptor blocker and NADPH-oxidase activity inhibitor (Siragy and Huang, 2008). (P)RR mediated stimulation of signal cascade (depicted in Figure 3) of transforming growth factor-β1 (TGF-β1) and connective tissue growth factor (CTGF) in renal glomeruli (Huang et al., 2011) and enhancement of renal production of the inflammatory cytokines- TNF-alpha and IL-1beta, independent of the effects of renal Ang-II (Matavelli et al., 2010), contributes to the development and progression of kidney disease in diabetes. Up-regulation of (P)RR expression by high glucose is mediated by both PKC-Raf-ERK and PKC-JNK-c-Jun signaling pathways. Also, nuclear factor-κB and activation protein-1 are involved in high-glucose-induced (P)RR up-regulation in rat mesangial cells (Huang and Siragy, 2010).

At 5–6 months of age, transgenic rats over expressing the human (P)RR gene nonspecifically developed glomerulosclerosis with proteinuria by three to seven times without elevating the blood pressure (Kaneshiro et al., 2007). Transgenic rats over expressing human (P)RR gene exclusively in smooth muscle cells developed hypertension at their 7 months of age (Burckle et al., 2006). (Pro)renin receptor mediated non-proteolytically activated prorenin contributes to the development and progression of nephropathy including proteinuria and glomerulosclerosis in diabetic animals with high plasma levels of prorenin by increasing angiotensin II tissue generation (Ichihara et al., 2004). An increased Ang-I content was observed in the heart of double transgenic mice over expressing human prorenin and angiotensinogen compared to the single-transgenic mice (Prescott et al., 2002). These results indicate how prorenin contribute to the generation of angiotensin peptides locally and tissue

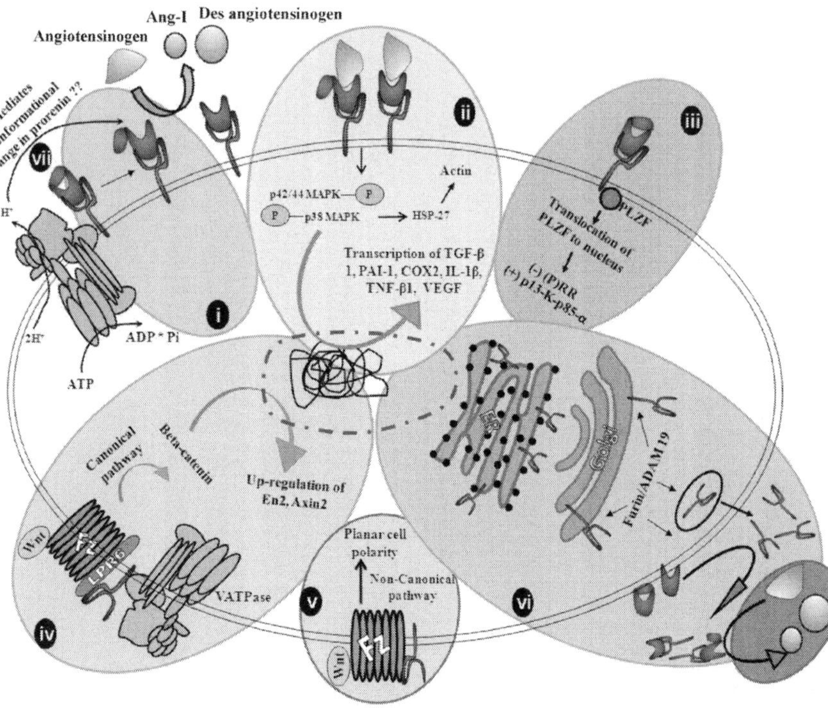

Fig. 3. Receptor associated prorenin system or RAPS mediated by prorenin and (pro)renin receptor [(P)RR] has set a new perception about the involvement of renin-angiotesin system in the pathophysiology of end-stage organ damage. Such nomenclature has been proposed due to the dual activation of tissue RAS (i) and RAS-independent signaling pathways (ii-vii). Augmentation of tissue RAS or RAPS via (P)RR initiates endocrine, paracrine or autocrine activities mediated by angiotensin II peptide. Binding of renin/prorenin to (P)RR initiates signal transduction via angiotensin II independent pathway by activating mitogen activated protein kinases (MAPKs) that induce expression of many regulatory proteins (ii). Translocation of promyelocytic zinc finger (PLZF) after prorenin binding to (P)RR leads to a short negative feed back loop that in turn suppresses (P)RR expression (iii). Also, (P)RR itself, independent of renin/prorenin, mediates Wnt-βcatenine (canonical, iv) and Ez/PCP (non-canonical) signaling pathways (v). (P)RR can be processed in the golgi apparatus by Furin or ADAM19 to its soluble form (vi). The shedded (P)RR through exocytosis can come outside of the cell and has been detected in human plasma and cultured cell medium. Shedded (P)RR binds (pro)renin. Prorenin bound to soluble (P)RR performs enzymatic activity (vi). V-ATPase participates in proton transport. C-terminal region of (P)RR is identical to the sequence of V-ATPase. This is not clear whether non-proteolytic activation of prorenin through conformational change after receptor binding is also mediated by the acidic environment created by membrane associated V-ATPase and (P)RR (vii).

damage after being taken up by tissues from circulation. (P)RR is up-regulated in kidneys of diabetic rats and renal mesangial cells exposed to high glucose concentration (Siragy and Huang, 2008; Huang and Syragi, 2010). Rapid phosphorylation at the serine residues of (P)RR in response to hyperglycemia up-regulates TGF-beta1-CTGF cascade (Huang et al., 2011), which initiates or augments kidney disease in diabetic rats. An increased Ang-I content was observed in the heart of double transgenic mice over expressing human prorenin in the liver and human angiotensinogen in the heart as compared to the single-transgenic mice (Prescott et al., 2002). These results indicate how prorenin contribute to the generation of angiotensin peptides locally and tissue damage after being taken up by tissues from circulation. Moreover, (P)RR by stimulating non-proteolytic activation of prorenin contribute to the development of renal and cardiac fibrosis in spontaneously hypertensive rats (SHRs) (Ichihara et al, 2006a,b). These data demonstrated the possible involvement of (P)RR in the pathogenesis of heart failure and kidney tissue damage. Association of (P)RR gene polymorphism with high blood pressure and left ventricular hypertrophy substantiate the important role of (P)RR in the pathogenesis of hypertension in humans (Hirose et al., 2011; Ott et al., 2011) *In vitro* and animal studies have shown that increased receptor expression could be linked to high blood pressure and to cardiac and glomerular fibrosis by activating mitogen-activated protein kinases and by upregulating gene expression of profibrotic molecules. Also, animal studies with angiotensin-II type 1a receptor deficiency showed that the (P)RR is involved in the development and progression of diabetic nephropathy through angiotensin-II independent pathway via activation of intracellular pathways (Ichihara et al., 2006c).

Association of prorenin and the (P)RR with the development of ocular pathology/diseases has been reported (Satofuka et al., 2006, 2007, 2008; Wikinson-Burka et al., 2011). Non-proteolytic activation of prorenin mediated by (P)RR is associated with retinal neovascularization in experimental retinopathy model of prematurity (Satofuka et al., 2007). Using the same model, the involvement of prorenin and (P)RR in the pathological angiogenesis, leukocyte accumulation and intracellular adhesion molecule-1 with vascular endothelial growth factor expression; retinal gene and protein expression of inflammatory mediators has also been demonstrated (Satofuka et al., 2006, 2008). RILLKKMPSV, a peptide sequence of rat prorenin prosegment, influences the vasculature, glia and neurons, and (pro)renin receptor expression in the retina (Wikinson-Burka et al., 2011).

4. Functions of (P)RR other than its involvement in renin-angiotensin system

The prototype of sequence homology between (P)RR from human and other species actually gave a clue regarding its plausible additional role in biological processes other than RAS. Two or more genes homologous to (P)RR have been found in *C. elegans* and *Drosophila melanogaster* that are phylogenetically distant from human. These species also express some components of RAS, which are not involved in homeostasis or electrolyte balance. Thus, (P)RR in these species may contribute to functions not related to RAS.

The C-terminal truncated fragment of (P)RR helps to assemble vacuolar H^+-proton adenosine triphosphatase (V-H^+-ATPase) (Ludwig et al., 1998). The (P)RR is also identical to endoplasmic reticulum-localized type 1 transmembrane adaptor precursor (CAPER) (Burckle & Bader, 2006; Campbell, 2006; Bader, 2007; Strausberg et al., 2002). Evolutionarily

V-H^+-ATPase is a highly conserved ancient enzyme in eukaryotic cells (Nelson et al., 2000) and this could be one of the most plausible reasons behind the sequence homology of C-terminus of human (P)RR with the evolutionarily close species like rat, mouse, chicken, drosophila, mosquito, zebra fish, frog and remote like *C. elegans* and bacteria *Ehrlichia chaffeensis*. (P)RR exists in truncated form composed of transmembarne region and the cytoplasmic tail that co-precipitates with V-ATPase may govern its function unrelated to RA system (Bader 2007). For this reason, (P)RR is also known as ATP6ap2 (adaptor protein type II vacuolar H^+-ATPase). The V-H^+-ATPase is expressed in the collecting ducts and distal tubules within the kidney, where it contributes to the urinary acidification as well as play pivotal role in endocytosis (Toei et al., 2010). Different subunits of V-ATPase perform different functions, notably, mutations in genes encoding C or D subunits in mice involved in embryonic lethality giving an evidence that V-ATPase plays an important role in development (Inoue et al, 1999; Miura et al, 2003), mutated B1 or A3 subunit involved in metabolic acidosis and osteoporosis in mice, respectively (Li et al., 1999; Finberg et al., 2005) and altered B1 or A4 subunit causes distal renal tubular acidosis in human (Karret, 1999; Smith, 2000). It is suggested that (P)RR and vacuolar H^+-ATPase are linked together in the kidney (Advani et al., 2009) while, for the assembly and function of vacuolar H^+-ATPase in the heart, (P)RR plays very pivotal roles (Kinouchi et al., 2010).

Recent evidences demonstrate that (P)RR is a component of the Wnt receptor complex (Cruciat et al., 2010). It is essential for *en2* expression because of its requirement in Wnt signaling. It also acts down stream of Wnts and upstream of β-catenin [Figures 3 (iv) and (v)]. Deletion of the cytoplasmic domain of (P)RR, which mediates renin signaling inside cell, showed no effect on Wnt receptor binding suggesting that (P)RR acts in a renin-independent manner as an adaptor between Wnt receptors and the V-ATPase complex. Moreover, malfunctioning (P)RR contributed to the abnormal tadpoles characterized by small heads, shortened tails, as well as defects in melanocyte and eye pigmentation at the early embryonic stage as (P)RR and V-ATPase are required to mediate Wnt signaling during antero-posterior patterning of *Xenopus*'s early central nervous system development (Cruciat et al., 2010). A homologue of (pro)renin receptor in Drosophila [d(P)RR], localized mainly to the plasma membrane, has an evolutionarily conserved role at the receptor level for activation of canonical and noncanonical Wnt/Fz (frizzled) signaling pathways [Figure 3 (v)]. Attenuation of d(P)RR affects Wg target genes in cultured cells and *in vivo* (Buechling et al., 2010). Over expressed d(P)RR interacts with Fz and Fz2 receptors which is required for planar cell polarity in Drosophila epithelia and also for convergent extension movements in *Xenopus* gastrulae. Small interfering RNAs targeting human (pro)renin receptor significantly reduced Wnt-responsive TopFlash reporter activity in HEK293T cells. Thus, (P)RR has a conserved role in mediating Wnt signaling in human (Buechling et al., 2010). This data is also consistent with the findings of Cruciat et al (2010) who demonstrated the developmental role of (P)RR in *Xenopus*. Further, asymmetric subcellular localization of frizzled, a seven-pass transmembrane receptor that acts in both wingless (Wg) and planar cell polarity (PCP), is prerequisite for the proper functioning of PCP signaling pathway (Hermle et al., 2010). It has been demonstrated that the function of VhaPRR, an accessory subunit of the vacuolar (V)-ATPase proton pump in Drosophila and also known as the VhaM8-9 because of its sequence homology with V-ATPase, is tightly associated with Fz but not to other PCP core proteins. Fz fails to localize asymmetrically in absence of VhaPRR. It also acts as the modulators of canonical Wnt signaling pathway in larval and adult wing tissues. VhaPRR knock down caused multiple wing hair and hair mispolarization phenotypes (Hermle et al., 2010). These indicate the association of (P)RR in non-canonical (Fz/PCP) signaling pathways. Recent evidences regarding the association of (P)RR with H+-ATPase and Wnt signaling pathway shedding light on the reason behind the connection of non-proteolytic activation of prorenin by (P)RR with glomerulosclerosis, fibrosis, proteinuria. Though *in vitro* studies suggested non-proteolytic activation of prorenin mediated by (P)RR, but it is yet to determine whether the activation is mediated only by the protein-protein interaction, or by the co-operation of (P)RR and V-H^+-ATPase, or only due to the acidic environment created as a result of proton transport *in vivo* [Figure 3 (vii)]. However, because Wnt signaling pathway promotes renal fibrosis, glomerulosclerosis and proteinuria, (He et al., 2009; Dai et al., 2009) it is possible that (P)RR might act in a combination of (P)RR-H^+-ATPase-Wnt signaling pathway. Thus, (P)RR is involved in the Wnt/β-catenin canonical and Wnt/PCP non-canonical pathways in conjunction with V-H^+-ATPase in a renin-independent fashion [Figures 3 (iv) and (v)].

Using zebra fish, the important association of (P)RR and V-ATPase in the development of brain and eye at the very early stage of embryonic development has been demonstrated (Amsterdam et al., 2004). A mutation in (P)RR is very lethal that causes death before the completion of embryonic stage by creating severe malformations of the central nervous system. In fact, while ACE is required to maintain fertility and ACE2 serves as a receptor for the SARS corona virus [causing factor for severe acute respiratory syndrome (SARS)], a single amino acid mutation in exon-4 of (P)RR mRNA associated with X-linked mental retardation and epilepsy (Ramser et al., 2005), and thus, (P)RR seems to be important for brain development and cognition. Also, another major finding (Contrepas et al., 2009) stated that (P)RR play essential role in neuronal cell differentiation. Other than the embryonic development, (P)RR gene polymorphism has been found to be associated with high blood pressure in Caucasian and Japanese male subjects (Hirose et al., 2011; Ott et al., 2011). Elevated blood pressure and increased heart rate in transgenic rats over expressing (P)RR in smooth muscles have been reported in their models (Burckle et al., 2006).

5. Inhibition of the activities of the components of renin angiotensin system: (pro)renin receptor as a new therapeutic target

Peptides mimicking the structural part of prorenin prosegment (pro-enzyme of renin) or N-terminal sequence of angiotensinogen containing the renin cleavage site were the first-generation of renin inhibitors (Boger et al., 1985; Hui et al., 1987; Bolis et al., 1987). Parenteral administration of these drugs efficiently reduced blood pressure by inhibiting renin activity in animals and in human being (Boger et al., 1985; Webb et al., 1985). However, due to their peptidic nature, these drugs had very poor oral bioavailability. Later, chemically modified CGP29287 achieved more attention as renin inhibitor due to its stability and longer duration of action when given orally (Wood et al., 1985). Further, development of other drugs like enalkiren (A 64662), CGP38560A, remikiren (Ro 425892) and zankiren (A 72517) with molecular weight of a tetra-peptide (Wood et al., 1994,1989; Maibaum et al., 2003) also failed to attract attention due to their low bioavailability (<2%), a short half-life and weak blood pressure lowering activity when administered orally (Wood et al., 1994; Nussberger et al., 2002; Rongen et al., 1995). On the other hand, an orally inactive peptide from snake venom established the important role of angiotensin converting enzyme (ACE) inhibitors in regulating blood pressure. This led to the development of Captopril, the first ACE inhibitor. Moreover, blood pressure lowering activity, to a great extent, depends on the inhibiting ability of plasma renin activity (PRA) and/or reducing plasma renin concentration (PRC). Thus, use of ACE inhibitors or angiotensin receptor blockers (ARBs) for inhibiting renin angiotensin system is not as effective as it should be because these inhibitors ultimately increase PRA or PRC (Mooser et al., 1990; Azizi et al., 2004). In addition, inhibition of ACE increases angiotensin I, which would be, via ACE-independent pathways by using cathepsins and tonins, converted into angiotensin II (Wolny et al., 1997; Hollenberg et al., 1998). Together these data indicate that direct renin inhibitors could be the superlative choice as an anti-hypertensive agent which would lower plasma renin activity.

Aliskiren, an octanamide, the first known representative of a new class of completely non-peptide, low-molecular weight, orally active transition-state renin inhibitor, that progressed to phase-III clinical trials (Wood et al., 2003). After oral dose of aliskiren (from 40 to 640 mg/day) in healthy volunteers, its plasma concentration increased dose dependently and the peak concentration reached after 3–6 hour with an average half life of 23.7 hour (Nussberger et al., 2002) making the compound suitable for once-daily administration. The oral bioavailability was 2.7%. Plasma steady-state concentrations were reached after 5–8 days of treatment. Aliskiren can inhibit enzymatic activities of receptor-bound renin and non-proteolytically activated prorenin, while it has no effect on the interaction of renin/prorenin with (P)RR. Also, aliskiren could not act as (P)RR blocker to inhibit renin/prorenin binding to (P)RR or failed to prevent (pro)renin signaling (Feldt et al., 2008b). Interestingly, when renin was incubated with aliskiren and then, allowed to bind to (P)RR, the binding affinity of renin to (P)RR decreased more than 1000 fold *in vitro* (Biswas et al., 2010b).

Also, an ideal blocker for (pro)renin receptor is indeed a necessity of time considering the direct association of (P)RR with increased blood pressure and its indirect involvement, via non-proteolytic activation of prorenin, in the pathogenesis of end-stage organ damage in hypertension, diabetes and ocular diseases. The efficacy of a peptidic blocker known as decoy peptide (R^{10P}IFLKRMPSI^{19P}) designed from the N-terminus of prorenin prosegment on the basis of the sequence of handle (I^{11P}FLKR^{15P}) region peptide was reported earlier for improving organ damage (Ichihara et al., 2004). Both human and rat decoy peptides inhibited the bindings of human and rat prorenins to their respective (P)RR expressed on the membranes of COS-7 cells with a similar K_i of 6.6 nM (Nurun et al., 2007). This peptide inhibited the bindings of not only prorenin but also renin to the preadsorbed receptors with the K_i values of 15.1 and 16.7 nM, respectively (Nabi et al., 2009b). Moreover, real-time bindings using surface plasmon resonance (SPR) technique in BIAcore assay system revealed evidence for the direct binding of native decoy peptide to the immobilized (P)RR with K_i of 3.5 nM (Nabi et al., 2009a, 2009b). The SPR technique displayed reduced resonance signal of prorenin binding to (P)RR while co-incubated with the decoy peptide.

The decoy proposition has also been tested *in vivo* using various transgenic models. Administration of HRP significantly inhibited increased levels of renal angiotensin II, the development of proteinuria and glomerulosclerosis in a model of diabetic nephropathy; rat

HRP completely prevented the development of diabetic nephropathy in heminephrectomized streptozotocin induced diabetic rats without affecting hyperglycemia (Takahashi et al., 2007). Urinary albumin excretion and the renal production of tumor necrosis factor-α and interleukine-β1 were decreased significantly when rat HRP was given directly into the renal cortical interstitium of diabetic rats (Matavelli et al., 2010). Prevention of the development of proteinuria, glomerulosclerosis, and complete inhibition of the activation of ERK1/2, p38, JnK in the kidney of diabetic angiotensin-II type-1a receptor-deficient mice was reported and thus, the role of (P)RR via angiotensin II independent pathway in association with prorenin was suggested (Ichihara et al., 2006c). Other investigators also confirmed the action of prorenin and (P)RR via angiotensin-II independent pathway (Huang et al., 2006; Muller et al., 2008; Feldt et al., 2008a, b). Moreover, HRP inhibits the development of retinal neovascularization by inhibiting non-proteolytic activation of prorenin caused by interaction with (P)RR in experimental retinopathy model of prematurity (Satofuka et al., 2007). Satofuka *et al.* using the same model, showed that the HRP suppressed the pathological angiogenesis, leukocyte adhesion and retinal expression of ICAM-1 and VEGF; also, reduced retinal gene and protein expression of inflammatory mediators (Satofuka et al., 2006, 2009). HRP also improved vascular disorder in a model of retinopathy of prematurity, but had detrimental effects on retinal neurons and glia. These effects occurred despite HRP not being detected in plasma. In young spontaneously hypertensive rats (SHR) under high salt-diet, HRP not completely but significantly attenuated glomerulosclerosis with proteinuria, cardiac hypertrophy with left ventricular fibrosis without affecting the development of hypertension (Ichihara et al, 2006 a, b). In addition, Susic et al made a further interesting observation by reporting reduced beneficial effects of decoy (PRAM-1) in SHR rat with normal diet (Susic et al., 2008).

On the contrary, many researchers are not satisfied about decoy's role as a fruitful (P)RR blocker (Batenburg et al., 2007; Muller et al., 2008; Feldt et al., 2008a, b; Mercure et al., 2009). Chronic HRP treatment did not improve target organ damage in renovascular Goldblatt hypertensive rats with high renin, prorenin and PRA that lead to Ang-II dependent target organ damage rather HRP counteracts the beneficial effects of aliskiren (van Esch et al., 2011). Also, HRP had no effects on the activation of signal transduction mediated by prorenin-(P)RR interaction (Feldt et al., 2008a). On the other hand, very recently, (P)RR siRNA technique and prolonged use of HRP or valsartan showed inhibition of rapid phosphorylation in the serine residues of (P)RR that ultimately suppressed inflammation in the kidneys (Huang et al., 2011). The concentration of HRP could not be measured in both blood and plasma of rats infused with either 0.1 or 1 mg/kg HRP per day, which suggested rapid metabolism of the peptide *in vivo* and this interpretation was supported by the finding that HRP was metabolized with a half-life of 5 minutes in EDTA-plasma at 37°C (Wikinson-Burka et al., 2011). Recycling of (P)RR between the cellular compartments and cell surface has been demonstrated earlier (Batenburge et al., 2007). Later, this annotation has been experimentally proved by the action of furin (Cousins et al., 2009) and ADAM19 (Yoshikawa et al., 2011), which till-to-date could be one of the most appropriate and acceptable explanation behind the useful execution of "decoy" as (P)RR blocker in some animal model or cell line while in other models, the "decoy" is not effective even at the same or sometimes higher concentration.

6. Conclusion and future direction

A sensitive enzyme-linked immunosorbent assay has been established to detect the level of soluble (P)RR in the medium of cultured cells and also in cell lysates (Kazal et al., 2011). It is now very important to set up such easily pursuable and sensitive method for the detection of (P)RR in human plasma. It may facilitate to diagnose specific disease or to measure degree of organ damage or to predict the end-stage organ damage. Three dimensional structure of (pro)renin receptor has to be resolved to clear the ambiguity of decoy hypothesis, to find out the binding site(s) of prorenin, renin and the decoy peptide within the molecule. Furthermore, a well accepted (P)RR blocker is now the demand of time to reduce the effects of (P)RR on end-stage organ damage. Thus, (P)RR, now-a-days, should be the novel target for developing new therapeutic approaches to ameliorate end-stage organ damage related disorders. However, considering the involvement of (P)RR in organ development specially in eye and brain development, more extensive studies should be performed before designing a (P)RR blocker.

7. References

Achard V., Boullu-Ciocca S., Desbriere R., Nguyen G., & Grino M. Renin receptor expression in human adipose tissue. *Am. J. Physiol - Regu. Physiol.*, vol. 292, no. 1, (January 2007), pp. (R274-R282).

Advani A., Kelly DJ., Cox AJ, White KE., Advani SL., Thai K., Connelly KA., Yuen D., Trogadis J., Herzenberg AM., Kuliszewski MA., Leong-Poi H., & Gilbert RE. The (Pro)renin receptor: site-specific and functional linkage to the vacuolar H+-ATPase in the kidney. *Hypertension*, vol. 54, no. 2, (August 2009), pp. (261-269).

Admiraal P.J., van Kesteren C.A., Danser A.H.J., Derkx F.H., Sluiter W., & Schalekamp M.A.D.H. Uptake and proteolytic activation of prorenin by cultured human endothelial cells. *J. Hypertens.*, vol. 17, no. 5, (May 1999), pp. (621-629).

Allen AM., Dosanjh JK., Erac M., Dassanayake S., Hannan RD., & Thomas WG. Expression of constitutively active angiotensin receptors in the rostral ventrolateral medulla increases blood pressure. *Hypertension*, vol. 47, no.6, (June 2006), pp. (1054–1061).

Amsterdam A., Nissen RM., Sun Z., Swindell EC., Farrington S., & Hopkins N. Identification of 315 genes essential for early zebra fish development. *Proc. Natl. Acad. Sci. USA*, vol. 10, no. 35, (August 2004), pp. (12792–12797).

Aronsson M., Almasan K., Fuxe K., Cintra A., Harfstrand A., Gustafsson JA., & Ganten D. Evidence for the existence of angiotensinogen mRNA in magnocellular paraventricular hypothalamic neurons. *Acta. Physiol. Scand.*, vol. 132, no. 4, (April 1988), pp. (585–586).

Azizi M., & Menard J. Combined blockade of the renin-angiotensin systemwith angiotensin-converting enzyme inhibitors and angiotensinII type1 receptor antagonists. *Circulation*, vol. 109, no. 21, (June 2004), pp. (2492–2499).

Bader M. The second life of the (pro)renin receptor. *Journal of Renin Angiotensin Aldosterone System*, vol. 8, no. 4, (December 2007), pp. (205–208).

Batenburg W.W., Krop M., Garrelds I.M., de Vries R., de Bruin RJ., Burcklé CA., Müller DN., Bader M., Nguyen G., Danser AH. Prorenin is the endogenous agonist of the (pro)renin receptor. Binding kinetics of renin and prorenin in rat vascular smooth muscle cells overexpressing the human (pro)renin receptor. *Journal of Hypertension*, vol. 25, no. 12, (December 2007), pp. (2441-2453).

Bickerton RK., & Buckley JP. Evidence for a central mechanism in angiotensin-induced hypertension. *Proc. Soc. Exp. Biol. Med.*, vol 106, no. , (1961), pp. (834–837).

Biswas K.B., Nabi A.N., Arai Y., Nakagawa T., Ebihara A., Ichihara A., Inagami T., & Suzuki F. Qualitative and quantitative analyses of (pro)renin receptor in the medium of cultured human umbilical vein endothelial cells. *Hypertens. Res.*, vol. 34, no. 6, (June 2011), pp. (735-739).

Biswas KB, Nabi AH, Arai Y, Nakagawa T, Ebihara A, Suzuki F. Species specificity of prorenin binding to the (pro)renin receptor in vitro. Front. Biosci., vol. 2, (June 2010a), pp. (1234-1240).

Biswas K.B., Nabi A.H., Arai Y., Nakagawa T., Ebihara A., Ichihara A., Watanabe T., Inagami T., & Suzuki F. Aliskiren binds to renin and prorenin bound to (pro)renin receptor in vitro. *Hypertens. Res.*, vol. 33, no. 10, (October 2010b), pp. (1053-1059).

Boger J., Payne L.S., Perlow D.S., Lohr N.S., Poe M., Blaine E.H., Ulm E.H., Schorn T.W., LaMont B.I., Lin, T.Y. *et al.* Renin inhibitors. Syntheses of subnanomolar, competitive, transition-state analogue inhibitors containing a novel analogue of statine. *J. Med. Chem.*, vol. 28, no. 12, (December 1985), pp. (1779–1790).

Bolis G., Fung AK., Greer J., Kleinert HD., Marcotte PA., Perun TJ., Plattner JJ. & Stein H. Renin inhibitors: dipeptide analogues of angiotensinogen incorporating transition-state, nonpeptidic replacements at the scissile bond. *J .Med. Chem.*, vol. 30, no. 10, (October 1987), pp. (1729–1237).

Buechling T., Bartscherer K., Ohkawara B., Chaudhary V., Spirohn K., Niehrs C., & Boutros M. Wnt/Frizzled signaling requires dPRR, the Drosophila homolog of the prorenin receptor. *Curr. Biol.*, vol. 20, no. 14, (July 2010), pp. (1263-1268).

Brecher A.S., Shier D.N., Dene H., Wang S.M., Rapp J.P., Franco S.R., & Mulrow P.J. Regulation of adrenal renin messenger ribonucleic acid by dietary sodium chloride. *Endocrinology*, vol. 124, no. 6, (June 1989), pp. (2907–2913).

Burckle C.A., Danser A.H.J., Muller D.N., Garrelds I.M., Gasc J.M., Popova E., Plehm R., Peters J., Bader M., & Nguyen G. Elevated blood pressure and heart rate in human renin receptor transgenic rats. *Hypertension*, vol. 47, no. 6, (March 2006), pp. (552-556).

Burckle, C., & Bader M. Prorenin and its ancient receptor. *Hypertension*, vol. 48, no. 4, (2006), pp. (549–551).

Campbell D.J. Critical review of prorenin and (pro)renin receptor research. *Hypertension*, vol. 51, no. 5, (March, 2008), pp. (1259–1264).

Campbell D.J., Kladis A., & Duncan A.M. Nephrectomy, converting enzyme inhibition and angiotensin peptides. *Hypertension*. Vol. 22, no. 4, (October, 1993), pp. (513-522).

Campbell D.J., & Valentijn A.J. Identification of vascular renin-binding proteins by chemical cross-linking: inhibition of binding of renin by renin inhibitors. *J. Hypertens.*, vol. 12, no. 8, (August 1994), pp. (879–890).

Cohen S., Taylor J.M., Murakami K., Michelakis A.M., & Inagami T. Isolation and characterization of renin-like enzymes from mouse submaxillary glands. *Biochemistry*, vol. 11, no. 23, (November 1972); pp. (4286-4293).

Cooper J.R., Bloom F.E., Roth R.H. (1996). *Biochemical Basis of Neuropharmacology*. Oxford Univ. Press, Oxford, UK.

Cousin C., Bracquart D., Contrepas A., Corvol P., Muller L., & Nguyen G. Soluble form of the (pro)renin receptor generated by intracellular cleavage by furin is secreted in plasma. *Hypertension*, vol. 53, no. 6, (June 2009), pp. (1077-1082).

Contrepas A., Walker J., Koulakoff A., Franek K.J., Qadri F., Giaume C., Corvol P., Schwartz C.E., & Nguyen G.A. Role of the (pro)renin receptor in neuronal cell differentiation. *Am. J. Physiol. Regul. Integr. Comp. Physiol.*, vol. 297, no. 2, (August 2009), pp. (R250-R257).

Dai C.., Stolz D.B., Kiss L.P., Monga S.P., Holzman L.B., & Liu Y. Wnt/beta-catenin signaling promotes podocyte dysfunction and albuminuria. *J. Am. Soc. Nephrol.*, vol. 20, no. 9, (September 2009), pp. (1997–2008).

Danser A.H.J., van Kats J.P., Admiraal P.J.J., Derkx F.H.M., Lamers J.M.J., Verdouw P.D., Saxena P.R., & Schalekamp M.A.D.H. Cardiac renin and angiotensins: uptake from plasma versus in situ synthesis. *Hypertension*. vol. 24, no. 1, (July 1994), pp. (37-48).

Danser A.H.J., Koning M.M.G., Admiraal P.J.J., Sassen L.M., Derkx F.H., Verdouw P.D., & Schalekamp M.A. Production of angiotensins I and II at tissue sites in the intact pig. *Am. J. Physiol.*, vol. 263, no. 2, (August 1992), pp. (H429–H437).

Derkx F.H.M., Tan-Tjiong H.L., Man in 't Veld A.J., Schalekamp M.P., & Schalekamp M.A.D.H. Activation of inactive plasma renin by plasma and tissue kallikreins. *Clin. Sci. (Lond).*, vol. 57, no. 4, (October 1979), pp. (351–357).

Derkx F.H.M., Tan-Tjiong L., Wenting G.J., Boomsma F., Man in 't Veld A.J and Schalekamp M.A.D.H. Asynchronous changes in prorenin and renin secretion after captopril in patients with renal artery stenosis. *Hypertension.*, vol. 5, no. 2, (March-April 1983), pp. (244–256).

Derkx F.H., Alberda A.T., de Jong F.H., Zeilmaker F.H., Makovitz J.W., Schalekamp & M.A. Source of plasma prorenin in early and late pregnancy: observations in a patient with primary ovarian failure. *J. Clin. Endocrinol. Metab.*, vol. 65, no. 2, (August 1987a), pp. (349-354).

Derkx F.H.M., Schalekamp M.P., & Schalekamp M.A.D.H. Two-step prorenin-renin conversion. Isolation of an intermediary form of activated prorenin. *J. Biol. Chem.*, vol. 262, no. 6, (February 1987b), pp. (2472–2477).

Derkx F.H.M., Deinum J., Lipovski M., Verhaar M., Fischli W., & Schalekamp M.A.D.H. Nonproteolytic "activation" of prorenin by active site-directed renin inhibitors as demonstrated by renin-specific monoclonal antibody. *J. Biol. Chem.*, vol. 267, no. 32, (November 1992), pp. (22837–22842).

Deschepper C.F., Mellon S.H., Cumin F., Baxter J.D., & Ganong W.F. Analysis by immunocytochemistry and *in situ* hybridization of renin and its mRNA in kidney, testis, adrenal, and pituitary of the rat. *Proc. Natl. Acad. Sci. U.S.A.*, vol. 83, no. 19, (October 1986), pp. (7552-7556).

Do Y.S., Sherrod A., Lobo R.A., Paulson R.J., Shinagawa T., Chen S.W., Kjos S., & Hsueh W.A. Human ovarian theca cells are a source of renin. *Proc. Natl. Acad. Sci. U. S. A.*, vol. 85, no. 6, (1March 988), pp. (1957-1961).

de Lannoy L.M., Danser A.H.J., van Kats J.P., Schoemaker R.G., Saxena P.R., & Schalekamp M.A.D.H. Renin-angiotensin system components in the interstitial fluid of the isolated perfused rat heart. Local production of angiotensin I. *Hypertension*, vol. 29, no. 6, (June 1997), pp. (1240–1251).

de Lannoy L.M., Danser A.H.J., Bouhuizen A.M.B., Saxena P.R., & Schalekamp M.A.D.H. Localization and production of angiotensin II in the isolated perfused rat heart. *Hypertension*, vol. 31, no. 5, (May 1998), pp. (1111–1117).

De Mello W.C. Influence of intracellular renin on heart cell communication. *Hypertension*, vol. 25, no. 6, (June 1995), pp. (1172–1177).

Deschepper C.F., Bouhnik J., & Ganong W.F. Colocalization of angiotensinogen and glial fibrillary acidic protein in astrocytes in rat brain. *Brain. Res.*, vol. 374, no. 1, (May 1986), pp. (195–198).

Dhanaraj V., Dealwis C.G., Frazao C., Badasso M., Sibanda B.L., Tickle I.J., Cooper J.B., Driessen H.P., Newman M., Aguilar C. X-ray analyses of peptide-inhibitor complexes define the structural basis of specificity for human and mouse renins. *Nature.*, vol. 357, no. 6378, (June, 1992), pp. (466-472).

Deinum J., Rønn B., Mathiesen E., Derkx F.H., Hop W.C., Schalekamp M.A. Increase in serum prorenin precedes onset of microalbuminuria in patients with insulin-dependent diabetes mellitus.*Diabetologia*, vol. 42, no. 8, (August 1999), pp. (1006-1010).

Doi Y., Franco S.R. & Mulrow P.J. Evidence for an extra renal source of inactive renin in rats. *Hypertension*, vol. 6, no. 5, (September 1984) pp. (6627–6632).

Dostal D.E., Rothblum K.N., Chernin M.I., Cooper G.R., & Baker K.M. Intracardiac detection of angiotensinogen and renin: a localized renin—angiotensin system in neonatal rat heart. *Am. J. Physiol.*, vol. 263, no. 4 Pt 1, (October 1992), pp. (C838–C850).

Du D., Kato T., Nabi A.H.M.N., Suzuki F., & Park E.Y. Expression of functional human (pro)renin receptor in silkworm (*Bombyx mori*) larvae using BmMNPV bacmid. *Biotechnology and Applied Biochemistry*, vol. 49, no. 3, (March 2008), pp. (195-202).

Dzau V.J., Ellison K.E., Brody T., Ingelfinger J., & Pratt R.E. A comparative study of the distributions of renin and angiotensinogen messenger ribonucleic acids in rat and mouse tissues. *Endocrinology*, vol. 120, no. 6, (June 1987), pp. (2334-2338).

Dzau V.J., Ingelfinger J., Pratt R.E., & Ellison K.E. Identification of renin and angiotensinogen messenger RNA sequences in mouse and rat brains. *Hypertension*, vol. 8, no. 6, (June 1986), pp. (544–548).

Dzau V.J., Slater E.E., & Haber, E. Complete purification of human renin. *Biochemistry*, vol. 18, no. 23, (November 1979), pp. (524-528).

Epstein A.N., Fitzsimons J.T., & Rolls B.J. Drinking induced by injection of angiotensin into the rain of the rat. *J. Physiol.*, vol. 210, no. 2, (September 1970), pp. (457–474).

Feldt S. Maschke U. Dechend R., Luft F.C. & Muller D.N. The putative (pro)renin receptor blocker HRP fails to prevent (pro)renin signaling. *J. Am. Soc. Nephrol.*, vol. 19, no. 4, (April 2008a), pp. (743-748).

Feldt S. Batenburg W.W., Mazak I., Maschke U., Wellner M., Kvakan H., Dechend R., Fiebeler A., Burckle C., Contrepas A., Danser A.H.J., Bader M., Nguyen G., Luft F.C., & Muller D.N. Prorenin and renin-induced extracellular signal-regulated kinase 1/2 activation in monocytes is not blocked by aliskiren or the handle-region peptide. *Hypertension*, vol. 51, no. 3, (March 2008b), pp. (682-688).

Figueired, A.F.S., Takii Y., Tsuji, H., Kato K. & Inagami T. Rat kidney renin and chatepsin D: purification and comparison of properties. *Biochemistry*, vol. 22, no. 24, (November 1985), pp. (5476-5481).

K.E. Finberg, G.A. Wagner, M.A. Baileyetal. The B1-subunit of the H^+ATPase is required for maximal urinary acidification. *Proc. Natl. Acad. Sci. USA*, vol.102, no.38, (September 2005), pp. (13616–13621).

Fischer-Ferraro C., Nahmod V.E., Goldstein D.J., & Finkielman S. Angiotensin and renin in rat and dog brain. *J. Exp. Med.*, vol. 133, no. 2, (February 1971), pp. (353–361).

Ganten D, Minnich JL, Granger P, Hayduk K, Brecht HM, Barbeau A, Boucher R, Genest J. Angiotensin-forming enzyme in brain tissue. *Science*, vol. 173, no. 991, (July 1971), pp. (64–65).

Ganten D., Ganten U., Kubo S., Granger P., Nowaczynski W., Boucher R., & Genest J. Influence of sodium, potassium and pituitary hormones on iso-renin in rat adrenal glands. *Am. J. Physiol.*, vol. 227, no. 1, (July 1974), pp. (224–229).

Ganten D., Hutchinson J.S., Schelling P., Ganten U., & Fischer H. The iso-renin angiotensin systems in extra-renal tissue. *Clin. Exp. Pharmacol. Physiol.*, vol. 3, no. 2, (March-April 1976), pp. (102-126).

Glorioso N., Atlas S.A., Laragh J.H., Jewelewicz R., & Sealey J.E. Prorenin in high concentrations in human ovarian follicular fluid. *Science*, vol. 233, no. 4771, (September 1986), pp. (1422-1424).

Goto M., Mukoyama M., Suga S.I., Matsumoto T., Nakagawa M., Ishibashi R., Kasahara M., Sugawara A., Tanaka I., & Nakao K. Growth-dependent induction of angiotensin II type 2 receptor in rat mesangial cells. *Hypertension*, vol. 30, no. 3, (September 1997), pp. (358–362).

Gross V., Schunck W.H., Honeck H., Milia A.F., Kärgel E., Walther T., Bader M., Inagami T., Schneider W., & Luft F.C. Inhibition of pressure natriuresis in mice lacking the AT receptor. *Kidney Int.*, vol. 57, no. 1, (January 2000), pp. (191–202).

He W., Dai C., Li Y., Zeng G., Monga S.P., Liu Y. Wnt/beta-catenin signaling promotes renal interstitial fibrosis. *J. Am. Soc. Nephrol.*, vol. 20, no. 4, (April 2009), pp. (765–776).

Hermle T., Saltukoglu D., Grünewald J., Walz G., & Simons M. Regulation of Frizzled-dependent planar polarity signaling by a V-ATPase subunit. *Curr. Biol.*, vol 20, no. 14, (July 2010), pp, (1269-1276).

Hirose T., Hashimoto M., Totsune K., Metoki H., Hara A., Satoh M., Kikuya M., Ohkubo T., Asayama K., Kondo T., Kamide K., Katsuya T., Ogihara T., Izumi S., Rakugi H., Takahashi K., & Imai Y. Association of (pro)renin receptor gene polymorphisms with lacunar infarction and left ventricular hypertrophy in Japanese women: the Ohasama study. *Hypertens. Res.*, vol. 34, no. 4, (April 2011), pp. (530-535).

Hirose T., Mori N., Totsune K., Morimoto R., Maejima T., Kawamura T., Metoki H., Asayama K., Kikuya M., Ohkubo T., Kohzuki M., Takahashi K., & Imai Y. Gene expression of (pro)renin receptor is upregulated in hearts and kidneys of rats with congestive heart failure. *Peptides*, vol. 30, no. 12, (December 2009), pp. (2316–2322).

Hirose S., Ohsawa T., Inagami T., & Murakami K., Brain renin from bovine anterior pituitary: isolation and properties. *J. Biol. Chem.*, vol. **257**, no. 11, (June 1982), pp. (6316-6321).

Hirose S., Yokosawa H., & Inagami T. Immunochemical identification of renin in rat brain and distinction from acid protease. *Nature*, vol. 274, no. 5669, (July 1978), pp. (392-393).

Ho M.M., & Vinson G.P. Transcription of (pro)renin mRNA in the rat adrenal cortex, and the effects of ACTH treatment and a low sodium diet. *Journal of Endocrinology*, vol. 157, no. 2, (May 1998), pp. (217–223).

Hokimoto S., Yasue H., Fujimoto K., Yamamoto H., Nakao K., Kaikita K., Sakata R., & Miyamoto E. Expression of angiotensin-converting enzyme in remaining viable myocytes of human ventricles after myocardial infarction. *Circulation*, vol. 94, no. 7, (October 1996), pp. (1513–1518).

Hollenberg N.K., Fisher N.D., & Price, D.A. Pathways for angiotensin II generation in intact human tissue: evidence from comparative pharmacological interruption of the renin system. *Hypertension*, vol. 32, no. 3, (September 1998), pp. (387-392).

Holm I., Ollo R., Panthier J.J., & Rougeon F. Evolution of aspartyl proteases by gene duplication: the mouse renin gene is organized in two homologous clusters of four exons. *EMBO J.*, vol. 3, no. 3, (March 1984), pp. (557-562).

Huang Y., Wongamorntham S., Kasting J., McQuillan D., Owens R.T., Yu L., Noble N.A., & Border W. Renin increases mesangial cell transforming growth factor-beta1 and matrix proteins through receptormediated, angiotensin II-independent mechanisms. *Kidney Int.*, vol. 69, no. 1, (January 2006), pp. (105-113).

Huang J., Matavelli L.C., & Siragy M. Renal (pro)renin receptor contribute to the development of diabetic kidney disease through transforming growth factor-β1 and connective tissue growth factor signaling cascades. *Clinical and Experimental Pharmacology and Physiology*, vol. 38, no. 4, (April 2011), pp. (215-221).

Huang J., & Siragy H,M. Regulation of (pro)renin receptor expression by glucose-induced mitogen-activated protein kinase, nuclear factor-kappaB, and activator protein-1 signaling pathways. *Endocrinology*, vol. 151, no. 7, (July 2010), pp. (3317-3325).

Hui K.Y., Carlson W.D., Bernatowicz M.S., & Haber, E. Analysis of structure-activity relationships in renin substrate analogue inhibitory peptides. *J. Med. Chem.*, vol. 30, no. 8, (August 1987), pp. (1287–1295).

Ichihara A., Hayashi M., Kaneshiro Y., Suzuki F., Nakagawa T., Tada Y., Koura Y., Nishiyama A., Okada H., Uddin M.N., Nabi A.H.M.N., Ishida Y., Inagami T., & Saruta T. Inhibition of diabetic nephropathy by a decoy peptide corresponding to the "handle" region for non-proteolytic activation of prorenin. *J. Clin. Invest.*, vol. 114, no. 8, (October 2004), pp. (1128–1135).

Ichihara A., Kaneshiro Y., & Takemitsu T. Contribution of nonproteolytically activated prorenin in glomeruli to hypertensive renal damage. *J. Am. Soc. Nephrol.*, vol. 17, no. 9, (September 2006a), pp. (2495-2503).

Ichihara A., Kaneshiro Y., Takemitsu T., Sakoda M., Suzuki F., Nakagawa T., Nishiyama A., Inagami T., & Hayashi M. Non proteolytic activation of prorenin contributes to development of cardiac fibrosis in genetic hypertension. *Hypertension*, vol. 47, no. 5, (May 2006b), pp. (894–900).

Ichihara A., Suzuki F., Nakagawa T., Kaneshiro Y., Takemitsu T., Sakoda M., Nabi A.H.M.N., Nishiyama A., Sugaya T., Hayashi M., & Inagami T. Prorenin receptor blockade inhibits development of glomerulosclerosis in diabetic angiotensin II type

1a receptor deficient mice. *J. Am. Soc. Nephrol.*, vol. 17, no. 7, (July 2006c), pp. (1950–1961).

Imagawa M., Chiu R., & Karin M. Transcription factor AP-2 mediates induction by two different signal-transduction pathways: protein kinase C and cAMP. *Cell*, vol. 51, no. 2, (October 1987), pp. (251-260).

Inagami T. & Murakami K. Prorenin. *Biomed. Res.*, vol. 1, no. , (1980), pp. (456-475).

Inagami T. & Murakami K. Pure renin: isolation from hog kidney and characterization. *J. Biol. Chem.*, vol. 252, no. 9, (May 1977), pp. (2978-2983).

Inoue H., Noumi, T. Nagata M., Murakami H., & Kanazawa H. Targeted disruption of the gene encoding the proteolipid subunit of mouse vacuolar H-ATPase leads to early embryonic lethality. *Biochimica et Biophysica Acta.*,vol. 1413, no. 3, (November 1999), pp. (130–138).

Ito M., Oliverio M.I., Mannon P.J., Best C.F., Maeda N., Smithies O., & Coffman T.M. Regulation of blood pressure by the type 1A angiotensin II receptor gene. *Proc. Natl. Acad. Sci. USA*, vol. 92, no. 8, (April 1995), pp. (3521–3525).

Itskovitz J., & Sealey J.E. Ovarian prorenin-renin-angiotensin system. *Obstet. Gynecol. Surv.*, vol. 42, no. 9, (September 1987), pp. (545-551).

Jones C.A., Petrovic N., Novak E.K., Swank R.T., Sigmund C.D. & Gross K.W. Biosynthesis of renin in mouse kidney tumor As4.1 cells. *Eur. J. Biochem.*, vol. 243, no. 1-2, (January 1997), pp. (181-190).

Jutras I., & Reudelhuber T.L. Prorenin processing by cathepsin B in vitro and in transfected cells. *FEBS Lett.*, vol. 443, no. 1, (January 1999), pp. (48-52).

Kaneshiro Y., Ichihara A., Sakoda M., Takemitsu T., Nabi A.H.M.N., Uddin M.N., Nakagawa T., Nishiyama A., Suzuki F., Inagami T., & Itoh H. Slowly progressive, angiotensin II-independent glomerulosclerosis in human-renin/prorenin-receptor-transgenic rats. *J. Am. Soc. Nephrol.*, vol. 18, no. 6, (June 2007), pp. (1789–1795).

Julie K.J., Toma I., Sipos A., Elliott J., Sarah M., Vargas L., & Peti-Peterdi J. The collecting duct is the major source of p rorenin in diabetes. *Hypertension*, vol 51, no. 6, (June 2008), pp. (1597–1604).

Kato T., Du D., Suzuki F., & Park E.Y. Localization of human (pro)renin receptor lacking the transmembrane domain on budded baculovirus of Autographa californica multiple nucleopolyhedrovirus. *Appl. Microbiol. Biotechnol.*, vol. 82, no. 3, (March, 2009), pp. (431-437).

Kato T., Kageshima A., Suzuki F., & Park E.Y. Expression and purification of human (pro)renin receptor in insect cells using baculovirus expression system. Protein Expr Purif., vol. 58, no. 2, (April, 2008), pp. (242-248).

Karet F.E., Finberg K.E., & Nelson R.D., Nayir A., Mocan H., Sanjad S.A., Rodriguez-Soriano J., Santos F., Cremers C.W., Di Pietro A., Hoffbrand B.I., Winiarski J., Bakkaloglu A., Ozen S., Dusunsel R., Goodyer P., Hulton S.A., Wu D.K., Skvorak A.B., Morton C.C., Cunningham M.J., Jha V., Lifton R.P.. Mutations in the gene encoding B1 subunit of H+-ATPase cause renal tubular acidosis with sensor in neural deafness. *Nature Genetics*, vol. 21, no. 1, (January 1999), pp. (84–90).

Kim W.S., Nakayama K., Nakagawa T., Kawamura Y., Haraguchi K., & Murakami K. Mouse submandibular gland prorenin-converting enzyme is a member of glandular kallikrein family. *J. Biol. Chem.*, vol. 266, no. 29, (October 1991), pp. (19283-19287).

Kobori H., Ozawa Y., Suzakietal Y., Prieto-Carrasquero M.C., Nishiyama A., Shoji, T., Cohen E.P., & Navar L.G. Young scholars award lecture. Intratubular angiotensinogen in hypertension and kidney diseases. *Am. J. Hypertens.*, vol. 19, no. 5, (May 2006), pp. (541–550).

Kikkawa Y., Yamanaka N., Tada J., Kanamori N., Tsumura K., & Hosoi K. Prorenin processing and restricted endoproteolysis by mouse tissue kallikrein family enzymes (MK1, MK9, MK13, and MK22). *Biochim. Biophys. Acta Protein Struc. Mol. Enzymol.*, vol. 1382, no. 1, (January 1998), pp. (55-64).

Kinouchi K., Ichihara A., Sano M., Sun-Wada G.H., Wada Y., Kurauchi-Mito A., Bokuda K., Narita T., Oshima Y., Sakoda M., Tamai Y., Sato H., Fukuda K., & Itoh H. The (pro)renin receptor/ATP6AP2 is essential for vacuolar H^+-ATPase assembly in murine cardiomyocytes. *Circ. Res.*, vol. 107, no. 1, (July 2010), pp. (30–34).

Leckie B.J., Bottrill A.R. A specific binding site for the prorenin propart peptide Arg^{10}-Arg^{20} does not occur on human endothelial cells. J Renin Angiotensin Aldosterone Syst., vol. 12, no. 1, (May 2011), pp. (36-41).

Lenz T., James G.D., Laragh J.H., & Sealey J.E. Prorenin secretion from human placenta perfused in vitro. American Journal of Physiology, Endocrinology and Metabolism. *Am. J. Physiol.*, vol. 260 no. 6, (June 1991), pp. (E876-E882).

Liang F., & Gardner D.G. Autocrine/paracrine determinants of strain-activated brain natriuretic peptide gene expression in cultured cardiac myocytes. *J. Biol. Chem.*, vol. 273, no. 23, (1998), pp. (14612–14619).

Li Y.P., Chen W., Liang Y., Li E., & Stashenko P. ATP6i-deficient mice exhibit severe osteopetrosis due to loss of osteoclast-mediated extracellular acidification. *Nature Genetics.*, vol. 23, no. 4, (December 1999), pp. (447–451).

Logan C.Y., & Nusse R. The Wnt signaling pathway in development and disease. *Annu. Rev. Cell Dev. Biol.*, vol. 20, (2004), pp. (781-810).

Ludwig J., Kerscher S., Brandt U., Pfeiffer K., Getlawi F., Apps D.K., & Schägger H. Identification and characterization of a novel 9.2-kDa membrane sector-associated protein of vacuolar proton-ATPase from chromaffin granules. *J. Biol. Chem.*, vol. 278, no. 18, (May 1998), pp. (10939–10947).

Maibaum J., & Feldman D.L. Renin inhibitors as novel treatments for cardiovascular disease. *Expert. Opin. Ther. Patents*, vol. 13, (2003), pp. (589–603).

Manne K., & Danser A.H.J. Circulating versus tissue renin-angiotensin system: On the origin of (pro)renin. *Curr. Hypertense. Rep.*, vol. 10, no., 2, (April 2008), pp. (112-118).

Matavelli L.C., Huang J., & Siragy H.M. (Pro)renin receptor contributes to diabetic nephropathy by enhancing renal inflammation. *Clin. Exp. Pharmacol. Physiol.*, vol. 37, no., 3, (March 2010), pp. (277-282).

Miura G.I., Froelick G.J., Marsh D.J., Stark K.L., & Palmiter R.D. The d subunit of the vacuolar ATPase (Atp6d) is essential for embryonic development. *Transgenic Research.*, vol. 12, no. 1, (February 2003), pp. (131–133).

Matoba T., Murakami K. & Inagami, T. Rat renin: purification and characterization. *Biochim. Biophys. Acta.*, vol. 526, no. 2, (October 1978), pp. (560-571).

Ménard J, Boger R.S., Moyse D.M., Guyene T.T., Glassman H.N., & Kleinert H.D. Dose-dependent effects of the renin inhibitor zankiren HCI after a single oral dose in mildly sodium-depleted normotensive subjects. *Circulation*, vol. 91, no. 2, (January 1995), pp. (330–338).

Mercure C., Prescott G., Lacombe M.J., Silversides D.W., & Reudelhuber T.L. Chronic increases in circulating prorenin are not associated with renal or cardiac pathologies. *Hypertension.*, vol. 53, no. 6, (June 2009), pp. (1062-1069).

Mooser V., Nussberger J., Jullierat L. Reactive hyperreninemia is a major determinant of plasma angiotensin II during ACE inhibition. *J. Cardiovasc. Pharmacol.*, vol. 15, no. 2, (February, 1990), pp. (276–282).

Murakami K., & Inagami T. Isolation of pure and stable renin from hog kidney. *Biochem. Biophys. Res. Commun.*, vol. 62, no. 3, (February, 1975), pp. (757-763).

Muller D.N., Klanke B., Feldt S., Cordasic N., Hartner A., Schmieder R.E., Luft F.C., & Hilgers K.F. (Pro)renin receptor peptide inhibitor "handle-region" peptide does not affect hypertensive nephrosclerosis in goldblatt rats. *Hypertension*, vol. 51, no. 3, (March 2008), pp. (676-681).

Nabi A.H.M.N., Biswas K.B., Nakagawa T., Ichihara A., Inagami T., & Suzuki F. 'Decoy peptide' region (RIFLKRMPSI) of prorenin prosegment plays a crucial role in prorenin binding to the (pro)renin receptor. *Int. J. Mol. Med.* Vol. 24, no. 1, (July 2009a), pp. (83-89).

Nabi A.H.M.N., Biswas K.B., Nakagawa T., Ichihara A., Inagami T., & Suzuki F. Prorenin has high affinity multiple binding sites for (pro)renin receptor. *Biochim. Biophys. Acta.*, vol. 1794, no. 12, (December 2009b), pp. 1838-1847.

Nabi A.N., Biswas K.B., Arai Y., Nakagawa T., Ebihara ., Islam L.N., Suzuki F. (Pro)renin receptor and prorenin: their plausible sites of interaction. *Front Biosci.*, January 2012, In press.

Nabi A.H.M.N., Kageshima A., Uddin M.N., Nakagawa T., Park E.Y., & Suzuki F. Binding properties of rat prorenin and renin to the recombinant rat renin/prorenin receptor prepared by a baculovirus expression system. *Int. J. Mol. Med.*, vol. 18, no. 3, (2006), pp. (483-488).

Nabi A.H.M.N., Biswas K.B., Arai Y., Uddin M.N., Nakagawa T., Ebihara A., Ichihara A., Inagami T., & Suzuki F. Functional characterization of the decoy peptide, R^{10P}IFLKRMPSI^{19P}. *Front. Biosci.*, vol. 2, (June, 2010), pp. (1211-1217).

Naruse, K.; Takii, Y.; Inagam, T., Immunohistochemical localization of luteinizing hormone producing cells of rat pituitary. Proc. Nat. Acad. Sci., U.S.A. vol. 78, no. , (1981), pp. (7579-7583).

Naruse M., & Inagami T., Markedly elevated specific renin level in the adrenal in genetically hypertensive rats. Proc. Natl. Acad. Sci. U.S.A.78, (1982) 3295-3299.

Nelson N., Perzov N., Cohen A., Hagai K., Padler V., Nelson H. The cellular biology of proton-motive force generation by V-ATPases. *J. Exp. Biol.*, vol. 203, no. 1, (2000), pp. (89–95).

Neri Serneri G.G., Boddi M., Coppo M., Chechi T., Zarone N., Moira M., Poggesi L., Margheri M., Simonetti I. Evidence for the existence of a functional cardiac renin-angiotensin system in humans. *Circulation*, vol. 94, no. 8, (October 1996), pp. (1886–1893).

Nguyen G. Increased cyclooxygenase-2, hyperfiltration, glomerulosclerosis, and diabetic nephropathy: put the blame on the (pro)renin receptor? *Kidney International*, vol. 70, no. 4, (August 2006), pp. (618–620).

Nguyen G., Delarue F., Berrou J., Rondeau E., Sraer J.D. Specific receptor binding of renin on human mesangial cells in culture increases plasminogen activator inhibitor-1 antigen. *Kidney Int.*, vol. 50, no. 6, (1996), pp. (1897-1903).

Nguyen G., Bouzhir L., Delarue F., Rondeau E., & Sraer J.D. Evidence of a renin receptor on human mesangial cells: effects on PAI1 and cGMP. *Nephrologie*, vol. 19, no. 7, (1998), pp. (411-416).

Nguyen G., Delarue F., Burckle C., Bouzhir L., Giller T. & Sraer J.D. Pivotal role of the renin/prorenin receptor in angiotensin II production and cellular responses to renin. *J. Clin. Invest.*, vol. 109, no. 11, (June 2002), pp. (1417-1427).

Nurun NA., Uddin MN., Nakagawa T., Iwata H., Ichihara A., Inagami T., & Suzuki F. Role of "handle" region of prorenin prosegment in the non-proteolytic activation of prorenin by binding to membrane anchored (pro)renin receptor. *Front. Biosci.*, vol. 12, (September 2007), pp. (4810-4817).

Nussberger J., Wuerzner G., Jensen C., & Brunner, H.R. Angiotensin II suppression in humans by the orally active renin inhibitor aliskiren (SPP100). Comparison with enalapril. *Hypertension*, vol. **39**, no. 1, (January 2002), pp. (e1-e8).

Oparil S., & Haber E. The renin–angiotensin system (first of two parts). *N. Engl. J. Med.*, vol. 291, no. 8, (August 1974a), pp. (389-401).

Oparil S., & Haber E. The renin–angiotensin system (second of two parts). *N. Engl. J. Med.*, vol. 291, no. 9, (August 1974b), pp. (446-457).

Ott C., Schneider M., Delles C., Schlaich M.P., Hilgers K.F., & Schmieder R.E. Association of (pro)renin receptor gene polymorphism with blood pressure in Caucasian men. *Pharmacogenetics and Genomics*, vol. 21, no. 6, (June 2011), pp. (347-349).

Pandey K.N., Melner N. M., Paramentier M., & Inagami T. Demonstration of renin activity in purified rat Leydig cells: evidence for the existence of endogenous inactive (latent) form of enzyme. *Endocrinol.*, vol. 115, no. 5, (November 1984), pp. (1753-17590).

Paul P., Wagner D., Metzger R., Ganten D., Lang R.E., Suzuki F., Murakami K., Burbach J.H.P., & Ludwig G. Quantification of renin mRNA in various mouse tissues by a novel solution hybridization assay. *J. Hypertens.*, vol. 6, no. 3, (March 1988), pp. (247-252).

Pickens P.T., Bumpus F.M., Lloyd AmM., Smeby R.R., & Page I.H. Measurements of renin activity in human plasma. *Circ. Res.*, vol. 17, no. 5, (November 1965), pp. (438-448).

Pitarresi T.M., Rubattu S., Heinrikson R. & Sealey J.E. Reversible cryoactivation of recombinant human prorenin. *J. Biol. Chem.*, vol. 267, no. 17, (June 1992), pp. (11753–11759).

Prescott G., Silversides D.W., & Reudelhuber T.L. Tissue activity of circulating prorenin. *Am. J. Hypertens.*, vol. 15, no. 3, (March 2002), pp. (280–285).

Prieto-Carrasquero M.C., Harrison-Bernard L.M., Kobori H., Ozawa Y., Hering-Smith K.S., Hamm L.L., Navar L.G. Enhancement of collecting duct renin in angiotensin II-dependent hypertensive rats. *Hypertension.*, vol. 44, no. 4, (August 2004), pp. (223–229).

Rohrwasser A., Morgan T., Dillon H.F., Zhao L., Callaway C.W., Hillas E., Zhang S., Cheng T., Inagami T., Ward K., Terreros D.A., & Lalouel J.M. Elements of a paracrine tubular renin angiotensin system along the entire nephron. *Hypertension.*, vol. 34, no. 6, (December 1999), pp. (1265–1274).

Rongen G.A., Lenders J.W.M., Smits P. & Thien T. Clinical pharmacy okinetics and efficacy of renin inhibitors. *Clin. Pharmacokinet.*, vol. 29, no. 1, (July 1995), pp. (6–14).

Ruzicka M., Yuan B., & Leenen F.H.M. Effects of enalapril versus losartan on regression of volume overload-induced cardiac hypertrophy in rats. *Circulation.*, vol. 90, no. 1, (July 1994), pp. (484-491).

Sadoshima J., Xu Y., Slayter H.S., & Izumo S. Autocrine release of angiotensin II mediates stretch-induced hypertrophy of cardiac myocytes in vitro. *Cell*, vol. 95, no. 5, (December 1993), pp. (977–984).

Saris J.J., Derkx F.H.M., de Bruin R.J.A., Dekkers D.H., Lamers J.M., Saxena P.M., Schalekamp M.A., & Danser A.H.J. High-affinity prorenin binding to cardiac man-6-P/IGF-II receptors precedes proteolytic activation to renin. *Am. J. Physiol.-Heart Circ. Physiol.*, vol. 280, no. 4, (April 2001), pp. (H1706–H1715).

Saris J.J., 't Hoen P.A., Garrelds I.M., Dekkers D.H., den Dunnen J.T., Lamers J.M., Danser A.H.J. Prorenin induces intracellular signalling in cardiomyocytes independently of angiotensin II. *Hypertension*, vol. 48, no. 4, (October 2006), pp. (564-571).

Satofuka S., Ichihara A., Nagai N., Yamashiro K., Koto T., Shinoda H., Noda K., Ozawa Y., Inoue M., Tsubota K., Suzuki F., Oike Y., & Ishida S. Suppression of ocular inflammation in endotoxin-induced uveitis by inhibiting nonproteolytic activation of prorenin. *Invest. Ophth. Vis. Sci.*, vol. 47, no. 6, (June 2006), pp. (2686-2692).

Satofuka S., Ichihara A., Nagai N., Koto T., Shinoda H., Noda K., Ozawa Y., Inoue M., Tsubota K., Itoh H., Oike Y., & Ishida S. Role of nonproteolytically activated prorenin in pathologic, but not physiologic, retinal neovascularization. *Invest. Ophth. Vis. Sci.*, vol. 48, no. 1, (January 2007), pp. (422-429).

Satofuka S., Ichihara A., Nagai N., Noda K., Ozawa Y., Fukamizu A., Tsubota K., Itoh H., Oike Y., & Ishida S. (Pro)renin receptor promotes choroidal neovascularization by activating its signal transduction and tissue renin-angiotensin system. *Am. J. Pathol.*, vol. 173, no. 6, (December 2008), pp. (1911–1918).

Satofuka S., Ichihara A., Nagai N., Noda K., Ozawa Y., Fukamizu A., Tsubota K., Itoh H., Oike, Y., & Ishida S. (Pro)renin receptor-mediated signal transduction and tissue renin-angiotensin system contribute to diabetes-induced retinal inflammation. *Diabetes*, vol. 58, no. 7, (July 2009), pp. (1625-1633).

Schefe J.H., Menk, M., Reinemund J., Effertz K., Hobbs R.M., Pandolfi P.P., Ruiz P., Unger T., Funke-Kaiser H. A novel signal transduction cascade involving direct physical interaction of the renin/prorenin receptor with the transcription factor promyelocytic zinc finger protein. *Circ. Res.*, vol. 99, no. 12, (December 2006), pp. (1355-1366).

Schieffer B., Wirger A., Meybrunn M., Seitz S., Holtz J., Riede U.N., & Drexler H. Comparative effects of chronic angiotensin-converting enzyme inhibition and angiotensin II type 1 receptor blockade on cardiac remodeling after myocardial infarction in the rat. *Circulation*, vol. 89, no. 5, (May 1994), pp. (2273-2282).

Schnermann J., & Briggs J.P. Function of the juxtaglomerular apparatus: control of glomerular hemodynamics and renin secretion. In: Alpern RJ, Hebert SC, editors. *The Kidney Physiology and Pathophysiology.* Burlington-San Diego-London: Elsevier Academic Press; 2008. pp. 589–626.

Schweda F., Friis U., Wagner C., Skott O., & Kurtz A. Renin release. *Physiology (Bethesda),* vol 22, (October 2007), pp. (310–319).

Sealey J.E., Catanzaro D.F., Lavin T.N., Gahnem, F., Pitarresi T., Hu L.F., & Laragh J.H. Specific prorenin/renin binding (ProBP). Identification and characterization of a novel membrane site. *Am. J. Hypertens.,* vol. 9, no. 5, (May 1996), pp. (491-502).

Sealey J.E., & Rubattu S. Prorenin and renin as separate mediators of tissue and circulating systems. *Am. J. Hypertens.,* vol. 2, no. 5, (May 1989), pp. (358-366).

Sealey J.E., Goldstein M., Pitarresi T., Kudlak T.T., Glorioso N., Fiamengo S.A., & Laragh J.H. Prorenin Secretion From Human Testis: No Evidence for Secretion of Active Renin or Angiotensinogen. J. Clin. Endocrinol. Metab., vol. 66, no. 5, (May 1988), pp. (974-978).

Shaw K.J., Do Y.S., Kjos S., Anderson P.W., Shinagawa T., Dubeau L., Hsueh W.A. Human decidua is a major source of renin. *J. Clin. Invest.,* vol. 83, no. 6, (June 1989), pp. (2085–2092).

Shinagawa, T., Do., Y.S., Baxter, J.D., Carili, C., Schilling, J. & Hseuh, W. Identification of an enzyme in human kidney that correctly processes prorenin. *Proc. Natl. Acad. Sci., U.S.A.,* vol. 87, no. 5, (March 1990), pp. (1927-1931).

Shinagawa T., Nakayama K., Uchiyama Y., Kominami E., Doi Y., Hashiba K., Yano K., Hseu W.A., & Murakami K. Role of cathepsin B as prorenin processing enzyme in human kidney. *Hypertens. Res.,* vo. 18, no. 2, (June 1995), pp. (131-136).

Strausberg R.L., Feingold E.A., Grouse L.H., Derge, J.G.; Klausner, R.D.; Collins, F.S.; Wagner, L.; Shenmen, C.M.; Schuler, G.D.; Altschul, S.F.; Zeeberg, B.; Buetow, K.H.; Schaefer, C.F.; Bhat, N.K.; Hopkins, R.F.; Jordan, H.; Moore, T.; Max, S.I.; Wang, J.; Hsieh, F.; Diatchenko, L.; Marusina, K.; Farmer, A.A.; Rubin, G.M.; Hong, L.; Stapleton, M.; Soares, M.B.; Bonaldo, M.F.; Casavant, T.L.; Scheetz, T.E.; Brownstein, M.J.; Usdin, T.B.; Toshiyuki, S.; Carninci, P.; Prange, C.; Raha, S.S.; Loquellano, N.A.; Peters, G.J.; Abramson, R.D.; Mullahy, S.J.; Bosak, S.A.; McEwan, P.J.; McKernan, K.J.; Malek, J.A.; Gunaratne, P.H.; Richards, S.; Worley, K.C.; Hale, S.; Garcia, A.M.; Gay, L.J.; Hulyk, S.W.; Villalon, D.K.; Muzny, D.M.; Sodergren, E.J.; Lu, X.; Gibbs, R.A.; Fahey, J.; Helton, E.; Ketteman, M.; Madan, A.; Rodrigues, S.; Sanchez, A.; Whiting, M.; Madan, A.; Young, A.C.; Shevchenko, Y.; Bouffard, G.G.; Blakesley, R.W.; Touchman, J.W.; Green, E.D.; Dickson, M.C.; Rodriguez, A.C.; Grimwood, J.; Schmutz, J.; Myers, R.M.; Butterfield, Y.S.; Krzywinski, M.I.; Skalska, U.; Smailus, D.E.; Schnerch, A.; Schein, J.E.; Jones, S.J.; Marra, M.A. Generation and initial analysis of more than 15,000 full-length human and mouse cDNA sequences", *Proc. Natl. Acad. Sci. USA,* vol. 99, no. 26, (December 2002), pp. (16899–16903).

Smith A.N., Skaug J., & Choateetal K.A. Mutations in ATP6N1B, encoding a new kidney vacuolar proton pump 116-kD subunit, cause recessive distal renal tubular acidosis with preserved hearing. Nature Genetics., vol. 26, no. 1, (September 2000), pp. (71–75).

Siragy H.M., & Carey R.M. The subtype-2 (AT) angiotensin receptor mediates renal production of nitric oxide in conscious rats. *J. Clin. Invest.*, vol. 100, no. 2, (1997), pp. (264–269).

Siragy H.M., & Huang J. Renal (pro)renin receptor upregulation in diabetic rats through enhanced angiotensin AT_1 receptor and NADPH oxidase activity. *Exp. Physiol.*, vol. 93, no. 5, (May 2008), pp. (709–714).

Susic D., Zhou X., Frohlich E.D., Lippton H., & Knight M. Cardiovascular effects of prorenin blockade in genetically hypertensive rats (SHR) on normal and high salt diet. *Am. J. Physiol. Heart Circ. Physiol.*, vol. 295, no. 3, (September 2008), pp. (H1117-H1121).

Sun Y., Cleutjens J.P.M., Diaz-Arias A.A., & Weber K.T. Cardiac angiotensin converting enzyme and myocardial fibrosis in the rat. *Cardiovasc. Res.*, vol 28, no. 9, (September 1994), pp. (1423–1432).

Suzuki F., Ludwig G., Hellmann W., Paul M., Lindpaintner K., Murakami K., & Ganten D. Renin gene expression in rat tissues: a new quantitative assay method for rat renin mRNA using synthetic cRNA. *Clin. Exp. Hyper. A.*, vol. 10, no. 2, (1987), pp. (345-359).

Suzuki F., Nakagawa T., Kakidachi H., Murakami K., Inagami T., & Nakamura Y. The dominant role of the prosegment of prorenin in determining the rate of activation by acid or trypsin: studies with molecular chimeras. Biochem. Biophys. Res. Commun., vol. 267, no. 2, (January 2000), pp. (:577-580).

Suzuki F., Hayakawa M., Nakagawa T., Uddin M.N., Ebihara A., Iwasawa A., Ishida Y., Nakamura Y., & Kazuo M. Human prorenin has 'gate and handle' regions for its non-proteolytic activation. *J. Biol. Chem.*, vol. 278, no. 25, (June 2003), pp. (22217–22222).

Tada, M.; Fukamizu, A.; Seo, M.S.; Takahashi, S.; Murakami, K. Renin expression in the kidney and brain is reciprocally controlled by captopril. *Biochem. Biophys. Res. Commun.*, vol. 159, no. 3, (March 1989), pp. (1065-1071).

Tada M., Takahashi S., Miyano M., & Miyake Y. Tissue-specific regulation of renin-binding protein gene expression in rats. *The J. Biochem.*, vol. 112, no. 2, (August 1992), pp. (175–182).

Takahashi H., Ichihara A., Kaneshiro Y., Inomata K., Sakoda M., Takemitsu T., Nishiyama A., & Itoh H. Regression of nephropathy developed in diabetes by (Pro)renin receptor blockade. *J. Am. Soc. Nephrol.*, vol 18, no. 7, (July 2007), pp. (2054-2061).

Takahashi K., Hiraishi K., Hirose T., Kato I., Yamamoto H., Shoji I., Shibasaki A., Kaneko K., Satoh, F., & Totsune K. Expression of (Pro)renin Receptor in the Human Brain and Pituitary, and Co-localisation with Arginine Vasopressin and Oxytocin in the Hypothalamus. *J. Neuroendocrinol.*, vol. 22, no. 5, (May 2010), pp. (453–459).

Takahashi S., Ohsawa T., Miura R., & Miyake Y. Purification and characterization of renin binding protein (RnBP) from porcine kidney. *The J. Biochem.*, vol. 93, no. 6, (June 1983), pp. (1583-1594).

Tigerstedt R., & Bergman P.G. Niere and Kreislauf. *Scand Arc Physiol.*, vol. **8**, no. , (1898), pp. (223-271).

Toei M., Saum R., & Forgac M. Regulation and isoform function of the V-ATPases. *Bochemistry*, vol. 49, no. 23, (June 2010), pp. (4715–4723).

van Esch J.H.M., van Veghel R, Garrelds I.M., Leijten F, Bouhuizen A. M., & Danser A.H.J. Handle Region Peptide Counteracts the Beneficial Effects of the Renin Inhibitor Aliskiren in Spontaneously Hypertensive Rats. *Hypertension*, vol. 57, no. 4, (February 2011), pp. (852-858).

Van Gilst W.H., de Graeff P.A., Kingma J.H., Wesseling H., & de Langen C.D.J. Captopril reduces purine loss and reperfusion arrhythmias in the rat heart after coronary artery ligation. *Eur. J. Pharmacol.*, vol. 100, no. 1, (April 1984), pp. 113-117.

van Kesteren C.A., Saris J.J., Dekkers D.H., Lamers J.M., Saxena P.R., Schalekamp M.A., Danser A.H.J. Cultured neonatal rat cardiac myocytes and fibroblasts do not synthesize renin or angiotensinogen: evidence for stretch-induced cardiomyocyte hypertrophy independent of angiotensin II. *Cardiovasc. Res.*, vol. 43, no. 1, (July 1999), pp. 148-156.

van Kesteren C.A.M., Danser A.H.J., Derkx F.H., Dekkers D.H., Lamers J.M., Saxena P.M., & Schalekamp M.A. Mannose 6-phosphate receptor-mediated internalization and activation of prorenin by cardiac cells. *Hypertension*, vol. 30, no. 6, (December 1997), pp. (1389–1396).

Wang P.H., Do Y.S., Macaulay L., Shinagawa T., Anderson P.W., Baxter J.D., & Hseuh W.A. (1991) Identification of renal cathepsin B as a human prorenin-processing enzyme. *J. Biol. Chem.*, vol. 266, no. 19, (July 1991), pp. (12633-12638).

Webb, D.J., Manhem, P.J., Ball, S.G., Inglis, G., Leckie, B.J., Lever, A.F., Morton, J.J., Robertson, J.I., Murray, G.D., Ménard, J., et al. A study of the renin inhibitor H142 in man. *J. Hypertens.*, vol. 3, no. 6, (December 1985), pp. (653–658).

Wilkinson-Berka, J.L., Heine R., Tan G., Cooper M.E., Hatzopoulos K.M., Fletcher E.L., Binger K.J., Campbell D.J., & Miller A.G. RILLKKMPSV influences the vasculature, neurons and glia, and (pro)renin receptor expression in the retina. *Hypertension*, vol. 55, no. 5, (June 2010), pp. (1454-1460).

Wilson C.M., Cherry M., Taylor B.A., & Wilson J.D. Genetic and endocrine control of renin activity in the submaxillary gland of the mouse. *Biochem. Genet.*, vol. 19, no. (5-6), (June 1981), pp. (509-523)

Wolny A., Clozel J.P., Rein J., Mory P., Vogt P., Turino M., Kiowski W. F., Ischli W. Functional and biochemical analysis of angiotensin II-forming pathways in the human heart. *Circ.Res.*, vol. 80, no. 2, (February 1997), pp. (219-227).

Wood J.M., Gulati N., Forgiarini P., Fuhrer W. & Hofbauer, K.G. Effects of a specific and long-acting renin inhibitor in the marmoset. *Hypertension*, vol. 7, no. 5, (October 1985), pp. (797–803).

Wood J.M., Criscione L., de Gasparo M., Bühlmayer P., Rüeger H., Stanton J.L., Jupp R.A. & Kay J. CGP 38 560: orally active, low-molecular-weight renin inhibitor with high potency and specificity. *J. Cardiovasc. Pharmacol.*, vol. 14, no. 2, (August 1989), pp. (221–226).

Wood J.M., Cumin F. & Maibaum J. Pharmacology of renin inhibitors and their application to the treatment of hypertension. *Pharmacol. Ther.*, vol. 61, no. 3, (1994) pp. (325-344).

Wood J.M., Maibaum J., Rahuel J., Grütter MG., Cohen NC., Rasetti V., Rüger H., Göschke R., Stutz S., Fuhrer W., Schilling W., Rigollier P., Yamaguchi Y., Cumin F., Baum HP., Schnell CR., Herold P., Mah R., Jensen C., O'Brien E., Stanton A., & Bedigian

MP. Structure-based design of aliskiren, a novel orally effective renin inhibitor. Biochem. Biophys. Res .Commun., vol. 308, no. 4, (September 2003), pp. (698-705).

Yokosawa H., Inagami T., & Haas E. Purification of human renin. *Biochem. Biophys. Res. Commun.*, vol. 83, no. 1, (July 1978), pp. (306-312).

Yokosawa H., Holladay L.A., Inagami T., Haas E. & Murakami K. Human renal renin: complete purification and characterization. *J. Biol. Chem.*, vol. 255, no. 8, (April 1980), pp. (3498-3502).

Yoshikawa A., Kusano Y.K., Kishi F., Kishi F,; Susumu T., Iida S., Ishiura S., Nishimura S., Shichiri M., & Senbonmatsu T. The (pro)renin receptor is cleaved by ADAM19 in the Golgi leading to its secretion into extracellular space. *Hypertens. Res.*, vol. 34, no. 5, (May 2011), pp. (599-605).

Yusuf S., on behalf of the SOLVD investigators. Effect of enalapril on survival in patients with reduced left ventricular ejection fractions and congestive heart failure. *N. Engl. J. Med.*, vol. 325, no. 5, (August 1991), pp. (293-302).

Zhang X., Dostal D.E., Reiss K., et al. Identification and activation of autocrine renin—angiotensin system in adult ventricular myocytes. *Am. J. Physiol.*, vol. 269, no. 5, (November 1995), pp. (H1791–H1802).

Zhang J., Noble N.A., Border W.A., Owens R.T., & Huang Y. Receptor-dependent prorenin activation and induction of PAI-1 expression in vascular smooth muscle cells. *Am. J. Physiol. Endocrinol. Metab.*, vol. 295, no. 4, (October 2008), pp. (E810–E819).

14

Cholesterol-Binding Peptides and Phagocytosis

Antonina Dunina-Barkovskaya
Belozersky Institute of Physico-Chemical Biology at Moscow Lomonosov State University Russia

1. Introduction

Phagocytosis is an important cellular process that in multicellular organisms ensures a defence against microbial invasion and removal of effete/apoptotic cells. Phenomenologically, phagocytosis is a process of internalization or engulfment by a cell of particles of a certain size (more than 0.5 µm) (Ofek et al., 1995; Pratten & Lloyd, 1986; Koval et al., 1998; Aderem & Underhill, 1999; Morrissette et al., 1999; Tjelle et al., 2000; May & Machesky, 2001; Djaldetti et al., 2002). After the contact of a particle with a phagocytozing cell, named "phagocyte" in the 19th century (*see* Heifets, 1982; Gordon, 2008), plasma membrane underneath the particle forms either invagination or extensions (pseudopods) surrounding the particle and eventually forms a vesicle (phagosome) that delivers the particle inside the cell.

The ability to phagocytoze is an integral feature of eucariotic cells, starting from single-celled animals to the higher vertebrates. In mammals, most of differentiated cells are able to phagocytoze to a certain extent; specialized cells named "professional phagocytes" (Rabinovitch, 1995) (monocytes, macrophages, neutrophils) do this most efficiently, but the activity of "non-professional" phagocytes is also very important both for anti-microbial defence and for tissue development, remodeling, and repair. For instance, macrophage-mediated phagocytosis plays a significant role in muscle tissue regeneration (Tidball & Wehling-Henricks, 2007). Senescent erythrocytes, mostly removed from circulation by macrophages, are also phagocytozed by epithelial cells of thyroid gland and urinary bladder (Aderem & Underhill, 1999). Fibroblasts incorporate solid particles - fragments of bone or prosthetic materials (Grinnell, 1984; Knowles et al., 1991). Lung epithelium cells can take up foreign particles inhaled with air (Kato et al., 2003; Saxena et al., 2008). Cells of retinal pigmented epithelium (RPE) phagocytoze and digest the shed outer segment membranes of rods (Rabinovitch, 1995; Aderem & Underhill, 1999; Krigel et al, 2010), and so on.

The vast diversity of the tasks and performances of the phagocytes may account for the fact that impairments in the phagocytic machinery accompany a number of serious illnesses, such as immunodeficiency (review of Lekstrom-Himes & Gallin, 2000), rheumatoid arthritis (Turner et al., 1973), retinal dystrophies (Gal et al., 2000), paroxysmal arrhythmia (James, 1994), cystic fibrosis and bronchiectasis (Vandivier et al., 2002). Macrophages are shown to be involved in promoting tumor angiogenesis, an essential step in the tumor progression to malignancy (Lin et al., 2006). Defective phagocytic clearance of apoptotic cells and macrophages as such are involved in the development of the aterosclerotic lesions that initiate acute thrombotic and vascular diseases, including myocardial infarction and stroke (Lucas & Greaves, 2001; Takahashi et al., 2002). Therefore, understanding the molecular mechanisms of phagocytosis is very important and should help to solve a number of medicinal problems and elaborate new approaches for regulation and control/correction of the phagocytic process.

By now, it is accepted that mechanism of phagocytosis implicates such processes as exocytosis, endocytosis, and adhesion (Aderem & Underhill, 1999; Botelho et al., 2000; Booth et al., 2001; Dunina-Barkovskaya, 2004; Lee at al., 2007; Fairn et al., 2010). A detailed list of the molecular participants that accomplish the initial membrane reorganization after the contact with the particle, subsequent formation and pinching-off of the phagosomal vesicle, and the components involved in the phagolysosome maturation and recycling has been created (Araki et al., 1996; Hackam et al., 1998; Morrissette et al., 1999; Garin et al., 2001; May & Machesky, 2001; Booth et al., 2001; Grinstein, 2010). This list includes receptors, membrane lipids, enzymes, cytoskeletal elements, ion-transporting systems (channels, exchanges, and pumps), and accessory cytoplasmic proteins required for membrane fusion, vesicle fission, and oxidative burst. However, molecular mechanism-based tuning of the phagocytic process in vivo and in vitro remains a challenge for the contemporary life sciences. This mini-review will briefly outline the role of cholesterol in the phagocytic process and consider cholesterol-binding peptides as potential tools for modulations and studies of the phagocytic process.

2. Cholestrol-dependence of the early stages of phagocytosis: What is cholesterol-dependent?

It has long been shown that the phagocytic process is cholesterol-dependent (Werb & Cohn, 1972) and very sensitive to sterols (Schreiber et al., 1975). Werb & Cohn, 1972, in their studies of the membrane composition changes following phagocytosis of latex particles showed that the ability to phagocytoze is regained several hours after the particle engulfment provided that the recovery medium contains cholesterol. Depleting plasma membrane cholesterol considerably inhibits phagocytosis (Peyron et al., 2000; Gatfield & Pieters, 2000). These observations raise a question: what molecular components accomplishing the phagocytic process are cholesterol-dependent?

2.1 Examples of cholesterol-dependence of phagocytic receptors

It is generally agreed that phagocytosis is triggered as a result of binding of cell membrane "phagocytic" receptors with their ligands on the particle surface. Ligand-receptor binding is followed by lateral clustering of the ligand–receptor complexes and by an unexplained way initiates a biochemical cascade leading to the actin polymerization at the sites of the vesicle formation. It is also assumed that the type of ligand (and the receptor involved) determines the "scenario" of the phagocytic process (Aderem & Underhill, 1999; Tjelle et al., 2000; Greenberg, 2001).

There are many reviews cataloging various types of phagocytic receptors and corresponding signaling cascades leading to the phagosome formation and detachment (Mosser, 1994;

Greenberg, 1995; Ofek et al., 1995; Aderem & Underhill, 1999; Astarie-Dequeker et al., 1999; Greenberg, 1999; Peyron, 2000; Tjelle et al., 2000; Greenberg, 2001; Djaldetti et al., 2002). In human phagocytes, the best studied are the receptors recognizing host serum immunoglobulin G (IgG) and complement C3 components (C3b and iC3b). These immune humoral factors are termed opsonins, and particles covered with opsonins are termed opsonized. Phagocytic receptor recognizing Fc-domain of IgG is termed FcγR, and that recognizing C3b and iC3b, complement 3 receptor, or CR3. In real life, however, phagocytes have to deal with non-opsonized particles, for example, in open wounds or in organs directly contacting with the environment (respiratory tract, gastro-intestinal tract) (Mosser, 1994; Ofek, 1995; Peyron, 2000; Djaldetti et al., 2002). In these cases an important role belongs to various receptors, such as mannose or beta-glucan receptors that bind integral components of the microorganism surface. To this group belong several receptors of macrophages, and in particular CD14, a receptor recognizing bacterial surface components, including lipopolysacharide (LPS); scavenger receptor A, as well as receptors CD36 and CD68 (macrosialin) that participate in phagocytosis of apoptotic cells. There is also phosphatidylserine receptor (PSR) recognizing phosphadidylserine that relocates from the inner to the outer monolayer of the plasma membrane of apoptotic cells (Fadok et al., 1992; Pradhan, 1997; Devitt et al., 1998; Giles et al., 2000; Li et al., 2003).

Although phagocytic receptors (as receptors in general, by definition) are considered "specialized", they are multispecific and multifunctional (Aderem & Underhill, 1999). This means that they recognize different ligands or certain molecular configurations ("pattern receptors") and can mediate other processes, such as endocytosis or adhesion. A striking example is CD36, multiligand scavenger receptor of class B. These ligands include thrombospondin-1, long-chain fatty acids, modified LDL, retinal photoreceptor outer segments, *Plasmodium falciparum* malaria-parasitized erythrocytes, sickle erythrocytes, anionic phospholipids, apoptotic cells, and collagens I and IV (Febbraio et al., 2001 and refs. therein). Another example is a complement receptor CR3 - also known as CD11b/CD18 and $\alpha_M\beta_2$, β_2-integrin - that functions not only as a membrane receptor recognizing iC3b but also as an adhesion molecule and binds diverse ligands, for example, intercellular adhesion molecule-1 (ICAM-1) (Ross & Větvicka, 1993; Ofek et al, 1995). Even Fcγ receptors may trigger either endocytosis or phagocytosis, depending on the size of the ligand-receptor cluster (Koval et al., 1998; Huang, 2006).

There are a number of works indicating that phagocytic processes involving certain phagocytic receptors are cholesterol-dependent. For example, Han et al., 1997 and Han et al., 1999 showed that lipoprotein lipids and cholesterol can upregulate the expression of the CD36 gene and protein. Moreover, according to Febbraio et al., 2001, CD36 colocalizes with caveolin-1 in specialized cholesterol- and sphingolipid-enriched microdomains of plasma membrane termed rafts. There are plenty of comprehensive reviews considering molecular structure, biophysics, and the roles of rafts in cell physiology (Simons & Ikonen, 1997; Brown & London, 2000; Pike, 2003; Lingwood & Simons, 2010, and references therein). In brief, rafts are defined as protein-lipid domains enriched with cholesterol and sphingomyelin. Rafts feature resistance to non-polar detergents (like Triton X-100), which points to strong interactions between molecules in a raft. At the same time, rafts are dynamic structures: they move laterally in the plane of the membrane, and big rafts can split into smaller ones, which in turn can fuse with each other. In artificial systems, in the absence of protein, rafts can be of micron size, while in cell membrane, raft size was reported to be several tens on nanometers. Cholesterol-sequestering agents (e.g., nystatin, filipin, β-cyclodextrin, etc.) or those interfering with its synthesis and metabolism (e.g., progesterone) prevent raft formation and inhibit cell processes in which rafts are involved, such as caveolar endocytosis. It is proposed that rafts may serve to concentrate signaling molecules and facilitate the integration of signaling cascades.

Another example of cholesterol- (and raft-) dependence concerns CR3-mediated phagocytosis. According to Peyron et al., 2000, nystatin and other cholesterol-sequestering agents (filipin, methyl-β-cyclodextrin) notably inhibit CR3-mediated phagocytosis of non-opsonized bacteria *Mycobacterium kansasii* by neutrophils. Moreover, phagocytosis is blocked if glycosylphosphatidylinositol- (GPI-) anchored proteins are removed with phosphatidyl inositol phospholipase C. The authors suggested that CR3-mediated phagocytosis of *Mycobacterium kansasii* requires binding of CR3 with GPI-ancored proteins localized in rafts. Once CR3 is not associated with rafts, it can mediate phagocytosis of zymozan or opsonized zymozan but not *Mycobacterium kansasii*. The observations suggest that the phagocytic "scenario" involving a given receptor, CR3, depends on the presence of cholesterol and interaction of the receptor with the lipid.

Fcγ-receptor-mediated phagocytosis is also cholesterol-dependent. Clustering of Fcγ receptors induced by binding to multiple opsonic ligands on a particle leads to phosphorylation of the Fcγ-immunoreceptor tyrosine-based activation motif (ITAM) by members of the Src family of kinases and followed by recruitment of the kinase Syk. Syk activation in turn initiates a signaling cascade, including activation of phosphatidylinositol 3-kinase (PI 3-kinase) and of the small GTPases Rac and Cdc42, which coordinate actin remodeling (Henry et al., 2004). As was reported (Kwiatkowska & Sobota, 2001; Katsumata et al., 2001; Kono et al., 2002; Kwiatkowska et al., 2003), one of the earliest signal events after the cross-binding of Fcγ receptors is lateral raft assembly that occurs before the activation of kinases of the Src family and independent of their activity. To trigger the lateral assembly of rafts, sufficient was the expression of ligand-binding monomer FcγR, without signal subunits carrying the activating fragment with tyrosine. Moreover, expression of the ligand-binding fragment of the receptor triggered fast mobilization of calcium. The authors (Kono et al., 2002; Kwiatkowska et al., 2003) suggested that lateral assembly of rafts is caused by ligand-binding subunits of the Fcγ-receptor and that it is the raft coalescence that triggers the signaling cascade of the biochemical reactions leading to rearrangements of membrane and cytoskeleton and eventually, to the formation of a membrane vesicle containing the particle. The importance of the integrity of the plasma membrane detergent-resistant microdomains for IgG-dependent phagocytosis was also shown by Marois et al., 2011. The authors also reported that phagocytosis of IgG-opsonized zymosan by human neutrophils required an extracellular influx of calcium that was blocked only by antibodies against FcγRIIIb. These data revive the question wheather FcγR may function as ligand-dependent channels (Young et al., 1983; Young et al., 1985). A quick local change of the ion channel activity at the site of the contact of the phagocyte with a particle remains a possible (but yet unexplored) step in the phagocytic signal cascade. In our hands, phagocytosis of non-opsonized beads by IC-21 macrophages is sensitive to methyl-β-cyclodextrin and carbenoxolon - glucocorticoid and a connexin channel blocker (Golovkina et al., 2009; Vishniakova et al., 2011).

Thus, even a brief overview of the very early stages of the phagocytic process shows that at least some of phagocytic receptors depend on cholesterol.

2.2 Phosphoinositides and lipid rafts in phagocytosis

It is recognized that successful phagosome formation requires local actin polymerization/depolymerization (Aderem & Underhill, 1999; Tjelle et al., 2000; May & Machesky, 2001; Greenberg, 2001; Lee at al., 2007) that dependes on the phosphatidylinsiotol metabolism. A detailed quantitative assessment of membrane remodeling during FcγR-mediated phagocytosis revealed marked changes in membrane composition that concerned the localization and metabolism of phosphoinositides (Botelho et al., 2000; Lee at al., 2007; Fairn et al., 2010). It was found in particular that at the onset of phagocytosis phosphatidylinositol (PI) 4,5-bisphosphate ($PI4,5P_2$) accumulates at sites where pseudopods are formed. Following the closure and fission of the phagosome, the phosphoinositide concentration drops, while phospholipase Cγ (PLCγ) is mobilized and local concentration of DAG increases. The authors suggested that this localized increase in $PI4,5P_2$ serves as a platform for the robust actin polymerization required for pseudopod extension. Recent works (Fairn et al., 2010; Grinstein, 2010) demonstrated in detail a highly localized sequence of changes in the level of several phosphoinositides as well as phosphatidylserine. The net changes in the content of these anionic phospholipids notably altered the surface charge of the membrane and caused the relocation of membrane-associated proteins due to electrostatic interactions. The authors hypothesize that modification of the membrane surface charge may play a role of an "electrostatic switch" that attracts or repulses proteins carrying polycationic or polyanionic motifs. Perhaps this hypothesis can be further extended: the role of such electrostatic switch may play a charged particle that touches the cell and triggers the phagocytic process.

Are there any correlations beween changes in the phosphoinisitide content and lipid raft assembly during phagocytosis? Lee at al., 2007 reported that there was an obvious clearance of the raft marker YFP-GPI from the base of forming phagosomes within minutes of particle contact and that this clearance resulted from the focal insertion of unlabeled endomembranes that are delivered focally by directed exocytosis. The authors interpreted these results as evidence against the raft involvement in the phagocytic process. However, these findings do not exclude the possibility that the raft formation preceded the observed changes in the phosphatidylinositide contents and possibly triggered these changes. The insertion of the new membranes should displace laterally and/or dilute the molecules and molecular ensembles (and rafts in particular) that initiated the phagocytic signaling cascade.

A number of works suggest close relations between rafts and phosphoinositides. For example, Hope & Pike, 1996 showed that polyphosphoinositide phosphatase, but not several other phosphoinositide-utilizing enzymes, is highly enriched in a low-density Triton-insoluble membrane fraction that contains caveolin; this fraction is also enriched in polyphosphoinositides, containing approximately one-fifth of the total cellular phosphatidylinositol $(4,5)P_2$. Defacque et al, 2002 suggested that $PI(4,5)P_2$ may exist in raft-like microdomains on latex bead phagosomes after isolation. On activation of cells with agonists or addition of ATP to the in vitro actin assay, PIPs are rapidly synthesized and may aggregate laterally into larger raft domains. The authors speculate that rafts may provide a platform for the proteins and lipids necessary for actin assembly to occur locally on the membrane of latex bead phagosomes.

Thus, the importance of phosphoinositides does not exclude the involvement of rafts in the phagocytosis mechanisms; in contrast, the dynamic functions of these lipid components of the plasma membrane appear to be highly coordinated in time and space throughout the course of the phagocytic process.

3. Cholesterol-binding sites in integral proteins involved in phagocytosis

Once phagocytosis is cholesterol-dependent and at least some of the phagocytic receptors aggregate in lipid rafts, it is important to know what structure(s) in these integral proteins makes them cholesterol-sensitive. Epand, 2006 reviewed the structural features of a protein that favour its association with cholesterol-rich domains. One of the best documented of these is certain types of lipidations; relatively new are the sterol-sensing domain (SSD) and the cholesterol recognition/interaction amino acid consensus (CRAC) domain. The latter was first described by Li & Papadopolous, 1998 for the peripheral-type benzodiazepine receptor (PBR). The CRAC sequence was formulated as -L/V-$(X)_{1\text{-}5}$-Y-$(X)_{1\text{-}5}$-R/K-; the presence of this site in C-terminus of PBR was necessary and sufficient for the cholesterol transport (Li & Papadopolous, 1998; Li et al., 2001). Moreover, the authors found a similar sequence in some other proteins known to interact with cholesterol (apolipoprotein A-I, some enzymes of steroid metabolism (e.g., P450scc (side-chain cleavage P450)), annexin, and some other proteins. Among them was caveolin - an integral protein accompanying caveolae, i.e., domains of plasma membrane containing lipid rafts and participating in caveolin-dependent endocytosis (Pelkmans et al., 2002).

A search for CRAC-like domains in the transmembrane areas of some integral proteins related to phagocytosis was performed by Cheshev et al., 2006. Apart from phagocytic receptor FcγR, the list of the molecules tested included proteins presumably involved in the regulation of the ionic composition in perimembrane cytoplasm during phagocytosis. Activation of phagocytes is accompanied by changes in the activity of the potassium channels of the IRK family (Eder, 1998; Arkett & Dixon, 1992; DeCoursey & Cherny, 1996; Colden-Stanfield, 2002). Ionotropic purinoreceptors P2X7 are found in monocytes/macrophages (North, 2002; Gudipatyet al., 2001); they cluster (Connon et al., 2003) and segregate in rafts (Torres et al., 1999). Gap-junction proteins are also known to accumulate in cholesterol-rich rafts (Schubert et al., 2002; Lin et al., 2003; Lin et al., 2004; Dunina-Barkovskaya, 2005). Connexin Cx43 is found in macrophages (Beyer & Steinberg, 1991; Beyer & Steinberg, 1993; Anand et al., 2008); colocalization of purinoreceptors P2X7 with connexin Cx43 in macrophages was reported (Beyer & Steinberg, 1991; Beyer & Steinberg, 1993; Fortes et al., 2004).

Figure 1 shows the fragments of the proteins aligned versus the CRAC sequence. Most of the proteins studied contained a conservative sequence of hydrophobic amino acids Val, Leu, Tyr, and Trp. This sequence can be described as follows: $(L/V)_{1\text{-}2}$-$(X)_{1\text{-}4}$-Y/F-$(X)_{1\text{-}3}$-W, which is close to the CRAC consensus (L/V-$(X)_{1\text{-}5}$-Y-$(X)_{1\text{-}5}$-R/K). In contrast to PBR that has cholesterol-binding consensus in a cytoplasmic C-terminus, in the integral proteins (Fig. 1) the consensus sequence was always localized in the transmembrane domain. It may account for a slight difference between the PBR cholesterol-binding consensus and the sequences found.

FcγRI (374/1)

```
          280       *       300         *         320         *         340
P12314 : LGLQLPTPVWFHVLFYLAVGIMFLVNTVLWVTIRKELKRKKKWDLEISLDSGHEKKVISS : 340
site   : ----------------------TVLNYYVW------------------------------ :
```

P2X₁ (399/2)

```
                *         20         *         40         *         60
P47824 : MARRLQDELSAFFFEYDTPRMVLVRNKKVGVIFRLIQLVVLVYVIGWVFVYEKGYQTSSD : 60
site   : -------------------------------------TVLNYYV-WR------------- :
```

P2X₂ (472/2)

```
                *         20         *         40         *         60
P49653 : MVRRLARGCWSAFWDYETPKVIVVRNRRLGFVHRMVQLLILLYFVWYVFIVQKSYQDSET : 60
site : : --------------------------------------TVLNYYVW-------------- :
```

P2X₇ (595/2)

```
                *         20         *         40         *         60
Q64663 : MPACCSWNDVFQYETNKVTRIQSVNYGTIKWILHMTVFSYVSFALMSDKLYQRKEPLISS : 60
site   : -----------------------------------TVLYYV-W----------------- :
```

IRK2, G-protein-dependent (427/2)

```
                *         80         *        100         *        120
P48049 : NVHHGNVQETYRYLSDLFTTLVDLKWRFNLLVFTMVYTVTWLFFGFIWWLIAYIRGDLDH : 120
site   : ---------------------------------TVLNYYVW------------------- :
```

MIP26 (263/6)

```
                *        200         *        220         *        240
P06624 : AGMNPARSFAPAILTRNFTNHWVYWVGPVIGAGLGSLLYDFLLFPRLKSVSERLSILKGS : 240
site   : ----------------TVLNYYVW------------------------------------ :
```

Cx32 (284/4)

```
                *         20         *         40         *         60
O18968 : MNWTGLYTLLSGVNRHSTAIGRVWLSVIFIFRIMVLVVAAESVWGDEKSSFICNTLQPGC : 60
site : : ------------------------------------TVLNYYVW---------------- :
```

Cx43 (382/4)

```
                *        140         *        160         *        180
P18246 : ANVDMHLKQIEIKKFKYGIEEHGKVKMRGGLLRTYIISILFKSVFEVAFLLIQWYIYGFS : 180
site : : -------------------------------------------------TVLNYYVW--- :
```

Fig. 1. Proposed cholesterol-binding sites in transmembrane domains of some proteins. Fragments of the proteins studied are highlighted with *grey colour*; a number at the left is the protein Uniport code (AC); numbers (and asterisks) above the fragments show the amino-acid residue numbers. Cholesterol-binding consensus (*TVLNYYVW*) is shown below the protein fragment (in the line marked "site"). Shown are the protein fragments (~60 amino-acid residues), in which the amino-acid sequence was comparable with the cholesterol-binding consensus. Identical or similar amino-acid residues are marked *black*. A square bracket below shows the position of transmembrane domains.

The presence in a transmembrane domain of the cholesterol-binding site that is potentially able to intract with cholesterol indicates that at least such interaction is possible. Binding of cholesterol at the level of the transmembrane domain of the receptor may not only account for segregation of the receptors in the lipid rafts at the early stages of phagocytosis (Kono et al., 2002; Kwiatkowska et al., 2003) but also favour an optimal FcγR configuration required for further interactions with kinases. As regards the channel proteins, binding of cholesterol may regulate the state of the ion-conducting pore. The channel molecules studied may be involved in the phagocytic process and their cholesterol-dependence may contribute to the cholesterol-dependence of the phagocytic process.

The presence of cholesterol-binding sequence in a number of integral proteins not only explains cholesterol-dependence of their functions. It seems very likely that experimental expression (and overexpression in particular) of integral proteins possessing cholesterol-binding sites may produce a cholesterol-sequestering effect in the transformed cells, similar to the effect of such cholesterol-sequestering agent as nystatin, methyl-β-cyclodextrin, etc. (Cheshev et al., 2006).

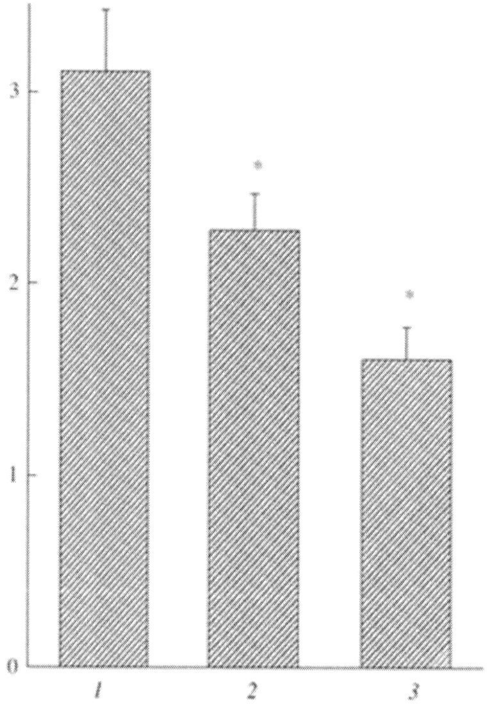

Fig. 2. Effect of cholesterol-binding peptide VLNYYVW on phagocytic activity of IC-21 macrophages (Dunina-Barkovskaya et al., 2007). Results of typical experiment are shown. *1*, Control; *2*, 1% DMSO; *3*, VLNYYVW (100 μg/ml) in DMSO (1%). *Ordinate*, mean number of beads per cell. *, $p < 0.05$ vs. control.

The effect of a cholesterol-binding peptide VLNYYVW corresponding to the consensus cholesterol-binding sequence (Fig.1) on phagocytic activity was tested on IC-21 macrophages (Dunina-Barkovskaya et al., 2007). Phagocytosis was assessed by fluorescent microscopy, using 2-µm non-opsonized fluorescent latex beads. The peptide wase dissolved in DMSO, which turned to affect phagocytic activity by itself: in the presence of 0.5–1.3% DMSO the number of beads per cell was lower by 20–30% than in the absence of DMSO. Peptide VLNYYVW (5–100 µg/ml) augmented the inhibitory action of DMSO (Fig. 2). This result suggests that cholesterol-binding peptide may indeed affect the phagocytic process. What is the mechanism of this effect remains to be determined. Peptides may interfere with the interactions between integral proteins and membrane cholesterol and thus hinder the segregation of the proteins in cholesterol-rich domains. Another possibility is a cholesterol-sequestering effect similar to that exerted by nystatin, methyl-β-cyclodextrin and analogous substances. The formation of non-functional rafts in the membranes, like what was observed in artificial systems by Epand et al., 2003 is also possible. The authors showed that peptide LWYIK, a fragment of cholesterol-binding sequence, induced raft formation in phosphatidylcholine–cholesterol bilayer membranes.

4. Conclusion

Phagocytosis is an important cellular process underlying the innate and acquired immunity and involved in tissue remodelling throughout development or repair. Phagocytosis is a multi-stage process that engages endocytosis, exocytosis, and adhesion mechanisms. Highly coordinated local and dynamic rearrangements of the membrane underneath the target particle result in its engulfment and intracellular processing. Phagocytosis is a cholesterol-dependent process. One of the reason of this cholesterol dependency is the formation of cholesterol-enriched domains in plasma membrane, where phagocytic receptors (FcγR, in particular) may cluster and form supramolecular complexes required to set off and perform a further cascade of biochemical reactions leading to the rearrangements of cytoskeleton and formation of a membrane vesicle containing the particle. A molecular basis for direct interaction between integral proteins and membrane cholesterol can be provided by a cholesterol recognition/interaction amino acid consensus (CRAC) -L/V-$(X)_{1-5}$-Y-$(X)_{1-5}$-R/K-, described for a cholesterol-binding site of the peripheral-type benzodiazepine receptor (Li & Papadopolous, 1998; Li et al., 2001). Alignment of this site with amino acid sequences of a phagocytic receptor FcγRI and some ionic channels that may be involved in the phagocytic process and/or are capable of clustering in rafts (e.g., purinoreceptors and connexins) revealed that most of the proteins studied possessed a relatively conservative hydrophobic amino-acid sequence (Val-Leu---Tyr---Trp) analogous to that in the PBR cholesterol-binding site. This sequence was always localized in a transmembrane domain of a protein (Cheshev et al., 2006). Functional activity of a cholesterol-binding peptide VLNYYVW was tested and confirmed on cultured macrophages IC-21. Cholesterol-binding peptides can thus be a useful tool for further investigations and possibly serve for correction of phagocytosis and other cholesterol-dependent processes.

5. Acknowledgments

I appreciate enthusiasm and support of my colleaugues Kh.S. Vishniakova and I.I.Kireev.

6. References

Aderem & Underhill, 1999 Aderem, A. & Underhill, D.M. (1999). Mechanisms of Phagocytosis in Macrophages, *Ann. Rev. Immunol.*, vol. 17, pp. 593–623.

Anand R.J., Dai S., Gribar S.C., Richardson W., Kohler J.W., Hoffman R.A., Branca M.F., Li J., Shi X.H., Sodhi C.P., Hackam D.J. (2008). A Role for Connexin43 in Macrophage Phagocytosis and Host Survival after Bacterial Peritoneal Infection, *J. Immunol.*, vol. 181 (12), pp. 8534–8543.

Araki, N., Johnson, M.T. & Swanson, J.A. (1996). A Role for Phosphoinositide 3-Kinase in the Completion of Macropinocytosis and Phagocytosis by Macrophages, *J. Cell Biol.*, vol. 135, pp. 1249–1260.

Arkett, S.A., Dixon, S.J. & Sims, S.M. (1992). Substrate Influences Rat Osteoclast Morphology and Expression of Potassium Conductances, *J. Physiol.*, vol. 458, pp. 633–653.

Astarie-Dequeker, C., N'Diaye, E.-N., Le Cabec, V., Rittig, M.G., Prandi, J., Maridonneau-Parini, I. (1999). *Infection and Immunity*, vol. 67, pp. 469–477.

Beyer, E.C. & Steinberg, T.H. (1991). Evidence That the Gap Junction Protein Connexin-43 Is the ATP-Induced Pore of Mouse Macrophages, *J. Biol. Chem.*, vol. 266 (13), pp. 7971-7974.

Beyer, E.C. & Steinberg, T.H. (1993). Connexins, Gap-Junction Proteins, and ATP-Induced Pores in Macrophages, *Progress in Cell Research,* vol. 3. Gap Junctions. Ed. Hall J.E., Zampighi G.A., Davis R.M. Amsterdam: Elsevier Sci. Publ., pp. 71–74.

Booth, J.W., Trimble, W.S. & Grinstein, S. (2001). Membrane Dynamics in Phagocytosis, *Seminars Immunol.*, vol. 13, pp. 357–364.

Botelho, R.J., M. Teruel, R. Dierckman, R. Anderson, A. Wells, J.D. York, T. Meyer, and S. Grinstein. (2000). Localized Biphasic Changes in Phosphatidylinositol-4,5-bisphosphate at sites of Phagocytosis, *J. Cell Biol.*, vol. 151, pp. 1353–1368.

Brown, D.A. & London, E. (2000). Structure and Function of Sphingolipid- and Cholesterol-Rich Membrane Rafts, *J. Biol. Chem.*, vol. 275, pp. 17221–17224.

Cheshev, D.A., Chekanov, N.N. & Dunina-Barkovskaya, A.Ya. (2006). Cholesterol1Binding Sites in Transmembrane Domains of Integral Membrane Proteins Involved in Phagocytosis and/or Capable of Clustering in Lipid Rafts, *Biologicheskie membrany* (Rus.), vol. 23 (1), pp. 69–73.

Colden-Stanfield, M. (2002). Clustering of Very Late Antigen-4 Integrins Modulates K^+ Currents to Alter Ca^{2+}-mediated Monocyte Function, *Am. J. Physiol. Cell Physiol.*, vol. 283 (3), pp. C990–C1000.

Connon, C.J., Young, R.D. & Kidd, E.J. (2003). P2X7 Receptors Are Redistributed on Human Monocytes after Pore Formation in Response to Prolonged Agonist Exposure, *Pharmacology*, vol. 67 (3), pp. 163–168.

DeCoursey, T.E. & Cherny, V.V. (1996). Voltage-Activated Proton Currents in Human THP-1 Monocytes, *J. Membr. Biol.*, vol. 152, pp. 131–140.

Defacque, H., Bos, E., Garvalov, B., Barret, C., Roy, Ch., Mangeat, P., Shin, Hye-Won, Rybin, V. & Griffiths, G. (2002). Phosphoinositides Regulate Membrane-dependent Actin Assembly by Latex Bead Phagosomes. *Molecular Biology of the Cell,* vol. 13, 1190–1202.

Devitt, A., Moffatt, O.D., Raykundalia, C., Capra, J.D., Simmons, D.L. & Gregory, C.D. (1998). Human CD14 Mediates Recognition and Phagocytosis of Apoptotic Cells, *Nature*, vol. 392, pp. 505–509.

Djaldetti, M., Salman, H., Bergman, M., Djaldetti, R. & Bessler, H. (2002). Phagocytosis--the Mighty Weapon of the Silent Warriors. *Microscopy Research and Technique*. vol. 57, pp. 421–431.

Dunina-Barkovskaya, A.Y. (2004). Phagocytosis - Three in One: Endocytosis, Exocytosis, and Adhesion, *Biologicheskie membrany* (Rus.), vol. 21 (4), pp. 245–272.

Dunina-Barkovskaya, A.Ya. (2005). Are Gap Junctions Protein–Lipid Rafts? *Biologicheskie membrany* (Rus.), vol. 22 (1), pp. 27–33.

Dunina-Barkovskaya, A.Ya., Vishniakova, Kh.S., Cheshev, D.A., Chekanov, N.N., Bujurina, I.M. (2007). Effects of DMSO and Cholesterol-binding Peptides on Phagocytic Activity of Cultured Macrophages IC-21, *Biologicheskie membrany* (Rus.), vol. 24 (6), pp. 451–456.

Eder, C. (1998). Ion channels in microglia, *J. Cell Biol.*, vol. 275, pp. C327–C342.

Epand, R.M. (2006). Cholesterol and the Interaction of Proteins with Membrane Domains, *Prog Lipid Res.*, vol. 45 (4), pp. 279–294.

Epand, R.M., Sayer, B.G. & Epand R.F. (2003). Peptide-induced Formation of Cholesterol-rich Domains, *Biochemistry*, vol. 42 (49), pp. 14677–14689.

Fadok, V.A., Voelker, D.R., Campbell, P.A., Cohen, J.J., Bratton, D.L. & Henson, P.M. (1992). Exposure of Phosphatidylserine on the Surface of Apoptotic Lymphocytes Triggers Specific Recognition and Removal by Macrophages, *J. Immunol.*, vol. 148, pp. 2207–2216.

Fairn, G.D., Ogata, K., Botelho, R.J., Stahl, Ph.D., Anderson, R.A., de Camilli, P., Meyer, T., Wodak, Sh. & Grinstein, S. (2010). An Electrostatic Switch Displaces Phosphatidylinositol Phosphate Kinases from the Membrane during Phagocytosis, *J. Cell Biol.*, vol. 187 (5), pp. 701–714.

Febbraio, M., Hajjar, D.P. & Silverstein, R.L. (2001). CD36: A Class B Scavenger Receptor Involved in Angiogenesis, Aterosclerosis, Inflammation and Lipid Metabolism, *J. Clin. Invest.*, vol. 108, pp. 785–791.

Fortes, F.S.A., Pecora, I.L., Persechini1, P.M., Hurtado, S., Costa, V., Coutinho,1S.R., Braga, M.B.M., Silva-Filho, F.C., Bisaggio, R.C., de Farias, F.P., Scemes, E., Campos de Carvalho, A.C. & Goldenberg, R.C.S. (2004). Modulation of Intercellular Communication in Macrophages: Possible Interactions between Gap Junctions and P2 Receptors, *J. Cell Sci.*, vol. 117, pp. 4717–4726.

Fuki, I.V., Meyer, M.E. & Williams K.J. (2000). Transmembrane and Cytoplasmic Domains of Syndecan Mediate a Multi-step Endocytic Pathway Involving Detergent-insoluble Membrane Rafts, *Biochem. J.*, vol. 351 (3), pp. 607–612.

Gal, A., Li, Y., Thompson, D.A., Weir, J., Orth, U., Jacobson, S.G., Apfelstedt-Sylla, E. & Vollrath, D. (2000). Mutations in MERTK, the Human Orthologue of the RCS Rat Retinal Dystrophy Gene, Cause Retinitis Pigmentosa, *Nat. Genet.*, vol. 26 (3), pp. 270–271.

Garin, J., Diez, R., Kieffer, S., Dermine, J.F., Duclos, S., Gagnon, E., Sadoul, R., Rondeau, C. & Desjardins, M. (2001). The Phagosome Proteome: Insight into Phagosome Functions, *J. Cell Biol.,* vol. 152, pp. 165–180.

Gatfield, J. & Pieters, J. (2000). Essential Role for Cholesterol in Entry of Mycobacteria into Macrophages, *Science,* vol. 288 (5471), pp. 1647–1650.

Giles, K.M., Hart, S.P., Haslett, C., Rossi, A.G. & Dransfield, I. (2000). An Appetite for Apoptotic Cells? Controversies and Challenges, *Br. J. Haematol.,* vol. 109, pp. 1–12.

Golovkina, M.S., Skachkov, I.V., Metelev, M.V., Kuzevanov, A.V., Vishniakova, Kh.S., Kireev, I.I. & Dunina-Barkovskaya, A.Ya. (2009). Serum-Induced Inhibition of the Phagocytic Activity of Cultured Macrophages IC-21. *Biologicheskie membrany* (Rus.), vol. 26 (5), pp. 379–386 [Translated version in: *Biochemistry (Moscow) Suppl. Series A: Membrane and Cell Biology* (2009), vol. 4 (3), pp. 412–419].

Gordon, S. (2008). Elie Metchnikoff: Father of Natural Immunity, *Eur. J. Immunol.,* vol. 38 (12), pp. 3257-3264.

Greenberg, S. (2001). Diversity in Phagocytic Signalling, *J. Cell Science,* vol. 114, pp. 1039–1040.

Greenberg, S. (1995). Signal Transduction of Phagocytosis, *Trend Cell Biol.,* vol. 5, pp. 93–99.

Greenberg, S. (1999). Modular Components of Phagocytosis, *J. Leukoc. Biol.,* vol. 66, pp. 712–717.

Grinnell, F. (1984). Fibroblast Spreading and Phagocytosis: Similar Cell Responses to Different-sized Substrata, *J. Cell Physiol.,* vol. 119, pp. 58–64.

Grinstein, S. (2010) Imaging Signal Transduction during Phagocytosis: Phospholipids, Surface Charge, and Electrostatic Interactions, *Am. J. Physiol. Cell Physiol.,* vol. 299 (5), C876–C881.

Gudipaty, L., Humphreys, B.D., Buel, G. & Dubyak, G.R. (2001). Regulation of P2X 7 nucleotide receptor function in human monocytes by extracellular ions and receptor density, *Am. J. Physiol. Cell Physiol.,* vol. 280, pp. C943–C953.

Hackam, D.J., Rotstein, O.D., Sjolin, C., Schreiber, A.D., Trible, W.S. & Grinstein, S. (1998). v-SNARE-Dependent Secretion is Required for Phagocytosis, *Proc. Natl. Acad. Sci. USA.,* vol. 95, pp. 11691–11696.

Han, J., Hajjar, D.P., Febbraio, M. & Nicholson, A.C. (1997). Native and Modified Low Density Lipoproteins Increase the Functional Expression of the Macrophage Class B Scavenger Receptor, CD36, *J. Biol. Chem.,* vol. 272, pp. 21654–21659.

Han, J., Hajjar, D.P., Tauras, J.M. & Nicholson, A.C. (1999). Cellular Cholesterol Regulates Expression of the Macrophage Type B Scavenger Receptor, CD36, *J. Lipid. Res.,* vol. 40, pp. 830–838.

Heifets, L. (1982). Centennial of Metchnikoff's Discovery, *J.Reticuloendothel. Soc.,* vol. 31 (5), pp. 381–391.

Henry, R.M., Hoppe, A.D., Joshi, N. & Swanson, J.A. (2004). The Uniformity of Phagosome Maturation in Macrophages, J. Cell Biol., vol. 164 (2), pp. 185–194.

Hope, H.R. & Pike, L.J. (1996). Phosphoinositides and Phosphoinositide-utilizing Enzymes in Detergent-insoluble Lipid Domains, *Mol. Biol. Cell,* vol. 7 (6), pp. 843–851.

Huang, Z.Y., Barreda, D.R., Worth, R.G., Indik, Z.K., Kim, M.K., Chien, P. & Schreiber, A.D. (2006). Differential Kinase Requirements in Human and Mouse Fcγ Receptor Phagocytosis and Endocytosis, *J. Leukoc. Biol.*, vol. 80 (6), pp. 1553–1562.

James, T.N. (1994). Normal and Abnormal Consequences of Apoptosis in the Human Heart from Postnatal Morphogenesis to Paroxysmal Arrhythmias, *Circulation*, vol. 90, pp. 556–573.

Kato, T., Yashiro, T., Murata, Y., Herbert, D.C., Oshikawa, K., Bando, M., Ohno, S. & Sugiyama, Y. (2003). Evidence that Exogenous Substances Can Be Phagocytized by Alveolar Epithelial Cells and Transported into Blood Capillaries, *Cell Tissue Res.*, vol. 311, pp. 47–51.

Katsumata, O., Hara-Yokoyama, M., Sautès-Fridman, C., Nagatsuka, Y., Katada, T., Hirabayashi, Y., Shimizu, K., Fujita-Yoshigaki, J., Sugiya, H. & Furuyama, S. (2001). Association of FcγRII with Low-density Detergent-resistant Membranes Is Important for Cross-linking-dependent Initiation of the Tyrosine Phosphorylation Pathway and Superoxide Generation, *J. Immunol.*, vol. 167 (10), pp. 5814-5823.

Knowles, G.C., McKeown, M., Sodek, J. & McCulloch, C.A. (1991). Mechanism of Collagen Phagocytosis by Human Gingival Fibroblasts: Importance of Collagen Structure in Cell Recognition and Internalization, *J. Cell Sci.*, vol. 98, pp. 551–558.

Kono, H., Suzuki, T., Yamamoto, K., Okada, M., Yamamoto, T. & Honda, Z. (2002). Spatial Raft Coalescence Represents an Initial Step in FcγR Signaling, *J.Immunol.*, vol. 169, pp. 193–203.

Koval, M., Preiterb, K., Adles, C., Stahl, P.D. & Steinberg, T.H. (1998). Size of IgG-opsonized Particles Determines Macrophage Response during Iternalization, *Exp. Cell Res.*, vol. 242, pp. 265–273.

Krigel, A., Felder-Schmittbuhl, M.P. & Hicks, D. (2010). Circadian-Clock Driven Cone-Like Photoreceptor Phagocytosis in the Neural Retina Leucine Zipper Gene Knockout Mouse, *Mol. Vis.*, vol. 16, pp. 2873–2881.

Kwiatkowska, K. & Sobota, A. (2001). The Clustered Fcγ Receptor II Is Recruited to Lyn-containing Membrane Domains and Undergoes Phosphorylation in a Cholesterol-dependent Manner, *Eur. J. Immunol.*, vol. 31 (4), pp. 989–998.

Kwiatkowska, K., Frey, J. & Sobota A. (2003). Phosphorylation of FcγRIIA Is Required for the Receptor-induced Actin Rearrangement and Capping: The Role of Membrane Rafts, *J. Cell Sci.*, vol. 116 (Pt 3), pp. 537–550.

Lee, W.L., Mason, D., Schreiber, A.D. & Grinstein S. (2007). Quantitative Analysis of Membrane Remodeling at the Phagocytic Cup. Molecular Biology of the Cell. Vol. 18, 2883–2892

Lekstrom-Himes, J.A. & Gallin, J.I. (2000). Immunodeficiency Diseases Caused by Defects in Phagocytes, *N. Engl. J. Med.*, vol. 343 (23), pp. 1703–1714.

Li, H. & Papadopoulos, V. (1998). Peripheral-type Benzodiazepine Receptor Function in Cholesterol Transport. Identification of a Putative Cholesterol Recognition/Interaction Amino Acid Sequence and Consensus Pattern, *Endocrinology*, vol. 139 (12), pp. 4991–4997.

Li, H., Yao, Z., Degenhardt, B., Teper, G. & Papadopoulos, V. (2001). Cholesterol Binding at the Cholesterol Recognition/Interaction Amino Acid Consensus (CRAC) of the

Peripheral-Type Benzodiazepine Receptor and Inhibition of Steroidogenesis by an HIV TAT-CRAC Peptide, *Proc. Natl. Acad. Sci. USA,* vol. 98, pp. 1267-1272.

Li, M.O., Sarkisian, M.R., Mehal, W.Z., Rakic, P. & Flavell R.A. (2003). Phosphatidylserine Receptor Is Required for Clearance of Apoptotic Cells, *Science,* vol. 302, pp. 1560-1563.

Lin, D., Lobell, S., Jewell, A. & Takemoto, D.J. (2004). Differential Phosphorylation of Connexin 46 and Connexin 50 by H_2O_2 Activation of Protein Kinase Cγ, *Mol. Vis.,* vol. 10, pp. 688-695.

Lin, D., Zhou, J., Zelenka, P.S. & Takemoto, D.J. (2003). Protein Kinase Cγ Regulation of Gap Junction Activity through Caveolin-1-containing Lipid Rafts, *Invest. Ophthalmol. Vis. Sci.,* vol. 44 (12), pp. 5259-5268.

Lin, E.Y,, Li, J.F., Gnatovskiy, L., Deng, Y., Zhu, L., Grzesik, D.A., Qian, H., Xue, X.N. & Pollard, J.W. (2006). Macrophages Regulate the Angiogenic Switch in a Mouse Model of Breast Cancer, *Cancer Res.,* vol. 66 (23), pp. 11238-11246.

Lingwood, D. & Simons, K. (2010). Lipid Rafts As a Membrane-Organizing Principle, *Science,* vol. 327, pp. 46-50.

Lucas, A.D. & Greaves, D.R. (2001). Atherosclerosis: Role of Chemokines and Macrophages, *Expert. Rev. Mol. Med.,* vol. 3 (25), pp. 1-18.

Marois, L., Paré, G., Vaillancourt, M., Rollet-Labelle, E. & Naccache, P.H. (2011) FcγRIIIb Triggers Raft-dependent Calcium Influx in IgG-mediated Responses in Human Neutrophils, *J. Biol. Chem.,* vol. 286 (5), pp. 3509-3519.

May, R.C. & Machesky, L.M. (2001). Phagocytosis and the Actin Cytoskeleton, *J. Cell Sci.,* vol. 114, pp. 1061--1077.

Morrissette, N.S., Gold, E.S., Guo, J., Hamermann, J.A., Ozinsky, A., Bedian, V. & Aderem, A.A. (1999). Isolation and Characterization of Monoclonal Antibodies Directed against Novel Components of Macrophage Phagosomes, *J. Cell Science,* vol. 119, pp. 4705-4713.

Mosser, D.M. (1994). Receptors on Phagocytic Cells Involved in Microbial Recognition, *Immunol. Ser.,* vol. 60, pp. 99-114.

North, R.A. (2002). Molecular Physiology of P2X Receptors, *Physiol. Rev.,* vol. 82 (4), pp. 1013-1067.

Ofek, I., Goldhar, J. & Sharon, N. (1995). Nonopsonic Phagocytosis of Microorganisms, *Ann. Rev. Microbiol.,* vol. 49, pp. 239-276.

Pelkmans, L., Puentener, D. & Helenius, A. (2002). Local Actin Polymerization and Dynamin Recruitment in SV40-induced internalization of Caveolae, *Science,* vol. 296, pp. 535-539.

Peyron, P., Bordier, C., N'Diaye, E.-N. & Maridonneau-Parini, I. (2000). Nonopsonic Phagocytosis of *Mycobacterium kansasii* by Human Neutrophils Depends on Cholesterol and Is Mediated by CR3 Associated with Glycosylphosphatidylinositol-Anchored Proteins, *J. Immunol.,* vol. 165, pp. 5186-5191.

Pike, L.J. (2003). Lipid Rafts: Bringing Order to Chaos, *Journal of Lipid Research,* vol. 44 (4), pp. 655-667.

Pradhan, D., Krahling, S., Williamson, P. & Schlegel, R.A. (1997). Multiple Systems for Recognition of Apoptotic Lymphocytes by Macrophages, *Mol.Biol. Cell*, vol.8, pp. 767–778.

Pratten, M.K. & Lloyd, J.B. (1986). Pinocytosis and Phagocytosis: The Effect of Size of a Particulate Substrate on Its Mode of Capture by Rat Peritoneal Macrophages Cultured In Vitro, *Biochim. et biophys. Acta,* vol. 881, pp. 307–313.

Rabinovitch, M. (1995). Professional and Nonprofessional Phagocytes: An Introduction, *Trends Cell Biol.*, vol. 5, pp. 85–87.

Ross, G.D. & Větvicka, V. (1993). CR3 (CD11b, CD18): A Phagocyte and NK Cell Membrane Receptor with Multiple Ligand Specificities and Functions, *Clin. Exp. Immunol.*, vol. 92 (2), pp.181–184.

Saxena, R.K., Gilmour, M.I. & Hays, M.D. (2008). Isolation and Quantitative Estimation of Diesel Exhaust and Carbon Black Particles Ingested by Lung Epithelial Cells and Alveolar Macrophages In Vitro, *Biotechniques*, vol. 44 (6), pp. 799–805.

Schreiber, A.D., Parsons, J., Mcdermot, P. & Cooper R.A. (1975) Effect of Corticosteroids on the Human Monocyte IgG and Complement Receptors, J. Clin. Invest. Vol. 56, pp. 1189–1197.

Schubert, A.L, Schubert, W., Spray, D.C. & Lisanti, M.P. (2002). Connexin Family Members Target to Lipid Raft Domains and Interact with Caveolin-1, *Biochemistry*, vol. 41 (18), pp. 5754--5764.

Simons, K. & Ikonen, E. (1997). Functional Rafts in Cell Membranes, *Nature*, vol. 387, pp. 569–572.

Takahashi, K., Takeya, M. & Sakashita, N. (2002). Multifunctional Roles of Macrophages in the Development and Progression of Atherosclerosis in Humans and Experimental Animals, *Med. Electron Microsc.*, vol. 35 (4), pp. 179–203.

Tidball, J.G. & Wehling-Henricks, M. (2007). Macrophages Promote Muscle Membrane Repair and Muscle Fibre Growth and Regeneration during Modified Muscle Loading in Mice In Vivo, *J. Physiol.*, vol. 578 (Pt 1), pp. 327–336.

Tjelle, T.E., Levdal, T. & Berg, T. (2000). Phagosome Dynamics and Function, *BioEssays*, vol. 22, pp. 255 – 263.

Torres, G.E., Egan, T.M. & Voigt, M.M. (1999). Hetero-oligomeric Assembly of P2X Receptor Subunits, *J. Biol. Chem.*, vol. 274, pp. 6653–6659.

Turner, R.A., Schumacher H. R. & Myers A.R. (1973). Phagocytic Function of Polymorphonuclear Leukocytes in Rheumatic Diseases, *J. Clin. Invest.*, vol. 52 (7), 1632–1635.

Vandivier, R.W., Fadok, V.A., Hoffman, P.R., Bratton, D.L., Penvari, C., Brown, K.K., Brain, J.D., Accurso, F.J. & Henson P.M. (2002). Elastase-Mediated Phosphatidylserine Receptor Cleavage Impairs Apoptotic Cell Clearance in Cystic Fibrosis and Bronchiectasis, *J. Clin. Invest.*, vol. 109, pp. 661–670.

Vishniakova Kh. S., Kireev I. I. & Dunina-Barkovskaya A.Ya. (2011). Effects of Cell Culture Density on Phagocytosis Parameters in IC-21 Macrophages, *Biochemistry (Moscow) Supplement Series A: Membrane and Cell Biology*, vol. 5 (4), pp. 355–363 [Original Russian Text in: *Biologicheskie membrany* (Rus.), vol. 28 (5), pp. 387–396].

Werb, Z. & Cohn, Z.A. (1972). Plasma Membrane Synthesis in the Macrophage Following Phagocytosis of Polystyrene Latex Particles, *J. Biol. Chem.*, vol. 247, pp. 2439–2446.

Yeung, T., Heit, B., Dubuisson, J.-F., Fairn, G.D., Chiu, B., Inman R., Kapus, A., Swanson, M., & Grinstein, S. (2006). Contribution of Phosphatidylserine to Membrane Surface Charge and Protein Targeting during Phagosome Maturation, *J. Cell Biol.*, vol. 185 (5), pp. 917–928.

Young J.D., Unkeless J.C., Cohn Z.A. (1985). Functional Ion Channel Formation by Mouse Macrophage IgG Fc Receptor Triggered by Specific Ligands. J. Cell Biochem., vol. 29, pp. 289–297.

Young J.D., Unkeless J.C., Young T.M., Mauro A., Cohn Z.A. (1983). Role for Mouse Macrophage IgG Fc Receptor as Ligand-Dependent Ion Channel, *Nature,* vol. 306 (5939), pp. 186–189.

Part 2

Studying Protein Interactions

15

One-by-One Sample Preparation Method for Protein Network Analysis

Shun-Ichiro Iemura and Tohru Natsume
Biomedicinal Information Research Center (BIRC),
National Institute of Advanced Industrial Science and Technology (AIST)
Japan

1. Introduction

Proteomics is the large-scale study of an organism's complete complement of proteins, and its relevant technologies have matured over recent years. Along with the development of mass spectrometry (MS), MS-based proteomics has emerged as an invaluable tool for large-scale identification and quantification of protein networks (Aebersold & Mann, 2003; Domon & Aebersold, 2006). Proteomic data is important for a wide range of research in basic and medical biology. In recent years, many large-scale projects have been performed and a huge amount of data has accumulated. However, because the data sets from individual projects often vary in quality, the value of proteomics for the wider scientific community is limited (Olsen & Mann, 2011).

One of the causes of this variation in proteomic data quality is thought to be the manual process of large-scale sample preparation. The sample preparation process for proteomic analysis consists of the several complicated steps. For example, sample preparation for protein interaction analysis using mammalian cells expressing a target protein typically requires 1×10^7-10^8 cells (one 10-cm or 15-cm tissue culture dish) (Blagoev et al., 2003; Burckstummer et al., 2006; Ewing et al., 2007). After cell recovery, steps such as cell lysis, purification of protein complexes, denaturation and modification of proteins, separation by gel electrophoresis, and enzymatic digestion are performed sequentially. In fact, many researchers and technicians are involved in laborious, repetitive work of large-scale sample preparation, in which they must handle tens of culture dishes at a time. In such a 'parallel sample preparation' process, during the preparation of a number of samples, the conditions undoubtedly differ between the first and last treated samples. Denaturation of the component proteins of complexes and proteolysis progress over time, and the denatured proteins are thought to be the cause of nonspecific binding. We came to realize that highly sensitive analysis could not be performed using the prevailing parallel sample preparation methods.

To optimize sample preparation conditions and improve sample quality, we considered that a 'one-by-one sample preparation' method would be useful. One-by-one sample preparation is the concept that one sample is finished at a time, followed by preparation of the next sample (Fig. 1). In this way, each sample can be prepared carefully under almost equal conditions; however, this method is not realistic for large-scale analysis, because of the large amount of human time and work involved.

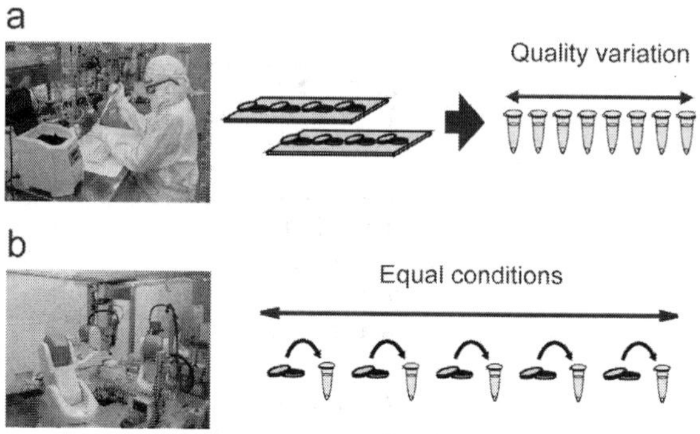

Fig. 1. Comparison of sample preparation processes. (a) Parallel preparation by the manual method. The quality of the samples was uneven. (b) One-by-one preparation. This method enables the preparation of samples under the same conditions.

To realize the one-by-one concept and perform a pilot feasibility study, a fully automated sample preparation system is required. However, in the proteomics field, partial automation for parallel preparation is usually only applied to save analysis time, to eliminate sample contamination, and to reduce human error (Alterovitz et al., 2006). Several semi-automated robots that are specialized in certain processes are commercially available, such as liquid dispenser robots, cell culture robots, and electrophoresis gel cutting robots. However, to develop a fully automated and highly precise system for sample preparation using commercial robots would be difficult, because these robots do not meet our specifications, or if they do, the integration of the robots from different vendors may prove difficult. Furthermore, robots for other multiple sample preparation processes have not yet been developed. To achieve a significant breakthrough, we need a versatile robotic system. Recently, high-performance and reliable multi-axis articulated vertical robots have been developed, and are used in various fields, such as the motor industry. The motion of these industrial robots is fast, precise, and flexible. Moreover, these robots are relatively easy to integrate with other robots and equipment. Although the robotic system requires considerable effort and patience to set up (Blow, 2008), once one of the designated conditions is determined, it becomes applicable in many other situations.

In this chapter, we assess the one-by-one sample preparation method compared with parallel preparation in protein network analysis, using an automated sample preparation system for liquid chromatography-tandem mass spectrometry (LC-MS/MS). This automated system is compatible with the single-step affinity purification technique using the Flag-tag system (Einhauer & Jungbauer, 2001), without sodium dodecylsulfate polyacrylamide gel electrophoresis (SDS-PAGE) separation. Affinity-purification is a technique for purification of physiological protein complexes using target proteins (bait proteins) fused with affinity tags, such as short epitope peptides (e.g., Flag and Myc) or tandem-affinity purification (TAP) tags (Kocher & Superti-Furga, 2007). The bait proteins are overexpressed in cells and are separated, together with the protein complexes, using affinity beads that bind to the tags. Finally, all component proteins are identified by LC-MS/MS. Using this system, we tested two Wnt signaling pathway (Rao & Kuhl, 2010) proteins, β-catenin and Axin1, as baits, and demonstrated that the one-by-one purification method using this system is highly sensitive and reproducible compared with the manual parallel purification method. The results indicate that gentle and equal preparation conditions are important for generating reliable data for large-scale protein-protein interaction network and for quantitative analysis.

2. Experimental procedures

2.1 Design and development of a robotic system for one-by-one sample preparation

The robotic system was manufactured using four 6-axis robots, FC03N (Kawasaki Heavy Industries, Ltd., Hyogo, Japan) and a 3-axis robot comprising three single-axis robots (IAI corporation, Shizuoka, Japan), with help from the Japan Support System, Co., Ltd. (Ibaraki, Japan) and Nikkyo Technos, Co., Ltd. (Tokyo, Japan). In low femtomole level analysis, the key to obtaining reliable data quickly is to minimize contaminants, such as chemicals, airborne particles, and keratin proteins. Chemicals cause background noise, which limit the sensitivity of MS by decreasing the signal to noise ratio (S/N). Airborne particles, including dust, cause the blockage of the flow path and the nano LC column. Keratin proteins also cause background noise, which disturbs the detection of low abundance of proteins. Therefore, because we needed to perform sample preparation in a super clean room, our automated robotic system was designed for clean room specification (ISO class 4).

2.2 Immobilization of Anti-Flag antibodies to magnetic beads

Anti-Flag M2 antibodies (Sigma-Aldrich, St. Louis, MO) were immobilized via covalent binding of the primary amine group with 1-Ethyl-3-[3-dimethylaminopropyl] carbodiimide hydrochloride (EDC; Thermo Fisher Scientific, Waltham, MA)–modified Magnosphere MS300 magnetic beads (JSR, Tokyo, Japan). The beads (10 mg) suspension was transferred into a 1.5 ml-microtube. The beads were washed twice with 1 ml of activation buffer (0.1 M 2-[*N*-morpholino]ethane sulfonic acid (MES), pH 6.0, 0.5 M NaCl) and were resuspended in 1 ml of activation buffer. EDC and *N*-hydroxysulfosuccinimide (Sulfo-NHS; Thermo Fisher Scientific, Waltham, MA) were then added to the beads suspension. The final concentrations of EDC and sulfo-NHS were 2 and 5 mM, respectively. The mixture was incubated for 15 min at room temperature (RT), placed on the magnet, and the supernatant was discarded. The antibody (100 μg/ml) in conjugation buffer (50 mM sodium phosphate, pH 7.4, 0.15 M NaCl) was added to the beads and the mixture was incubated for 3 hr at 4 °C. After incubation, the supernatant was discarded and quenching buffer (20 mM HEPES-NaOH, pH 7.5, 0.15 M NaCl, 50 mM ethanolamine) was added. After quenching for 2 hr at 4 °C, the beads were washed three times with 1 ml of washing buffer (50 mM Tris-HCl, pH 8.0, 0.5 M NaCl, 0.1% Triton X-100) and twice with storage buffer (20 mM HEPES-NaOH, pH 7.5, 0.15 M NaCl, 0.5% digitonin). The antibody-immobilized beads were stored in 1 ml of storage buffer at 4 °C.

2.3 Cell culture and transfection

HEK293T cells (approximately 5.0×10^6 cells per 10-cm dish) were seeded in Dulbecco's modified Eagle's medium (DMEM; Invitrogen, San Diego, CA) containing 10% heat-inactivated fetal bovine serum (FBS; Invitrogen) the day before transfection. The cells were transfected with human β-catenin or human Axin1 cDNA, using Lipofectamine 2000 (Invitrogen) according to the manufacturer's protocol. The cells were collected 24 h after transfection.

2.4 Cell collection and lysis

The culture medium was discarded from the 10-cm dish, and the HEK293T cells expressing a bait protein were scraped into 1 ml of cold phosphate buffered saline (PBS) and transferred into a 1.5 ml-microtube. After centrifugation at low speed (3,000 rpm) for 1 min at 4 °C, the supernatant was discarded, and 1.0 ml of lysis buffer (20 mM HEPES, pH 7.5, 150 mM NaCl, 50 mM NaF, 1 mM Na_3VO_4, 0.5% digitonin, 1 mM $MgCl_2$, 1 mM PMSF, 5 μg/ml leupeptin, 5 μg/ml aprotinin and 3 μg/ml pepstatin A) was added. The cells were lysed by gently mixing for a short time with a vortex mixer (parallel method) or with a pipette tip (one-by-one method). In this step, we chose the vortexing in the parallel method because we thought, in reality, this way had to be adopted in large-scale sample treatment. The lysate was centrifuged at high speed (15,000 rpm) for 10 min at 4 °C, and the cleared lysate was transferred into a microtube containing the anti-Flag antibody immobilized magnetic beads.

2.5 Immunoprecipitation

The supernatant was incubated with the magnetic beads at 4 °C for 10 min with a rotator (parallel method) or the 6-axis robot (one-by-one method; 10 times mixing → interval: 4 min at 4 °C → 10 times mixing → interval: 4 min at 4 °C). After incubation, the beads were washed twice with 1 ml of wash buffer (10 mM HEPES, pH 7.5, 150 mM NaCl, 0.1% Triton X-100). The protein complexes containing the bait protein were then mixed with 100 μl of Flag peptide (0.5 mg/ml, SIGMA) in wash buffer for 5 min at 4 °C using a mixer (parallel method) or a 'protein complexes elution device' (Fig. 2a) (one-by-one method). The eluted fraction was transferred to a new microtube.

2.6 Limited proteolysis with lysyl endopeptidase C (Lys-C)

To concentrate the purified proteins and to exchange the buffer, trichloroacetic acid (TCA) precipitation was performed. Sodium deoxycholate (DOC) was added to a final concentration of 0.1%. After mixing, TCA was added to a final 10% concentration and the solution was precipitated at 0 °C for 30 min. The protein precipitate was collected by centrifugation (15,000 rpm for 10 min at 4 °C). The supernatant was carefully removed, 1 ml of acetone (precooled at -30 °C) was added to the pellet, and vortexing was carried out until the pellet became unstuck from the bottom of the tube. The proteins were collected by centrifugation (15,000 rpm for 5 min at 4 °C) and the supernatant was removed. The pellet was redissolved in 10 μl extraction buffer (0.1 M Tris-HCl, pH 8.8, 0.05% n-octyl glucopyranoside, 7M guanidine hydrochloride) using the microtube mixer. After the proteins were dissolved almost completely, 40 μl of digestion buffer (0.1 M Tris-HCl, pH 8.8,

0.05% n-octyl glucopyranoside) was added and mixed. Finally, 0.1 µg of lysyl endopeptidase (Lys-C; Wako, Osaka, Japan) was added and the mixture was incubated over night at 37 °C.

2.7 Western blotting

HEK293T cells were transfected with human β-catenin or human Axin1 cDNA, or pcDNA3 vector (as a negative control) as described in section 2.3. The purified proteins (from the immunoprecipitation step, section 2.5) were separated by electrophoresis on 10% SDS-PAGE and transferred onto Polyvinylidene difluoride (PVDF) membranes. The membranes were blocked with 2% BSA in TBS-T for 1 h at RT, followed by incubation with each primary antibody for 1 h at RT. After incubation with the secondary antibody for 1 h at RT, protein bands were detected with an ECL detection kit.

2.8 Direct nanoflow liquid chromatography tandem mass spectrometry system (DNLC-MS/MS)

All samples were diluted 10-fold with 0.1% formic acid and analyzed (2 µl) by DNLC system (Natsume et al., 2002) coupled to a QSTAR XL (AB Sciex, Foster City, CA). Peptides were separated on a C18 reversed-phase column packed with Mightysil C18 (particle size 3 µm; Kanto Chemical, Tokyo, Japan) at a flow rate of 100 nl/min by a 40-min linear gradient from 5% to 40% acetonitrile in 0.1% formic acid, and were sprayed on-line to the mass spectrometer. MS and MS/MS spectra were obtained in an Information Dependent Acquisition (IDA) mode. Up to two precursor ions above the intensity threshold of 50 counts with a charge state from 2 to 3 were selected for MS/MS analyses (1.0 sec) from each survey scan (0.5 sec). The MS and MS/MS scan ranges were m/z 400-1500 and 100-1500, respectively.

2.9 Data analysis

Peak lists were created by scripts of Analyst QS 1.1 Software (AB Sciex) using the following parameters: 0.1 amu Mass tolerance for combining MS/MS spectra, 2 cps MS/MS export threshold, 5 Minimum number of MS/MS ions for export, 50% Centroid height percentage, and 0.05 amu Centroid merge distance. All MS/MS spectra were queried against the National Center for Biotechnology Information (NCBI) non-redundant database (human; January 25, 2011; 137,349 sequences) using an in-house Mascot server (version 2.2.1; Matrix Science, London, UK). Search parameters were as follows: MS and MS/MS tolerance of 250 ppm and 0.5 Da, respectively; enzymatic specificity allowing for 1 missed cleavage site and K cleavage (enzyme: Lys-C/P); no fixed modification; and variable modification of N-acetyl (protein N terminus) and phosphorylations (Ser, Thr, and Tyr). Proteins that were identified by two or more peptides with a peptide expectation value of $p < 0.05$ were considered as reliable identifications.

3. Results

3.1 Automated robotic system for one-by-one sample preparation

To perform precise one-by-one sample purification for protein network analysis, we designed and developed a robotic system for fully automated sample preparation from cell collection to limited proteolysis with Lys-C. This system consists of four 6-axis industrial robots, one 3-axis robot, high- and low-speed centrifuges, a CO_2 incubator, and other components, as illustrated in detail in Fig. 2.

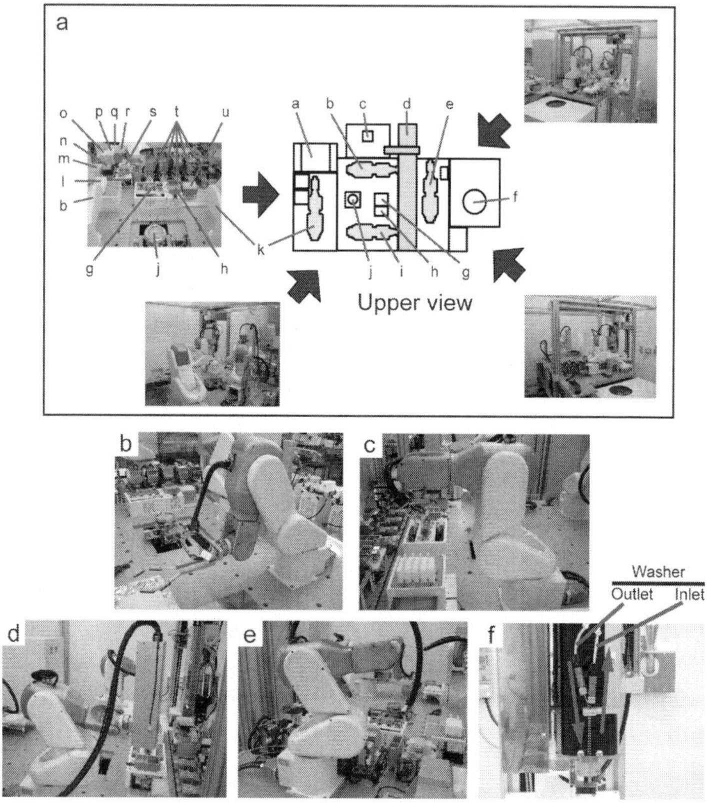

Fig. 2. Layout of the automated robotics for one-by-one sample preparation system.
(a) A schematic upper view diagram of the system and four photographs showing different views indicated by arrows. a: CO_2 incubator; b: 6-axis robot No. 2; c: low-speed centrifuge; d: 3-axis robot; e: 6-axis robot No. 4; f: high-speed centrifuge; g: microtube carriers for low-speed centrifuge; h: buffers position (lysis buffer and phosphate-buffered saline); i: 6-axis robot No. 3; j: culture dish stage; k: 6-axis robot No. 1; l: cell scrapers specialized for this system; m: pipette tips (2-200 μl); n: pipette tips (0.1-10 μl); o: protein complexes elution device; p: incubator (4 °C); q: microtube rack; r: incubator (37 °C); s: reagents rack (elution buffer, TCA, etc.); t: microtube capper/decapper (temperature-controlled); u: pipette tip (200-1,000 μl). (b) 6-axis robot No. 1: culture dish-carrying robot. (c) 6-axis robot No. 2: scraping and tube-carrying robot. (d) 6-axis robot No. 3: dispenser robot. (e) 6-axis robot No. 4: microtube-carrying robot. (f) 3-axis robot: micro-dispenser robot. A washer is attached to this robot.

The features of this system are: (i) The system is optimized for sample preparation from 10-cm culture dishes, and the process operates under gentle conditions to decrease protein denaturation and degradation compared to manual operation. The scraping robot (6-axis robot No. 2) can collect cells gently in a single scraping motion (Fig. 3a and 3b). In addition, a microtube delivery robot (6-axis robot No. 4) can mix the magnetic beads immobilized on the anti-Flag M2 antibody with cell extracts at intervals that will not over-mix or create a foam. Moreover, the elution of the protein complexes in the 'protein complexes elution device' (Fig. 2a) is performed by moving the beads backwards and forwards in the elution buffer between two magnets (Fig. 3c-e). The solution is not mixed vigorously; therefore, this procedure is expected to prevent the denaturation of the eluted protein. (ii) This system allows rapid purification of the protein complexes. One sample, from cell scraping to elution of protein complexes, can be prepared in 40 min. The manual parallel treatment of 20 samples takes more than 120 min. (iii) The one-by-one system can operate 24 hours a day, automatically, generating approximately 500 samples per month.

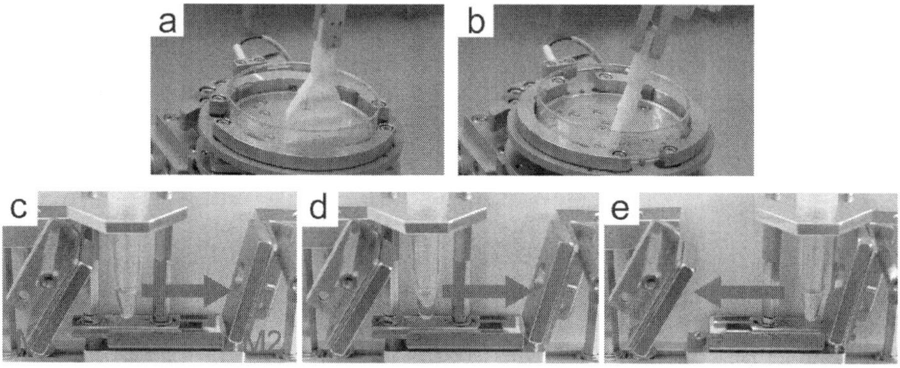

Fig. 3. Automated one-by-one sample preparation system. (a and b) Cell collection on the dish stage. (c-e) Process for elution of the protein complexes in the 'protein complexes elution device' (Fig. 1a). M1 and M2: magnets.

3.2 Comparison of parallel and one-by-one methods for the sample preparation by western blot analysis

To evaluate one-by-one sample preparation, we chose β-catenin and Axin1 as bait proteins because they are well-studied proteins that play key roles in the Wnt signaling pathway, and because, to date, many partners that interact with them have been identified (Daugherty & Gottardi, 2007; H. Huang & He, 2008; S.M. Huang et al., 2009). Furthermore, it is difficult to analyze β-catenin and Axin1-interacting proteins using affinity purification and LC-MS/MS, because these bait proteins are likely to be degraded, not only by the ubiquitin-proteasome system, but also nonspecifically by various proteases during the purification steps, even if protease inhibitors are added. Therefore, we expected that the gentle one-by-one purification method would allow these proteins to remain intact to the greatest extent possible, and would permit the identification of more interacting partner proteins.

We first compared the bait proteins (β-catenin and Axin1) from parallel preparation with those of one-by-one preparation. Flag-tagged β-catenin or Axin1 was expressed in HEK293T cells, purified by the parallel and the one-by-one method, and analyzed by western blotting (Fig. 4). In the case of parallel preparation, both β-catenin and Axin1 were found to be degraded. In particular, Axin1 degradation tended to be fast, and a protein band of approximately 120 kDa, corresponding to the intact form, was almost absent in some cases. On the other hand, in samples prepared by the one-by-one method, degradation of the bait proteins was significantly reduced. Interestingly, for Axin1, only one prominent band of the size of the intact protein was detected in most cases. These data indicated that the one-by-one method minimizes protein denaturation and degradation during sample preparation compared to the parallel method.

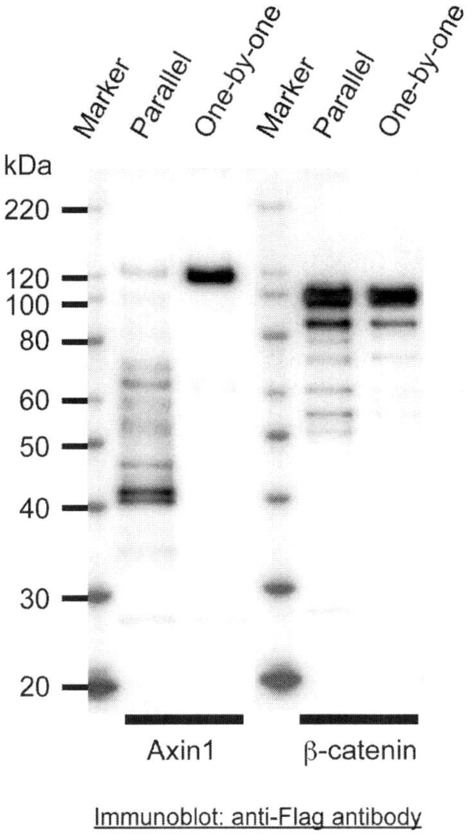

Fig. 4. Comparison of bait protein (β-catenin and Axin1) purification quality. Flag-tagged β-catenin or Axin1 proteins were expressed in HEK293T cells, purified by the parallel or one-by-one methods until the elution steps, and analyzed by western blot analysis. One-by-one: automated one-by-one method; Parallel: manual parallel method.

3.3 Comparison of parallel and one-by-one methods for the sample preparation by protein network analysis

Next, we compared the component proteins interacting with the bait proteins (β-catenin and Axin1) prepared by the parallel and one-by-one methods. Each bait protein was expressed in HEK293T cells and purified with its binding partner proteins. These proteins were then digested with Lys-C and analyzed by a DNLC-MS/MS system (Natsume et al., 2002). The identified proteins that interact with β-catenin and Axin1, excluding nonspecific binding, are listed in Table 1. As expected, the one-by-one preparation method showed better detection sensitivity and reproducibility compared with the parallel method.

Bait: β-catenin

Name[a]	Symbol[a]	Parallel[b] reproducibility (n = 10)	One-by-one[b] reproducibility (n = 10)
Adenomatous polyposis coli	APC	10 (100%)	10 (100%)
Adenomatosis polyposis coli 2	APC2	0	8 (80%)
Axin 1	AXIN1	10 (100%)	10 (100%)
Axin 2	AXIN2	1 (10%)	9 (90%)
Beta-transducin repeat containing	BTRC | FBXW11	1 (10%)	10 (100%)
Cadherin 1, type 1	CDH1	1 (10%)	9 (90%)
Cadherin 2, type 1	CDH2	2 (20%)	10 (100%)
Casein kinase 1, alpha 1	CSNK1A1	10 (100%)	10 (100%)
Catenin, alpha 1	CTNNA1	10 (100%)	10 (100%)
Catenin, alpha; 1 or 2	CTNNA1 | CTNNA2	10 (100%)	10 (100%)
Catenin, alpha; 1 or 3	CTNNA1 | CTNNA3	10 (100%)	10 (100%)
Catenin, beta interacting protein 1	CTNNBIP1	8 (80%)	10 (100%)
Catenin, delta 1; isoform 1B	CTNND1	0	10 (100%)
Cathepsin A	CTSA	9 (90%)	10 (100%)
Cullin 1	CUL1	2 (20%)	10 (100%)
Ezrin	EZR	0	8 (80%)
Family with sequence similarity 123B	FAM123B	10 (100%)	10 (100%)
F-box and WD repeat domain containing 11	FBXW11	1 (10%)	10 (100%)
Galactosidase, beta 1	GLB1	8 (80%)	10 (100%)
Glycogen synthase kinase 3 alpha	GSK3A	2 (20%)	10 (100%)
Glycogen synthase kinase 3; alpha or beta	GSK3A | GSK3B	6 (60%)	10 (100%)
Glycogen synthase kinase 3 beta	GSK3B	4 (40%)	10 (100%)
HMG-box transcription factor TCF-3	TCF7L1	7 (70%)	10 (100%)
Lymphoid enhancer-binding factor 1	LEF1	8 (80%)	10 (100%)

Bait: β-catenin

Name[a]	Symbol[a]	Parallel[b] reproducibility (n = 10)	One-by-one[b] reproducibility (n = 10)
S-phase kinase-associated protein 1	SKP1	2 (20%)	10 (100%)
Transcription factor 7 (T-cell-specific, HMG-box); isoform 1	TCF7	3 (30%)	10 (100%)
Transcription factor 7-like 2	TCF7L2	10 (100%)	10 (100%)

Bait: Axin1

Name[a]	Symbol[a]	Parallel[b] reproducibility (n = 10)	One-by-one[b] reproducibility (n = 10)
Adenomatous polyposis coli	APC	0	10 (100%)
Beta-catenin	CTNNB1	0	10 (100%)
Casein kinase 1, alpha 1	CSNK1A1	1 (10%)	10 (100%)
Glycogen synthase kinase 3 beta	GSK3B	2 (20%)	10 (100%)
Macrophage erythroblast attacher	MAEA	0	10 (100%)
WD repeat domain 26; isoform b	WDR26	0	8 (80%)

Table 1. Comparison of identified proteins and their reproducibility from samples prepared by parallel and one-by-one methods (analyzed by MS). [a]Protein names and Symbols refer to the Entrez Gene database. The proteins identified by a common peptide sequence are indicated by 'or' in the Name column, and '|' in the Symbol column. The identified proteins exclude nonspecific proteins (Table 2). [b]The samples were prepared independently by the parallel or the one-by-one method and analyzed by the DNLS-MS/MS system.

In the analysis of the one-by-one preparation β-catenin, we identified membrane proteins (Cadherins 1 and 2), peripheral membrane proteins (δ-catenin and Ezrin), the Skp1- Cullin-F-box-protein (SCF) E3 ubiquitin ligase complex (BTRC/FBXW11, Skp1, and Cullin1) and other component proteins (Adenomatosis polyposis coli 2 (APC2) and Axin2) using the one-by-one method, whereas some of these proteins were not identified by the parallel method. The reproducibility increased from below 20% (parallel preparation, n = 10) to above 80% (one-by-one preparation, n = 10). In the analysis of Axin1, the one-by-one method dramatically increased the precision of the identification of well-known interaction partners, such as Adenomatous polyposis coli (APC), δ-catenin, Glycogen synthase kinase 3β (GSK3β), and Casein kinase 1, whereas no specific interactions were identified using the parallel method (Table 1). This improvement is probably the result of the minimal degradation of Axin1 (Fig. 4). Furthermore, we found two new interacting partners: MAEA and WDR26. To confirm these interactions, Flag-tagged Axin1 was expressed in HEK293T cells and the cell extracts were subjected to immunoprecipitation with anti-Flag antibody, followed by western blotting with anti-MEAE or anti-WDR26 antibody. As shown in Fig. 5, both MAEA and WDR26 were found to form a complex with Axin1. Further work is required to determine the biological relevance of these interactions.

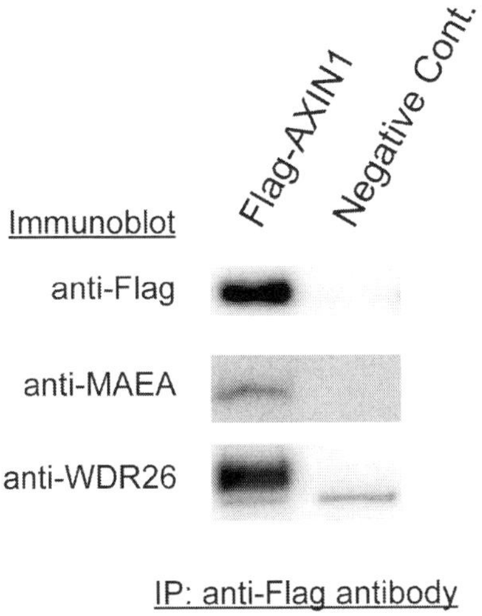

Fig. 5. Interaction of Axin1 with MAEA and WDR26. HEK293T cells were transfected with Flag-Axin1 or an empty vector (pcDNA3) as a negative control (Negative cont.). Expressed protein complexes were purified by the automated one-by-one methods until the elution step and analyzed by western blot analysis.

4. Discussion

Sample preparation is one of the most important processes for MS-based proteomics, such as large-scale protein-protein interaction networks and quantitative analyses. In affinity purification, although the single Flag-tag purification MS approach is useful and raises the possibility of identification of low abundant and transient interacting proteins, the problem is that this approach leads to a high false positive rate (Chen & Gingras, 2007). To overcome this problem, several protocols have been devised (Burckstummer et al., 2006; Selbach & Mann, 2006), and computational data processing to remove nonspecific proteins is performed during large-scale analysis (Ewing et al., 2007; Ho et al., 2002; Gavin et al., 2002). However, because it is possible to reliably identify low amounts of true interacting proteins by improving the signal-to-noise ratio in LC-MS/MS, we considered that reproducibly decreasing the level of nonspecific noise proteins in single-step purification samples would be a valid approach. Therefore, we empirically developed and optimized the conditions for sample preparation, and using this methodology, found more than fifty significant protein-protein interactions (Hirano et al., 2005; Kitajima et al., 2006; Iioka et al., 2007; Komatsu et al., 2007; Nishiyama et al., 2009; Kaneko et al., 2009; Komatsu et al., 2010). In spite of this useful methodology, we realized the limitations of the existing preparation method in large-scale analysis, because we found that the amount of true interactors, as well as nonspecific proteins, in manually parallel-prepared samples varied. The ultimate solution for this problem was to use a one-by-one purification method. In addition, because this preparation process needs to be automated to prepare samples under precisely equal conditions, we designed and developed a fully automated robotic sample preparation system for LC-MS/MS.

In a validation study using the Wnt signaling pathway proteins, β-catenin and Axin1, the rate of protein degradation was significantly higher in the parallel preparation compared with the one-by-one preparation. This higher protein degradation in parallel preparation is probably caused by the manual scraping of cells and increased preparation time. In parallel preparation, manual scraping of cells involves several rapid strokes, which may increase the cells' susceptibility to damage and increase the level of proteolytic enzymes released from subcellular compartments. The proteases, similarly to nonspecific binding proteins, are likely to attach to and degrade the purified protein complexes over time, and these degraded and denatured proteins are thought to cause nonspecific binding.

In contrast to manual parallel preparation, an important feature of the one-by-one system is the careful and brief sample preparation. The association rate of nonspecific proteins is thought to be slower than that of specific binding proteins; therefore, the careful and rapid one-by-one method reduces nonspecific protein associations. In fact, as shown in Table 2, the number of nonspecific proteins precipitated using the one-by-one method was significantly lower than that by the parallel method. Using the one-by-one method, this decrease was accompanied by a remarkable increase in known interactors, because the signal-to-noise ratio was increased in combination with the prevention of protein degradation. Although it was previously reported that single-affinity tag purifications brought an increase in nonspecific binding proteins (Chen & Gingras, 2007), we have found that the single-step one-by-one purification using anti-Flag antibody immobilized magnetic beads is valuable because of its considerable reduction in nonspecific binding proteins under optimized conditions.

Name[a]	Symbol[a]	Parallel[b]	One-by-one[b]
Actin, alpha 1, skeletal muscle | Actin, alpha 2, smooth muscle, aorta | Actin, beta | Actin, alpha, cardiac muscle 1 | Actin, gamma 1 | Actin, gamma 2, smooth muscle, enteric	ACTA1 | ACTA2 | ACTB | ACTC1 | ACTG1 | ACTG2	2	2
Actin, alpha 1, skeletal muscle | Actin, alpha 2, smooth muscle, aorta | Actin, beta | Actin, gamma 2, smooth muscle, enteric	ACTA1 | ACTA2 | ACTC1 | ACTG2	1	1
ATPase family AAA domain-containing protein 3A | ATPase family AAA domain-containing protein 3B	ATAD3A | ATAD3B	6	ND
Complement component 1, q subcomponent binding protein	C1QBP	2	ND

Name[a]	Symbol[a]	Parallel[b]	One-by-one[b]
DEAH (Asp-Glu-Ala-His) box polypeptide 9	DHX9	4	ND
Eukaryotic translation elongation factor 1 alpha 1 | Eukaryotic translation elongation factor 1 alpha 2 | Eukaryotic translation elongation factor 1 alpha 1 pseudogene 5	EEF1A1 | EEF1A2 | EEF1A1P5	5	ND
Eukaryotic translation initiation factor 4A1	EIF4A1	6	2
Eukaryotic translation initiation factor 4A1 | Eukaryotic translation initiation factor 4A2	EIF4A1 | EIF4A2	2	ND
Histone cluster 1, H1c | Histone cluster 1, H1d | Histone cluster 1, H1e	HIST1H1C | HIST1H1D | HIST1H1E	2	ND
Heat shock protein 90kDa alpha (cytosolic), class A member 1	HSP90AA1	7	2
Heat shock protein 90kDa alpha (cytosolic), class A member 1 | Heat shock protein 90kDa alpha (cytosolic), class A member 2	HSP90AA1 | HSP90AA2	3	1
Heat shock protein 90kDa alpha (cytosolic), class A member 1 | Heat shock protein 90kDa alpha (cytosolic), class A member 2 | Heat shock protein 90kDa alpha (cytosolic), class B member 1 | heat shock protein 90kDa alpha (cytosolic), class B member 2 (pseudogene)	HSP90AA1 | HSP90AA2 | HSP90AB1 | HSP90AB2P	2	2
Heat shock protein 90kDa alpha (cytosolic), class A member 1 | Heat shock protein 90kDa alpha (cytosolic), class B member 1	HSP90AA1 | HSP90AB1	2	2
Heat shock protein 90kDa alpha (cytosolic), class B member 1	HSP90AB1	4	1
Heat shock protein 90kDa alpha (cytosolic), class B member 1 | Heat shock protein 90kDa alpha (cytosolic), class B member 3 (pseudogene)	HSP90AB1 | HSP90AB3P	3	1
Heat shock 70kDa protein 1A | Heat shock 70kDa protein 1B	HSPA1A | HSPA1B	13	5
Heat shock 70kDa protein 1A | Heat shock 70kDa protein 1B | Heat shock 70kDa protein 1-like	HSPA1A | HSPA1B | HSPA1L	4	3

Name[a]	Symbol[a]	Parallel[b]	One-by-one[b]
Heat shock 70kDa protein 5 (glucose-regulated protein, 78kDa)	HSPA5	19	8
Heat shock 70kDa protein 8	HSPA8	12	9
Heat shock 60kDa protein 1 (chaperonin)	HSPD1	21	10
Nucleolin	NCL	13	2
Nucleophosmin (nucleolar phosphoprotein B23, numatrin)	NPM1	3	1
Poly(A) binding protein, cytoplasmic 1	PABPC1	3	ND
Poly (ADP-ribose) polymerase 1	PARP1	15	4
Ribosomal protein L10a	RPL10A	3	ND
Ribosomal protein L11	RPL11	4	ND
Ribosomal protein L12	RPL12	4	ND
Ribosomal protein L13	RPL13	2	ND
Ribosomal protein L18	RPL18	3	1
Ribosomal protein L22	RPL22	2	1
Ribosomal protein L23	RPL23	2	1
Ribosomal protein L23a	RPL23A	5	ND
Ribosomal protein L24	RPL24	2	1
Ribosomal protein L28	RPL28	2	1
Ribosomal protein L29	RPL29	4	ND
Ribosomal protein L3	RPL3	5	ND
Ribosomal protein L30	RPL30	2	ND
Ribosomal protein L31	RPL31	3	ND
Ribosomal protein L35	RPL35	2	ND
Ribosomal protein L37a	RPL37A	2	ND
Ribosomal protein L38	RPL38	3	ND
Ribosomal protein L4	RPL4	5	2
Ribosomal protein L5 | Ribosomal protein, large, P0	RPL5 | RPLP0	5	2
Ribosomal protein L6	RPL6	6	ND
Ribosomal protein L7a	RPL7A	3	2
Ribosomal protein L8	RPL8	2	ND
Ribosomal protein L9	RPL9	3	1
Ribosomal protein, large, P0	RPLP0	2	ND
Ribosomal protein, large, P2	RPLP2	4	2
Ribosomal protein S11	RPS11	2	ND
Ribosomal protein S12	RPS12	2	ND
Ribosomal protein S13	RPS13	5	1
Ribosomal protein S15	RPS15	2	ND
Ribosomal protein S16	RPS16	4	ND
Ribosomal protein S19	RPS19	4	2
Ribosomal protein S20	RPS20	3	ND
Ribosomal protein S23	RPS23	2	ND

Name[a]	Symbol[a]	Parallel[b]	One-by-one[b]
Ribosomal protein S24	RPS24	2	ND
Ribosomal protein S25	RPS25	4	ND
Ribosomal protein S27a | Ubiquitin A-52 residue ribosomal protein fusion product 1 | Ubiquitin B | Ubiquitin C	RPS27A | UBA52 | UBB | UBC	3	2
Ribosomal protein S3	RPS3	5	2
Ribosomal protein S3A	RPS3A	7	2
Ribosomal protein S4, X-linked	RPS4X	4	1
Ribosomal protein S4, X-linked | Ribosomal protein S4, Y-linked 1 | Ribosomal protein S4, Y-linked 2	RPS4X | RPS4Y1 | RPS4Y2	3	ND
Ribosomal protein S5	RPS5	3	ND
Ribosomal protein S6	RPS6	4	ND
Ribosomal protein S7	RPS7	4	2
Ribosomal protein S8	RPS8	3	ND
Ribosomal protein S9	RPS9	4	1
Ubiquitin A-52 residue ribosomal protein fusion product 1	UBA52	3	1

Table 2. Comparison of nonspecific proteins identified from samples prepared by parallel and one-by-one methods (analyzed by MS). [a]The nonspecific proteins co-purified with β-catenin (n = 10) and Axin 1 (n = 10) using each method were categorized according to the criteria reported by Chen and Gingras. Protein Symbols and Names refer to the NCBI Gene database. Proteins identified by a common peptide sequence are indicated by '|' in the Name, Symbol columns. [b]Total number of the identified peptides. ND: not detected.

5. Conclusion

We have described a one-by-one sample preparation method for MS-based high-precision protein network analysis. To perform a pilot feasibility study of the one-by-one method, we designed and developed a fully automated robotic system. This system makes it possible to prepare samples under equally fast and gentle conditions. To clarify the importance of the one-by-one method, we compared protein complexes prepared by the automated one-by-one system with manual parallel preparation using β-catenin and Axin1 as baits, which are well-characterized Wnt signaling pathway proteins. One-by-one purification resulted in a sharp decrease in proteolytic degradation of purified proteins and in nonspecific binding proteins, allowing the reproducible identification of known interaction partners, as well as novel component proteins. These results suggest that one-by-one sample preparation by the automated system is useful for obtaining reliable data for high-precision analysis of protein identification and quantification for large-scale protein network analysis compared with manual parallel preparation.

We expect that this system will allow highly sensitive analyses of protein interactions using various types of cells, such as embryonic stem (ES), neuronal, and primary cells, which are limited in supply. Furthermore, we envision that this system could be used for qualitative and quantitative protein interaction network studies including chemical proteomics (Rix & Superti-Furga, 2009).

In future work, we will develop a multi-purpose robotic system that can be flexibly customized. Finally, our goal is to develop an automated robotic system that can operate not only in affinity purification, but also in general proteomics.

6. Acknowledgment

We thank H. Shibuya (Medical Research Institute, Tokyo Medical and Dental University) for providing the Flag-tagged human β-catenin and Axin1 cDNAs, and Y. Hioki, K. Koike, K. Nishimura, T. Asano and H. Kusano for technical assistance. This work was supported by a 'Development of Basic Technology to Control Biological Systems Using Chemical Compounds' grant from the New Energy and Industrial Technology Development Organization (NEDO), Japan.

7. References

Aebersold, R. & Mann, M. (2003). Mass spectrometry-based proteomics. *Nature*, 422 (6928), 198-207.

Alterovitz, G., Liu, J., Chow, J. & Ramoni, M. F. (2006). Automation, parallelism, and robotics for proteomics. *Proteomics*, 6 (14), 4016-4022.

Blagoev, B., Kratchmarova, I., Ong, S. E., Nielsen, M., Foster, L. J. & Mann, M. (2003). A proteomics strategy to elucidate functional protein-protein interactions applied to EGF signaling. *Nat. Biotechnol.*, 21 (3), 315-318.

Blow, N. (2008). Lab automation: tales along the road to automation. *Nat. Methods*, 5 (1), 109-112.

Burckstummer, T., Bennett, K. L., Preradovic, A., Schutze, G., Hantschel, O., Superti-Furga, G. & Bauch, A. (2006). An efficient tandem affinity purification procedure for interaction proteomics in mammalian cells. *Nat. Methods*, 3 (12), 1013-1019.

Chen, G. I. & Gingras, A. C. (2007). Affinity-purification mass spectrometry (AP-MS) of serine/threonine phosphatases. *Methods*, 42 (3), 298-305.

Daugherty, R. L. & Gottardi, C. J. (2007). Phospho-regulation of Beta-catenin adhesion and signaling functions. *Physiology (Bethesda)*, 22, 303-309.

Domon, B. & Aebersold, R. (2006). Mass spectrometry and protein analysis. *Science*, 312 (5771), 212-217.

Einhauer, A. & Jungbauer, A. (2001). The FLAG peptide, a versatile fusion tag for the purification of recombinant proteins. *J. Biochem. Biophys. Methods*, 49 (1-3), 455-465.

Ewing, R. M., Chu, P., Elisma, F., Li, H., Taylor, P., Climie, S., McBroom-Cerajewski, L., Robinson, M. D., O'Connor, L., Li, M., Taylor, R., Dharsee, M., Ho, Y., Heilbut, A., Moore, L., Zhang, S., Ornatsky, O., Bukhman, Y. V., Ethier, M., Sheng, Y., Vasilescu, J., Abu-Farha, M., Lambert, J. P., Duewel, H. S., Stewart, I. I., Kuehl, B., Hogue, K., Colwill, K., Gladwish, K., Muskat, B., Kinach, R., Adams, S. L., Moran, M. F., Morin, G. B., Topaloglou, T. & Figeys, D. (2007). Large-scale mapping of human protein-protein interactions by mass spectrometry. *Mol. Syst. Biol.*, 3, 89.

Gavin, A. C., Bosche, M., Krause, R., Grandi, P., Marzioch, M., Bauer, A., Schultz, J., Rick, J. M., Michon, A. M., Cruciat, C. M., Remor, M., Hofert, C., Schelder, M., Brajenovic, M., Ruffner, H., Merino, A., Klein, K., Hudak, M., Dickson, D., Rudi, T., Gnau, V., Bauch, A., Bastuck, S., Huhse, B., Leutwein, C., Heurtier, M. A., Copley, R. R., Edelmann, A., Querfurth, E., Rybin, V., Drewes, G., Raida, M., Bouwmeester, T., Bork, P., Seraphin, B., Kuster, B., Neubauer, G. & Superti-Furga, G. (2002). Functional organization of the yeast proteome by systematic analysis of protein complexes. *Nature*, 415 (6868), 141-147.

Hirano, Y., Hendil, K. B., Yashiroda, H., Iemura, S., Nagane, R., Hioki, Y., Natsume, T., Tanaka, K. & Murata, S. (2005). A heterodimeric complex that promotes the assembly of mammalian 20S proteasomes. *Nature*, 437 (7063), 1381-1385.

Ho, Y., Gruhler, A., Heilbut, A., Bader, G. D., Moore, L., Adams, S. L., Millar, A., Taylor, P., Bennett, K., Boutilier, K., Yang, L., Wolting, C., Donaldson, I., Schandorff, S., Shewnarane, J., Vo, M., Taggart, J., Goudreault, M., Muskat, B., Alfarano, C., Dewar, D., Lin, Z., Michalickova, K., Willems, A. R., Sassi, H., Nielsen, P. A., Rasmussen, K. J., Andersen, J. R., Johansen, L. E., Hansen, L. H., Jespersen, H., Podtelejnikov, A., Nielsen, E., Crawford, J., Poulsen, V., Sorensen, B. D., Matthiesen, J., Hendrickson, R. C., Gleeson, F., Pawson, T., Moran, M. F., Durocher, D., Mann, M., Hogue, C. W., Figeys, D. & Tyers, M. (2002). Systematic identification of protein complexes in Saccharomyces cerevisiae by mass spectrometry. *Nature*, 415 (6868), 180-183.

Huang, H. & He, X. (2008). Wnt/beta-catenin signaling: new (and old) players and new insights. *Curr. Opin. Cell Biol.*, 20 (2), 119-125.

Huang, S. M., Mishina, Y. M., Liu, S., Cheung, A., Stegmeier, F., Michaud, G. A., Charlat, O., Wiellette, E., Zhang, Y., Wiessner, S., Hild, M., Shi, X., Wilson, C. J., Mickanin, C., Myer, V., Fazal, A., Tomlinson, R., Serluca, F., Shao, W., Cheng, H., Shultz, M., Rau, C., Schirle, M., Schlegl, J., Ghidelli, S., Fawell, S., Lu, C., Curtis, D., Kirschner, M. W., Lengauer, C., Finan, P. M., Tallarico, J. A., Bouwmeester, T., Porter, J. A., Bauer, A. & Cong, F. (2009). Tankyrase inhibition stabilizes axin and antagonizes Wnt signalling. *Nature*, 461 (7264), 614-620.

Iioka, H., Iemura, S., Natsume, T. & Kinoshita, N. (2007). Wnt signalling regulates paxillin ubiquitination essential for mesodermal cell motility. *Nat. Cell Biol.*, 9 (7), 813-821.

Kaneko, T., Hamazaki, J., Iemura, S., Sasaki, K., Furuyama, K., Natsume, T., Tanaka, K. & Murata, S. (2009). Assembly pathway of the Mammalian proteasome base subcomplex is mediated by multiple specific chaperones. *Cell*, 137 (5), 914-925.

Kitajima, T. S., Sakuno, T., Ishiguro, K., Iemura, S., Natsume, T., Kawashima, S. A. & Watanabe, Y. (2006). Shugoshin collaborates with protein phosphatase 2A to protect cohesin. *Nature*, 441 (7089), 46-52.

Kocher, T. & Superti-Furga, G. (2007). Mass spectrometry-based functional proteomics: from molecular machines to protein networks. *Nat. Methods*, 4 (10), 807-815.

Komatsu, M., Waguri, S., Koike, M., Sou, Y. S., Ueno, T., Hara, T., Mizushima, N., Iwata, J. I., Ezaki, J., Murata, S., Hamazaki, J., Nishito, Y., Iemura, S., Natsume, T., Yanagawa, T., Uwayama, J., Warabi, E., Yoshida, H., Ishii, T., Kobayashi, A., Yamamoto, M., Yue, Z., Uchiyama, Y., Kominami, E. & Tanaka, K. (2007). Homeostatic levels of p62 control cytoplasmic inclusion body formation in autophagy-deficient mice. *Cell*, 131 (6), 1149-1163.

Komatsu, M., Kurokawa, H., Waguri, S., Taguchi, K., Kobayashi, A., Ichimura, Y., Sou, Y. S., Ueno, I., Sakamoto, A., Tong, K. I., Kim, M., Nishito, Y., Iemura, S., Natsume, T., Ueno, T., Kominami, E., Motohashi, H., Tanaka, K. & Yamamoto, M. (2010). The selective autophagy substrate p62 activates the stress responsive transcription factor Nrf2 through inactivation of Keap1. *Nat. Cell Biol.*, 12 (3), 213-223.

Natsume, T., Yamauchi, Y., Nakayama, H., Shinkawa, T., Yanagida, M., Takahashi, N. & Isobe, T. (2002). A direct nanoflow liquid chromatography-tandem mass spectrometry system for interaction proteomics. *Anal. Chem.*, 74 (18), 4725-4733.

Nishiyama, M., Oshikawa, K., Tsukada, Y., Nakagawa, T., Iemura, S., Natsume, T., Fan, Y., Kikuchi, A., Skoultchi, A. I. & Nakayama, K. I. (2009). CHD8 suppresses p53-mediated apoptosis through histone H1 recruitment during early embryogenesis. *Nat. Cell Biol.*, 11 (2), 172-182.

Olsen, J. V. & Mann, M. (2011). Effective representation and storage of mass spectrometry-based proteomic data sets for the scientific community. *Sci. Signal.*, 4 (160), pe7.

Rao, T. P. & Kuhl, M. (2010). An updated overview on Wnt signaling pathways: a prelude for more. *Circ. Res.*, 106 (12), 1798-1806.

Rix, U. & Superti-Furga, G. (2009). Target profiling of small molecules by chemical proteomics. *Nat. Chem. Biol.*, 5 (9), 616-624.

Selbach, M. & Mann, M. (2006). Protein interaction screening by quantitative immunoprecipitation combined with knockdown (QUICK). *Nat. Methods*, 3 (12), 981-983.

16

Live In-Cell Visualization of Proteins Using Super Resolution Imaging

Catherine H. Kaschula[1,2], Dirk Lang[3] and M. Iqbal Parker[1,2]
[1]*Department of Medical Biochemistry, University of Cape Town, Anzio Road, Observatory, Cape Town,*
[2]*International Centre for Genetic Engineering and Biotechnology, Wernher and Beit South, Anzio Rd, Observatory, Cape Town,*
[3]*Department of Human Biology, University of Cape Town, Anzio Road, Observatory, Cape Town, South Africa*

1. Introduction

Fluorescence microscopy is a non-invasive technique that allows for the dynamic recording of molecular events in live cells, tissues and animals and is based on the principle that fluorescently-labelled material can be illuminated at one wavelength and emit light or fluoresce at another wavelength. The live material is selectively labelled with a fluorescent probe (the fluorophore) to generate a fluorescent image which is detected and recorded through the objective of a microscope. At the one end of the scale: positron emission tomography (PET), magnetic resonance spectroscopy (MRI) and optical coherence spectroscopy (OCT) provide real-time images from live animal or human subjects with resolutions up to about 1 mm, 100 μm and 10 μm respectively (Fernandez-Suarez and Ting 2008) (see Figure 1). At the other end of the scale: electron microscopy provides near molecular-level spatial resolution down to a few nanometers, but here cells must be fixed, which is invasive and prevents dynamic imaging. The most widely used fluorescent imaging methods in research are confocal and wide field microscopy which can provide resolutions down to a few hundred nanometers (ie can resolve intracellular organelles and track proteins in live cells). With the recent emergence of new far field fluorescence imaging techniques, it is now possible to achieve a higher level of resolution down to 10 nm to resolve single synaptic vesicles or pairs of interacting proteins (Fernandez-Suarez and Ting 2008; Huang, Babcock, and Zhuang 2010). In this chapter, we will focus on the relatively new field of far field or super resolution fluorescence imaging. The current limitations in terms of spatial and temporal resolution will be discussed together with recent fluorescent probe technology. A few applications of these techniques which have led to new discoveries will be presented.

2. The spatial resolution limit

In light microscopy, resolution is fundamentally limited by the properties of diffraction (Abbe 1873) or the "spreading out" of a light wave when it passes through a small aperture or is focused to a focal point. The diffraction barrier, which was first described by Ernst

Abbe in 1873, describes the inability of a lens-based optical microscope to discern details that are closer together than half the wavelength of light (Toomre and Bewersdorf 2010). As a result, a minutely small object that emits light will be detected as a finite-sized spot, the size of which is referred to as the point-spread function (PSF) (Figure 2). The PSF is

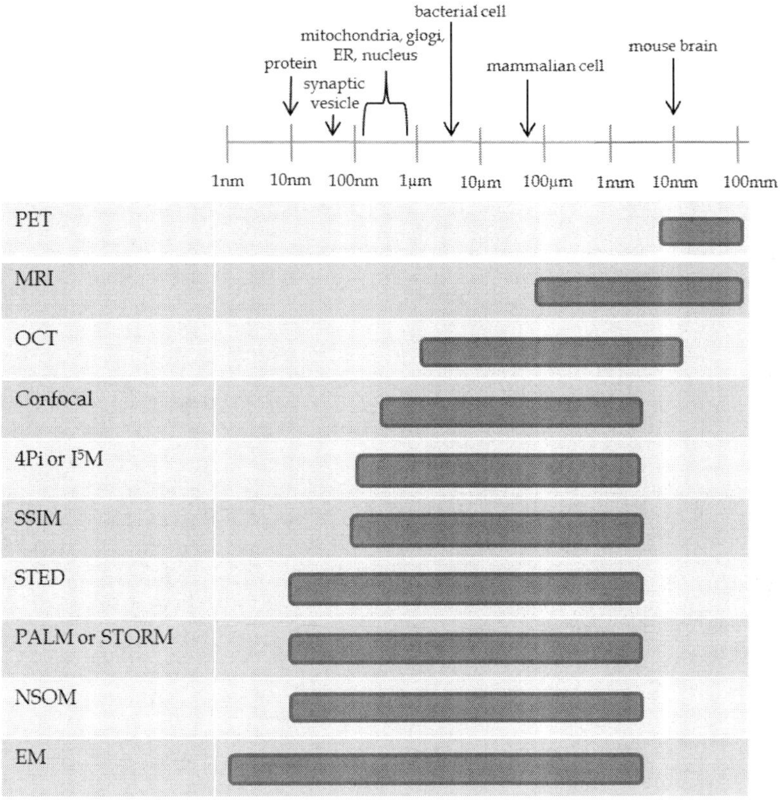

Fig. 1. **Comparison of the spatial resolutions of biological imaging techniques**. The size scale is logarithmic and approximate sizes of biological features are displayed. In addition, the spatial resolutions are estimates and are given for the focal plane. ER, endoplasmic reticulum; PET, positron-emission microscopy; MRI, magnetic resonance imaging; OCT, optical coherence tomography; SSIM, saturated structured-illumination microscopy; STED, stimulated emission depletion; PALM, photoactivated localization microscopy; STORM, stochastic optical reconstruction microscopy; NSOM, near-filed scanning optical microscopy; EM, electron microscopy. Adapted from (Fernandez-Suarez and Ting 2008).

elongated in shape along the optical axis due to the nature of the non-symmetrical wavefront that emerges from a conventional objective lens (Heilemann 2010). According to the diffraction barrier: the resolution obtainable with a wide field microscope is 200 - 250 nm in the x and y directions and 500 - 700 nm in the z-direction (Toomre and Bewersdorf 2010) which is suitable resolution to view organelles and proteins.

Imaging techniques such as multiphoton fluorescence microscopy and confocal laser microscopy have gently pushed the diffraction limit by using a focused laser beam to reduce the focal point size. In addition the confocal microscope uses a spatial pinhole to eliminate out-of-focus light thicker than the focal plane. 4Pi microscopy and I^5M is another branch of microscopy that makes use of two opposing objective lenses to sharpen the PSF along the optical axis through interference of the counter-propagating wavefronts (Heilemann 2010). Although all of the above mentioned methods improve resolution (down to about 100 nm), they are still fundamentally limited by diffraction (Huang, Babcock, and Zhuang 2010).

Diffraction-limited resolution applies only to light that has propagated for a distance substantially larger than its wavelength (i.e. in the far field). In 1992, the first super-resolution image of a biological sample was obtained using near-field scanning optical microscopy (NSOM) (Betzig and Trautman 1992). Here the excitation source or detection probe is placed near the sample to obtain resolutions in the 20 - 120 nm range. Although NSOM has been used to study the nanoscale organisation of several membrane proteins, it cannot be used for intracellular imaging as the probe or excitation source needs to be within tens of nanometers of the target object.

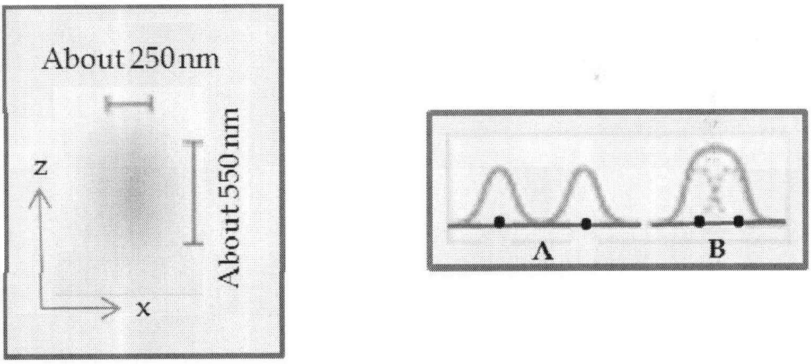

Fig. 2. **Diffraction-limited resolution of conventional light microscopy**. The focal spot of a typical objective with a high aperture is depicted by the ellipse with a width of about 250 nm in the x-direction and about 550 nm in the z-direction. The image of a point emitter imaged through the objective (the point spread function), has similar widths which define the diffraction-limited resolution. Two objects separated by a distance > the resolution limit are resolvable and appear as two separate entities in the image (i.e. in **A**) whereas images closer together than the resolution limit appear unresolvable (i.e. in **B**). Adapted from (Huang, Babcock, and Zhuang 2010).

3. Far field super resolution imaging

Features that are spectrally different are not challenged by diffraction. Likewise, Abbe's barrier does not prevent determining the coordinates of a molecule down to 1 nm with great precision (Kural et al. 2005) if there is no similar marker molecule within $\lambda/2n$ (the diffraction limited region). Overcoming the diffraction limit has been achieved by discerning groups of labelled features within a distance $<\lambda/2n$. This has been realised by modulating the emissions of fluorescent probes (i.e. transitions between bright and dark states) within a diffraction-limited region (Hell 2007).

One class of super-resolution techniques use patterned illumination to spatially modulate the fluorescence behaviour of a *population of molecules* within a diffraction-limited region, such that not all of them emit simultaneously. Microscopies utilizing this technique include stimulated emission depletion (STED), RESOLFT technology and saturated structured illumination microscopy (SSIM).

Another other class of super-resolution techniques uses photoswitching or other mechanisms to activate *individual molecules* within a diffraction-limited region. Images are then reconstructed with subdiffraction limit resolution from the measured positions of individual fluorophores. Microscopies utilizing this technique include stochastic optical reconstruction microscopy (STORM), photoactivated microscopy (PALM), and fluorescence photoactivation localization microscopy (FPALM).

4. Super-resolution imaging of a *population of molecules*

These techniques apply patterned light to a sample to manipulate its fluorescence emission. This spatial modulation can be applied in either a positive or negative manner. In the negative case, patterned light is applied to supress the population of molecules that can fluoresce. In the positive case, the light field used to excite the sample is patterned. In both of these techniques, the spatial information encoded into the illumination pattern allows neighbouring fluorophores to be distinguished from each other, leading to enhanced spatial resolution.

4.1 Principles of STED microscopy

In STED microscopy, fluorescence emission of a cluster of fluorophores is selectively "turned off" or quenched. The sample is illuminated with an excitation laser pulse which is immediately chased by a red-shifted pulse or STED beam (see Figure 3). The STED pulse quenches the fluorophores that reside within the excited state everywhere except those close to the zero intensity position to give a doughnut emission profile. When the two pulses are superimposed, only molecules close to the zero of the STED beam fluoresce, thereby lowering the PSF and increasing resolution. This approach offers improved resolution given a strong depletion light source, low scattering from the sample and good photostability of the fluorophores. In biological samples, STED images have achieved a resolution down to 20 nm in the case of organic dyes and 50 - 70 nm in the case of fluorescent proteins (Fernendez-Suarez and Ting 2008; Huang, Babcock, and Zhuang 2010).

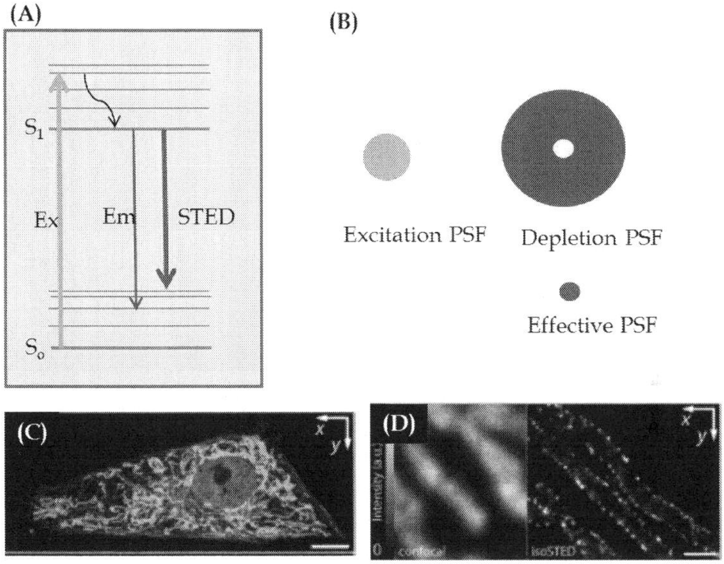

Fig. 3. **Super-resolution imaging with STED microscopy.** (**A**) A fluorophore can enter the first excited state S_1, following the absorption of a photon of appropriate energy. Fast relaxation to the vibrational ground state of S_1 can cause the emission of fluorescence to occur. The key principle of STED microscopy is that this excited state is locally depopulated by inducing stimulated emission. (**B**) A first laser that excites fluorophores into the S_1 state is overlaid with a depletion laser which has a doughnut-shaped intensity profile, where the area of zero-intensity scales with the irradiation intensity of the depletion beam. The resulting "effective" PSF represents the remaining area where fluorescence emission is still observed, and which is well below the diffraction limit. (**C**) Overview of the mitochondrial network of a PtK2 (kangaroo rat) cell. The mitochondria was labelled with antibodies against the translocase of the outer membrane of mitochondria (TOM) complex (green) and the microtubule cytoskeleton was labelled with antibodies against β-tubulin (red). The nucleus was stained with DAPI (blue). Scale bar 10 μm (**D**) Mitochondria labelled for the outer membrane with antibodies specific for the TOM complex imaged with a confocal microscope (left) and isoSTED nanoscope (right). Scale bar 500 nm. Both (**C**) and (**D**) are reprinted with permission (Schmidt et al. 2009). Copyright 2009 American Chemical Society.

4.2 Principles of SIM microscopy

SIM microscopy utilizes a positive sinusoidal pattern of excitation light by combining two light beams. A final image is computationally reconstructed from multiple snapshots collected by scanning and rotating the pattern. Spatial modulation from the excitation pattern brings about enhanced spatial resolution (see Figure 4).

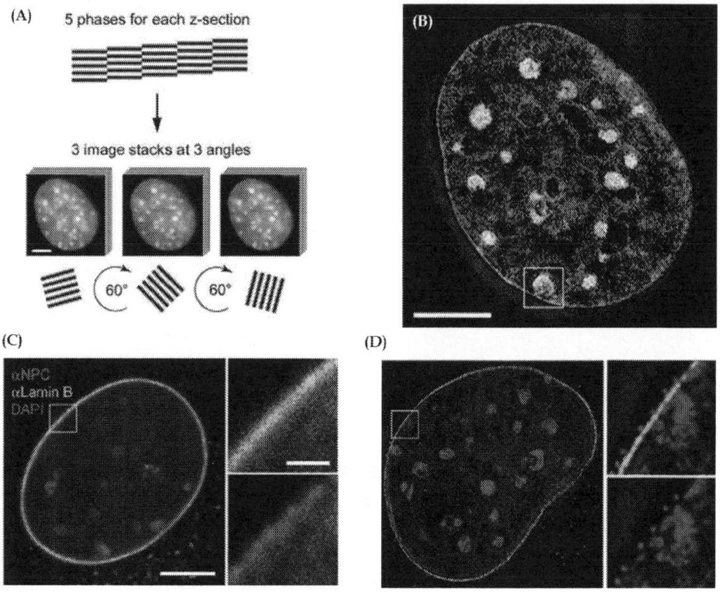

Fig. 4. **Super-resolution imaging with SIM spectroscopy.** Cross section through a DAPI-stained C2C12 cell nucleus aacquired with structured illumination. Five phases in the sine wave pattern were recorded at each z-position (in **A**), allowing the shifted components to be separated and returned to their proper location in space. Three image stacks are recorded with the diffraction grating rotated to three positions 60° apart. The cross section is reconstructed to give the 3D SIM image (in **B**). Scale bar 5 μm. (**C**) Cross section of a confocal image of the nucleus stained for DNA (blue), lamin B (green), and the nuclear pore complex (red), The right panes show the magnified images of the boxed region. (**D**) 3D SIM image of a similarly stained nucleus. Reprinted with permission (Schermelleh et al. 2008).

4.3 Video rate super-resolution images of live cells using STED and SIM

Temporal resolution refers to the precision of a measurement with respect to time, which is critical for the dynamic imaging in living cells. There is an interplay between temporal and spatial resolution due to the finite speed of light and the time taken for the photons to reach the detector. During this timeframe, the system may have undergone a change, thus the longer the light has to travel, the lower the temporal resolution. Video rate STED imaging (28 frames per second) with 62 nm spatial resolution has been demonstrated in a field of view of about 5 μM^2, allowing the motion of individual synaptic vesicles in a dendritic spine to be followed (Westphal et al. 2008). This was achieved by increasing laser intensity (to 400 mW per cm^2) and reducing the number of photons collected per imaging cycle (resulting in increased spatial resolution). This situation is not ideal as the high laser intensities not damaging to living cells. Replacing the pulsed STED lasers with continuous wave lasers permits faster scanning and higher time resolution (Willig et al. 2007). In this regard, a 70 μM^2 image of an endoplasmic reticulum took only 0.19s to acquire (Moneron et al. 2010).

SIM is good for live-cell applications that require a large field of view but not very high spatial resolution as this technique is limited by how fast the illumination pattern can be modulated and the rate of camera speed (Huang, Babcock, and Zhuang 2010).

5. Super-resolution fluorescence microscopy by *single-molecule switching*

Fluorescent probes with photoswitchable properties have been developed to modulate the fluorescence emission profile of individual fluorophores such that only an optically resolvable subset of fluorophores are activated at any moment, allowing their localization with high accuracy. Over the course of multiple activation cycles, the positions of numerous fluorophores are determined and used to construct a high-resolution image. This is the basis of PALM, FPALM and STORM super-resolution spectroscopy. Here spatial resolution is dependent on the precision of a molecules' position which in turn is related to the number of photons which are detected. For example, in the absence of background, if 10 000 photons are collected from a single fluorophore before it bleaches or is turned off, its position can be determined to 2 nm precision (Yildiz et al. 2003).

Being able to localize a single molecule does not directly translate to super-resolution imaging as the labelled biological sample may contain thousands of fluorophores within a diffraction limited region. The fluorescence emissions of the fluorophores will overlap such that the overall image will appear as a blur. However, if the fluorescence emission from these molecules is controlled such that only one molecule is emitting at a time, individual molecules can be imaged and localized.

6. Fluorescent probes used in super-resolution imaging

STORM and PALM microscopy are made possible only though the use of fluorescent probes. Despite the high specifications required for these probes, a large number of switchable fluorophores are available. These probes must firstly have a fluorescent state that emits light at one wavelength and a dark state that does not emit light at this wavelength. Secondly, in order to achieve high precision of localization, the probes should emit a large number of photons before entering the dark phase. Thirdly, because only one fluorophore is activated within a diffraction-limited area at any time, the fluorophores within the dark state should remain as such to ensure high precision localization of the activated fluorophore. A low spontaneous rate of activation of the fluorophores in the dark state is also desired (i.e. spontaneous activation by thermal energy) (Huang, Babcock, and Zhuang 2010). Currently available probes range from organic dyes to fluorescent proteins. Some of these will be discussed below.

6.1 Fluorescent proteins

There are two classes of fluorescent proteins used in super-resolution imaging: those that convert from a dark to a bright fluorescent state upon irradiation (called photo-activatable proteins), and those whose fluorescence wavelength shifts upon irradiation (also called photoshiftable fluorescent proteins). All known photoshiftable proteins shift their wavelength emission irreversibly, whereas other non-photoshiftable fluorescent proteins emit both reversibly and irreversibly (Fernandez-Suarez and Ting 2008; Lukyanov et al. 2005).

EosFP is the most commonly used irreversible photoswitchable fluorescent protein which exhibits both a high contrast and brightness (Wiedenmann et al. 2004). This protein emits strong green fluorescence (516 nm) that changes to red (581 nm) upon near UV irradiation because of a photo-induced modification involving a break in the peptide backbone next to the chromophore (see Figure 5). This protein was used successfully to perform single-particle tracking of membrane proteins in live COS7 cells at an imaging speed of 20 frames per second using PALM (Manley et al. 2008). The main disadvantage of monomeric EosFP however, is that the chromophore formation occurs only at temperatures below 30 °C, which limits its use in mammalian cells (Wiedenmann et al. 2004). Even the brightest photoswitchable fluorescent proteins are still much dimmer than many of the small molecule organic fluorophores. For example EosFP provides about 490 collected photons per molecule (Schroff et al. 2007) whereas the switchable fluorophore pair Cy3-Cy5 provides about 6000 collected photons per molecule per switching cycle which lasts about 200 cycles (Bates et al. 2007; Bates, Blosser, and Zhuang 2005).

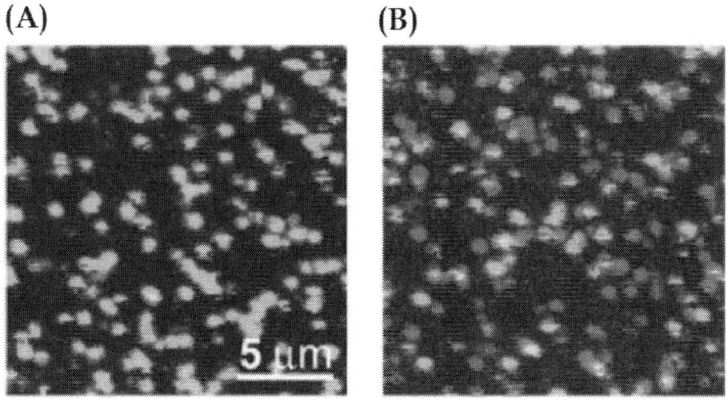

Fig. 5. **Single-molecule spectroscopy of EosFP immobilized on a BSA surface**. Confocal images were taken at 488 nm excitation (**A**) and 400 nm excitation (**B**). Reprinted with permission (Wiedenmann et al 2004)

Reversible fluorescent proteins are advantageous in super-resolution imaging as the same fluorophore can be imaged multiple times. Reversible photoswitching is a prerequisite in RESOLFT imaging, in which each molecule is switched on and off many times in order to reconstruct a subdiffraction image. The best known reversible fluorescent protein is the naturally occurring Dronpa (Ando, Mizuno, and Miyawaki 2004) and its variants of which Padron is one of them (Andresen et al. 2008).

6.2 Organic dyes

There are three main classes of non-genetically encoded probes that have been used in super-resolution imaging, namely inorganic quantum dots, reversible photoswitches and irreversible photocaged fluorophores.

Fluorescent molecules suitable for STED imaging need to have a high quantum yield and slow fluorescence decay, in which case ATTO or DY dyes are ideal. For RESOLFT imaging, the photoswitches FP595 and futyl fulgides are useful (Fernandez-Suarez and Ting 2008). The small molecule analogues to the reversible photoactivatable proteins (i.e. Dronpa) are photochromic probes which include rhodamines and diarylethenes and photoswitchable cyanines. These dyes have higher contrast ratios and higher extinction coefficients than their fluorescent protein counterparts, resulting in a larger number of photons collected per molecule. The photoswitchable cyanines have been used in both PALMIRA and STORM imaging (Bates et al. 2007; Huang et al. 2008) (see Figure 6 below). Cy5 is best used in combination with a secondary chromophore (or activator) that facilitates the switching. For example when Cy5 is paired with Cy3, the same red laser that excites Cy5 is also used to switch the dye to a stable dark state. Subsequently, exposure to green laser light converts Cy5 back to the fluorescent state, and this recovery depends on the close proximity of the secondary dye Cy3 (Bates, Blosser, and Zhuang 2005). Cy3 has also been found to facilitate switching of other cyanines which has greatly increased the amount of colours that are available for STORM imaging and has allowed for the simultaneous visualization of microtubules and clathrin-coated pits in fixed mammalian cells with 20 - 30 nm lateral resolution (see Figure 7) (Bates et al. 2007). The availability of several colours of photoswitchable cyanine dyes gives these fluorophores more diverse application than the photoswitchable fluorescent proteins of which only a few colours are available. Photoswitchable rhodamines are also an important class of photoswitches as they are membrane permeable which enables their use for live-cell imaging, compared to the cyanine dyes which are not.

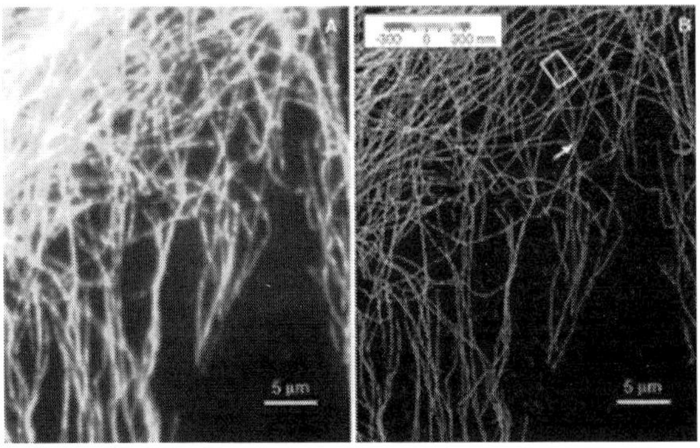

Fig. 6. **A three dimensional STORM image of microtubules in a BS-C-1 cell.**
(**A**) Conventional immunofluorescence imaging of microtubules. (**B**) The 3D STORM image of the same area using Cy3 and Alexa 647 photoswitchabe cyanine pair. A red laser (657 nm) was used to image Alexa 647 molecules and deactivate them to the dark state; a green laser (532 nm) was used to reactivate Alexa 646 in a Cy3-dependent manner. Reprinted with permission (Huang et al. 2008).

Another important class of dyes are the irreversible caged fluorophores such as the caged Q-rhodamine (Gee, Weinberg, and Kozlowski 2001) although these compounds have not been used for super-resolution imaging of biological samples.

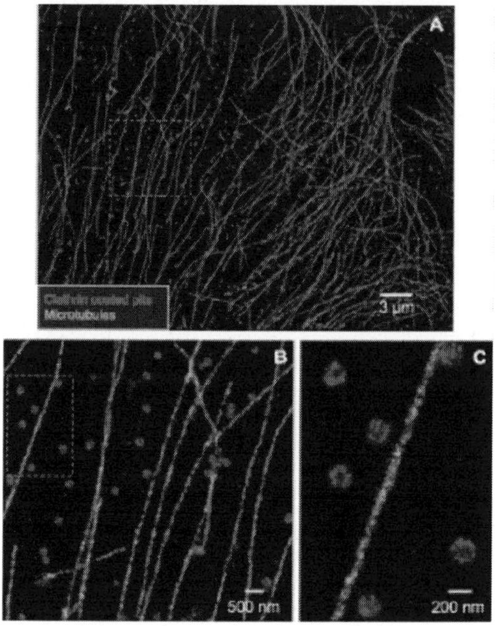

Fig. 7. **Two-colour STORM imaging of microtubules and clathrin coated pits in a mammalian cell.** (**A**) STORM image of a large area of a BS-C-1 cell. The microtubules were immunostained with Cy2 and Alexa 647, and those for clathrin with Cy3 and Alexa 647. The 457 nm and 532 nm laser pulses were used to selectively activate the two pairs of fluorophores. Each localization was false coloured according to the following code: green for 457 nm activation and red for 532 nm activation. (**B**) Enlarged STORM image of the boxed area. (**C**) Further magnification of the boxed area. Reprinted with permission (Bates et al. 2007).

7. Site-specific targeting of fluorophores to cellular proteins

Although non-genetically encoded probes generally show increased brightness and photostability compared to their fluorescent protein counterparts, they have their disadvantages. The lack of genetic encoding means that these probes require targeting to the biomolecule of interest inside the cell. These probes have been traditionally targeted using antibodies although their application is not widespread. Antibodies are not membrane permeable, and hence are not useful for labelling living cells intracellulary. Antibody staining also usually results in a low labelling efficiency and the large size of antibodies contributes to uncertainty in the spatial relationship between the label and target (Fernandez-Suarez and Ting 2008).

Some current approaches to site-specific labelling of biomolecules in living cells has been reviewed by Fernandez-Suarez and Ting (Fernandez-Suarez and Ting 2008). One method involves fusion of a peptide that recruits a small molecule to the protein of interest (Martin et al. 2005; Lata et al. 2006). Other methodologies use proteins to recruit the small molecule tag (Marks, Braun, and Nolan 2004; Bonasio et al. 2007) which can improve the specificity of binding due to the larger interaction surface although the increased size of this protein can perturb protein/enzyme function. In a combination method which seeks to achieve high labelling specificity with minimal perturbation to the protein target, a peptide recognition sequence has been used comprising an enzyme to catalyse the attachment of the probe to the sequence (Fernandez-Suarez et al. 2007).

8. Perspectives on emerging applications of super-resolution microscopy in live cells

The major technological principles of super-resolution microscopy (SIM, STED and PALM/STORM) have now matured to the extent that they have been implemented in commercially available systems that are relatively easy to use and within reach for well-established research laboratories. Thus, it is likely that we are standing at the beginning of an era of groundbreaking discoveries, fuelled by a multitude of applications of these novel imaging approaches to the challenging questions in cell biology.

Substantial potential for super-resolution imaging exists, for example, in understanding the structural basis of signal transduction within cells. Aspects of the organization and function of lipid rafts or microdomains in the cell membrane have been controversially discussed in the past, and imaging with resolution on the nanometer scale now allows addressing questions such as the molecular composition and dynamics of putative signaling complexes (Lang and Rizzoli 2010) (see Figure 8), the dynamic cytoskeletal changes underlying cell motility and migration, the way plasmamembrane structures are linked to and interact with the cytoskeleton (Ahmed 2011), or how cells interact with substrate molecules.

Our understanding of how cells communicate *in vitro* or even in the context of live tissues is set to benefit substantially from super-resolution technologies. The STED approach has been used to analyse the subcellular distribution of Na-K-ATPase in neurons (Blom et al. 2011) and to map synaptic spines in live brain tissue (Nagerl and Bonhoeffer 2010) (see Figure 9). Protein localization in chemical synapses has been investigated using STORM imaging (Dani et al. 2010). In the context of immunology, super-resolution imaging has been applied to study the composition of the immunological synapse (Dani et al. 2010) (See Figure 10) and it is now well within reach to visualize the dynamic process of how viral particles interact with immune cells, as recently shown (Felts et al. 2010). It is even possible to map GFP-tagged proteins in live multicellular organisms, as has been demonstrated in the nematode C. elegans, using STED (Rankin et al. 2011).

Super-resolution microscopy will also enhance our ability to study molecular interactions, based on signal colocalization, FRET analysis or the genetic engineering of constructs that emit fluorescence when two interaction partners are in close proximity, as has been demonstrated (Ahmed 2011).

Beyond the study of proteins, super-resolution imaging, particularly STED due to its high temporal resolution and the fact that it is based on the simultaneous imaging of a number of fluorophores in a given volume, has the potential of becoming a powerful tool to study cell physiology using diffusible fluorescent indicator dyes, e.g. for Ca^{2+} (Nagerl and Bonhoeffer 2010). Single-molecule super-resolution approaches using such dyes have been employed to visualize single ion channels (Patterson et al. 2010; Wiltgen, Smith, and Parker 2010).

In summary, each of the different approaches to super-resolution microscopy holds enormous potential in addressing key questions in current cell biology. However, they also have their characteristic advantages and drawbacks. SIM is a widefield technique easily implemented and not very demanding in terms of specimen preparation and labelling that can be used for multichannel fluorescence detection and is reasonably suitable for imaging of dynamic processes (image acquisition rates upward of 10 frames/s are possible), but it has comparatively low resolution upward of 50 nm. PALM, STORM and their derivatives are widefield fluorescence microscopy-based techniques that are currently achieving the highest resolution (in the range of 20 nm) and allow for multi-channel fluorescence imaging, but are largely confined to analysis of static or relatively slow processes in the order of minutes and in thin monolayers of cells or tissue sections. STED is a confocal laser scanning-based technique, allowing for imaging of fast dynamic processes in the range of milliseconds and analysis of relatively thick tissue slices with high lateral resolution. STED as well as PALM/STORM techniques have very specific requirements with regard to specimen preparation and labelling and the potential of these techniques is still limited to some extent by the availability of suitable fluorophores.

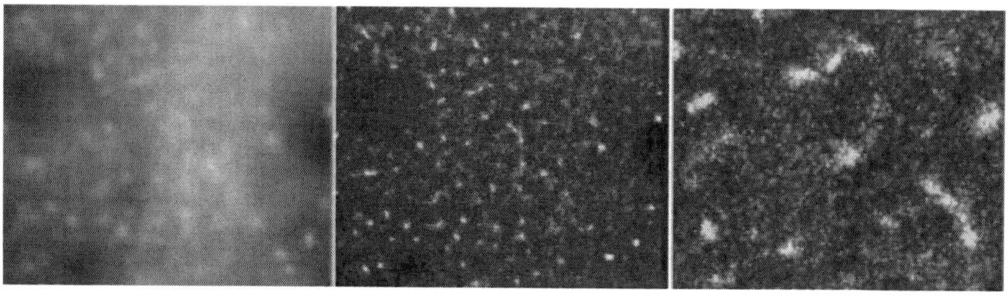

Fig. 8. **Application of TIRF and PALM imaging**. Demonstration of clusters of transferrin receptor (labeled with PalmCherry, red) and clathrin light chain (labeled with PAGFP, green) in the cell membrane by TIRF-microscopy (left) and PALM (middle; right, magnified view) (Lang and Rizzoli 2010).

As super-resolution microscopy techniques become established tools in cell biology research, a future challenge will be to design multimodal imaging approaches that combine the strengths of the different techniques. There is also a need to develop more fluorophores that are suitable for live-cell labelling, have sufficient quantum yield and provide a palette of spectral ranges suitable for the sensitive and simultaneous labelling of multiple cellular components. The drive towards a more sophisticated microscope, light source and computing hardware is still likely to lead to substantial improvements in the theoretically unlimited resolution beyond the diffraction barrier, and will enhance the capability of the systems for temporal resolution and 3-dimensional imaging.

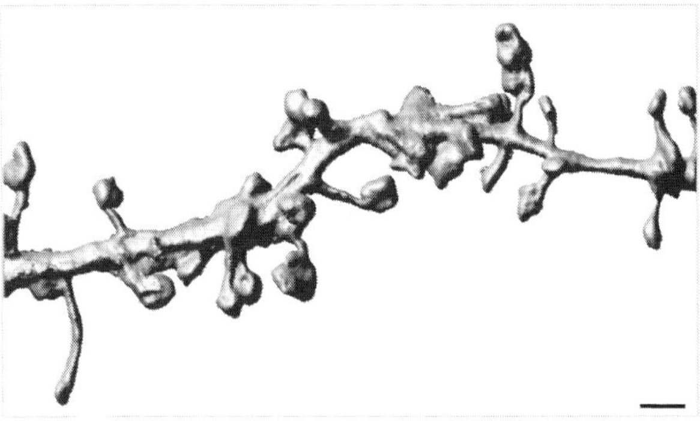

Fig. 9. **Application of STED imaging**. STED-based 3-dimensional reconstruction of dendritic spines genetically tagged with GFP. Scale bar: 1µm (Nagerl and Bonhoeffer 2010).

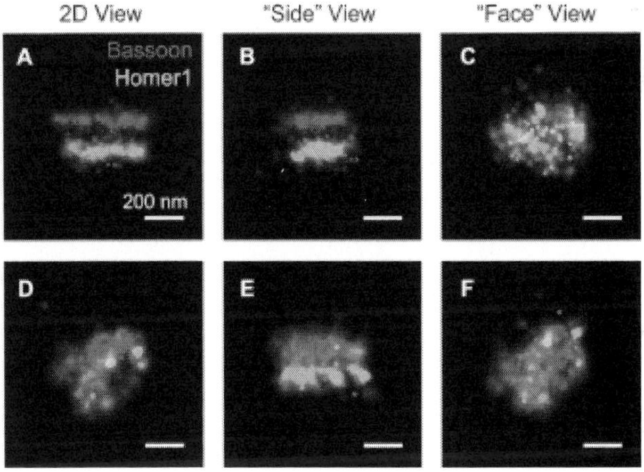

Fig. 10. **Application of STORM imaging**. STORM imaging of the pre-synaptic protein Bassoon and post-synaptic Homer1 using STORM super-resolution microscopy (Dani et al. 2010).

9. Conclusion

With the development of super-resolution imaging techniques it is now possible to image live cells down to tens of nanometers. STED imaging has allowed video rate tracking of synaptic vesicles in a dendritic spine down to 62 nm spatial resolution (Westphal et al. 2008), whereas STED, STORM and PALM have allowed cellular structures to be imaged in 3D and multiple colours. With such improved resolution, protein pairs have been visualised that contradict previous reports (Shroff et al. 2007) demonstrating the power of visualizing biomolecules of high resolution. For further improvements in spatial and temporal resolution, increased computational methods as well as fluorophores and site-specific live cell labelling are required.

10. Acknowledgments

This work was supported by the South African Research Chairs Initiative (SARCHI) of the Department of Science and Technology, the National Research Foundation (NRF), and research grants from the Medical Research Council (MRC) of South Africa and the University of Cape Town (UCT).

11. References

Abbe, E. 1873. "Beitrage zur Theorie des Mikroscops und der mikroskopischen Wahrnehmung." *Arch. Mikr. Anat.* no. 9:413-468.

Ahmed, S. 2011. "Nanoscopy of cell architecture: the actin-membrane interface." *Bioarchitecture* no. 1:32-38.

Ando, R., H. Mizuno, and A. Miyawaki. 2004. "Regulated fast nucleocytoplasmic shuttling observed by reversible protein highlighting" *Nature Biotechnology* no. 306 (5700):1370-1373.

Andresen, M., A.C. Stiel, J. Folling, D. Wenzel, A. Schonle, A. Egner, C. Eggeling, S.W. Hell, and S. Jakobs. 2008. "Photoswitchable fluorescent proteins enable monochromatic mulilabel imaging and duel color fluorescence nanoscopy." *Nature Biotechnology* no. 26 (9):1035-1040.

Bates, M., T.R. Blosser, and X. Zhuang. 2005. "Short-Range Spectroscopic Ruler Based on a Single-Molecuke Opical Switch." *Physical Review Letters* no. 94:108101.

Bates, M., B. Huang, G.T. Dempsey, and X. Zhuang. 2007. "Multicolor Super-Resolution Imaging with Photo-Switchable Fluorescent Probes." *Science* no. 317:1749-1753.

Betzig, E., and J.K. Trautman. 1992. "Near-field optics: microscopy, spectroscopy, and surface modification beyond the diffraction limit." *Science* no. 257 (5067):189-195.

Blom, H., D. Ronnlund, L. Scott, Z. Spicarova, J. Widengren, A. Bondar, A. Aperia, and H. Brismar. 2011. "Spatial distribution of Na+ K+ ATPase in dendritic spines dissected by nanoscale superresolution STED microscopy." *BMC Neuroscience* no. 12:16.

Bonasio, R., C.V. Carman, E. Kim, P.T. Sage, K.R. Love, T.R. Mempel, T.A. Springer, and U.H. von Andrian. 2007. "Specific and covalent labeling of a membrane protein with organic fluorochromes and quantum dots." *Proceedings of the National Academy of Science USA* no. 104 (37):14753-14758.

Dani, A., B. Huang, J. Bergan, C. Dulac, and X. Zhuang. 2010. "Superresolution imaging of chemical synapses in the brain." *Neuron* no. 68 (5):843-856.

Felts, R.L., K. Narayan, J.D. Estes, D. Shi, C.M. Trubey, J. Fu, L.M. Hartnell, G.T. Ruthel, D.K. Schneider, K. Nagashima, J.W. Jr. Bess, S. Bavari, B.C. Lowekamp, D. Bliss, J.D. Lifson, and S. Subramaniam. 2010. "3D visualization of HIV transfer at the virological synapse between dendritic cells and T cells." *Proceedings of the National Academy of Science USA* no. 107 (30):13336-13341.

Fernandez-Suarez, M., H. Baruah, L. Martinez-Hernandez, K.T. Xie, J.M. Baskin, C.R. Bertozzi, and A.Y. Ting. 2007. "Redirecting lipoic acid ligase for cell surface protein labeling wioth small molecule probes." *Nature Biotechnology* no. 25 (12):1483-1487.

Fernandez-Suarez, M., and A.Y. Ting. 2008. "Fluorescent probes for super-resolution imaging in living cells." *Nature Reviews Molecular Cell Biology* no. 9:292-944.

Gee, K.R., E.S. Weinberg, and D.J. Kozlowski. 2001. "Caged Q-rhodamine dextran: a new photoactivated fluorescent tracer." *Bioorganic and Medicinal Chemistry Letters* no. 11:2181-2183.

Heilemann, M. 2010. "Fluorescence Microscopy Beyond the Diffraction Limit." *Journal of Biotechnology* no. 149:243-251.

Hell, S.W. 2007. "Far-Field Optical Nanoscopy." *Science* no. 316:1153-1158.

Huang, B., H. Babcock, and X. Zhuang. 2010. "Breaking the Diffraction Barrier: Super-Resolution Imaging of Cells." *Cell* no. 143 (7):1047-1058.

Huang, B., W. Wang, M. Bates, and X. Zhuang. 2008. "Three-dimensional Super-Resolution Imnaging by Stochastic Optical Reconstruction Microscopy." *Science* no. 319:810-813.

Kural, C., H. Kim, S. Syed, G. Goshima, V.I. Gelfand, and P.R. Selvin. 2005. "Keneisin and SDynein Move a Peroxisome in Vivo: A Tug-of-War or Coordinated Movement." *Science* no. 308:1469-1472.

Lang, T., and S.O. Rizzoli. 2010. "Membrane protein clusters at nanoscale resolution: more than pretty pictures." *Physiology* no. 25 (2):116-124.

Lata, S., M. Gavutis, R. Tampe, and J. Piehler. 2006. "Specific and stable fluorescence labeling of histidibne-tagged proteins for dissecting mulit-protein complex formation " *Journal of the American Chemical Society* no. 128:2365-2372.

Lukyanov, K.A., D.M. Chudakov, S. Lukyanov, and V.V. Verkusha. 2005. "Photoactivatable Fluorescent Proteins." *Nature Reviews: Moleculaer Cell Biology* no. 6:885-891.

Manley, S., J.M. Gillette, G.H. Patterson, H. Schroff, H.F. Hess, E. Betzig, and J. Lippincott-Schwartz. 2008. "High-density mapping of Single Molecule Trajectories with Photoactivated Localization Microscopy." *Nature Methods* no. 5 (2):155-157.

Marks, K.M., P.D. Braun, and G.P. Nolan. 2004. "A general approach for chemical labelling and rapid spatially controlled protein inactivation." *Proceedings of the National Academy of Science USA* no. 101:9982-9987.

Martin, B.R., B.N. Giepmans, S.R. Adams, and R.Y. Tsien. 2005. "Mammalian cell-based optimization of the biarsenical-binding tetracysteine motif for improved fluorescence and affinity." *Nature Biotechnology* no. 23 (10):1308-1314.

Moneron, G., R. Medda, B. Hein, A. Giske, V. Westphal, and S.W. Hell. 2010. "Fast STED Microscopy with Continuous Wave Fiber Lasers." *Optics Express* no. 18 (2):1302-1308.

Nagerl, U.V., and T. Bonhoeffer. 2010. "Imaging Living Synapses at the Nanoscale by STED Microscopy." *The Journal of Neuroscience* no. 30 (28):9341-9346.

Patterson, G., M. Davidson, S. Manley, and J. Lippincott-Schwartz. 2010. "Superresolution imaging using single-molecule localization." *Annual Reviews in Physical Chemistry* no. 61:345-367.

Rankin, B.R., G. Moneron, C.A. Wurm, J.C. Nelson, A. Walter, D. Schwarzer, J. Schroeder, D.A. Colon-Ramos, and S.W. Hell. 2011. "Nanoscopy in a living multicellular organism expressing GFP." *Biophysical Journal* no. 100 (12):L63-65.

Schermelleh, L., P.M. Carlton, S. Haase, L. Shoa, L. Winito, P. Kner, B. Burke, M.C. Cardoso, D.A. Agard, M.G.L. Gustafsson, H. Leonhardt, and J.W. Sedat. 2008.

"Subdiffraction multicolor imaging of the nuclear periphery with 3D structured illumination microscopy." *Science* no. 320:1332-1336.

Schmidt, R., C.A. Wurm, C.A. Punge, A. Egner, S. Jakobs, and S.W. Hell. 2009. "Mitochondrial Cristae Revealed with Focused Light." *Nano Letters* no. 9 (6):2508-2510.

Shroff, H., C.G. Galbraith, J.A. Galbraith, H. White, J.M. Gillette, S. Olenych, M.W. Davidson, and E. Betzig. 2007. "Dual-color superresolution imaging of genetically expressed probes within individual adhesion complexes." *PNAS* no. 104 (51):20308-20313.

Toomre, D., and J. Bewersdorf. 2010. "A New Wave of Cellular Imaging." *The Annual Review of Cell and Developmental Biology* no. 26:285-314.

Westphal, V., S.O. Rizzoli, M.A. Lauterbach, D. Kamin, R. Jahn, and S.W. Hell. 2008. "Video-Rate Far-Field Optical Nanoscopy Dissects Synaptic Vesicle Movement." *Science* no. 320:246-249.

Wiedenmann, J., S. Ivanchenko, F. Oswald, F. Schmidtt, C. Rocker, A. Salih, K-D. Spindler, and G.U. Nienhaus. 2004. "EosFP, a fluorescent marker protein with UV-inducible green-to-red fluorescence conversion." *PNAS* no. 101 (45):15905-15910.

Willig, K.I., B. Harke, R. Medda, and S.W. Hell. 2007. "STED Microscopy with Continuous Wave Beams." *Nature Methods* no. 4 (11):915-918.

Wiltgen, S.M., I.F. Smith, and I. Parker. 2010. "Superresolution ,localization of single functional IP3R channels utilizing Ca2+ flux as a readout." *Biophysical Journal* no. 99 (2):437-446.

Yildiz, A., J.N. Forkey, S.A. McKinney, H. Taekjip, Y.E. Goldman, and P.R. Selvin. 2003. "Myosin V Walks Hand-Over-Hand: Singled Flurophore Imaging with 1.5nm Localization." *Science* no. 300 (5628):2061-2065.

Approaches to Analyze Protein-Protein Interactions of Membrane Proteins

Sabine Hunke* and Volker S. Müller
Molekulare Mikrobiologie, Universität Osnabrück, Osnabrück, Germany

1. Introduction

About one quarter of an organismal genome encodes membrane proteins that play key roles in signal transduction, transport, energy recruitment and virulence traits of bacterial pathogens (Jones 1998; Krogh et al. 2001). The significance of membrane proteins is reflected by the fact that about 60% of all pharmaceuticals target membrane proteins (Bakheet and Doig 2009; Yildirim et al. 2007).

It can be estimated that most membrane proteins function in complexes (Fig. 1) (Daley 2008). Protein-protein interactions (PPIs) within these complexes can either be direct (primary interaction) or indirect (secondary interaction). Direct interactions occur either by homo-oligomerisation as determined for bacterial two-component systems (Gao and Stock 2009) (Fig. 1A) or by hetero-oligomerisation as shown for transport systems like ATP-binding cassette (ABC) transporters (Figs. 1B and 1C). Indirect interactions exist in large complexes as exemplified in energy producing systems such as the photosystem, bacterial surface appendages such as flagella, or secretion systems that even span two membrane systems in Gram-negative bacteria (Fig. 1D) (Jordan et al. 2001; Erhardt, Namba, and Hughes 2010). These high affinity, stable PPIs are important to form stable functional complexes (Jura et al. 2011). In addition, low affinity, transient PPIs are needed for proteins that regulate the activity of a stable complex and have been described for the interaction between e.g. ABC protein and inhibitory $EIIa^{Glc}$ (Blüschke et al. 2007; Blüschke, Volkmer-Engert, and Schneider 2006), substrate binding protein and ABC transporter (Locher, Lee, and Rees 2002) or accessory proteins in two-component systems (Heermann and Jung 2010; Buelow and Raivio 2010; Zhou et al. 2011).

Thus, there is high demand for techniques to screen for interactions partners of and to characterize the interaction with a specific membrane protein. However, due to the hydrophobic nature of membrane proteins application of classical approaches is far more challenging than for soluble proteins (Daley 2008). Recent reviews summarize and discuss approaches to investigate PPIs for soluble proteins (Lalonde et al. 2008; Miernyk and Thelen 2008). Here, we give an overview on the current techniques used to determine and characterize PPIs of membrane proteins.

---

* Corresponding Author

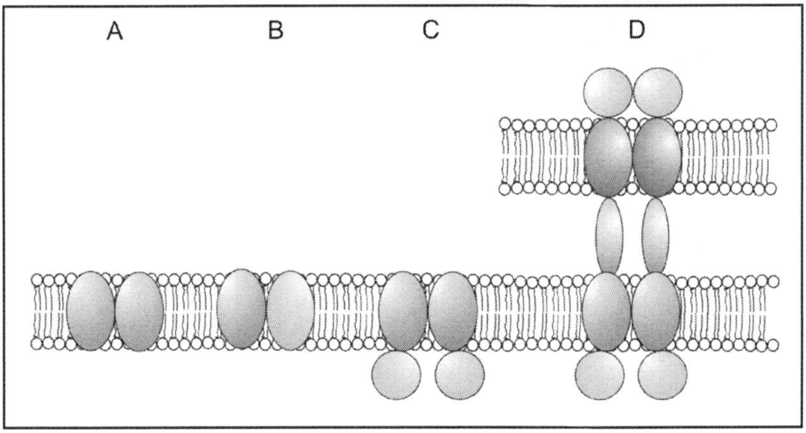

Fig. 1. Protein-protein interaction (PPIs) of membrane proteins.
Membrane proteins can be assembled as homo (A) or hetero (B) oligomers. (C) In addition, PPIs of membrane proteins exist with peripheral proteins on either site of the membrane. (D) Some membrane proteins are part of huge multi-protein complexes that can even span two membranes such as secretion systems in Gram-negative bacteria (Filloux 2011).

2. Determination of membrane protein-protein interactions

Several genetic and biochemical techniques have been developed to determine PPIs of membrane proteins. In general, these approaches use either protein fragment complementation assays (PCA) or combine affinity purification with mass spectrometry analysis (AP-MS). PCAs are based on the reconstituted interaction of a protein function (reporter) by fusing two proteins of interest to complementary fragments of the reporter protein (Ladant and Karimova 2000; Remy and Michnick 2007). AP-MS analysis allows the determination of indirect interactions in a complex. A major advantage of these techniques is that the protocols can be applied to almost any cell type or organism. Noteworthy, all AP-MS approaches need a rather large amount of material and a suitable affinity tag. Both types of screening methods, PCA and AP-MS, enable high-throughput analysis. However, the use of high-throughput approaches may compass high levels of false-positive results and consequently, novel PPI partners identified have to be confirmed by alternative methods (Lalonde et al. 2008; Miernyk and Thelen 2008).

2.1 Genetic systems to analyze membrane protein-protein interactions in eukaryotes

Genetic systems established to investigate PPIs in eukaryotes use PCA. Based on classic yeast two-hybrid systems (Fields and Song 1989) that are limited because the fusion proteins have to be translocated into the nucleus to activate a reporter gene, new adequate methods have been developed to overcome this limitation and allow now the analysis of membrane protein PPIs in eukaryotes.

2.1.1 Protein-fragment complementation assays

Johnsson and Varsharvsky invented the split-ubiquitin yeast two-hybrid system (SU-YTH), a system using the endogenous mechanism of cleavage of ubiquitin by ubiquitin-specific proteases (UBPs) (Johnsson and Varshavsky 1994; Johnsson and Varshavsky 1994). Ubiquitin is the recognition marker for UBPs and can be separated into the C-terminal (Cub) and the N-terminal (Nub) part when both prey and bait are in close proximity, These two parts fused to a bait and prey protein are able to reassociate spontaneously to a quasi-native ubiquitin-molecule which can be recognized by UBPs. These proteases cleave the C-terminal attached reporter polypeptide from Cub and thereby enable the reporter transcription factor to translocate into the nucleus and to activate the reporter genes.

A second PCA represents the dihydrofolate reductase (DHFR) strategy which is also called survival selection strategy (Ear and Michnick 2009). Bait and prey proteins are fused to corresponding fragments of a modified DHFR, insensitive to methotrexate which is reconstituted and active if both interaction partners are in close proximity. The proliferation of the cells is depend on DHFR which catalyzes the reduction of dihydrofolate to tetrahydropholate during the synthesis of nucleotides and several amino acids and can be inhibited by methotexate (Remy, Campbell-Valois, and Michnick 2007; Pelletier, Campbell-Valois, and Michnick 1998). Thus only cell carrying interacting fragments of the mutated DHFR which are reassembled due to the interaction of the bait and prey proteins are able to proliferate and survive in the presence of the inhibitor methotrexate.

Besides the DHFR, different reporter enzymes can be used for a PCA strategy e.g. yeast cyosine deaminase (OyCD) (Ear and Michnick 2009) or fluorescent proteins (see chapter 3.6.3) extensively reviewed (Michnick et al. 2011).

2.1.2 Reverse Ras recruitment system (reverse RRS)

An alternative method to SU-YTH is the reverse Ras recruitment system (reverse RRS) (Hubsman, Yudkovsky, and Aronheim 2001) that is based on Ras recruitment system (RRS) in yeast (Broder, Katz, and Aronheim 1998). Growth of yeast depends on cAMP. cAMP is generated by adenylate cyclase which is activated by Ras which itself is activated by Cdc25 (Cannon, Gibbs, and Tatchell 1986). In contrast to cytoplasmic Ras, membrane-bound Ras complement a temperature-sensitive mutant in Cdc25 (Aronheim et al. 1997; Aronheim 1997) (Petitjean, Hilger, and Tatchell 1990). PPI between a membrane protein and its interaction partner results in Ras translocation and allows cell growth at elevated temperature (Aronheim 2001). Thus, reverse RRS can be used to screen for a soluble protein as PPI partner for a membrane protein

2.2 Genetic systems to analyze membrane protein-protein interactions in bacteria

PCA can also be used as genetic systems to analyze membrane protein-protein interactions in bacteria. PCAs based on a protein function directly involved in transcription are restricted to determine the interaction of soluble proteins (bacteriophage lambda repressor λcI, *E. coli* LexA repressor, DNA loop formation, RNA polymerase recruitment) (Ladant and Karimova 2000). In contrast, PCAs based on metabolism or signaling cascades can be adapted for membrane proteins (bacterial mDHFR survival assay, BACTH).

2.2.1 Murine dihydrofolate reductase (mDHFR)

A PCA based on the essential DHFR has also been established to screen for PPI in bacteria. Prokaryotic DHFR but not murine DHFR (mDHFR) is inhibited by trimethoprim (Appleman et al. 1988). PPI of the two proteins of interest fused to mDHFR fragments allow *E. coli* to grow on media supplemented with trimethoprim (Remy, Campbell-Valois, and Michnick 2007), allowing a positive selection.

2.2.2 Bacterial adenylate cyclase two-hybrid assay (BACTH)

The bacterial adenylate cyclase two-hybrid assay (BACTH) is well established to investigate PPIs for membrane proteins in bacteria (Fig. 2) (Karimova, Dautin, and Ladant 2005). BACTH is based on the reconstituted interaction of two *Bordetella pertussis* adenylate cyclase fragments (T18 and T25) resulting in elevated levels of cAMP (Karimova et al. 1998). cAMP is a key signaling molecule in *E. coli* that activates the catabolite activator protein (CAP) resulting in transcriptional activation of metabolic operons such as those for lactose and maltose (Deutscher, Francke, and Postma 2006). Consequently, PPI of two membrane proteins fused to adenylate cyclase fragments results in fermentation of lactose or maltose which can easily be detected on either indicator (MacConkey maltose or X-Gal plates) or selection media (minimal media supplemented with either lactose or maltose as carbon source) (Karimova, Dautin, and Ladant 2005). In addition, BACTH allows quantification of the PPI by measuring the activity of the lactose cleaving β-galactosidase (Robichon et al. 2011).

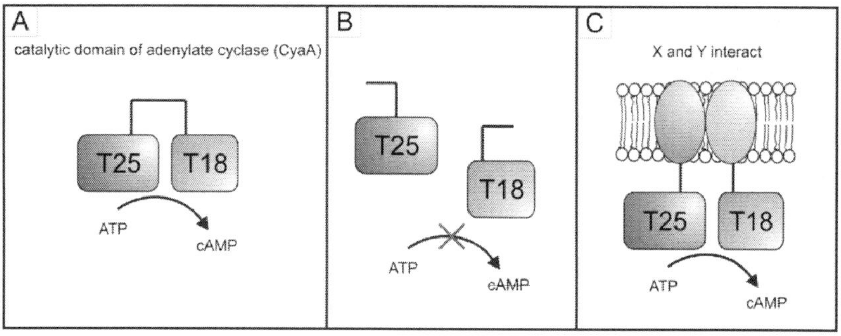

Fig. 2. The bacterial adenylate cyclase two-hybrid assay (BACTH). (A) The adenylate cyclase (CyaA) of *Bordetella pertussis* synthesizes cAMP in *E. coli*. cAMP activates the catabolite activator protein resulting in target gene expression. (B) Coexpression of two CyaA fragments (T25, T18) does not result in protein fragment complementation (PCA). (C) Fusion of T25 and T18 to interacting membrane proteins results in PCA and cAMP production (Karimova et al. 1998; Karimova, Dautin, and Ladant 2005).

2.3 Co-immunoprecipitation

Co-immunoprecipitation (co-IP) is the classical method to screen for and the proof of PPIs. On the one hand, co-IP is a highly specific, yet relatively simple technique that allows the identification of two or more proteins *in vivo* (Miernyk and Thelen 2008). On the other hand, co-IP requires an antibody with high specificity for the protein of interest. However, highly specific antibodies for membrane proteins are even more difficult to obtain than for soluble proteins. Therefore, when using co-IP for membrane proteins false-positive results are reduced by pre-clearing solubilized membrane proteins by the addition of immobilized protein A or protein G (protein A/G) (Vaidyanathan et al. 2010). The pre-cleared supernatant is then incubated with the primary antibodies and protein-antibody complexes are formed. The protein-antibody complexes are recovered using immobilized protein A/G. Using co-IP mainly strong PPIs as found in complexes can be detected, but transient interactions are rather difficult to be determined.

2.4 Tandem affinity purification (TAP)

One AP-MS technique is the tandem affinity purification (TAP) method (Xu et al. 2010; Puig et al. 2001; Rigaud, Pitard, and Levy 1995). TAP allows the identification of direct interactions in a protein complex and uses the fusion of a TAP tag to the either the N or the C terminus of the target protein (Xu et al. 2010). The TAP tag consists of a calmodulin-binding domain (CBD), a cleavage site for the tobacco etch virus (TEV), and the IgG binding units of the protein A of *Staphylococcus aureus*. Protein complexes containing a TAP tagged protein can be purified by two very specific purification steps. In the first purification step the TAP tagged complex is bound to an IgG column. Elution occurs by cleaving off the protein complex from the column using the TEV cleavage site of the TAP tag. In the second purification step, the TAP-tagged complex is bound to calmodulin beads. After EGTA elution the complex can be further analyzed with respect to the interaction partners of the TAP-tagged bait protein. When using mild detergent the TAP method can also be adapted for membrane proteins. However, the TAP system is considered to be inefficient in identifying transient interactions (Xu et al. 2010). A very recent review describes the application and limits of TAP in detail (Xu et al. 2010).

2.5 Chemical cross-linking and mass spectrometry techniques

2.5.1 Protein Interaction Reporter (PIR) technology

IP-based affinity purification methods always require the genetically introduction of a tag fused to a target protein of interest. The overexpression of the fusion proteins can lead to improper intercellular localization and by this to false positives (Bouwmeester et al. 2004). Furthermore, the co-elution of potential interaction partners of a target protein (see chapter 2.3 and 2.4) is often negatively affected during the purification of the target proteins afterwards. The protein interaction reporter (PIR) technology established by Xiaoting Tang and James E. Bruce overcomes these limitations by the use of new design of crosslinker. These cross-linkers include two reactive groups to cross link potential interaction partners, two labile bonds and a mass encoded reporter containing an affinity tag. The labile bonds can be cleaved afterwards by UV irradiation prior to the identification of interaction partners. The applications of the PIR technology have been extensively reviewed (Hoopmann, Weisbrod, and Bruce 2010; Tang and Bruce 2010; Yang et al. 2010).

2.5.2 Membrane strep–protein interaction experiment (SPINE)

Membrane-SPINE is an improved technique based on the Strep-protein interaction experiment (SPINE) adapted to membrane proteins (Herzberg et al., 2007). It combines the fixation of protein complexes in a cell by formaldehyde cross-linking *in vivo* with the specific purification of a Strep-tagged target membrane protein (Müller et al. 2011). Due to its small size formaldehyde can easily penetrate membranes and create an effective snap shot of the interactome of a living cell (Fig. 3). Thus not only the target protein but also cross-linked potential interaction partner can be co-eluted (Fig. 3 B) and identified afterwards by Mass spectrometry (Fig. 3 E) or immunoblot analysis (Fig. 3 D). By using Membrane-SPINE it is possible to monitor not only permanent protein-protein interactions but also transient interactions occurring during signal transduction (Müller et al. 2011).

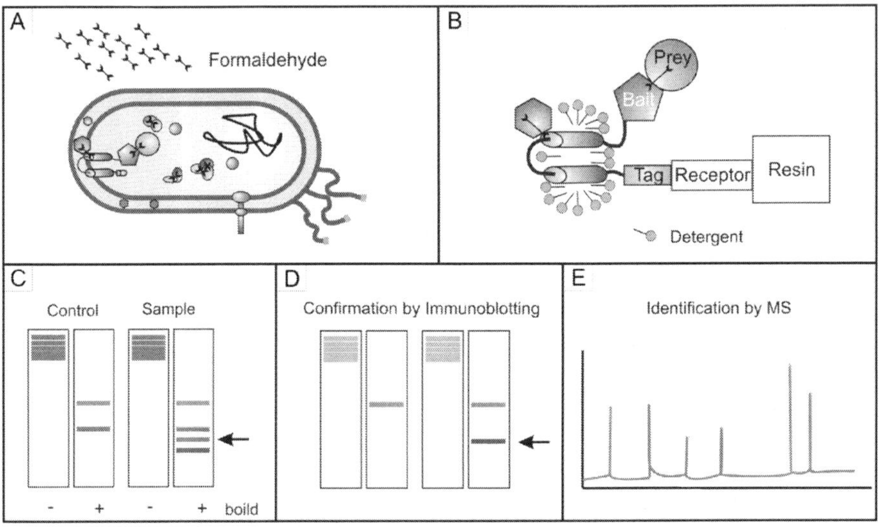

Fig. 3. Membrane-SPINE. (A) Protein complexes of a living cell are fixed by formaldehyde cross-linking. (B) After detergent solubilization the strep-tagged membrane protein is purified via a strep-tactin resin. (C) Co-eluted interaction partners of the strep-tagged membrane protein can be separated after boiling through SDS-PAGE and finally confirmed by immunoblotting (D) or identified by MS analysis (E) (Müller et al. 2011).

2.6 *In silico* prediction of membrane protein-protein interactions

The recent progress in the fields of bioinformatics has culminated in the development of powerful tools for the prediction of protein-protein interactions *in silico*. The growing amount of data of protein interactions and protein sequence information have been successfully used for the prediction of new protein interaction networks by the homogenous protein mapping method (Saeed and Deane 2008) or co-evolution analysis (Skerker et al. 2008). By using a computational approach Skerker and colleagues compared by sequence alignment nearly 1300 pairs of histidine kinases (HK) and response regulators (RR) of two-component systems of almost 200 sequenced bacterial genomes to identify the structural basis determining the interaction between HK and RR (Skerker et al. 2008). In the same line, Procaccini and coworkers identified a molecular interaction code between HK and RR by comparing 8998 paired SK/RR sequences were of 769 fully sequenced bacterial genomes which results in a highly specific preference between both interaction partners. Based on this code, it was possible to predict clusters of cross-talk candidates between non-cognate signaling partners (Procaccini et al., 2011). However, all these predictions can strongly contradict the situation *in vivo* which emphasizes the importance of the methods presented within this review for verification.

3. Characterization of membrane protein-protein interactions

Different methods are available to characterize PPIs between membrane proteins. They allow the investigation of functional interaction, kinetics, and affinities between membrane proteins. Moreover, the dynamics of the interaction, the interface between the proteins and the cellular localization can be determined.

3.1 Reconstitution of membrane proteins for functional studies

For detailed biochemical investigation of membrane proteins the incorporation of the purified proteins into a lipid bilayer is essential. This technique is also known as reconstitution and results in proteoliposomes that allow the characterization of membrane proteins without the influence of other membrane components (Rigaud 2002; Rigaud and Levy 2003; Paternostre, Roux, and Rigaud 1988). The importance of reconstitution results from the observation that many membrane proteins are only fully active when incorporated into a lipid bilayer (Rigaud 2002; Fleischer et al. 2007).

Basis for any successful reconstitution are the quality of the purified membrane protein, the lipid requirement of the membrane protein and the ratio between lipid and protein (Geertsma et al. 2008; Knol, Sjollema, and Poolman 1998). The usage of mild detergents, such as n-dodecyl-β-maltoside (DDM) or Triton X-100, is highly recommended to reduce dissociation during purification and to keep by this the complex active (Geertsma et al. 2008). The addition of chemical chaperones (glycerol, salt), phospholipids or ligands can further stabilize membrane protein complexes during the purification procedure (Geertsma et al. 2008).

To reconstitute a protein function into proteoliposomes, detergent-solubilized and purified membrane protein is mixed with detergent-destabilized lipid vesicles. In order to generate membrane vesicles and to incorporate the membrane protein into these vesicles the detergent has to be removed. Several techniques exist (dilution, dialysis, SEC) but when using mild detergents that have in general a low critical micellar concentration (CMC) the adsorption of the detergent onto polystyrene beads is the method of choice.

After reconstitution of a membrane protein into proteoliposomes several controls have to be performed: the morphology and residual permeability of the proteoliposomes have to be proved; the incorporation efficiency and the orientation of the membrane proteins have to be determined to allow kinetic studies; and the functionality of the reconstituted membrane proteins has to be proved by activity assays. We have observed during our studies with different kind of membrane proteins that not always the highest solubilization and purification efficiency results in the most active protein (Hunke et al., 1997; Fleischer et al., 2007). Notably, when working with the same membrane protein but from different organisms we had to change the detergent (Fleischer et al. 2007; Müller et al. 2011). We and others, use proteoliposomes not only to analyze the functional interaction between a membrane protein (reconstituted sensor kinase) and a soluble partner (cognate response regulator) (Fleischer et al. 2007; Jung, Tjaden, and Altendorf 1997) but also in order to investigate the impact of different conditions on this interaction (Fleischer et al. 2007).

Together, the knowledge on the physical background of lipid-detergent systems and the mechanisms of proteoliposome formation results in a number of basic principles in membrane protein reconstitution (Rigaud and Levy 2003; Silvius 1992) that allowed the establishment of general protocols (Geertsma et al. 2008).

3.2 Native electrophoretic techniques

Blue native electrophoresis (BNE; also known as Blue Native PAGE) has been developed to purify active membrane protein complexes from mitochondria (Schägger and von Jagow 1991). Therefore, membrane proteins are solubilzed using a mild neutral detergent and insoluble proteins are removed by centrifugation. Solubilized proteins are then mixed with the anionic dye Coomassie Brilliant Blue (CBB) G-250 which binds to protein surfaces (Compton and Jones 1985). Binding of CBB G-250 results in a negative charge shift and allows a protein complex to migrate into a non-denaturating polyacrylamid gel. Additional separation methods in the second dimension (2D) as e.g. denaturating SDS-PAGE can be used to separate single proteins of a membrane protein complex in order to estimate the native mass or to identify proteins of one membrane protein complex (Wittig and Schägger 2009). Because CBB interferes with the activity of proteins and the analysis of fluorescent-labeled proteins, clear-native electrophoresis (CNE) and high-resolution CNE (hrCNE) have been established (Wittig, Karas, and Schägger 2007). CNE uses no dye and proteins migrate according to their intrinsic p*I* (Wittig and Schägger 2009). The two disadvantages of CNE, proteins with a p*I*>7 are lost and smearing of the membrane proteins over the gel, can be partially compensated when applying hrCNE that uses detergent micelles to induce the charge shift (Wittig and Schägger 2008).

Extended recent reviews on BNE, CNE and hrCNE summarize the power of these techniques for the identification and characterization of PPIs for membrane proteins and explain the protocol in detail (Wittig and Schägger 2008, 2009; Krause 2006; Miernyk and Thelen 2008).

3.3 Far-Western Blot

Far-Western Blot analysis is an *in vitro* method to proof and to identify direct PPIs between two proteins (Edmondson and Roth 2001; Wu, Li, and Chen 2007).

When using Far-Western blot analysis to verify PPI, one protein is immobilized on a continuous membrane sheet and the blot is incubated with a purified second protein, the bait protein. Afterwards, the blot is treated as a normal immunoblot using an antibody against the bait protein. When using Far-Western blot analysis to identify a PPI between a bait protein and a prey protein in a cell lysate, the cell lysate is transferred to a continuous membrane sheet instead of a purified protein. Then, the blot is incubated with the bait protein and finally treated as an immune blot against the bait protein.

Although this method is most suitable for soluble proteins it can also be adapted to membrane proteins. However, detergent interferes with the immunoblot procedure (Zhou et al. 2011). Thus, when using Far-Western blot analysis for membrane proteins either the purified membrane protein or the membrane fractions should be immobilized to the continuous membrane sheet as the prey. In other words, Far-Western blot analysis allows the identification of a PPI between a membrane protein (in a membrane fraction) and a soluble bait protein without the need of purifying a membrane protein by detergent treatment.

Detailed protocols for the Fat-Western blot procedure are given by Edmondson & Roth (2001) and Wu et al. (2007).

3.4 SPOT-analysis

SPOT-analysis is an *in situ* screening technique developed for the identification of interacting epitopes (Frank 1992, 2002; Volkmer 2009). For SPOT-analysis a peptide array is generated by coupling single amino acids step by step first on a continuous membrane sheet then on the first amino acid and so on (Frank 2002; Reimer, Reineke, and Schneider-Mergener 2002; Wenschuh et al. 2000). By this, SPOT-analysis allows the rapid and parallel synthesis of different synthetic peptides that can be analyzed simultaneously. Even more important, SPOT-analysis permits the substitutional analysis of an epitope without the need of mutagenesis and purification (Volkmer 2009). As for Far-Western blot analysis the peptide array is first incubated with the bait protein and finally developed as normal immunoblot versus the bait protein. However, this elegant technique is limited on hydrophilic peptides and cannot be used to characterize PPIs of TMSs. Never the less, different groups have used SPOT-analysis to screen for and to identify epitopes in hydrophilic domains of membrane proteins important for the interaction with soluble proteins (Zhou et al. 2011; Blüschke, Volkmer-Engert, and Schneider 2006).

Generation of peptide arrays in macro- and micro-array format is described in detail by the groups of Frank and Schneider-Mergener (Frank 2002; Reimer, Reineke, and Schneider-Mergener 2002; Wenschuh et al. 2000).

3.5 Surface Plasmon Resonance (SPR)

The most elegant approach to quantify binding kinetics, thermodynamics and concentrations in PPIs *in vitro* is the surface plasmon resonance (SPR) technology. This technique needs both proteins to be purified.

When using SPR for soluble proteins, one purified protein is bound to a gold-coated surface of a chip. To obtain the background, the chip is floated with the buffer the second protein is purified with and the refractive index of the solvent near the gold surface is measured. In the next step the second purified protein in its buffer is floated and the refractive index is again measured. PPI is determined by the changes in refractive index.

Because membrane proteins cannot be bound to chip surface as efficient as soluble proteins, specific chips have been designed that allow the capture of proteoliposomes (Maynard et al. 2009). A detailed protocol for this approach is given by (Hodnik and Anderluh 2010). Technologies that allow the analysis of membrane proteins directly on a chip are still in develop (Maynard et al. 2009). Nevertheless SPR can already be utilized to analyze PPI between a membrane protein and a soluble protein. Therefore, the soluble protein is immobilized to a classical SPR chip and floated with proteoliposomes containing the membrane protein. This experimental setup has successfully been used to characterize the transient interaction between the membrane integral sensor kinase KdpD and the scaffolding protein UspC (Heermann et al. 2009).

3.6 Imaging technologies

Imaging technologies allow the characterization of PPIs in the native environment of proteins *in vivo* and in real time. Thus, they are excellent tools to study mechanisms in protein function.

In general, imaging technologies use genetic fusions between the protein of interest and a fluorescent protein (Fig. 4). Genetic fusions are easily applicable for any protein including membrane proteins. However, for membrane proteins it has to be taken into account that fluorescent proteins like the green fluorescent protein (GFP) are only folded correctly in the cytosol. Consequently, fusions between membrane proteins and fluorescent proteins should only be performed at those domains of a membrane protein known to be localized inside the cell.

Initial experiments to analyze PPIs with imaging technologies are co-localization studies of two-labeled proteins in order to determine their cellular distribution. Subsequently, a variety of imaging technologies can be used to characterize PPIs of membrane proteins in more detail (Lalonde et al. 2008; Schäferling and Nagl 2011).

3.6.1 Fluorescence resonance energy transfer (FRET)-based techniques

Fluorescence (or Förster) resonance energy transfer (FRET) is a biophysical method detecting energy transfer from a donor fluorophor to an acceptor fluorophor (Fig. 4A) (reviewed in (Masi et al. 2010; Schäferling and Nagl 2011). The principle was first described by Theodor Förster, 1948. The basis of FRET is the correct donor-acceptor pair. The emission wavelength of the donor fluorophor has to be in the range of the excitation wavelength for the acceptor fluorophor. When the two fluorophores are in sufficient proximity (2-8 nm) excitation of the donor induces energy emission that can be absorbed by the acceptor resulting in a characteristic energy emission of the acceptor. Well established donor acceptor pairs in cell biology are the combination of the cyan fluorescent proten (CFP) with the yellow fluorescent protein (YFP), GFP with rhodamine, fluoresceinisothiocyanate and Cy3, and CFP with the fluorescein arsenical helix binder (FlAsH) (Hoffmann et al. 2005).

To analyze the distance and dynamics of membrane proteins, cell lines are co-transfected (bacteria are co-transformed) with two vectors carrying a CFP -bait protein and an YFP-prey protein fusion. The FRET signal reflects the PPI between bait and prey and is determined by fluorescence microscopy. Recently, FRET has additionally been demonstrated as a tool for high-throughput screening of PPIs in living mammalian cells (Banning et al. 2010). Therefore, FRET measurement was combined with fluorescence activated cell sorting (FACS). To do so, the human cell line 293T was co-transfected with a vector carrying a fusion between the human immunodeficiency virus (HIV) Vpu accessory protein and YFP and a second vector carrying a fusion between a cDNA library and CFP. Cells were sorted for a positive FRET signal and PPIs proofed by co-IP. However, an average of more than 50% false positives was estimated which is comparable with Y2H screens (Banning et al. 2010).

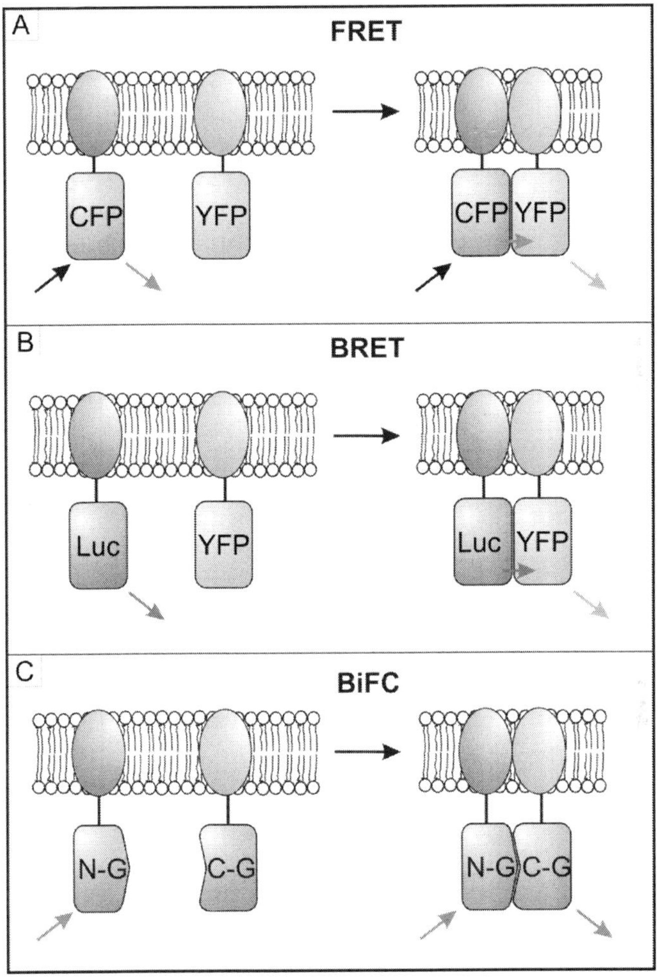

Fig. 4. Comparison of fluorescence resonance energy transfer (FRET), Bioluminescence Resonance Energy Transfer (BRET) and bimolecular fluorescence complementation (BiFC) (A) FRET: A donor fluorophore (here CFP) is fused to protein (orange) and an acceptor fluorophore (here YFP) is fused to a second protein (green). When the two proteins are in sufficient proximity fluorescence energy transfer can be monitored. (B) BRET: As in FRET, energy transfer between a donor and an acceptor is determined, but the donor is a protein that emits light (here luciferase). (C) BiFC: A fluorescent protein (here GFP) is split in two halves. Interaction of the two proteins fused to these two halves results in protein fragment complementation (PCA).

Fluorescence lifetime imaging microscopy (FLIM) is a FRET-based technique established to identify sub-cellular distributions of specific post-translational changes in protein targets (Peltan et al. 2006). In contrast to FRET, FLIM measurement determines the relaxation time of the acceptor flourophor and not the emission quantity (Biskup et al. 2007; Wouters 2006). As a consequent, FLIM measurement is independent from fluorophore concentrations and therefore the FRET-based method of choice to investigate dynamics in PPIs (Lalonde et al. 2008).

Total internal reflection fluorescence (TIRF) microscopy is a FRET-based approach used to study processes close to or at cell membranes (Mattheyses, Simon, and Rappoport 2010). In principle, TIRF results as the light beam propagates first through glass with a high refractive index and then through water with a low refractive index. As a consequence, the direction of the light beam is altered and an evanescent field is generated. Therefore, TIRF microscopy stimulates only fluorophores very close to the cover slip resulting in a minimized background fluorescence and reduced cellular photo-damage (Mattheyses, Simon, and Rappoport 2010; Lam et al. 2010).

General extended reviews on fluorescence microscopy techniques are given by Waters, North and Masi et al., (North 2006; Waters 2009; Masi et al. 2010). For FLIM background we refer to Lalonde et al. (2008). A detailed protocol and trouble-shooting for FLIM is given by Periasamy (Sun, Day, and Periasamy 2011). Detailed reviews on the physical basis of TIRF and advanced applications are given by Axelrod and Rappoport (Mattheyses, Simon, and Rappoport 2010; Axelrod 2003; Axelrod 2008).

3.6.2 Bioluminescence Resonance Energy Transfer (BRET)

Bioluminescence resonance energy transfer (BRET) is a variation of FRET using an autofluorescent protein as a donor (Fig. 4B) (Xia and Rao 2009; Pfleger and Eidne 2006). Consequently, excitation of the donor is not required. The most popular used BRET pair is a combination of coelenterazine emitting energy around 400 nm and a variant of GFP, termed GFP2 (Jensen et al. 2002).

3.6.3 Bimolecular Fluorescence Complementation (BiFC)

Bimolecular fluorescence complementation (BiFC) is fluorescence technique based on PCA (Fig. 4C). Two halves of a fluorescence protein, in general GFP (N-GFP and C-GFP), are fused to either the bait or the prey protein. PPI of bait and prey protein results in a fluorescent signal that can be monitored by fluorescence microscopy. However, BiFC cannot be used for dynamic studies because half-life time of the N-GFP and C-GFP interaction was estimated to be 10 years (Magliery et al. 2005).

3.7 Site-directed chemical cross-linking

Site-directed chemical cross-linking is a powerful tool to characterize the distance and the dynamics of specific amino acid pairs in and between membrane proteins both *in vivo* and *in vitro* (Kaback et al. 2011; Bordignon, Grote, and Schneider 2010). In many cases, homo-bifunctional sulfhydryl cross-linkers are used. These have variable spacer-arm length ranging from 5 to 50 Å. Because of their hydrophobic spacer arms, many cross-linkers are membrane-permeable and thus ideal to perform cross-linking with membrane proteins as exemplified for the maltose ABC transporter (Bordignon, Grote, and Schneider 2010). To prevent unspecific inter-molecular cross-linking, cross-linkers with a maximum spacer-arm length of 25 Å should be chosen.

Ideally, the native cysteine residues within proteins are first substituted by other amino acids (Ala or Ser) to allow specificity in site-directed chemical cross-linking. Hereafter, cysteine insertion mutagenesis is performed. The functionality of first the cysteine-free and then the mono-cysteine proteins has to be confirmed after each substitution step (Hunke et al. 2000; Hunke and Schneider 1999). Finally, the cross-linking procedure is performed either *in vivo* (Shiota et al. 2011), or *in vitro* using crude membranes or the reconstituted system (Hunke et al. 2000; Daus et al. 2007). When using the reconstituted system, substrates or inhibitors can be added during the cross-linking procedure providing information about the dynamics within a complex (Daus et al. 2007).

Comprehensive background and application for site-directed chemical cross-linking is given by the two major suppliers for cross-linking agents Pierce (http://www.piercenet.com/files/1601673_Crosslink_HB_Intl.pdf) and Molecular Probes (www.invitrogen.com/site/us/en/home/References/Molecular-Probes-The-Handbook.html).

3.8 Site-directed spin labeling electron paramagnetic resonance spectroscopy (EPR)

Site-directed spin labeling (SDSL) electron paramagnetic resonance (EPR) spectroscopy is a biophysical method introduced by Wayne L. Hubbell (Altenbach et al. 1990; Altenbach et al. 1989) that allows the determination not only of distances in and between macromolecules but also their dynamics (reviewed in Berliner et al., 2002; Klare & Steinhoff, 2010). In addition, EPR spectroscopy techniques provide a high time resolution and are independent on the protein size (reviewed in Klare & Steinhoff, 2010 and in this issue by Klare, 2012).

Spin labels are introduced at two cysteine residues in the otherwise cysteine-free complex and are excited by a strong microwave pulse. The most frequently used spin label is the methanethiosulfonate spin label (1-oxyl-2,2,5,5-tetramethyl-D3-pyrroline-3-yl)methanethiosulfonate (MTSL). The physical principle that the intensity of the dipolar interaction between the two spin labels is inversely proportional to the cube of their distance, allows the calculation of the distance between the two spin labeled residues (Fajer, Brown, and Song 2007; Klare and Steinhoff 2010; Klare 2012).

EPR spectroscopy methods (exchange EPR, dipolar continuous wave EPR) cover a distance range up to 2 nm. Moreover, dipolar continuous wave (CW) EPR spectroscopy yields information on the sidechain mobility as well as the accessibility and polarity of the microenvironment of a spin label at single labeled proteins (Bordignon and Steinhoff 2007).

Pulse dipolar EPR methods in particular double electron–electron resonance (DEER) spectroscopy allows the determination of a distance ranges from 2-8 nm in PPIs (Pannier et al. 2000). DEER uses two microwave frequencies resulting in two spin populations. Thereby, one spin population influences the echo amplitude of the second spin population

(Fajer, Brown, and Song 2007). An open-source software (DEER Analysis 2011) for extracting distance distributions from DEER data sets has been provided by the ETH-Zurich (http://www.epr.ethz.ch/software/index). DEER gave a deeper inside the transmembrane signaling mechanism of rhodopsin (Altenbach et al. 2008; Knierim et al. 2008), sensory rhodopsin (Holterhues et al. 2011; Klare et al. 2011), the maltose ABC transporter (Grote et al. 2008; Grote et al. 2009) and the KtrAB potassium transporter (Hänelt et al. 2010). Thus, during recent years and on the basis of crystallographic data, DEER has been established as the state of the art technique to allow description of signal and transport mechanisms (Bordignon, Grote, and Schneider 2010; Klare and Steinhoff 2010).

4. Conclusions and outlook

The advances in genome, proteome and *in silico* analysis have identified membrane proteins with no assigned function. Moreover, it became evident that most membrane proteins function in complexes that are composed of several subunits (Daley 2008). Elucidation of the identification and characterization of PPIs of integral membrane proteins is the challenging task of today's research. During recent years new methods have emerged that offer new opportunities to determine partner, kinetics and thermodynamics in membrane protein PPIs. Fluorescence techniques allow now the investigation of the location and interaction of membrane proteins *in vivo*. The application of EPR techniques has just started to allow a deeper inside into the mechanisms in membrane protein PPIs. Combination of the techniques presented here will allow in the future to elucidate the mechanism of signal transmission and substrate transport from one side of a membrane to the other.

5. Acknowledgment

This work was financially supported by the Deutsche Forschungsgemeinschaft. We are grateful to Michael Hensel for critical reading the manuscript.

6. References

Altenbach, C, SL Flitsch, HG Khorana, and WL. Hubbell. 1989. Structural studies on transmembrane proteins. 2. Spin labeling of bacteriorhodopsin mutants at unique cysteines. *Biochemistry* 28 (19):7806-7812.

Altenbach, Christian, Ana Karin Kusnetzow, Oliver P. Ernst, Klaus Peter Hofmann, and Wayne L. Hubbell. 2008. High-resolution distance mapping in rhodopsin reveals the pattern of helix movement due to activation. *Proc. Natl. Acad. Sci. U.S.A.* 105 (21):7439-7444.

Altenbach, Christian, T. Marti, H. Gobind Khorana, and Wayne L. Hubbell. 1990. Transmembrane Protein Structure: Spin Labeling of Bacteriorhodopsin Mutants. *Science* 248 (1088-1092).

Appleman, J R, N Prendergast, T J Delcamp, J H Freisheim, and R L Blakley. 1988. Kinetics of the formation and isomerization of methotrexate complexes of recombinant human dihydrofolate reductase. *J. Biol. Chem.* 263 (21):10304-10313.

Aronheim, A, E Zandi, H Hennemann, S.J. Elledge, and M. Karin. 1997. Isolation of an AP-1 repressor by a novel method for detecting protein-protein interactions. *Mol. Cell Biol.* 17 (6):3094-3102.

Aronheim, A. 1997. Improved efficiency sos recruitment system: expression of the mammalian GAP reduces isolation of Ras GTPase false positives. *Nucleic. Acids Res.* 25 (16):3373-3374.

Aronheim, Ami. 2001. Protein recruitment systems for the analysis of protein-protein interactions. *Methods* 24 (1):29-34.

Axelrod, D. 2003. Total internal reflection fluorescence microscopy in cell biology. *Biophotonics B* 361:1-33.

Axelrod, Daniel. 2008. Chapter 7: Total Internal Reflection Fluorescence Microscopy. In *Methods in Cell Biology*, edited by J. C. Dr. John and Dr. H. William Detrich, III: Academic Press.

Bakheet, Tala M., and Andrew J. Doig. 2009. Properties and identification of human protein drug targets. *Bioinformatics* 25 (4):451-457.

Banning, Carina, Jörg Votteler, Dirk Hoffmann, Herwig Koppensteiner, Martin Warmer, Rudolph Reimer, Frank Kirchhoff, Ulrich Schubert, Joachim Hauber, and Michael Schindler. 2010. A Flow Cytometry-Based FRET Assay to Identify and Analyse Protein-Protein Interactions in Living Cells. *PLoS ONE* 5 (2):e9344.

Biskup, C., T. Zimmer, L. Kelbauskas, and B. Hoffmann. 2007. Multi-dimensional fluorescence lifetime and FRET measurements. *Microsc. Res. Tech.* 70:442–451.

Blüschke, Bettina, Viola Eckey, Britta Kunert, Susanne Berendt, Heidi Landmesser, Michael Portwich, Rudolf Volkmer, and Erwin Schneider. 2007. Mapping putative contact sites between subunits in a bacterial ATP-binding cassette (ABC) transporter by synthetic peptide libraries. *J. Mol. Biol.* 369 (2):386-399.

Blüschke, Bettina, Rudolf Volkmer-Engert, and Erwin Schneider. 2006. Topography of the surface of the signal-transducing protein EIIAGlc that interacts with the MalK subunits of the maltose ATP-binding cassette transporter (MalFGK2) of *Salmonella typhimurium. J. Biol. Chem.* 281 (18):12833-12840.

Bordignon, Enrica, Mathias Grote, and Erwin Schneider. 2010. The maltose ATP-binding cassette transporter in the 21st century - towards a structural dynamic perspective on its mode of action. *Mol. Microbiol.* 77 (6):1354-1366.

Bordignon, Enrica, and H.J. Steinhoff. 2007. Membrane Protein Structure and Dynamics Studied by Site-directed Spin Labeling ESR. In *Biological Magnetic Resonance*, edited by M. A. Hemmingo and L. J. Berliner. New York: Springer.

Bouwmeester, T, A Bauch, H Ruffner, PO Angrand, G Bergamini, K Croughton, C Cruciat, D Eberhard, J Gagneur, S Ghidelli, C Hopf, B Huhse, R Mangano, AM Michon, M Schirle, J Schlegl, M Schwab, MA Stein, A Bauer, G Casari, G Drewes, AC Gavin, DB Jackson, G Joberty, G Neubauer, J Rick, B Kuster, and G. Superti-Furga. 2004. A physical and functional map of the human TNF-alpha/NF-kappa B signal transduction pathway. *Nat. Cell Biol.* 6 (2):97-105.

Broder, Yehoshua C., Sigal Katz, and Ami Aronheim. 1998. The Ras recruitment system, a novel approach to the study of protein-protein interactions. *Curr. Biol.* 8 (20):1121-1130.

Buelow, Daelynn R. , and Tracy L. Raivio. 2010. Three (and more) component regulatory systems & auxiliary regulators of bacterial histidine kinases. *Mol. Microbiol.* 75 (3):547-566.

Cannon, J.F., J.B. Gibbs, and K Tatchell. 1986. Suppressors of the ras2 mutation of Saccharomyces cerevisiae. *Genetics* 113 (2):247-264.

Compton, S.J., and C.G. Jones. 1985. Mechanism of dye response and interference in the Bradford protein assay. *Anal. Biochem.* 151 (2):369-374.

Daley, Daniel O. 2008. The assembly of membrane proteins into complexes. *Curr. Opin. Struct. Biol.* 18 (4):420-424.

Daus, M.L., M. Grote, P. Müller, M. Doebber, S. Herrmann, H.J. Steinhoff, E. Dassa, and Schneider E. 2007. ATP-driven MalK dimer closure and reopening and conformational changes of the "EAA" motifs are crucial for function of the maltose ATP-binding cassette transporter (MalFGK2). . *J Biol Chem* 282 (22387-22396).

Deutscher, Josef, Christof Francke, and Pieter W. Postma. 2006. How phosphotransferase system-related protein phosphorylation regulates carbohydrate metabolism in bacteria. *Microbiol. Mol. Biol. Rev.* 70 (4):939-1031.

Ear, Po Hien, and Stephen W. Michnick. 2009. A general life-death selection strategy for dissecting protein functions. *Nat Meth* 6 (11):813-816.

Edmondson, Diane G., and Sharon Y. Roth. 2001. Identification of protein interactions by far western analysis. In *Curr. Protoco. Mol. Biol.*: John Wiley & Sons, Inc.

Erhardt, Marc, Keiichi Namba, and Kelly T. Hughes. 2010. Bacterial Nanomachines: The Flagellum and Type III Injectisome. *Cold Spring Harbor Perspectives in Biology* 2 (11).

Fajer, P.G., L. Brown, and L. Song. 2007. Practical pulsed dipolar ESR (DEER). *in: Hemminga M.A. and Berliner L.J. (2007) ESR spectroscopy in membrane biophysics. Series: Biological Magnetic Resonance, Springer: New York* 27:95-128.

Fields, Stanley, and Ok-kyu Song. 1989. A novel genetic system to detect protein–protein interactions. *Nature* 340 (6230):245-246.

Filloux, Alain. 2011. Protein secretion systems in *Pseudomonas aeruginosa*: an essay on diversity, evolution and function. *Front. Microbiol.* 2.

Fleischer, Rebecca, Ralf Heermann, Kirsten Jung, and Sabine Hunke. 2007. Purification, reconstitution, and characterization of the CpxRAP envelope stress system of *Escherichia coli*. *J. Biol. Chem.* 282 (12):8583-8593.

Frank, R. 1992. SPOT-synthesis: an easy technique for the positionally addressable, parallel chemical synthesis on a membrane support. *Tetrahedron* 48:9217-9232.

Repeated Author. 2002. The SPOT-synthesis technique. Synthetic peptide arrays on membrane supports - principles and applications. *J. Immunol. Methods* 267 (1):13-26.

Gao, Rong, and Ann M. Stock. 2009. Biological Insights from Structures of Two-Component Proteins. *Annual Rev. Microbiol.* 63 (1):133-154.

Geertsma, E.R., N.A.B.N. Mahmood, G.K. Schuurman-Wolters, and B. Poolmann. 2008. Membrane reconstitution of ABC transporters and assays of translocator function. . *Nature Prot.* 3 (2):256-266.

Grote, Mathias, Enrica Bordignon, Yevhen Polyhach, Gunnar Jeschke, Heinz-Jürgen Steinhoff, and Erwin Schneider. 2008. A Comparative Electron Paramagnetic

Resonance Study of the Nucleotide-Binding Domains Catalytic Cycle in the Assembled Maltose ATP-Binding Cassette Importer. *Biophys. J.* 95 (6):2924-2938.

Grote, Mathias, Yevhen Polyhach, Gunnar Jeschke, Heinz-Jürgen Steinhoff, Erwin Schneider, and Enrica Bordignon. 2009. Transmembrane Signaling in the Maltose ABC Transporter MalFGK2-E. *J. Biol. Chem.* 284 (26):17521-17526.

Hänelt, Inga, Dorith Wunnicke, Meike Müller-Trimbusch, Marc Vor der Brüggen, Inga Kraus, Evert P. Bakker, and Heinz-Jürgen Steinhoff. 2010. Membrane Region M2C2 in Subunit KtrB of the K+ Uptake System KtrAB from Vibrio alginolyticus Forms a Flexible Gate Controlling K+ Flux. *J. Biol. Chem.* 285 (36):28210-28219.

Heermann, Ralf, and Kirsten Jung. 2010. Stimulus perception and signaling in histidine kinases. In *Bacterial Signaling*: Wiley-VCH Verlag GmbH & Co. KGaA.

Heermann, Ralf, Arnim Weber, Bettina Mayer, Melanie Ott, Elisabeth Hauser, Günther Gabriel, Torsten Pirch, and Kirsten Jung. 2009. The Universal Stress Protein UspC Scaffolds the KdpD/KdpE Signaling Cascade of Escherichia coli under Salt Stress. *J. Mol. Biol.* 386 (1):134-148.

Hodnik, V, and G. Anderluh. 2010. Capture of intact liposomes on biacore sensor chips for protein-membrane interaction studies. *Methods Mol Biol.* 627 (201-211).

Hoffmann, C., G. Gaietta, M. Bünemann, S.R. Adams, S. Oberdorff-Maass, B. Behr, J.-P. Viladarga, R. Y. Tsien, M. H. Ellisman, and M. J. Lohse. 2005. A FlAsH-based FRET approach to determine G protein–coupled receptor activation in living cells. *Nat Methods* 2 (3):171-176.

Holterhues, Julia, Enrica Bordignon, Daniel Klose, Christian Rickert, Johann P Klare, Swetlana Martell, Lin Li, Martin Engelhard, and Heinz-Jürgen Steinhoff. 2011. The Signal Transfer from the Receptor NpSRII to the Transducer NpHtrII Is Not Hampered by the D75N Mutation. *Biophys. J.* 100 (9):2275-2282.

Hoopmann, MR, CR Weisbrod, and JE Bruce. 2010. Improved strategies for rapid identification of chemically cross-linked peptides using protein interaction reporter technology. *J. Proteome Res.* 9 (12):6323-6333.

Hubsman, Monika, Guennady Yudkovsky, and Ami Aronheim. 2001. A novel approach for the identification of protein.protein interaction with integral membrane proteins. *Nucleic Acids Res.* 29 (4):e18.

Hunke, S., M. Mourez, M. Jéhanno, E. Dassa, and E. Schneider. 2000. ATP modulates subunit-subunit interactions in an ATP-binding-cassette transporter (MalFGK2) determined by site-directed chemical cross-linking. . *J. Biol. Chem.* 275 (15526-15534).

Hunke, Sabine, and Erwin Schneider. 1999. A Cys-less variant of the bacterial ATP binding cassette protein MalK is functional in maltose transport and regulation. *FEBS Lett.* 448 (1):131-134.

Jensen, A.A., J.L. Hansen, S.P. Sheikh, and H. Bräuner-Osborne. 2002. Probing intermolecular protein-protein interactions in the calcium-sensing receptor homodimer using bioluminescence resonance energy transfer (BRET). *Eur J Biochem.* 269 (20):5076-5087.

Johnsson, N, and A Varshavsky. 1994. Split ubiquitin as a sensor of protein interactions in vivo. *Proc. Natl. Acad. Sci. USA* 91 (22):10340-10344.

Johnsson, N, and A. Varshavsky. 1994. Ubiquitin-assisted dissection of protein transport across membranes. *EMBO J.* 13 (11):2686-2698.

Jones, DT. 1998. Do transmembrane protein superfolds exist? . *FEBS Lett* 423:281-285.

Jordan, Patrick, Petra Fromme, Horst Tobias Witt, Olaf Klukas, Wolfram Saenger, and Norbert Krausz. 2001. Three-dimensional structure of cyanobacterial photosystem I at 2.5[thinsp][angst] resolution. *Nature* 411 (6840):909-917.

Jung, Kirsten, Britta Tjaden, and Karlheinz Altendorf. 1997. Purification, Reconstitution, and Characterization of KdpD, the Turgor Sensor of Escherichia coli. *J. Biol. Chem.* 272 (16):10847-10852.

Jura, Natalia, Xuewu Zhang, NicholasÂ F Endres, MarkusÂ A Seeliger, Thomas Schindler, and John Kuriyan. 2011. Catalytic control in the EGF receptor and its connection to general kinase regulatory mechanisms. *Mol. Cell* 42 (1):9-22.

Kaback, H., Irina Smirnova, Vladimir Kasho, Yiling Nie, and Yonggang Zhou. 2011. The Alternating Access Transport Mechanism in LacY. *J. Mem. Biol.* 239 (1):85-93.

Karimova, Gouzel, Nathalie Dautin, and Daniel Ladant. 2005. Interaction network among *Escherichia coli* membrane proteins involved in cell division as revealed by bacterial two-hybrid analysis. *J. Bacteriol.* 187 (7):2233-2243.

Karimova, Gouzel, Josette Pidoux, Agnes Ullmann, and Daniel Ladant. 1998. A bacterial two-hybrid system based on a reconstituted signal transduction pathway. *Proc. Natl. Acad. Sci. U.S.A.* 95 (10):5752-5756.

Klare, J.P. 2012. Site-directed Spin Labeling and Electron Paramagnetic Resonance (EPR) Spectroscopy: A Versatile Tool to Study Protein-Protein Interactions. *InTech.*

Klare, J.P., and H.-J. Steinhoff. 2010. Site-directed spin labeling and pulse dipolar electron paramagnetic resonance. *Encyclopedia of Analytical Chemistry*

Klare, Johann P., Enrica Bordignon, Martin Engelhard, and Heinz-Jürgen Steinhoff. 2011. Transmembrane signal transduction in archaeal phototaxis: The sensory rhodopsin II-transducer complex studied by electron paramagnetic resonance spectroscopy. *Euro. J. Cell Biol.* 90 (9):731-739.

Knierim, Bernhard, Klaus Peter Hofmann, Wolfgang Gärtner, Wayne L. Hubbell, and Oliver P. Ernst. 2008. Rhodopsin and 9-Demethyl-retinal Analog. *J. Biol. Chem.* 283 (8):4967-4974.

Knol, Jan, Klaas Sjollema, and Bert Poolman. 1998. Detergent-Mediated Reconstitution of Membrane Proteins†. *Biochem.* 37 (46):16410-16415.

Krause, Frank. 2006. Detection and analysis of protein–protein interactions in organellar and prokaryotic proteomes by native gel electrophoresis: (Membrane) protein complexes and supercomplexes. *Electrophoresis* 27 (13):2759-2781.

Krogh, Anders, Björn Larsson, Gunnar von Heijne, and Erik L. L. Sonnhammer. 2001. Predicting transmembrane protein topology with a hidden markov model: application to complete genomes. *J. Mol. Biol.* 305 (3):567-580.

Ladant, Daniel, and Gouzel Karimova. 2000. Genetic systems for analyzing protein-protein interactions in bacteria. *Res. Microbiol.* 151 (9):711-720.

Lalonde, Sylvie, David W. Ehrhardt, Dominique Loqué, Jin Chen, Seung Y. Rhee, and Wolf B. Frommer. 2008. Molecular and cellular approaches for the detection of protein–protein interactions: latest techniques and current limitations. *Plant J.* 53 (4):610-635.

Lam, Alice D., Sahar Ismail, Ray Wu, Ofer Yizhar, Daniel R. Passmore, Stephen A. Ernst, and Edward L Stuenkel. 2010. Mapping Dynamic Protein Interactions to Insulin Secretory Granule Behavior with TIRF-FRET. *Biophys. J.* 99 (4):1311-1320.

Locher, Kaspar P., Allen T. Lee, and Douglas C. Rees. 2002. The *E. coli* BtuCD structure: A framework for ABC transporter architecture and mechanism. *Science* 296 (5570):1091-1098.

Magliery, T.J., C.G.M. Wilson, W.L. Pan, and D. Mishler. 2005. Detecting protein–protein interactions with a green fluorescent protein fragment reassembly trap: scope and mechanism. *J. Am. Chem. Soc.* 127:146-157.

Masi, A, R Cicchi, A Carloni, FS Pavone, and A. Arcangeli. 2010. Optical methods in the study of protein-protein interactions. . *Adv Exp Med Biol* 674:33-42.

Mattheyses, Alexa L., Sanford M. Simon, and Joshua Z. Rappoport. 2010. Imaging with total internal reflection fluorescence microscopy for the cell biologist. *J. Cell Sci.* 123 (21):3621-3628.

Maynard, Jennifer A., Nathan C. Lindquist, Jamie N. Sutherland, Antoine Lesuffleur, Arthur E. Warrington, Moses Rodriguez, and Sang-Hyun Oh. 2009. Surface plasmon resonance for high-throughput ligand screening of membrane-bound proteins. *Biotech. J.* 4 (11):1542-1558.

Michnick, Stephen W., Po Hien Ear, Christian Landry, Mohan K. Malleshaiah, and Vincent Messier. 2011. Protein-fragment complementation assays for large-scale analysis, functional dissection and dynamic studies of protein–protein interactions in living cells *Meth. Mol. Biol.* 756 (6):395-425.

Miernyk, Jan A., and Jay J. Thelen. 2008. Biochemical approaches for discovering protein-protein interactions. *Plant J.* 53 (4):597-609.

Müller, Volker Steffen, Peter Ralf Jungblut, Thomas F. Meyer, and S. Hunke. 2011. Membrane-SPINE: An improved method to identify protein-protein interaction partners of membrane proteins *in vivo*. *Proteomics* 11 (10):2124-2128.

North, A.J. 2006. Seeing is believing? A beginners' guide to practical pitfalls in image acquisition. *J. Cell Biol.* 172:9-18.

Pannier, M., S. Veit, A. Godt, G. Jeschke, and H.W. Spiess. 2000. Dead-time free measurement of dipole-dipole interactions between electron spins. . *J. Magn. Res.* 142 (331-340).

Paternostre, MT, M Roux, and JL. Rigaud. 1988. Mechanisms of membrane protein insertion into liposomes during reconstitution procedures involving the use of detergents. 1. Solubilization of large unilamellar liposomes (prepared by reverse-phase evaporation) by triton X-100, octyl glucoside, and sodium cholate. *Biochem.* 27 (8):2668-2677.

Pelletier, Joelle N., F.-X. Campbell-Valois, and Stephen W. Michnick. 1998. Oligomerization domain-directed reassembly of active dihydrofolate reductase from rationally designed fragments. *Proc. Natl. Acad. Sci. U.S.A.* 95 (21):12141-12146.

Peltan, I.D., A.V. Thomas, I. Mikhaienko, D.K. Strickland, B.T. Hymann, and von Arnim C.A.F. 2006. Fluorescence lifetime imaging microscopy (FLIM) detects stimulus-dependent phosphorylation of the low density lipoprotein receptor-related protein (LRP) in primary neurons. . *Biochem. Biophys. Res. Commun.* 349:34-30.

Petitjean, A, F Hilger, and K. Tatchell. 1990. Comparison of thermosensitive alleles of the CDC25 gene involved in the cAMP metabolism of *Saccharomyces cerevisiae. Gene* 124 (4):797-806.

Pfleger, K.D., and K.A. Eidne. 2006. Illuminating insights into protein-protein interactions using bioluminescence resonance energy transfer (BRET). *Nat Methods* 3 (3):165-174.

Puig, O. , F. Caspary, G. Rigaut, B. Rutz, E. Bouveret, E. Bragado-Nilsson, M. Wilm, and B. Seraphin. 2001. The tandem affinity purification (TAP) method: a general procedure of protein complex purification. *Methods* 24:218-229.

Reimer, U., U. Reineke, and J Schneider-Mergener. 2002. Peptide arrays: from macro to micro. *Curr. Opin. Biotechnol.* 13 (4):315-320.

Remy, Ingrid, F. X. Campbell-Valois, and Stephen W. Michnick. 2007. Detection of protein-protein interactions using a simple survival protein-fragment complementation assay based on the enzyme dihydrofolate reductase. *Nat. Protocols* 2 (9):2120-2125.

Remy, Ingrid, and Stephen W. Michnick. 2007. Application of protein-fragment complementation assays in cell biology. *Bio Tech* 42 (2):137-145.

Rigaud, J.-L. 2002. Membrane proteins: functional and structural studies using reconstituted proteoliposomes and 2-D crystals. *Brazil. J. Med. Biol. Res.* 35:753-766.

Rigaud, J.L., and D. Levy. 2003. Reconstitution of membrane proteins into liposomes. . *Methods Enzymol.* 372 (65-86).

Rigaud, J.L., B. Pitard, and D. Levy. 1995. Reconstitution of membrane proteins into liposomes: application to energy-transducing membrane proteins. *Biochim. Biophys. Acta* 1231:223-246.

Robichon, Carine, Gouzel Karimova, Jon Beckwith, and Daniel Ladant. 2011. Role of leucine zipper motifs in association of the *Escherichia coli* cell division proteins FtsL and FtsB. *J. Bacteriol.* 193 (18):4988-4992.

Saeed, Ramazan, and Charlotte Deane. 2008. An assessment of the uses of homologous interactions. *Bioinformatics* 24 (5):689-695.

Schäferling, M, and S. Nagl. 2011 Förster resonance energy transfer methods for quantification of protein-protein interactions on microarrays. *Methods Mol Biol.* 723:303-320.

Schägger, Hermann, and G. von Jagow. 1991. Blue native electrophoresis for isolation of membrane protein complexes in enzymatically active form. *Anal. Biochem.* 199 (2):223-231.

Shiota, Takuya, Hide Mabuchi, Sachiko Tanaka-Yamano, Koji Yamano, and Toshiya Endo. 2011. *In vivo* protein-interaction mapping of a mitochondrial translocator protein Tom22 at work. *Proc. Natl. Acad. Sci. U.S.A* 108 (37):15179-15183.

Silvius, JR. 1992. Solubilization and functional reconstitution of biomembrane components. *Annu. Rev. Biophys. Biomol. Struct.* 21 (323-348).

Skerker, Jeffrey M., Barrett S. Perchuk, Albert Siryaporn, Emma A. Lubin, Orr Ashenberg, Mark Goulian, and Michael T. Laub. 2008. Rewiring the specificity of two-component signal transduction systems. *Cell* 133 (6):1043-1054.

Sun, Y., R.N. Day, and A. Periasamy. 2011. Investigating protein-protein interactions in living cells using fluorescence lifetime imaging microscopy. *Nat Protoc* 6 (9):1324-1340.

Tang, Xiaoting, and James E. Bruce. 2010. A new cross-linking strategy: protein interaction reporter (PIR) technology for protein-protein interaction studies. *Mol. BioSyst.* 6 (6):939-947.

Vaidyanathan, Ravi, Steven M. Taffet, Karen L. Vikstrom, and Justus M. B. Anumonwo. 2010. Regulation of cardiac inward rectifier potassium current (IK1) by synapse-associated protein-97. *J. Biol. Chem.* 285 (36):28000-28009.

Volkmer, Rudolf. 2009. Synthesis and application of peptide arrays: quo vadis SPOT technology. *Chem Bio Chem* 15 (9):1431-1442.

Waters, J.C. 2009. Accuracy and precision in quantitative fluorescence microscopy. *J. Cell Biol.* 185:1135-1148.

Wenschuh, Holger , Rudolf Volkmer-Engert, Margit Schmidt, Marco Schulz, Jens Schneider-Mergener, and Ulrich Reineke. 2000. Coherent membrane supports for parallel microsynthesis and screening of bioactive peptides. *Peptide Science* 55 (3):188-206.

Wittig, Ilka, M Karas, and Hermann Schägger. 2007. High resolution clear native electrophoresis for in-gel functional assays and fluorescence studies of membrane protein complexes. *Mol. Cell. Proteomics* 6 (4):1215-1225.

Wittig, Ilka, and Hermann Schägger. 2008. Features and applications of blue-native and clear-native electrophoresis. *Proteomics* 8 (19):3974-3990.

Repeated Author. 2009. Native electrophoretic techniques to identify protein–protein interactions. *Proteomics* 9 (23):5214-5223.

Wouters, F.S. 2006. The physics and biology of fluorescence microscopy in the life sciences. *Contemp. Phys.* 47:239-255.

Wu, Y, Q Li, and X.Z. Chen. 2007. Detecting protein-protein interactions by Far western blotting. *Nat Protoc* 2 (12):3278-3284.

Xia, Z., and J. Rao. 2009. Biosensing and imaging based on bioluminescence resonance energy transfer. *Curr Opin Biotechnol* 20 (1):37-44.

Xu, Xiaoli, Yuan Song, Yuhua Li, Jianfeng Chang, Hua zhang, and Lizhe An. 2010. The tandem affinity purification method: An efficient system for protein complex purification and protein interaction identification. *Protein Expr. Purif.* 72 (2):149-156.

Yang, Li, Xiaoting Tang, Chad R. Weisbrod, Gerhard R. Munske, Jimmy K. Eng, Priska D. von Haller, Nathan K. Kaiser, and James E. Bruce. 2010. A Photocleavable and Mass Spectrometry Identifiable Cross-Linker for Protein Interaction Studies. *Anal.Chem.* 82 (9):3556-3566.

Yildirim, M. A., K. I. Goh, M. E. Cusick, A. L. Barabasi, and M. Vidal. 2007. Drug-target network. *Nat. Biotechnol.* 25 (1119-1126).

Zhou, Xiaohui, Rebecca Keller, Rudolf Volkmer, Norbert Krauß, Patrick Scheerer, and Sabine Hunke. 2011. Structural insight into two-component system inhibition and pilus sensing by CpxP. *J. Biol. Chem.* 286:9805-9814.

18

Relating Protein Structure and Function Through a Bijection and Its Implications on Protein Structure Prediction

Marco Ambriz-Rivas[1], Nina Pastor[2] and Gabriel del Rio[1]
[1]*Universidad Nacional Autónoma de México, Instituto de Fisiología Celular*
[2]*Universidad Autónoma del Estado de Morelos, Facultad de Ciencias*
México

1. Introduction

Proteins are studied by measuring different properties, typically the chemical structure and biochemical activity. Given that these measurements are done on the same protein molecule, they must be related. Despite the fact that this relationship exists, the mathematical nature of this relationship has remained elusive to our understanding, and is not commonly considered in the so called "structure-function relationship problem of proteins" (Punta & Ofran, 2008). While this is a fundamental problem in biochemistry and biology, that is, to establish a procedure that allows scientists to reliably relate protein structure and protein activity, the likelihood to succeed in this enterprise depends on our ability to understand the mere nature of this relationship. The possibility to effectively relate structure and activity has motivated years of research in different areas in biology, including biophysics, molecular biology, biochemistry, bioinformatics, and computational biology, among others. Although great advances have been achieved from these different areas of expertise, the question remains unsolved. That is, there is no general procedure that may have proven to effectively relate protein structure and activity. However, recent results in the prediction of protein three-dimensional structure (from now on referred simply as 3D structure) are addressing this problem with a fresh look, revealing a new aspect of this relationship that may explain why this particular problem has remained elusive. The present work reviews the general concepts being used to predict protein 3D structure with emphasis on the contribution of these methods to unravel the structure-activity relationship of proteins.

We divide this review in three sections. In the first section, we will present a mathematical view on the evolution of the concept about the 3D structure-activity relationship in proteins. The second section presents the general concepts behind template-based modelling and *ab initio* methods for the prediction of protein 3D structure. There, we will describe how these approaches have contributed to our current understanding of the 3D structure-activity relationship of proteins. Finally, we will review new methods for protein 3D structure prediction and how these may contribute to unravel the 3D structure-activity relationship of proteins.

2. Evolution of the 3D structure-activity paradigm from a mathematical perspective

Back in 1936 Mirsky and Pauling (Mirsky & Pauling, 1936) proposed that protein activity, or its function within a biological context, should be determined by its 3D structure. Considering that the characterization of protein activity has frequently been cumbersome, the possibility to determine it by simply looking at the 3D structure of proteins could be considered an impulse to establish this relationship. Yet, determination of protein 3D structure has not been an easy treat either. Perhaps the main motivation to establish this relationship consists in the possibility to design new devices capable of reproducing the highly efficient capabilities of proteins (Drexler, 1994; Robson, 1999; Balzani *et al.*, 2000) or to simply engineer proteins in order to adapt these for industrial use (Zaks, 2001; Huisman & Gray, 2002; Straathof *et al.*, 2002; Luetz *et al.*, 2008). Ultimately, establishing the 3D structure-activity relationship of proteins may serve to test our level of understanding of these molecules.

Hitherto, the approximation most frequently used to solve this relationship is to consider knowledge-based classification schemes. Such schemes are based on the existence of a given set of proteins with known activity; from that knowledge, it has been possible to identify new proteins sharing similar activity, from protein sequence comparisons. Although quite useful to classify the ever-increasing number of new protein sequences generated nowadays, this type of approaches has a limited ability to assist researchers in the design of protein activity (see next section). Alternatively, the activity of a protein is commonly analyzed from the knowledge of its 3D structure using biophysical methods (Neet & Lee, 2002; Chollet & Turcatti, 1999). In either case, previous knowledge of both protein 3D structure and activity is required to establish this relationship, indicating our current limitation in understanding this problem from basic principles. Even when new enzymatic activities have been designed "from scratch" (Siegel *et al.*, 2010) the active site residues are nestled within previously known protein folds. It has been possible to design completely novel folds, such as Top7, from scratch (Kuhlman *et al.*, 2003), but this refers to the sequence-3D structure relationship, which is not the main focus of this review.

We propose that one of the reasons for this limited understanding of the 3D structure-activity relationship of proteins is the absence of knowledge as to what type of mathematical relationship this one is. As we will show, determining the nature of this relationship may lead researchers to analyze this relationship with a new perspective and may accelerate the full understanding of it.

To explain this, let us first formally describe the 3D structure-activity relationship of proteins as a postulate:

P1: Protein activity depends on its 3D structure.

That is, protein activity may be represented as a mathematical relation of the protein 3D structure. Since both activity and 3D structure can always be measured on a given protein, that is they come in pairs, we postulate that this relation may be represented by a mathematical function. To further describe this postulate, let us define:

D1: Protein activity is defined as the capacity of proteins to interact with other molecules resulting in a change (on the interacting molecule or the environment) that is measurable (e.g., the chemical transformation of glucose to glucose 6-phosphate).

D2: Protein 3D structure is defined by two sets: the set of amino acid residues included in the protein and the set of physical interactions between these residues in the 3D space.

D3. A mathematical function is a particular class of relation between sets and it describes the dependence between the elements of these sets: an independent variable (an element in one of the sets) and the dependent variable (another element in the other set). In other words, for a given value of the independent variable there is one value of the dependent variable.

Postulate **P1** then refers to a mathematical function between two features of proteins: the activity and the 3D structure. The activity is usually expressed as a quantity (kinetic constants such as the Michaellis-Menten constant Km) and the structure may be represented by a quantity also, for instance the fold classification; yet, such quantities have not been easily related, so a new set of measurements is needed to evaluate **P1** (see below for a further discussion on this aspect). To do so, the question we want to address first is: what type of mathematical function is this? Basically, there are three types of mathematical functions:

D4: Injections. In mathematics, this refers to one-to-one relations: given two sets S (3D Structure) and A (protein Activity), there is at least one element in S related with one element in A (see Figure 1A and 1B). Therefore, there can be elements of the set A that do not have a matching partner in set S (Figure 1B).

D5: Surjections. This is defined as a mathematical function where given two sets S and A, there is an association of at least one element in S with an element in A (see Figure 1A and 1C). Therefore, there can be elements of the set A that have one or more relations with elements in set S.

D6: Bijections. These are defined as mathematical functions where for every element in set S there is exactly one element in set A associated to it. They occur when both an injection and a surjection relation exist (see Figure 1A).

In all these cases (injections, surjections and bijections), the mathematical function f might be reversible: given $f: S \rightarrow A$, then it is possible to find a function g such that $g: A \rightarrow S$. However, only in the case of bijections the reversibility of the association is a necessary condition of the function.

Expressing these concepts in terms of the 3D structure-activity relationship of proteins, we may say that this relationship presents the properties of injections. For a long time biochemists have characterized the activities of proteins; however, for some time many activities were known but no protein 3D structures were associated to them; more recently, with the advent of DNA sequencing, many protein sequences and 3D structures are known for which no activity has being assigned yet (Norin & Sundström, 2002). However, given postulate P1, we must expect that for each protein there have to be both an activity and a 3D structure associated to it; consequently, the currently unknown 3D structures or activities of proteins will be measured eventually.

Alternatively, most of the current approaches to study the 3D structure-activity relationship of proteins treat this as a surjection: the evolution theory postulates that protein activity or 3D structure has been conserved in different species (orthologous proteins); thus this is a case of a one-to-many (one function-many structures) relation. Additionally, in protein evolution the term "convergence" refers to the cases where different 3D structures of proteins have evolved to share a similar activity; conversely, an alternative example are single-domain moonlighting proteins, where one 3D structure is associated with multiple activities, albeit, using different molecular surfaces (Jeffery, 1999, 2003, 2009; Copley, 2003). In any case though, the one-to-many association prevails as much as we group together 3D structures or activities that are not identical. That is, to the best of our knowledge, there are no two proteins with identical activities reported so far with perfectly different 3D structures, nor are there two proteins with identical 3D structures with perfectly different activities. Take for instance the triose-phosphate isomerase proteins; these are proteins with a high degree of sequence-3D structure similarity, sharing similar but not identical activities (see Table 1). In the case of moonlighting proteins, there is no evidence that the two different activities may be performed in the same protein having exactly the same 3D structure, yet the structure may be slightly altered to accomplish different activities (Bateman *et al.*, 2003; Krojer *et al.*, 2002).

The need to move from considering similar to identical activity or 3D structure in the structure-activity relationship of proteins is important to improve our understanding of this relationship. On the one hand, it is convenient to assume similarity in 3D structure or activity of proteins in the discovery phase of biology (i.e., accelerated discovery of new proteins) because this assumption allows for the classification of new proteins into known families of proteins with known activity. Alternatively, provided the existence of an activity assay, it is possible to identify new proteins with such activity and presumably related in their 3D structure. However, after this initial phase of discovery, full understanding of the activity or 3D structure of a protein requires more detailed analysis both experimentally and theoretically. For the theoretical part, here we claim that in order to gain a better understanding of the 3D structure-activity relationship of proteins it is necessary to be precise in the terms used to relate these properties.

From this analysis we noted that since the 3D structure-activity relationship of proteins presents features of both injections and surjections, thus it may be best represented by a bijection. Furthermore, assuming that the injective feature is only a temporal one, and the surjective feature exists if and only if the definition of activity or 3D structure is not precise, we may conclude that the best way to analyze the 3D structure-activity relationship of proteins is as a bijection, where we postulate that for any given protein there is always one activity related to a given 3D structure. This approach necessarily implies that one has to come up with a rigorous and precise definition for both 3D structure and activity. Herein lies the challenge.

This conclusion leads us to the following scenario: let us assume that there is a set S with every possible 3D structure of proteins, and a set A with every possible measurable activity of proteins; then, for a given protein 3D structure in set S there is exactly one protein activity in set A; conversely, for a given protein activity in set A, there is exactly one protein 3D structure in set S. In this scenario, there are no identical activities in set A, neither there are identical structures in set S. To formally express this:

$$A = f(S) \tag{1}$$

Now, in order to express this relation in numerical terms, let us define the 3D structure as a matrix (e.g., adjacency matrix) and activity as a vector (e.g., list of critical residues for protein activity). Choosing this set of critical residues is a convenient pick since it has been reported that proteins sharing high 3D structural similarity do not share the same set of critical residues (Cota E *et al.*, 2000; Rivera MH *et al.*, 2003), yet some critical residues are indeed shared between homologue proteins (Zhang Z & Palzkill T, 2003). Thus, representing 3D structure as a matrix (M) and activity as a vector of critical residues (C) provides us with a way to express this relation formally and look for mathematical tools to define the mathematical function inherent to these quantities. Thus:

$$C = f(M) \tag{2}$$

In other words, given a set of contacts between the residues of a protein (3D structure), our problem is to find a mathematical transformation of this matrix into a vector containing the critical residues for the protein function. If the mathematical function relating M and C is a bijection, then it must be possible to transform the vector C back into the matrix M. In order to find the mathematical function involved in this transformation, having access to multiple 3D structures and multiple sets of critical residues for several proteins is required.

Our analysis has several implications for the analysis of the 3D structure-activity relationship. In the present review, we will discuss only those relevant for the prediction of protein 3D structure. That is encouraged by the emergence of new approaches for the prediction of protein 3D structure that are based on the notion that the 3D structure-activity relationship is a bijection. However, these approaches have been developed in the absence of the current mathematical context, as we will describe below; embracing this bijection may provide the basis to improve the current methods of protein 3D structure prediction.

3. Current methods for protein 3D structure prediction

In this section we will summarize the ideas behind them and the kind of relationship that they assume between 3D structure and activity. This review does not attempt to cover in detail these methodologies, but to present the basic aspects of them in the context of postulate P1. For detailed descriptions of these methodologies, there are other reviews published elsewhere (Jones & Thornton, 1993; Martí-Renom *et al.*, 2000; Osguthorpe, 2000; Hardin *et al.*, 2002; Koretke *et al.*, 2002; Zhang, 2002; Godzik, 2003).

3.1 General considerations

Despite of the diversity of approaches to perform structural predictions, they all share a common design. The two key components of any method are the model generator and a quality evaluator (Figure 2).

1. Model generators refer to algorithms that create native-like protein 3D structures. There are two ways to generate such structures: knowledge-based strategies that depend on the available structures in databases and *ab initio* strategies (also known as physics-based), which consider physics principles to generate structures. Typically, model generators produce many alternative 3D structures that are potential solutions to the native structure of the protein.

2. Quality evaluators. These algorithms aim to evaluate the quality of the models produced by the model generators, in order to select the best models; i.e., those resembling the known native-like structure of proteins. Like the model generators, quality evaluators can be knowledge-based or *ab intio*.

It is important to keep in mind that these methodologies have limitations, especially if they are used to gain insights into the relation between the 3D structure and activity of poorly characterized proteins. Knowledge-based model generators and evaluators assume surjective relations between structure and activity, since the common idea of modellers of protein 3D structures is to assist in the grouping of protein structures based on similar attributes (Gerstein & Hegyi, 1998; Domingues *et al.*, 2000; Skolnick *et al.*, 2000). Therefore, in these cases knowledge of the protein 3D structure may provide inaccurate information about the activity (Martin *et al.*, 1998). On the other hand, *ab initio* methods do not take into account the 3D structure-activity relation to perform predictions. With this kind of predictions, it is unlikely to get precise information about the activity of the protein from its 3D structure (Baker & Sali, 2001, and the results from CASP9).

Often the 3D structure is used to interpret the activity and rarely the other way around (Gherardini & Helmer-Citterich, 2008), thus it is not surprising that the current methods of protein 3D structure prediction do not address the prediction of 3D structure from the activity of the protein. In spite of this limitation, current methodologies for protein 3D structure prediction have been important in the development of the ideas about protein 3D structure determinants and their relationship with activity. Consequently, in the next two sections we will describe briefly the current methods for protein structure predictions, their features and limitations to elucidate protein activity.

3.2 Template-based modeling

This kind of predictions uses a protein of known 3D structure as a template to build the model of a protein whose 3D structure is unknown (target). The most critical part of this methodology is to identify adequate template(s) for the target. Accordingly, template-based modelling is classified in two main areas: homology modelling and fold recognition.

The idea behind homology modelling is that similar sequences have similar 3D structures (Doolittle, 1981, 1986; Chothia & Lesk, 1986). In this regard, the quality of a 3D model for a target protein depends strongly on the percentage of sequence identity between the target and template; the greater the identity, the more accurate the model will be. Likewise, below 30% of identity between the target and template proteins (sometimes referred as the "twilight zone"; Doolittle, 1986), several false templates may be identified for the target protein (Sander & Schneider, 1991; Rost, 1999). In that case, templates should be searched with fold recognition algorithms (Rost, 1999; see below). Templates can be found by searching databases of proteins with known 3D structure (e.g. the Protein Data Bank) with sequence alignment tools like BLAST (Altschul *et al.*, 1990, 1997) or FASTA (Pearson & Lipman, 1988). Then, models of the target protein are built from the templates, taking into account changes that must be introduced like insertions and deletions in the template (indels), side chain conformations of non-conserved residues, possible rearrangements in the backbone, among others (Jones & Thirup, 1986; Bruccoleri & Karplus, 1987; Vásquez, 1996).

Afterwards, the quality of the resulting models is evaluated (Laskowski *et al.*, 1993; Hooft *et al.*, 1996; Wallner & Elofsson, 2003; Ginalski *et al.*, 2003).

On the other hand, fold recognition methodologies identify proteins sharing similar 3D structures even if they do not have any obvious sequence similarity (Jones & Thornton, 1993; Godzik, 2003). Fold recognition can be performed in two ways. The first involves the enhancement of homology detection (Fischer & Eisenberg, 1996; Jaroszewski *et al.*, 1998; Rychlewski *et al.*, 2000), by using sequence profiles compiled from protein sequences that are compatible with the target. Two examples of this approach are PSI-BLAST (Altschul, 1997) and hidden Markov models (Durbin *et al.*, 1998). Accuracy of prediction is increased further if structural information (e.g. secondary structure) is incorporated in the profiles (Di Francesco *et al.*, 1997a, 1997b). The second approach is termed "threading" (Jones *et al.* 1992; Godzik & Skolnick, 1992). Here, the target sequence is forced to adopt the 3D structure of a potential target. Then the quality of the model is evaluated with a structure-based score. If the model has a high score, there is confidence that the target adopt a similar 3D structure as the template, otherwise the model is discarded. Once the template(s) is (are) found, a 3D-structural model of the target protein is built following the steps described in homology modelling after the initial template identification.

Template-based modelling has been recognized as the most accurate approach for protein structure prediction, especially if the identity between target and template is high (Chothia & Lesk, 1986; Sali *et al.*, 1995; Cozzetto *et al.*, 2009). However, as any model, these need to be tested in their ability to reproduce a biologically relevant feature, such as the activity. Since these methods assume a surjection for the structure-activity relationship, there are limitations imposed by such assumption, which are more notorious in the cases of low sequence similarity between the target and template proteins. One example of the limitation induced by the surjection conjecture in the structure-activity relationship of proteins is the TIM barrel fold, a common 3D-structure present in enzymes with very different activities such as oxidoreductases, hydrolases, lyases and isomerases (Greene *et al.*, 2007). Likewise, the opposite situation is common: proteins with very similar activities and structurally unrelated. For instance, both chymotrypsin and subtilisin are serine-proteases with the same catalytic triad in the active site even thought they have completely different 3D-structures (Wallace *et al.*, 1996).

Furthermore, even when there is a clear similarity between target and template sequences, there can be measurable structural differences. The most common example is loop structure. Precise prediction of loop regions is usually hard to accomplish since they tend to exhibit higher sequence variability and often have insertions and deletions relative to templates (Martí-Renom *et al.*, 2000). Loops though, play an important role conferring specificity to the protein activity. Another less frequent situation is when there are visible differences in active sites of related proteins. This can lead to inaccurate modelling of the structure of target proteins (Moult, 2005). One way to improve the modelling of loops would be to evaluate the predicted activity of the model.

The information summarized above provides a general notion about the relationships that template-based modelling assumes. One-to-many relations between protein structure and activity are quite common with this kind of predictions. Thus, it is frequent to misrelate the activity of a protein from the knowledge of its fold alone (Martin *et al.*, 1998). It is often necessary to use other resources to predict the activity more accurately, as the use of local structural features of proteins in active sites (see Gherardini & Helmer-Citterich, 2008 for more details). Such tools work with the traditional approaches for predictions: knowledge-based like the 3D-templates (Wallace *et al.*, 1996); or physics based, for example the identification of clefts and pockets in protein structures (Laskowski *et al.*, 1996; Binkowski *et al.*, 2003). These methods provide a theoretic framework to understand the 3D structure-activity relation in a one-way path: the prediction of activity from structure.

3.3 *Ab initio* modeling

Template based modelling can provide insights into the 3D structure and activity of poorly characterized proteins. In terms of generating reliable models it has an intrinsic limitation: it requires a protein of known 3D structure in order to produce a model. This may not be a problem in many situations, but there are proteins without any detectable template (more than half of the sequenced proteins in known genomes, see Yura *et al.*, 2006). In such cases the alternative is *ab initio* modelling, also known as template-free modelling (Osguthorpe, 2000; Hardin *et al.*, 2002; Koretke *et al.*, 2002). The premise of these modelling methods is that the protein sequence determines the native structure, which has the global minimum potential energy among all the alternative conformations. In other words, *ab initio* methods assume that sequence alone would be sufficient to model the structure of proteins. For this reason, *ab inito* methods are adjured to predict the structure folds that were previously unknown.

Ab initio methods carry out a large-scale search for protein structures that have a particularly low energy for a given amino acid sequence. The two critical parts of these predictors are the conformational search strategy and the energy evaluation method (known as energy potential). To perform a fast and efficient search of the conformational space, *ab initio* methods use sophisticated algorithms suited to solve combinatorial problems since it is impossible to systematically explore all the conformations of a polypeptide chain. Monte Carlo algorithms (Simons *et al.*, 1999; Ortiz *et al.*, 1999), genetic algorithms (Pedersen & Moult, 1997a, 1997b), zipping and assembly (Ozcan *et al.*, 2007) and molecular dynamics (Duan & Kollman, 1998; Shaw DE *et al.* 2010) are among the most frequently used methods to explore the conformational space of protein structures. Likewise, the energy potential is crucial to evaluate and select models of the target protein. Energy potentials can be of two kinds: molecular mechanics potentials, that are derived from physical-chemical calculations (Brooks *et al.*, 1983; Pearlman *et al.*, 1995) and knowledge-based potentials are constructed from the statistical analysis of the available structures in databases (Sippl, 1990; Koretke *et al.*, 1998; Kuhlman and Baker, 2000).

Ab initio predictions usually consume a great deal of time and computer power. Recent methods make simplifications on the protein 3D structure in order to keep an acceptable speed (Helles, 2007). One of the solutions is to reduce the number of atoms that represent the protein 3D structure in order to simplify the model generation process (Kolinski, 2004; Lee *et al.*, 1999). An alternative to speed up calculations is to consider fragment assembly strategies (Simons *et al.*, 1999; Jones & Thirup, 1986). The idea with this approach is to split the structure into smaller fragments composed by many residues. Fragments are selected from a knowledge-based database on the basis of structural compatibility with the target sequence and secondary structure propensities. The assembly of such substructures is determined by the energy potential and the conformation searching strategy. There are also multi-scale methods, like those of Cecilia Clementi, which change the resolution of the model depending on the questions that want to be asked of the protein (Shehu *et al.*, 2009).

Template-free modelling has experienced much progress since the first blind prediction experiment known as "Critical Assessment of Techniques for Protein Structure Prediction" (CASP) took place in the early 90's (Bourne, 2003; Moult, 2005). However, despite of the considerable efforts the accuracy of *ab initio* predictions is still very low, compared to template-based modelling. That is, models generated with *ab initio* methods may have very large deviations from the experimental structures. In other cases, the 3D structure of the model can be completely wrong (this is actually a common situation). These limitations have hindered the practical use of *ab intio* modeling for the inference of the 3D structure-activity relationship on the target proteins (Baker & Sali, 2001; the results from CASP9).

Finally, *ab initio* predictions do not take into account the relation between 3D structure and activity explicitly, therefore they provide little reliable information about this relationship. On the other hand, they assume that proteins fold autonomously to the 3D structure with the minimum free energy (this is the case for most globular proteins), but there are cases where this assumption may be unjustified, as in the case of protein folding under kinetic control. For example, it has long been recognized that transmembrane proteins do not adopt their final, functional 3D-structure unassisted, but they need a translocation machinery to insert into the membrane and fold (Elofsson & von Heijne, 2007). Hence, the use of these strategies is inadequate for transmembrane proteins. Nonetheless, the ROSETTA method (originally developed for globular proteins) has been adapted to predict transmembrane proteins, with limited success (Yarov-Yarovoy *et al.*, 2006). Additionally, *ab initio* predictions are unsuited for natively unstructured proteins (proteins that do not have a defined, unique structure), because they perform their activities as many alternative, rapidly interchanging conformations that correspond to multiple energy minima (Radivojac *et al.*, 2007).

Despite of these disadvantages, *ab initio* predictions sometimes provide insights about protein activity. For example, in the fourth CASP experiment, the ROSETTA method was able to predict the structure of a couple target proteins that are structurally related to proteins of known 3D structure that were missed by fold recognition methods (Bonneau *et al.*, 2001; Baker & Sali, 2001). Interestingly, the activities of the target proteins were similar, even thought there was no significant sequence identity between the proteins. A second example is the signalling protein Frizzled, whose critical residues for activity (previously characterized) were clustered together in the predicted structure in a surface patch likely to be involved in key protein-protein interactions (Baker & Sali, 2001). From these examples, it can be concluded that *ab initio* methods are more effective to gain information about the activity if they are combined with knowledge-based approaches (carrying on their limitations).

3.4 Concluding remarks about the current methods for protein 3D structure prediction

The available methodologies for the 3D structure prediction of proteins have provided useful insights about the relation between 3D structure and activity, and helped to construct the current paradigm. However, further refinement of these methods may assist to fully relate protein 3D structure with activity. In this review we propose that such refinement may come from the recognition of the bijective nature of the 3D structure-activity relationship. For instance, knowledge-based methods imply a surjective relationship between activity and 3D structure. Consequently, predicting details on the activity of a modelled 3D structure of a protein can be hard, since there are examples of folds associated with many activities and *vice versa*. Furthermore, *ab initio* methods do not consider the structure-activity relationship, therefore the information they provide about the activity is commonly inaccurate. Additionally, template-free methods assume that proteins fold autonomously into a stable, minimum energy conformation, limiting their applicability in proteins that do not have these features because they fold under kinetic control.

In summary, it is necessary to develop methods that take into account the bijective nature of the 3D structure-activity relationship, in order to improve the usefulness and reliability perhaps, of protein 3D structure prediction methods. In the following section we will describe the available methodologies that take into account this bijection.

4. Emerging methods for protein structure prediction based on the bijective nature of protein 3D structure and activity

The previous section outlined the current status in the protein 3D structure prediction field, its strengths and weaknesses with regard to activity inference. It is evident that current methodologies still have limitations to exploit the usefulness of the 3D structure-activity relationship. Fortunately, new methodologies have been developed that take into account the bijective nature between the 3D structure and activity of proteins. This section will discuss the principles behind these methods and their capabilities.

4.1 Relevance of critical residues in the 3D structure-activity relation

These methods are based on the concept of critical residues, which are defined as those residues that upon mutation abolish the activity of a protein. Such definition depends on the experimental procedure used to measure the activity of the protein, but generally speaking, residues are considered critical if they tolerate few if any mutations (Loeb *et al.*, 1989; Rennell *et al.*, 1991; Terwilliger *et al.*, 1994; Huang *et al.*, 1996; Axe *et al.*, 1998). Therefore, an experimentally determined critical residue may be either important to maintain the 3D structure of a protein or critical for the interaction with another molecule, or both. Thus, these residues constitute a key piece of knowledge that can be exploited to relate activity and 3D structure. Not surprisingly, methods have been developed to predict critical residues from protein sequence and/or 3D structure (Elcock, 2001; del Sol Mesa *et al.*, 2003; Glaser *et al.*, 2003; Thibert *et al.*, 2005; Cusack *et al.*, 2007).

Additionally, critical residues may provide a useful way to quantify structural features of proteins and relate them with the activity of a protein. As we mentioned earlier, there are no reports of two proteins with identical 3D structures with perfectly different activities and *vice versa* (please note that correctly representing both 3D structure and activity is one of the biggest challenges, and therefore, a Cartesian representation of the protein may not be the best to distinguish identical 3D structures). Hence, it is expected that proteins with similar, yet strictly different 3D structures, will have different sets of critical residues. If that is the case, the set of critical residues for a given protein should reflect its unique 3D structural and activity properties. Such assumption provides the framework for methodologies that are based in the bijective relation between 3D structure and activity.

In the next two sections, we will describe the available bijective approaches. To simplify, they are classified in two categories: phylogeny and structure-based methods. The usefulness of these methodologies to relate 3D structure and activity will also be discussed.

4.2 Phylogeny-based approaches

The idea behind phylogenetic methods is to exploit the evolutionary information that can be extracted from the analysis of the sequences of related proteins. To do so, it is necessary to identify a group of similar protein sequences and to construct a multiple sequence alignment with them. There are two types of information that can be extracted from the alignments: sequence conservation and sequence correlation.

The first property refers to the frequency of a specific amino acid at a given position in the alignment; residues occurring at high frequencies at particular positions are considered conserved residues. Sequence conservation is related to the direct evolutionary pressure to maintain the physical-chemical characteristics of some positions in order to retain the activity and/or 3D structure of a family of homologous proteins. Therefore, highly conserved residues are regarded as critical to retain the 3D structure and activity of the protein. In the literature, there are many reports of methods to calculate conservation (see Valdar *et al.*, 2002 and Sadowski & Jones, 2009 for comprehensive reviews).

Residue correlation (also known as co-evolution or co-variation) is defined as concerted patterns of variation between two or more different positions in a multiple sequence alignment of homologous proteins (Altschuh *et al.*, 1987). Such co-variating residues are proposed to correspond to compensatory substitutions that maintain the structural stability or functional properties of proteins throughout their evolutionary history. It has been observed that correlated residues tend to be in physical contact (Altschuh *et al.*, 1988); thus, this feature was proposed to be useful in residue contact predictions (Göbel *et al.*, 1994; Pazos *et al.*, 1997; Olmea & Valencia, 1997).

Critical residues predicted with phylogenetic approaches can be exploited to improve structural predictions. For example, the method reported by the group of Valencia (Olmea *et al.*, 1999) uses sequence conservation and correlation as part of a structure quality evaluator for a fold recognition structure predictor. The authors of this work report that the method is capable to distinguish correct models from incorrect models generated by the TOPITS threading algorithm (Rost, 1995). However, the accuracy of the algorithm decreases for large proteins, thus restricting its applicability.

Another exciting application of sequence correlation and co-variation is the design of new proteins (a field that strongly depends on 3D structure prediction tools). An illustrative example of protein engineering is the use of the Statistical Coupling Analysis method (SCA; Lockless & Ranganathan, 1999), which was used to design a novel artificial protein sequence with the same 3D structure and activity as natural WW domain proteins (Socolich *et al.*,

2005; Russ *et al.*, 2005). In order to design the protein, the method took into account the critical residues of the protein as well as their patterns of conservation and correlation (Socolich *et al.*, 2005). Furthermore, the methodology has been used recently to design a light-modulated chimerical enzyme (Lee *et al.*, 2008).

Ultimately, conserved residues will only capture the common critical residues for a set of homologous proteins, and will most likely miss the critical residues specific for the activity and 3D structure of each protein in that set. In that sense, conserved residues may be useful to score common structural features of proteins but may not be useful to evaluate the different 3D structure and biological activity of each homologous protein. To do so, a new method has recently being described that is now reviewed.

4.3 Methods based on structural information

A complementary approach to identify critical residues is to consider only 3D structural properties of proteins. One of the most recent approaches to study the 3D structure of proteins is graph theory (Vendruscolo *et al.*, 2002; Greene & Higman, 2003; Thibert *et al.*, 2005; Cusack *et al.*, 2007; Montiel Molina *et al.*, 2008), a theoretical approximation that has been used to characterize other biological systems, such as metabolism, genetic regulation and protein-protein interaction networks (Jeong *et al.*, 2000, 2001; Del Rio *et al.*, 2009). Under this view, protein 3D structure is modelled as a graph (network), which is defined by one set of nodes that represent the amino acid residues in a protein, and a set of edges that can be considered as molecular interactions between any two residues (nodes). The criterion to link two residues by an edge is based on maximum distances among the atoms of residues (Vendruscolo *et al.*, 2002; Greene & Higman, 2003; Thibert *et al.*, 2005; Cusack *et al.*, 2007; Milenković *et al.*, 2009).

Graph theory provides the mathematical basis to study the topological properties of networks derived from the protein structure. One useful concept of this field to characterize networks is network centrality, which measures the relative importance of nodes in the network. Thus, centrality can be used to predict critical residues (Thibert *et al.*, 2005; Cusack *et al.*, 2007) or to study topological features of protein structures (Vendruscolo *et al.*, 2002; Greene & Higman, 2003). Some of the most common centralities used to study networks derived from protein structures are betweenness and closeness, which relate nodes through the shortest paths among all the nodes in the graph (Freeman, 1977).

Centrality is reliable when it comes to predict critical residues (Chea & Livesay, 2007), but how can these be used to predict 3D structural and functional features? We have recently reported a tool named "JAMMING" to facilitate this task. The method predicts critical residues using betweenness or closeness centrality (Cusack *et al.*, 2007). We have shown that JAMMING may be used to identify protein structures involved in ligand binding by screening thousands of conformations generated from protein 3D structures in the unbound form; such functional conformers were found by a scoring system that matches critical residues with central residues (Montiel Molina *et al.*, 2008). Our results show that critical residues for a molecular interaction are preferentially found as central residues of protein structures in complex with a ligand. Therefore, the tool helps to relate the activity of the protein (binding to a molecule) with its structural properties (the conformers).

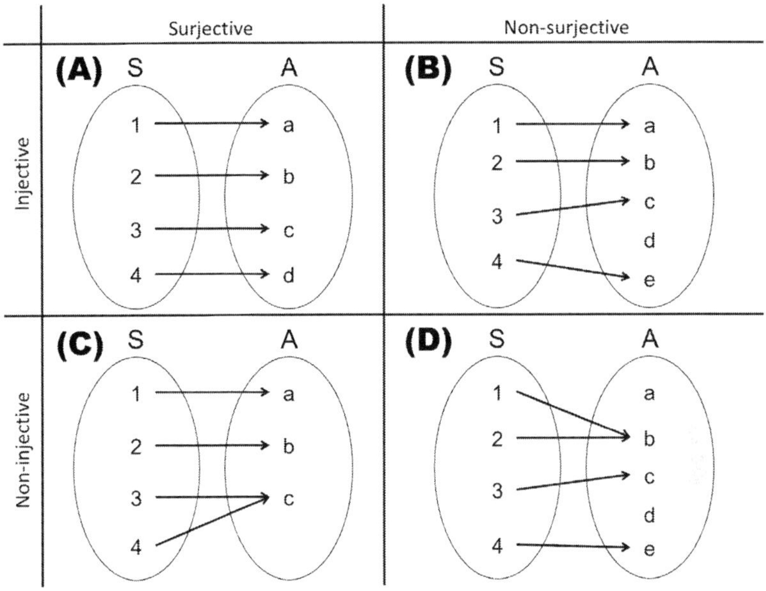

Fig. 1. Examples of injective and surjective functions
A) Injective and surjective (bijection). B) Injective and non-surjective. C) Non-injective and surjective. D) Non-injective and non-surjective.

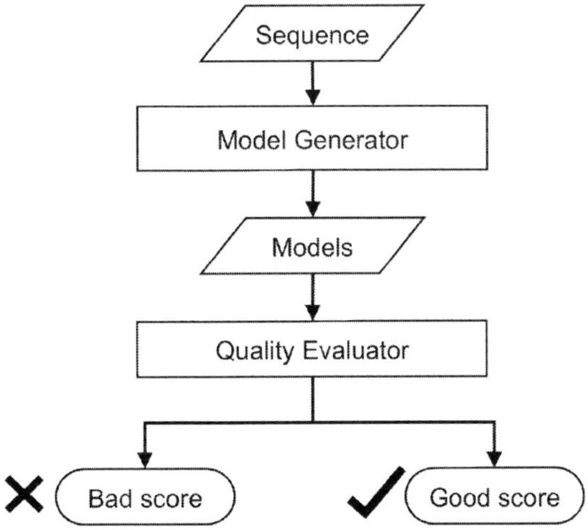

Fig. 2. Flowchart of structural prediction methods. The protein sequence is the input of the model generator algorithm. As a result, the generator produces multiple models that are assessed by the quality evaluator. Finally, the best scoring models are selected, whereas the models with bad scores are discarded.

Species	PDB	Identity[1] [%]	RMSD[1] [A]	K_m [mM]	K_{cat} [1/s]	References[2]
Homo sapiens	1wyi	100	0.0	0.34	16320	Gracy, 1975
Oryctolagus cuniculus	1r2t	98	0.4	0.42	8670	Krietsch, 1975a
Gallus gallus	8tim	89	0.8	0.47	4300	Xiang *et al.*, 2004
Saccharomyces cerevisiae	1ypi	53	1.0	1.27	16700	Krietsch, 1975b
Trypanosoma brucei	1tpf	53	1.1	0.19	6000	Kursula *et al.*, 2002
Leishmania mexicana	1amk	50	1.6	0.30	4170	Kohl *et al.*, 1994
Escherichia coli	1tre	46	1.4	1.03	9000	Alvarez *et al.*, 1998
Vibrio marinus	1aw2	42	1.4	1.90	7000	Alvarez *et al.*, 1998

[1] Sequence identities and RMSDs were calculated with the program DaliLite (Holm & Park, 2000) using 1wyi as the first molecule in all comparisons.
[2] References originally reporting the values for Km and Kcat.

Table 1. Structural and functional features of triose-phosphate isomerases from different species.

5. Conclusions

The structure-activity paradigm has travelled a long way since the first efforts to characterize the 3D structure and biological activity of proteins were performed back in the 1930's. Traditionally, the relationship between 3D structure and activity has been considered as a surjection to assist in the classification of the known proteins. Consequently, knowledge-based classification schemes, although useful to give sense to an ever-increasing list of known protein sequences, may not provide the basis to understand the subtleness of protein activity and structure in nature. In a similar fashion, most of the current methods for protein 3D structure prediction are unable to provide better insights about the activity of a protein of unknown structure (especially if it does not have a close homologue).

In this review, we propose that the relation between structure and activity may be modelled by a bijection. Critical residues provide a way to relate the structure and the activity of proteins, especially in the situation where structure and activity are represented by a bijection. Current methodologies based on the bijective 3D structure-activity relationship unnoticeably provided novel tools to explore the subtle determinants of protein activity, structure and their interaction. We claim that the incorporation of these methods into the traditional tools for protein structure prediction will improve the usefulness of the structural predictions to understand the details on the evolution of protein activity.

6. Acknowledgment

We want to acknowledge the technical assistance of Dra. Maria Teresa Lara Ortiz and the IT core facility of the Instituto de Fisiología Celular; Dr. Alejandro Fernández for his fruitful discussions on this subject and reading of the manuscript. This work was supported in part by one grant from CONACyT (82308) and two grants from the Universidad Nacional Autónoma de México to GDR, including the Macroproyecto: Tecnologías para la Universidad de la Información y la Computación and PAPIIT IN205911, and CONACyT (102182 and 133294) to NP.

7. References

Altschuh D, Lesk AM, Bloomer AC, Klug A. (1987). Correlation of co-ordinated amino acid substitutions with function in viruses related to tobacco mosaic virus. J Mol Biol. 193(4):693-707.

Altschuh D, Vernet T, Berti P, Moras D, Nagai K. (1988). Coordinated amino acid changes in homologous protein families. Protein Eng. 2(3):193-199.

Altschul SF, Gish W, Miller W, Myers EW, Lipman DJ. (1990). Basic local alignment search tool. J Mol Biol. 215(3):403-410.

Altschul SF, Madden TL, Schäffer AA, Zhang J, Zhang Z, Miller W, Lipman DJ. (1997). Gapped BLAST and PSI-BLAST: a new generation of protein database search programs. Nucleic Acids Res. 25(17):3389-402.

Alvarez M, Zeelen JP, Mainfroid V, Rentier-Delrue F, Martial JA, Wyns L, Wierenga RK, Maes D. (1998). Triose-phosphate isomerase (TIM) of the psychrophilic bacterium Vibrio marinus. Kinetic and structural properties. J Biol Chem. 273(4):2199-2206.

Axe DD, Foster NW, Fersht AR. (1998).A search for single substitutions that eliminate enzymatic function in a bacterial ribonuclease. Biochemistry. 37(20):7157-7166.

Baker D, Sali A (2001). Protein structure prediction and structural genomics. Science. 2001. 294(5540):93-6.

Balzani V V, Credi A, Raymo FM, Stoddart JF. (2000). Artificial Molecular Machines. Angew Chem Int Ed Engl39(19):3348-3391.

Bateman OA, Purkiss AG, van Montfort R, Slingsby C, Graham C, Wistow G. (2003). Crystal structure of eta-crystallin: adaptation of a class 1 aldehyde dehydrogenase for a new role in the eye lens. Biochemistry. 42(15):4349-56.

Binkowski TA, Adamian L, Liang J. (2003). Inferring functional relationships of proteins from local sequence and spatial surface patterns. J Mol Biol. 332(2):505-26.

Bonneau R, Tsai J, Ruczinski I, Baker D. (2001). Functional inferences from blind *ab initio* protein structure predictions. J Struct Biol. 134(2-3):186-90.

Bourne PE. (2003).CASP and CAFASP experiments and their findings. Methods Biochem Anal. 44:501-7.

Brooks BR, Bruccoleri RE, Olafson BD, States DJ, Swaminathan S, Karplus M. (1983). CHARMM: A program for macromolecular energy, minimization, and dynamics calculations. J Comp Chem. 4(2):187–217.

Bruccoleri RE, Karplus M. (1987). Prediction of the folding of short polypeptide segments by uniform conformational sampling. Biopolymers. 26(1):137-168.

Chea E, Livesay DR. (2007). How accurate and statistically robust are catalytic site predictions based on closeness centrality? BMC Bioinformatics. 8:153.

Chollet A, Turcatti G. (1999). Biophysical approaches to G protein-coupled receptors: structure, function and dynamics. J Comput Aided Mol Des. 13(3):209-219

Chothia C, Lesk AM. (1986). The relation between the divergence of sequence and structure in proteins. EMBO J. 5(4):823-6.

Copley SD. (2003). Enzymes with extra talents: moonlighting functions and catalytic promiscuity. Curr Opin Chem Biol. 7(2):265-72.

Cota E, Hamill SJ, Fowler SB, Clarke J. (2000). Two proteins with the same structure respond very differently to mutation: the role of plasticity in protein stability. J Mol Biol. 302(3):713-25.

Cozzetto D, Kryshtafovych A, Fidelis K, Moult J, Rost B, Tramontano A. (2009) Evaluation of template-based models in CASP8 with standard measures. Proteins 77 Suppl 9:18-28.

Cusack MP, Thibert B, Bredesen DE, del Rio G (2007) Efficient identification of critical residues based only on protein structure by network analysis. PLoS ONE 2(5):e421.

Del Rio G, Koschützki D, Coello G (2009) How to identify essential genes from molecular networks? BMC Syst. Biol. 3:102.

del Sol Mesa A, Pazos F, Valencia A (2003) Automatic methods for predicting functionally important residues J Mol Biol. 326(4):1289-1302.

Di Francesco V, Garnier J, Munson PJ. (1997a). Protein topology recognition from secondary structure sequences: application of the hidden Markov models to the alpha class proteins. J Mol Biol. 267(2):446-463.

Di Francesco V, Geetha V, Garnier J, Munson PJ.(1997b). Fold recognition using predicted secondary structure sequences and hidden Markov models of protein folds. Proteins. Supplement 1:123-128.

Domingues FS, Koppensteiner WA, Sippl MJ. (2000). The role of protein structure in genomics. FEBS Lett. 476(1-2):98-102.

Doolittle RF. (1981). Protein Evolution. Science. 214(4525):1123-1124.

Doolittle RF. (1986). Of URFs and ORFs: a primer on how to analyze derived amino acid sequences. University Science Books, Mill Valley, CA, USA.

Drexler KE. (1994). Molecular nanomachines: physical principles and implementation strategies. Annu Rev Biophys Biomol Struct. 23:377-405.

Duan Y, Kollman PA. (1998). Pathways to a protein folding intermediate observed in a 1-microsecond simulation in aqueous solution. Science. 282(5389):740-744.

Durbin R, Eddy S, Krogh A and Mitchison G. (1998). Biological Sequence Analysis: Probabilistic Models of Proteins and Nucleic Acids. Cambridge: Cambridge University Press.

Elcock AH (2001) Prediction of functionally important residues based solely on the computed energetics of protein structure J Mol Biol. 312(4):885-896.

Elofsson A, von Heijne G. (2007). Membrane protein structure: prediction versus reality. Annu Rev Biochem. 76:125-140.

Fischer D, Eisenberg D. (1996). Protein fold recognition using sequence-derived predictions. Protein Sci. 5(5):947-55.

Freeman LC. (1977). A set of measures of centrality based on betweenness. Sociometry 40:35-41.

Gerstein M, Hegyi H. (1998). Comparing genomes in terms of protein structure: surveys of a finite parts list. FEMS Microbiol Rev. 22(4):277-304.

Gherardini PF, Helmer-Citterich M. (2008). Structure-based function prediction: approaches and applications. Brief Funct Genomic Proteomic. 7(4):291-302.

Ginalski K, Elofsson A, Fischer D, Rychlewski L. (2003). 3D-Jury: a simple approach to improve protein structure predictions. Bioinformatics. 19(8):1015-8

Glaser F, Pupko T, Paz I, Bell RE, Bechor-Shental D, Martz E, Ben-Tal N. (2003). ConSurf: identification of functional regions in proteins by surface-mapping of phylogenetic information. Bioinformatics. 19(1):163-4.
Göbel U, Sander C, Schneider R, Valencia A. (1994). Correlated mutations and residue contacts in proteins. Proteins. 18(4):309-317.
Godzik A, Skolnick J. (1992). Sequence-structure matching in globular proteins: application to supersecondary and tertiary structure determination. Proc Natl Acad Sci USA. 89(24):12098-12102.
Godzik A. (2003). Fold recognition methods. Methods Biochem Anal. 44:525-546.
Gracy RW. (1975).Triosephosphate isomerase from human erythrocytes. Methods Enzymol. 41:442-447.
Greene LH, Higman VA (2003). Uncovering network systems within protein structures. J Mol Biol. 334(4):781-91.
Greene LH, Lewis TE, Addou S, Cuff A, Dallman T, Dibley M, Redfern O, Pearl F, Nambudiry R, Reid A, Sillitoe I, Yeats C, Thornton JM, Orengo CA. (2007). The CATH domain structure database: new protocols and classification levels give a more comprehensive resource for exploring evolution. Nucleic Acids Res. 35:D291-D297.
Hardin C, Pogorelov TV, Luthey-Schulten Z. (2002). *Ab initio* protein structure prediction. Curr Opin Struct Biol. 12(2):176-181.
Holm L, Park J. (2000). DaliLite workbench for protein structure comparison. Bioinformatics. 16(6):566-567.
Hooft RW, Vriend G, Sander C, Abola EE. (1996). Errors in protein structures. Nature. 381(6580):272.
Huang W, Petrosino J, Hirsch M, Shenkin PS, Palzkill T. (1996).Amino acid sequence determinants of beta-lactamase structure and activity. J Mol Biol. 258(4):688-703.
Huisman GW, Gray D. (2002). Towards novel processes for the fine-chemical and pharmaceutical industries. Curr Opin Biotechnol. 13(4):352-358.
Jaroszewski L, Rychlewski L, Zhang B, Godzik A. (1998). Fold prediction by a hierarchy of sequence, threading, and modeling methods. Protein Sci. 7(6):1431-40.
Jeffery CJ. (1999). Moonlighting proteins. Trends Biochem Sci. 24(1):8-11.
Jeffery CJ. (2003). Moonlighting proteins: old proteins learning new tricks. Trends Genet. 19(8):415-7.
Jeffery CJ. (2009). Moonlighting proteins - an update. Mol. BioSyst. 5:345-350.
Jeong H, Mason SP, Barabasi AL, Oltvai ZN (2001). Lethality and centrality in protein networks. Nature. 411 (6833), 41-42.
Jeong H, Tombor B, Albert R, Oltvai ZN, Barabasi AL (2000). The large-scale organization of metabolic networks. Nature. 407 (6804), 651-654.
Jones D, Thornton J. (1993). Protein fold recognition. J Comput Aided Mol Des. 7(4):439-456.
Jones DT, Taylor WR, Thornton JM. (1992). A new approach to protein fold recognition. Nature. 358(6381):86-89.
Jones TA, Thirup S.(1986). Using known substructures in protein model building and crystallography. EMBO J. 5(4):819-822.
Kohl L, Callens M, Wierenga RK, Opperdoes FR, Michels PA. (1994).Triose-phosphate isomerase of Leishmania mexicana mexicana. Cloning and characterization of the gene, overexpression in Escherichia coli and analysis of the protein. Eur J Biochem. 220(2):331-338
Kolinski A. (2004). Protein modeling and structure prediction with a reduced representation. Acta Biochim Pol. 51(2):349-71.

Koretke KK, Luthey-Schulten Z, Wolynes PG. (1998). Self-consistently optimized energy functions for protein structure prediction by molecular dynamics. Proc Natl Acad Sci U S A. 95(6):2932-2937.

Koretke KK, Luthey-Schulten Z, Wolynes PG. (2002). Ab initio protein structure prediction. Curr Opin Struct Biol. 12(2):176-181.

Krietsch WK. (1975a). Triosephosphate isomerase from rabbit liver. Methods Enzymol. 41:438-442.

Krietsch WK. (1975b). Triosephosphate isomerase from yeast. Methods Enzymol. 41:434-438.

Krojer T, Garrido-Franco M, Huber R, Ehrmann M, Clausen T. (2002). Crystal structure of DegP (HtrA) reveals a new protease-chaperone machine. Nature. 416(6879):455-459.

Kuhlman B, Baker D (2000) Native protein sequences are close to optimal for their structures. Proc. Natl. Acad. Sci. USA 97:10383-10388.

Kursula I, Partanen S, Lambeir AM, Wierenga RK. (2002).The importance of the conserved Arg191-Asp227 salt bridge of triosephosphate isomerase for folding, stability, and catalysis. FEBS Lett. 518(1-3):39-42.

Kuhlman B, Dantas, G, Ireton GC, Varani G, Stoddard BL, Baker D. (2003). Design of a novel globular protein fold with atomic-level accuracy. Science 302:1364-1368.

Laskowski RA, Luscombe NM, Swindells MB, Thornton JM. (1996). Protein clefts in molecular recognition and function. Protein Sci. 5(12):2438-2452.

Laskowski RA, Moss DS, Thornton JM. (1993). Main-chain bond lengths and bond angles in protein structures. J Mol Biol. 231(4):1049-1067.

Lee J, Liwo A, Scheraga HA. (1999).Energy-based de novo protein folding by conformational space annealing and an off-lattice united-residue force field: application to the 10-55 fragment of staphylococcal protein A and to apo calbindin D9K. Proc Natl Acad Sci USA. 96(5):2025-30.

Lee J, Natarajan M, Nashine VC, Socolich M, Vo T, Russ WP, Benkovic SJ, Ranganathan R. (2008). Surface sites for engineering allosteric control in proteins. Science. 322(5900):438-442.

Lockless SW, Ranganathan R. (1999). Evolutionarily conserved pathways of energetic connectivity in protein families. Science. 286(5438):295-299.

Loeb DD, Swanstrom R, Everitt L, Manchester M, Stamper SE, Hutchison CA 3rd. (1989). Complete mutagenesis of the HIV-1 protease. Nature. 340(6232):397-400.

Luetz S, Giver L, Lalonde J. (2008). Engineered enzymes for chemical production. Biotechnol Bioeng. 101(4):647-653.

Martí-Renom MA, Stuart AC, Fiser A, Sánchez R, Melo F, Sali A. (2000). Comparative protein structure modeling of genes and genomes. Annu Rev Biophys Biomol Struct. 29:291-325.

Martin AC, Orengo CA, Hutchinson EG, Jones S, Karmirantzou M, Laskowski RA, Mitchell JB, Taroni C, Thornton JM. (1998). Protein folds and functions. Structure. 6(7):875-884.

Milenković T, Filippis I, Lappe M, Pržulj N, (2009) Optimized Null Model for Protein Structure Networks. PLoS ONE 4: e5967.

Mirsky AE, Pauling L. (1936). On the Structure of Native, Denatured, and Coagulated Proteins. Proc Natl Acad Sci USA. 22(7):439-447.

Montiel Molina HM, Millan-Pacheco C, Pastor N, del Rio G (2008) Computer-Based Screening of Functional Conformers of Proteins. PLoS Comput Biol. 4(2):e1000009.

Moult J. (2005). A decade of CASP: progress, bottlenecks and prognosis in protein structure prediction. Curr Opin Struct Biol. 15(3):285-289.

Neet KE, Lee JC. (2002).Biophysical characterization of proteins in the post-genomic era of proteomics. Mol Cell Proteomics. 1(6):415-420

Norin M, Sundström M. (2002). Structural proteomics: developments in structure-to-function predictions. Trends Biotechnol. 20(2):79-84.

Olmea O, Rost B, Valencia A. (1999). Effective use of sequence correlation and conservation in fold recognition. J Mol Biol. 293(5):1221-1239.

Olmea O, Valencia A. (1997). Improving contact predictions by the combination of correlated mutations and other sources of sequence information. Fold Des. 2(3):S25-S32.

Ortiz AR, Kolinski A, Rotkiewicz P, Ilkowski B, Skolnick J. (1999). Ab initio folding of proteins using restraints derived from evolutionary information. Proteins. Supplement 3:177-185

Osguthorpe DJ. (2000). Ab initio protein folding. Curr Opin Struct Biol. 10(2):146-152.

Ozkan SB, Wu GA, Chodera JD, Dill KA (2007) Protein folding by zipping and assembly. Proc. Natl. Acad. Sci. USA 104: 11987-11992.

Pazos F, Helmer-Citterich M, Ausiello G, Valencia A. (1997).Correlated mutations contain information about protein-protein interaction. J Mol Biol. 271(4):511-523.

Pearlman DA, Case DA, Caldwell JW, Ross WS, Cheatham TE III, DeBolt S, Ferguson D, Seibel G, Kollman P. (1995) AMBER, a package of computer programs for applying molecular mechanics, normal mode analysis, molecular dynamics and free energy calculations to simulate the structural and energetic properties of molecules. Comp Phys Commun. 91:1-41.

Pearson WR, Lipman DJ.(1988). Improved tools for biological sequence comparison. Proc Natl Acad Sci USA. 85(8):2444-2448.

Pedersen JT, Moult J. (1997a). Ab initio protein folding simulations with genetic algorithms: simulations on the complete sequence of small proteins. Proteins. Supplement 1:179-184.

Pedersen JT, Moult J. (1997b). Protein folding simulations with genetic algorithms and a detailed molecular description. J Mol Biol. 269(2):240-259.

Punta M, Ofran Y. (2008). The rough guide to in silico function prediction, or how to use sequence and structure information to predict protein function. PLoS Comput Biol. 4(10):e1000160.

Radivojac P, Iakoucheva LM, Oldfield CJ, Obradovic Z, Uversky VN, Dunker AK. (2007). Intrinsic disorder and functional proteomics. Biophys J. 92(5):1439-56.

Rennell D, Bouvier SE, Hardy LW, Poteete AR. (1991). Systematic mutation of bacteriophage T4 lysozyme. J Mol Biol. 222(1):67-88.

Rivera MH, López-Munguía A, Soberón X, Saab-Rincón G. (2003). Alpha-amylase from Bacillus licheniformis mutants near to the catalytic site: effects on hydrolytic and transglycosylation activity. Protein Eng. 16(7):505-514.

Robson B. (1999). Beyond proteins. Trends Biotechnol. 17(8):311-315.

Rost B. (1995). TOPITS: threading one-dimensional predictions into three-dimensional structures. Proc Int Conf Intell Syst Mol Biol. 3:314-421.

Rost B. (1999). Twilight zone of protein sequence alignments. Protein Eng. 12(2):85-94.

Russ WP, Lowery DM, Mishra P, Yaffe MB, Ranganathan R. (2005). Natural-like function in artificial WW domains. Nature. 437(7058):579-83.

Rychlewski L, Jaroszewski L, Li W, Godzik A. (2000). Comparison of sequence profiles. Strategies for structural predictions using sequence information. Protein Sci. 9(2):232-41.

Sadowski MI, Jones DT. (2009) The sequence-structure relationship and protein function prediction. Curr. Opin. Struct. Biol. 19:357-362.

Sali A, Potterton L, Yuan F, van Vlijmen H, Karplus M. (1995). Evaluation of comparative protein modeling by MODELLER. Proteins. 23(3):318-326.

Sander C, Schneider R. (1991). Database of homology-derived protein structures and the structural meaning of sequence alignment. Proteins. 9(1):56-68.

Shaw DE, Maragakis P, Lindorff-Larsen K, Piana S, Dror RO, Eastwood MP, Bank JA, Jumper JM, Salmon JK, Shan Y, Wriggers W. (2010) Atomic-level characterization of the structural dynamics of proteins. Science 330: 341-346.

Shehu A, Kavraki LE, Clementi C. (2009) Multiscale characterization of protein conformational ensembles. Proteins 76:837-851.

Simons KT, Bonneau R, Ruczinski I, Baker D. (1999). Ab initio protein structure prediction of CASP III targets using ROSETTA. Proteins. Supplement 3:171-176.

Sippl MJ. (1990). Calculation of conformational ensembles from potentials of mean force. An approach to the knowledge-based prediction of local structures in globular proteins. J Mol Biol. 213(4):859-83.

Skolnick J, Fetrow JS, Kolinski A. (2000). Structural genomics and its importance for gene function analysis. Nat Biotechnol. 2000 Mar;18(3):283-287.

Socolich M, Lockless SW, Russ WP, Lee H, Gardner KH, Ranganathan R. (2005). Evolutionary information for specifying a protein fold. Nature. 437(7058):512-518.

Straathof AJ, Panke S, Schmid A. (2002). The production of fine chemicals by biotransformations. Curr Opin Biotechnol. 13(6):548-556.

Terwilliger TC, Zabin HB, Horvath MP, Sandberg WS, Schlunk PM. (1994). In vivo characterization of mutants of the bacteriophage f1 gene V protein isolated by saturation mutagenesis. J Mol Biol. 236(2):556-71.

Thibert B, Bredesen DE, del Rio G (2005). Improved prediction of critical residues for protein function based on network and phylogenetic analyses. BMC Bioinformatics. 6:213.

Valdar WS. (2002). Scoring residue conservation. Proteins. 48(2):227-241.

Vásquez M. (1996). Modeling side-chain conformation. Curr Opin Struct Biol. 6(2):217-21.

Vendruscolo M, Dokholyan NV, Paci E, Karplus M (2002). Small-world view of the amino acids that play a key role in protein folding. Phys Rev E. 65(6 Pt 1):061910.

Wallace AC, Laskowski RA, Thornton JM. (1996). Derivation of 3D coordinate templates for searching structural databases: application to Ser-His-Asp catalytic triads in the serine proteinases and lipases. Protein Sci. 5(6):1001-1013.

Wallner B, Elofsson A. (2003).Can correct protein models be identified? Protein Sci. 12(5):1073-1086.

Xiang J, Jung JY, Sampson NS. (2004).Entropy effects on protein hinges: the reaction catalyzed by triosephosphate isomerase. Biochemistry. 43(36):11436-11445.

Yarov-Yarovoy V, Schonbrun J, Baker D. (2006).Multipass membrane protein structure prediction using Rosetta. Proteins. 62(4):1010-1025.

Yura K, Yamaguchi A, Go M. (2006) .Coverage of whole proteome by structural genomics observed through protein homology modeling database. J Struct Funct Genomics. 7(2):65-76.

Zaks A. (2001). Industrial biocatalysis. Curr Opin Chem Biol. 5(2):130-6.

Zhang H (2002). Protein Tertiary Structures: Prediction from Amino Acid Sequences. In: ENCYCLOPEDIA OF LIFE SCIENCES. John Wiley & Sons, Ltd: Chichester http://www.els.net/ [doi:10.1038/npg.els.0006101].

Zhang Z, Palzkill T. (2003). Determinants of binding affinity and specificity for the interaction of TEM-1 and SME-1 beta-lactamase with beta-lactamase inhibitory protein. J Biol Chem. 278(46):45706-45712.

19

Protein-DNA Interactions Studies with Single Tethered Molecule Techniques

Guy Nir, Moshe Lindner and Yuval Garini
Physics Department and institute of Nanotechnology,
Bar Ilan University, Ramat Gan,
Israel

1. Introduction

The last decade has seen a leap forward in the understanding of molecular and cellular mechanisms with the development of advanced techniques for observing, manipulating and imaging single molecules. In contrast to conventional biochemical techniques which yield information derived from population averages, single molecule techniques give access to the dynamics and properties of individual biomolecules in situ.

One of the problems in studying single molecules is the need to observe and measure the molecule for a large enough period of time and hence different approaches were developed. Among the various experimental single molecule techniques, one of the most convenient to implement is Tethered Particle Motion (TPM) (Schafer, Gelles *et al.* 1991; Nelson, Zurla *et al.* 2006; Zurla, Franzini *et al.* 2006; Jeon & Metzler 2010). In the most common procedure of TPM approach, a bead is attached to the DNA in one end, while the other end is immobilized to a glass surface (Figure 1). The Brownian motion of the nano-bead can be imaged through a microscope with an array detector such as a charged coupled detector (CCD) camera that captures the position of the bead in time and space. The bead positions are analyzed by using single particle motion (SPT) algorithms and its distribution is calculated. It is directly related to the expected conformations of the DNA when it is treated as a polymer with a given nominal length and stiffness.

The DNA contains a long chain of nucleic acids that contains all of our genetic information and a stretched human DNA molecule is about 2 meters long. This long molecule is divided to 46 chromosomes (in human cells) that are packed into a human cell that is only 10-100 μm large, which necessitates that the DNA structure will be highly regulated. Moreover, the DNA packaging has to ensure the appropriate functioning of the DNA-related processes such as transcription and replication.

All the processes that involve DNA, mainly remodeling, transcription and replication, are performed by a set of proteins that interact with the DNA in different ways. Therefore, these proteins are crucial and the understanding of their interaction patterns with the DNA is of great interest. DNA-protein interactions are therefore a subject of an ongoing research mediated by different methodologies. Better understanding of the interaction mechanisms could lead to new diagnostics methods, the discovery of new drug targets and altogether can affect mankind health.

The mechanisms described above can be studied with single molecule techniques in great details and provide information that in many cases is undetectable when using ensemble techniques. One example is the motor enzymes, such as myosin V and kinesin, that proceed in nanometric steps (Yildiz, Forkey *et al.* 2003; Yildiz, Tomishige *et al.* 2004). A detailed description of their translocation mechanism requires labeling and tracking of single enzymes in high accuracy. Observing single labeled biomolecules avoids large-population averaging and allows deciphering each step-size. It also allows distinguishing transient kinetic steps in a multistep reaction and identifying rare or short-lived conformational states.

In this book chapter we will:

1. Describe single tethered-molecule detection methods that can be used for DNA-protein studies while emphasizing Tethered Particle Motion (TPM).
2. Review few key findings on relevant proteins.
3. Compare and summarize the method presented here.
4. Provide a fascinating glance to the dazzling near-future capabilities which are based on immerging single molecule detection methods for studying DNA-protein interactions.

2. Single tethered-molecule detection methods for DNA-protein studies

Recent progress in technology, especially in the fields of photonics and nanotechnology and their combination together with the profound knowledge obtained during the last decades in molecular biology has enabled the development of single-molecule applications. Here we will describe the following well-established methods: TPM, magnetic tweezers, optical tweezers, AFM and FRET. These methods have been used to perform cutting-edge experiments as we will show later in this chapter.

2.1 Tethered particle motion

2.1.1 General description of the method

TPM is an optical method that relies on physical models and biochemical approaches to detect and observe biophysical properties, such as the dynamic variations in conformations induced by enzymes acting on biopolymers. In TPM, the biopolymer of interest, i.e. linear double-stranded DNA (dsDNA), is attached at one end to a glass surface and hence held fixed, while the other end is labeled with a marker that can be a fluorescent tag or a metal bead and is free to diffuse in a restricted volume due to the anchoring of the other end (Figure 1). The bead positions that reflect the end-to-end distance of the biopolymer are recorded at different time-intervals and are than analyzed according to physical models to retrieve biophysical properties of the biopolymer, the enzymes acting on it and the nature of their interactions. One advantage of the method relies in the fact that unlike other single molecule techniques, TPM is a force-free technique, meaning that no external force is used to alter the studied molecules natural structure, which might lead to a more reliable measurement.

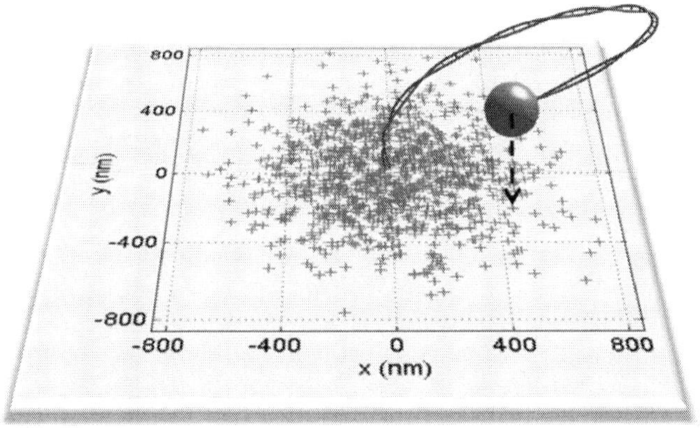

Fig. 1. TPM principles. A small bead is attached to one end of a polymer while its other end is attached to the surface. The bead performs Brownian motion in the solution constrained by the polymer. The bead positions (red crosses) are measured with an optical setup with an accuracy of few nanometers.

2.1.2 Physical model

Many biological processes such as replication, transcription and gene regulation require accessibility to the DNA. The DNA's contour in cells is regulated by DNA-bending proteins such as histones in eukaryotic cells (Shin, Santangelo *et al.* 2007) or histone-like proteins in prokaryotic cells (Rouviere-Yaniv 1987). Other proteins can bind or interact with the DNA and can modify its mechanical properties. Such interactions might alter the conformation of the DNA. TPM can detect such changes, but because they are not being measured directly, but only through the end-to-end distribution (as depicted from the bead measurement), it is necessary to use a model that describes the dependence of the DNA conformation on its basic physical parameters in space.

The DNA is known to be a rather stiff polymer that can be described to a good accuracy by the Worm-like-Chain (WLC) model. WLC is derived from the equivalent Freely Jointed Chain (FJC). The equivalent chain described by the model has the same mean square end-to-end distance $\langle R^2 \rangle$ and the same contour length, L, as the actual polymer, but it is described by N freely jointed effective bonds of length b (figure 2). This effective bond length b, is called Kuhn length. Accordingly we can write:

$$Nb = L \tag{1}$$

and its mean square end-to-end distance is

$$\langle R^2 \rangle = Nb^2 = bL \,. \tag{2}$$

Actually, one conventionally defines a ``persistence length'', l_p, in terms of how rapidly the direction of the polymer changes as a function of the contour length. Let us define the angle θ between a vector that is tangent to a certain polymer element and a tangent vector at a distance L along the polymer. It can be shown that the expectation value of the cosine of the angle falls off exponentially with distance,

$$\langle \cos\theta \rangle = e^{-L/l_p} \tag{3}$$

where the triangular brackets denote the average over all starting polymer-element positions. For DNA, the persistence length is twice the Kuhn length.

According to the WLC theory, for "naked" (proteins-free) double stranded DNA,

$$\langle R^2 \rangle \cong 2l_p L = bL \quad \text{for } L >> l_p . \tag{4}$$

The persistence length of DNA in normal conditions is equal to 50 nm (Rubinstein & Colby 2003). If the polymer is free to assume any configuration in three dimensions, it can be shown that the probability distribution function (PDF) for its projection length along one-dimension (x or y axis) is a Gaussian:

$$P_{1D}(x)dx = \sqrt{\frac{3}{4\pi L l_p}} \cdot \exp\left(-\frac{3x^2}{4Ll_p}\right) dx . \tag{5}$$

The PDF of the two-dimensional projection length of the polymer on a plane in Polar coordinates can also be calculated and it is found to be the Rayleigh distribution:

$$P_{2D}(r)dr = \frac{3}{4\pi L l_p} \cdot \exp\left(-\frac{3r^2}{4Ll_p}\right) \cdot 2\pi r \cdot dr \tag{6}$$

where $r = \sqrt{x^2 + y^2}$.

Although real polymer chains are subjected to self-avoidance, meaning that due to short range repulsive forces, monomers of the chain can't cross themselves which leads to an excluded volume, it is accustomed to treat DNA molecules as ideal chains (also called phantom polymers), since the probability of crossing is non negligible only for long DNA (>40 μm) (Strick, Allemand *et al.*), which are not usually used in single-molecule experiments (Slutsky 2005).

2.1.3 Marker considerations

The DNA conformation changes randomly in the solution and its end-to-end distance is measured in TPM by finding the position of the attached marker. Different types of markers can be used and with respect to their detection method, one can distinguish fluorescent beads, scattering beads and beads that are detected by normal transmission or contrast enhancement methods. A fluorescent marker might be small but suffers from quenching and bleaching and is not recommended for long-time observations. Polystyrene microspheres

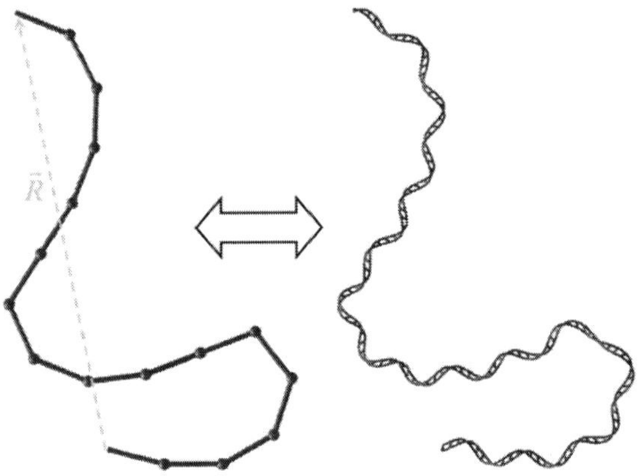

Fig. 2. Left: FJC model. Each segment is equal to b, the Kuhn length (~100 nm). $\bar{R}$ is the end-to-end vector. A real dsDNA is shown on the right and it is equivalent to the polymer in the FJC model.

are also used as markers normally with phase contrast microscopy, but it requires a rather large micron-size particle, which may lead to inaccurate analysis in TPM. If the size is too large, the measured position of the bead may be dominated by the free rotation of the marker around its tethering point. This motion is also affected by the position of the bead with respect to the surface, and therefore, the measured distribution may be dominated or severely influenced by the marker size. This will lead to errors when trying to extract the polymer properties from the bead distribution, and the bead-size effect cannot be easily compensated for. It is therefore better to work with smaller beads. The actual size that does not affect the distribution depends on the polymer contour length and persistence length. The problem was treated intensively in the literature (Segall, Nelson *et al.* 2006), and it was shown that the bead size will not affect the distribution significantly, as long as the parameter called the excursion number, is smaller than unity:

$$N_r = R / \sqrt{Ll_p / 3} < 1 . \tag{7}$$

The excursion number is defined as the ratio of the marker radius R to the radius of gyration of DNA which depends on the polymer persistence length l_p and contour length L.

Therefore, it is better to use a smaller bead size, and a more suitable solution is a small metal nano-bead. Such a bead has a significant plasmon scattering which results in an intense signal that can be easily detected by a CCD.

For DNA with a contour length of $L = 925nm$, a known persistence length of $l_p = 50nm$ for bare DNA and a gold bead with $r = 40$ nm gives $N_r \sim 0.32$ which meets the criterion (Segall, Nelson et al. 2006; Lindner, Nir et al. 2009).

Another advantage is the short exposure time that can be used, which still achieving a high-enough signal to noise for analyzing the bead position. If the exposure-time is too long, the bead motion broadens the image spot and the analyzed distribution of the bead position is skewed. Although this effect can be corrected (Destainville & Salomé 2006; Wong & Halvorsen 2006) it increases the error and should be avoided. We showed that with a gold nano-bead of 80 nm diameter, good results are achieved even at short exposure times as 1 ms.

2.1.4 Standard experimental set-up

Figure 3 (Nir, Lindner *et al.* 2011) shows a possible implementation of the experimental setup for TPM . It consists of a dark field (DF) microscope unit (Olympus BX-RLA2, Tokyo, Japan) with a x50 objective lens (NA=0.8) and an EM-CCD camera (Andor DU-885, Belfast, Northern Ireland) with a pixel size of 8x8 µm and a maximal pixel read-out rate of 35 MHz (Lindner, Nir *et al.* 2009).

The DF setup improves the signal to noise that is achieved in the measurement because it ensures that only the light that is scattered from the bead is collected by the objective lens, while eliminating the illumination background light.

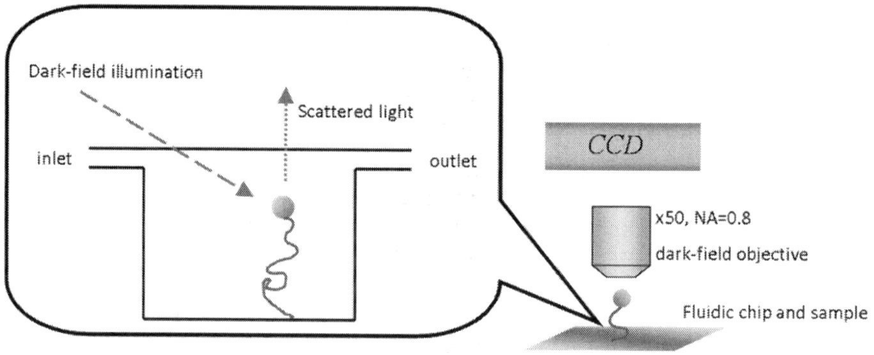

Fig. 3. The experimental setup. A dark-field microscope unit is used to track the metallic-bead position. A gold nano-bead is attached to DNA molecule and its position is tracked by the microscope and the CCD. When a protein interacts with the DNA, its biophysical properties are modified and can be tracked by the system.

2.1.5 Biochemical procedures

For constructing the dsDNA tethers it is common to use a PCR reaction to amplify desired dsDNA fragments from λ phage DNA. One end of the DNA is normally linked to digoxigenin (DIG) for a further attachment to an anti-DIG coated surface, this way the DNA is anchored to the surface. The other DNA end is biotin-linked for attaching a neutravidin conjugated nanobead. The modifications are done by using modified primers in the PCR reaction.

The tethering procedure is done in a flow cell and monitored with the microscope. First, a passivation procedure is required to reduce unspecific binding to the glass surface. Some common passivation reagents might be: Bio-Rad non-fat dry milk, Polyethylene Glycol (PEG) and Bovine Serum Albumin (BSA). After a proper incubation time (depends on the different passivation reagents) the buffer is washed and the surface is coated with anti-DIG. After one hour of incubation one should wash again and introduce the DNA to the solution. After incubation of one hour and another wash, the nano beads are introduced. After ~ 30 minutes of incubation and another wash the tethering procedure is complete (Selvin & Ha 2008; Lindner, Nir *et al.* 2009; Zimmermann, Nicolaus *et al.* 2010; Lindner, Nir *et al.* 2011).

2.1.6 Conducting experiments

First, each bead is measured a couple of times. The scattering plot is tested for circular symmetry. If it is found symmetric (figure 4), the persistence length is calculated (see section 2.1.6, data analysis). Measurement of the interaction of the bead with a protein continues only if its persistence length is ~50 ±5 nm. The next step would be to add an enzyme of interest, record the DNA's persistence length and realize if analyze the changes due to the enzyme acting on the DNA. Few beads are measured, few times each, in order to provide reliable statistic information. Each measurement consists of approximately ~2000 frames.

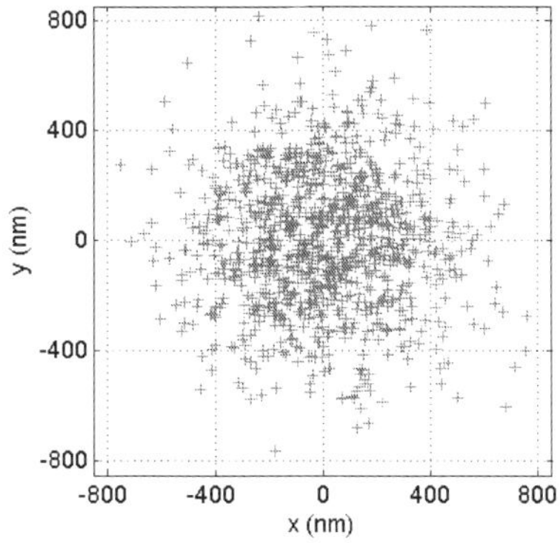

Fig. 4. *XY* projection. Each cross sign indicates a recording of the bead position at a time-point. Due to the symmetric polymer configurations distribution, the bead's distribution (representing the DNA end-to end vector) should be circular-symmetric centered at the DNA anchor point.

2.1.7 Data analysis

The data is analyzed with an SPT software package usually developed by the labs conducting such experiments with Matlab (The Mathworks, Natick, MA). For extraction of the DNA persistence length from the distribution, first, there should be an extraction of the bead position coordinates, $x(t), y(t)$ (2D projection) for each image (t). Then the radial distribution P(r) can be calculated and fitted to the expected distribution according to the Freely Joint Chain model which gives the Rayleigh distribution (Equation 6).

2.2 3D TPM

2.2.1 General description

3D TPM is an extended TPM method that was lately developed (Lindner, Nir *et al.* 2011). Instead of measuring only the 2D projection of the bead, it allows to measure the actual position of the bead in 3D.

The method relies on Total Internal Reflection (TIR), and employs the evanescent wave that is exponentially decreasing in the z-direction due to the TIR. Because the intensity depends on the bead height above the surface, the position of the bead in 3D can be determined.

2.2.2 Physical model

The intensity of an evanescent field decreases exponentially with the distance from the surface,

$$I = I_0 \exp(-z / d) \tag{8}$$

where I_0 is the intensity at the surface, z is the distance from surface, and d is the penetration depth. The penetration depth of the TIR illumination depends on a few fundamental parameter of the optical setup and can be tuned by changing the incident angle of the beam on the surface (Figure 5) according to:

$$d = \frac{\lambda_0}{4\pi\sqrt{n_i^2 \sin^2 \theta_i - n_t^2}} . \tag{9}$$

Here λ_0 is the wavelength of light in vacuum, n_i and n_t are the indexes of refraction of the materials above and below the surface and θ_i is the incident angle. By tuning the incident angle to be in the range of a few degrees above the critical angle, a penetration depth of 100–200 nm can be achieved, and axial distances in the range of 0–500 nm can be measured. Such a range is satisfying for measuring a 1-μm polymer with a persistence length of 50 nm; note that the end-to-end distance of such a polymer is rarely larger than 400 nm.

2.2.3 Standard experimental set-up

The experimental setup is very similar to a standard TPM setup, only with the addition of a diode laser and an equilateral prism to allow the creation of the evanescent wave (see figure 5).

2.2.4 Experimental procedures

A method that relies on the signal intensity to calculate the axial distance from the surface requires calibration in order to find d (Equation 9). We developed two calibration methods that rely on the actual system itself and do not require adding further optics to the setup. One is based on the 3D distribution of tethered beads (Volpe, Brettschneider *et al.* 2009). The distribution is measured with TIR illumination and the persistence length is calculated from the planar x and y distributions. Then, the distribution along z is fitted to the simulation results with a single parameter (the penetration depth d). The second method is based on measuring the free diffusion of the beads. The principle is similar to the fluorescence correlation spectroscopy method described by Harlepp *et al.* (Harlepp, Robert *et al.* 2004).

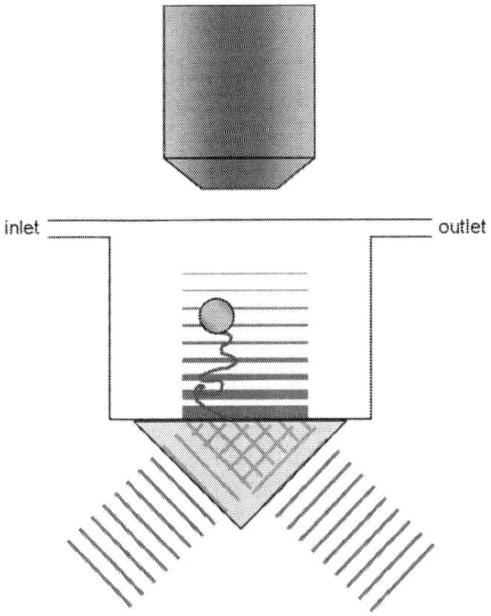

Fig. 5. 3D TPM illumination and detection. In order to achieve an evanescent wave, the light hits a prism above the critical angle. The evanescent wave decays exponentially into the sample while the bead's intensity and location is recorded. The bead position in along z is deduced according to Equation 8.

2.3 Single molecule force techniques

The following methods were thoroughly described before and hence will be briefly described here. Few key findings with these methods will be described as well.

So far we have only discussed a single molecule force-free technique. Although TPM allows us to observe a biopolymer in its natural form, it lacks the ability to manipulate the biopolymer. Many of the biological process such as DNA twisting, replication and cell migration are force driven. Another interesting fact is that biopolymers such as DNA, RNA, titin (the protein responsible for passive elasticity in the skeletal (Bustamante, Chemla *et al.* 2004)), etc posses "spring activity". If we combine these two facts, we can stretch and twist single biomolecules, hence examining their interactions with enzymes and regulating reaction coordinates as a function of load.

2.3.1 Magnetic tweezers

In magnetic tweezers a polymer of interest (such as DNA) is usually tethered to a glass surface while the other end is attached a magnetic microsphere which is pulled away from the surface with a magnet (Figure 6). The upper bound for force measurements in micromanipulation experiments is the tensile strength of a covalent bound, on the order of

1000 pN. The smallest measurable force is set by the Langevin force which is responsible for the Brownian motion of the sphere. Because of its random nature, the Langevin force is a noise density in force which is simply written as

$$f_n = \sqrt{4k_B T 6\pi\eta r} \tag{10}$$

where η is the viscosity of the medium and r is the radius of the particle. For a ~1 μm diameter microsphere in water, $f_n \sim 0.017\ pN/\sqrt{Hz}$. In between those two extremes lies the forces typical of the molecular scale, which is of order $k_B T/nm \sim 4pN$ (Strick, Allemand *et al.*). This is roughly the stall force of a single-molecular motor such as myosin (4 pN; (Finer, Simmons *et al.* 1994)) or RNA-polymerase (15–30 pN; (Yin, Landick *et al.* 1994; Wang 1998)). Applying forces in this regime on the magnetic microsphere, allows the delicate or aggressive stretching and twisting of the biopolymer, hence opening the door for manipulating single DNA-protein interactions (Manosas, Lionnet *et al.*; Bouchiat, Wang *et al.* 1999; Bustamante, Smith *et al.* 2000; Bustamante, Bryant *et al.* 2003; Neuman & Nagy 2008).

The most basic magnetic tweezers setup consists of a pair of permanent magnets, a flow cell, a magnetic bead and a CCD camera (Figure 6). For delicate manipulation, the magnet can be connected to a piezo-stage which allows bringing the magnets closer (for a stronger force) or away from the magnetic bead (for a weaker force) in nanometric steps.

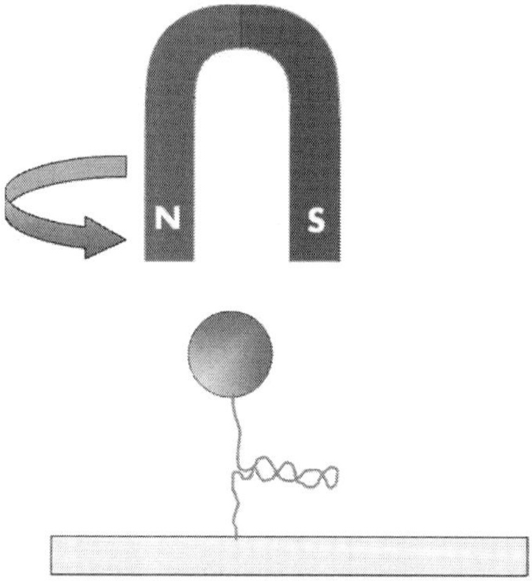

Fig. 6. Principle of magnetic tweezers. A magnetic force pulls the magnetic bead towards the magnet as a function of distance, stretching the DNA. The magnet can also be rotated, spinning the magnetic bead and twisting the DNA.

Figure 7 shows a typical plot that demonstrates how changing the force acting on the sphere allows the researchers to produce force-extension curves, meaning how much force is applied on the sphere to stretch a biopolymer to a certain extension.

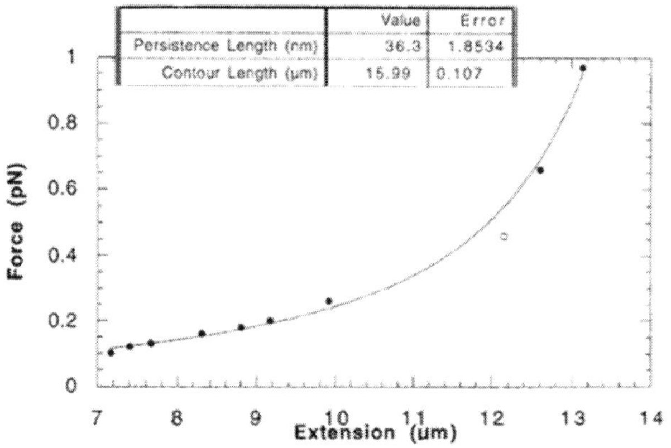

Fig. 7. Force-extension curve. Increasing the magnetic force results in further extension of the DNA. Reprinted with permission from (Haber & Wirtz 2000). Copyright [2000]. American Institute of Physics.

The Brownian fluctuations of the tethered sphere are equivalent to the motion of a damped pendulum of length $l = \langle z \rangle$ (Strick, Allemand *et al.*). Pulling the bead along z direction gives rise to a magnetic force, F. Its longitudinal, δz^2, and transverse fluctuations, δx^2, can be characterized as a spring with an effective stiffness $k_z = \partial_z F$ and $k_x = F / l$. By the equipartition theorem they satisfy:

$$\delta z^2 = \frac{k_B T}{k_z} = \frac{k_B T}{\partial_z F} \tag{11a}$$

and

$$\delta x^2 = \frac{k_B T}{k_x} = \frac{k_B T l}{F}. \tag{11b}$$

Thus by tracking the sphere fluctuations it is possible to extract the force pulling the sphere (and the biopolymer).

2.3.2 Optical tweezers

The type of experiments usually done with optical tweezers are stretching experiments, where one stretches a biomolecules of interest, for example, DNA, and follows how these manipulations alter the DNA conformation or how do DNA binding proteins respond to the load applied on the DNA. It is also used to follow the changes in extension or force as a result of a biochemical process of the DNA with proteins. The optical tweezers are implemented by creating an optical trap which is implemented by concentrating a laser beam to a diffracted-limited spot through a high Numerical Aperture (NA) objective lens. When light hits a transparent dielectric object, such as a polystyrene microsphere, it has two important optical outcomes (figure 8). The first one is reflection of the impinging light, which pushes away the microsphere (scattering force). The second one is refraction. The momentum of the light impinging on the microsphere is changed due to interaction with the microsphere. The momentum change of the light must be compensated by an equal and opposite change in momentum of the sphere, resulting in attraction of the sphere towards the center of the light spot (Williams 2002). To establish a stable trap, the force attracting the bead towards the light-spot must overcome the scattering force. The trap stiffness can be tuned by adjusting the laser intensity and focus. High NA objective lens is efficient at concentrating the light and stabilizes the trap. It is common to use high-power laser (>1 W) such as Nd:YAG and its derivatives that emit at the near-IR. The high-power reduces the spatial fluctuations of the trapped bead allowing a more stabilized trap and the near-IR wavelengths reduces the damage to the biomolecules that are in use (Neuman, Chadd *et al.* 1999). The detection of the microsphere is usually done with a quadrant photo-diode (QPD) which features high temporal resolution and leads to nanometric accuracy in detecting the deflection of the sphere (Rohrbach & AU - Stelzer 2002).

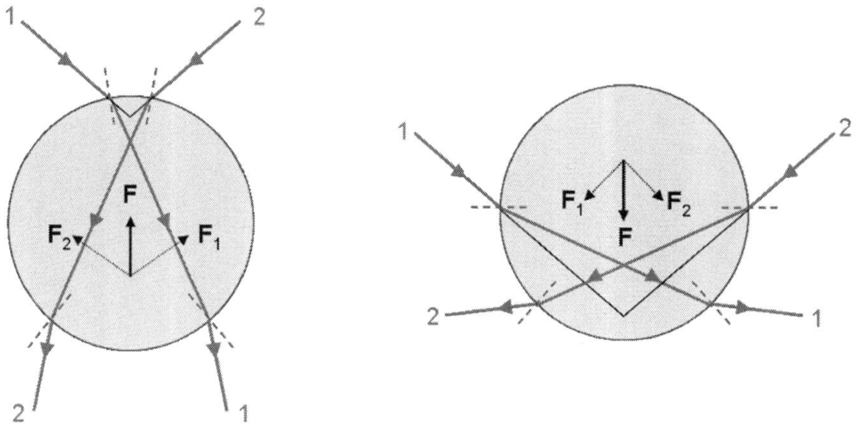

Fig. 8. Interaction of light with a transparent dielectric microsphere. Left: The microsphere is located beneath the center of the beam. When light hits the sphere it is reflected (not shown) and refracted according to Snell's law and the force acting on the object has to obey momentum conservation theory. Therefore, the net refraction force will point towards the center of the beam, pulling it up. Right: The microsphere is located above the center of the beam and the net refraction force is pointing towards the center of the beam, pulling it down.

An optical trap has to be calibrated for proper evaluating of the trap stiffness. The most common techniques treat the sphere as a linear spring, where the spring constant is determined by the sphere Brownian motion and the force obeys to Hooke's law ($f = -kx$). The sphere's position is calibrated by moving the sphere a known distance and recording the signal at that position. It can also be tuned by analyzing the frequency response of the fluctuations (Bustamante, Macosko *et al.* 2000). More profound and detailed explanations can be found in (Neuman & Nagy 2008; Selvin & Ha 2008).

As was mentioned at the beginning of this section, the most common experiments with optical tweezers are stretching experiments, which produce force-extension curves. A dsDNA molecule in solution bends and curves locally according to thermal fluctuations, which is of course an entropic-driven behavior, influenced by the DNA elasticity. According to the Worm-like Chain (WLC) model that was briefly explained in section 2.1.2, the persistence length of a dsDNA in solution containing physiological salt conditions is 50 nm (Rubinstein & Colby 2003). The WLC model is well-suited to describe the entropic behavior of dsDNA in the regime of low and intermediate forces.

According to the model, the force F required to induce an extension x of the end-to-end distance of a polymer with a contour length L and persistence length l_p is given by: (Bustamante, Smith *et al.* 2000):

$$\frac{F \cdot l_p}{k_B T} = \frac{1}{4\left(1 - x/L\right)^2} + \frac{x}{L} - \frac{1}{4} \qquad (12)$$

where k_B is the Boltzmann constant and T is the temperature.

2.3.3 Atomic force microscopy with tethered molecules

The Atomic Force Microscope (AFM) (Binnig, Quate *et al.* 1986; Martin, Williams *et al.* 1987) is another force-based technique that allows the stretching of individual biomolecules (Li, Wetzel *et al.* 2006; Perez-Jimenez, Garcia-Manyes *et al.* 2006; Neuman & Nagy 2008) and measure their elastic respond with sub-nanometer accuracy and picoNewton respond. Unlike previous-discussed single-molecule force spectroscopy methods, AFM is efficient also at high forces, which opened the door for exploring the properties of bio-entities in high-energy conformational states. For example, it enabled the detection of sub-Angstrom conformational changes of a single Dextran molecule (Walther, Bruji *et al.* 2006) and plot the unfolding force-histogram of a modular polyprotein (Li, Oberhauser *et al.* 2000).

In a typical experiment, a biomolecule of interest, say a protein, is linked to a flat surface that is mounted to on a piezoelectric stage. When the protein is approached by an AFM tip which is supported by a flexible cantilever, it might adsorb to the tip. When the tip is retreated from the surface it stretches the protein. The stretch bends the cantilever, which results in deflection of the laser beam impinging on the cantilever and recorded in a photo detector. If the cantilever elastic properties are known, it is possible to extract the force acting on the protein (Fisher, Oberhauser *et al.* 1999) (see figure 9). Low-force stretching can be modeled by the WLC model and the force acting on a polymer can be extracted in the same way done with optical tweezers according to equation 12. However, it was already mentioned that the great advantage in the AFM force spectroscopy technique is the ability to apply hundreds of picoNewtons . Moreover, force-extension curves for small or single-fold proteins are difficult to interpret because of non-specific interactions that might arise between the cantilever tip and the surface.

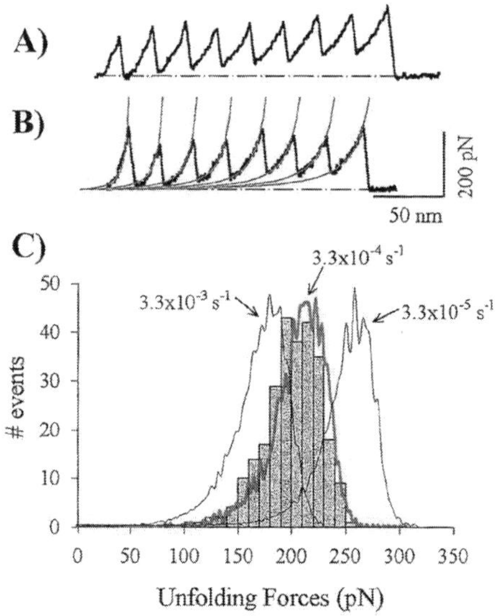

Fig. 9. Force–extension relationships for recombinant poly(I27) measured with AFM techniques. (A and B) Stretching of single I27GLG12 (A) or I27RS8 polyproteins (B) gives force–extension curves with a saw-tooth pattern having equally spaced force peaks. The saw-tooth pattern is well described by the WLC equation (continuous lines). (C) Unfolding force frequency histogram for I27RS8. The lines correspond to Monte Carlo simulations of the mean unfolding forces (n=10,000) of eight domains placed in series by using three different unfolding rate constants, k_u^0, an unfolding distance, Δx_u, of 0.25 nm, and a pulling rate of 0.6 nm/ms. Reprinted with permission from (Carrion-Vazquez, Oberhauser *et al.* 1999). PNAS.

To overcome this limitation and to utilize this technique advantages, modular proteins were engineered and formed "beads-on-a-string" structure (see figure 10). This structure is composed of tandem repeats of the same domain. At high forces (>100 pN), the domains are unfolded one-by-one, where each unfolding event is characterized by a saw-tooth in the force-extension curve and is explained as the elongation of the "string" due to the unfolding of a "bead" (domain) (Oberhauser, Marszalek *et al.* 1998). Each unfolding event can be fitted to the WLC model to recover the persistence length and the contour length of the polymer.

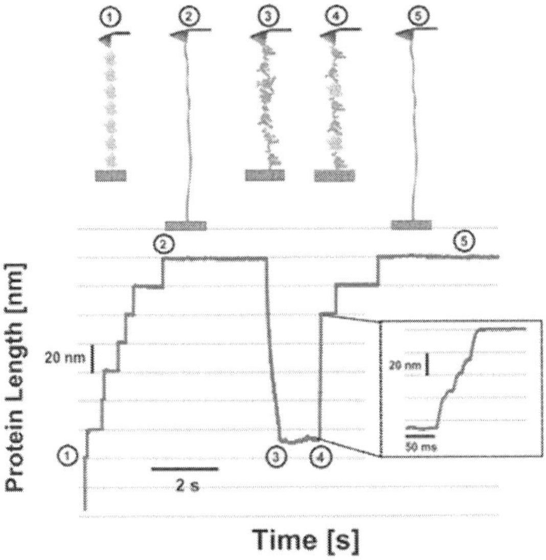

Fig. 10. 'Beads-on-a-string' stretching. At stage 1 the proteins are found at their native form (yellow). At stage 2 unfolding of the entire polyprotein occurs through pulling with the cantilever. For this protein (ubiquitin), it happens in 20 nm steps. At step 3 the force is quenched and the proteins maintain a collapsed form (gray). At step 4 refolding occur for some proteins along the chain and at step 5 the experiment is repeated by pulling again and causing another complete unfolding Reprinted with permission from (Garcia-Manyes, Dougan *et al.* 2009). PNAS.

2.4 Fluorescence resonance energy transfer with tethered molecules

Single-molecule FRET first introduced by Ha *et al.* (Ha, Enderle *et al.* 1996) is quite different from the other techniques discussed before. FRET aims to study the localization two entities in a biomolecule with a nanometer spatial resolution. This is done by measuring the non-radiative energy transfer from one fluorescent dye (donor) to another fluorescent dye (acceptor). The efficiency of energy transfer, E, depends on the donor-acceptor distance, R:

$$E = \frac{1}{1 + (R/R_0)^6} \tag{13}$$

Where R_0 is the distance when 50% of non-radiating energy transfer occurs (E=0.5) and is a function of the dyes properties (Selvin & Ha 2008). Due to the great sensitivity of this method to distance, it is applicable to a distance range of 3-8 nm. FRET is many times referred to as a spectroscopic ruler (figure 11). A biological molecule can be fluorescently-labeled in two sites, and intra-molecular dynamic motions of these sites relative to each other can be detected due to the energy transfer.

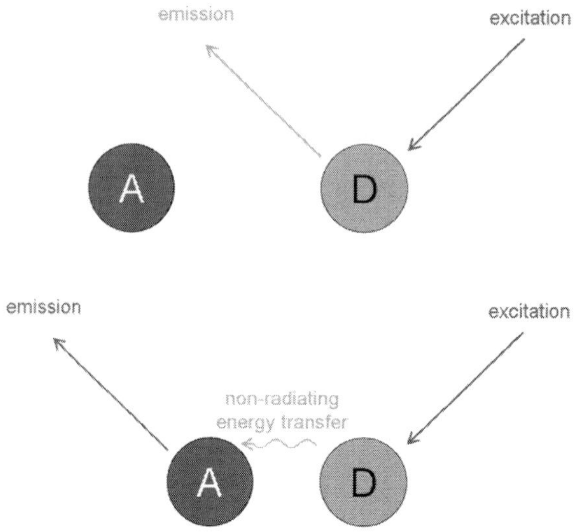

Fig. 11. FRET as a spectroscopic ruler. Up: The donor and acceptor dyes are too far for non-radiating energy transfer. Excitation of the donor results in its own emission. Down: The donor and acceptor are close enough for non-radiating energy transfer. Excitation of the donor might result in emission of the acceptor (other factors such as dipole-dipole orientation also affect the energy transfer).

Single molecule FRET (smFRET) has two major advantages over ensemble FRET. The first one derives from the fact that single-molecule experiments avoid averaging. smFRET allows the distinction between sub-populations and hence enables characterizing conformational states of biomolecules that result from dynamic and stochastic fluctuations. The second great advantage is the ability to observe in real-time dynamic events, an information that might be lost in ensemble FRET due to events being unsynchronized between different molecules.

In order to measure FRET, two detectors are needed, one for each dye. When there is a need for fast measurements and the signal is low, the preferred choice would be an Avalanche Photo-diode (APD), however choosing an Electron-Multiplying Charged Coupled Device (EMCCD) will enable visualizing hundreds of single-molecules simultaneously (Ha 2001). In smFRET the Signal-to-Noise Ratio (SNR) is relatively low (like most of the single-molecule experiments),so in order to decrease the auto fluorescence of the sample (noise) recorded by the detector, Total Internal Reflection Fluorescence microscopy (TIRF) is an appropriate solution. It ensures that only fluorescent dyes close to the surface are excited. smFRET can be applied to study surface-tethered dual-labeled DNA (McKinney, Declais *et al.* 2003) or the interactions between proteins and single tethered DNA (Myong, Rasnik *et al.* 2005; Myong, Bruno *et al.* 2007) or surface-tethered RNA (Arluison, Hohng *et al.* 2007).

3. Few key findings on relevant proteins

In this section we will show how the single molecule techniques that we described above, are employed to study a variety of biological processes such as: DNA bending by HU and IHF, DNA twisting by DNA gyrase, DNA replication by Φ29 DNA polymerase, refolding of ubiquitin and DNA unwinding by Hepatitis C virus NS3 helicase.

3.1 A DNA remodeling protein - HU

DNA-protein interactions are crucial also in bacteria cells where nucleoid-associated proteins (NAPs) together with macromolecular crowding effects play a major role in maintaining the architecture of the bacterial chromosome. NAPs ability to control the DNA structure is prominent for their role as regulators of DNA translocations (Krawiec & Riley 1990; Johnson, Johnson *et al.* 2005; Thanbichler, Wang *et al.* 2005; Luijsterburg, Noom *et al.* 2006; Stavans & Oppenheim 2006). Few of them were studied during the last decade with single molecule techniques and revealed new insights describing the dynamics of these protein-DNA/RNA interactions.

More than a few studies were performed on Integration Host Factor (IHF) protein of E. coli, which is involved in the integration of the bacteriophage λ DNA into the *E. coli* chromosome. In one of the studies, the local bending of a single 25 nm long DNA molecule caused by single IHF binding event was detected (Dixit, Singh-Zocchi *et al.* 2005). The experimental setup consisted of single linear dsDNA tethered at one end to a glass surface and to a microsphere at the other end. The tethered DNA had one consensus sequence for IHF binding. By optically monitoring the microsphere movement relative to the glass surface with evanescent microscopy, it was possible to detect conformational variations in the tether length. When IHF was introduced to the sample solution, the microsphere was pulled closer to the surface implying that the DNA bends and therefore adopts a more compact shape.

Another protein that belongs to the NAPs family is called Histone-like protein initially identified and characterized in *E.coli* strain U93 (HU). Ensemble studies revealed that the protein binds and bends DNA (Rouvière-Yaniv, Yaniv *et al.* 1979; Rouviere-Yaniv 1987; Pinson, Takahashi *et al.* 1999). Single molecule experiments refined this observation. They showed that while at relatively low HU concentrations the protein does compact the DNA, at high HU concentrations, it stretches the DNA (Sagi, Friedman *et al.* 2004; van Noort, Verbrugge *et al.* 2004; Skoko, Yoo *et al.* 2006; Xiao, Johnson *et al.* 2010). Some of these studies were performed with magnetic tweezers, (Figure 7) on a 50 nm dsDNA. Low concentrations of HU reduced the end-to-end distance in more than 50% but upon increasing the HU concentration, the persistence length increased up to ~150 nm. Similar results were also measured with a TPM setup (Figure 12). In comparison to the tweezers method, it has the advantage that force is not applied on the DNA and therefore the interaction occurs at the DNA natural form (Nir, Lindner *et al.* 2011). The bimodal effect of HU was recently explained by a model that assumes that the DNA is made of rigid segments and flexible joints (Rappaport & Rabin 2008). The model distinguishes two possible bending patterns along the polymer. If two neighboring segments are unoccupied by proteins, the bending angle θ is small, leading to the normal persistence length of DNA. When a protein occupies a segment without a neighboring protein, the spontaneous curvature increases, but when proteins occupy both neighboring segments, the spontaneous curvature is reduced again. The model therefore predicts that the DNA contains bent joints (large spontaneous curvature) and unbent joints (small spontaneous curvature) along the same DNA strand. These findings contribute to our understanding that the DNA flexibility is a more localized term when DNA-bending proteins are involved. It also raises a discussion of the contradicting role of HU. It might be, that even a low HU concentration (could be just a few nM) is efficient for chromosome condensation while higher HU concentrations might provide a degree of freedom for the interplay of a more rigid form versus chromosome condensation during different cell phases.

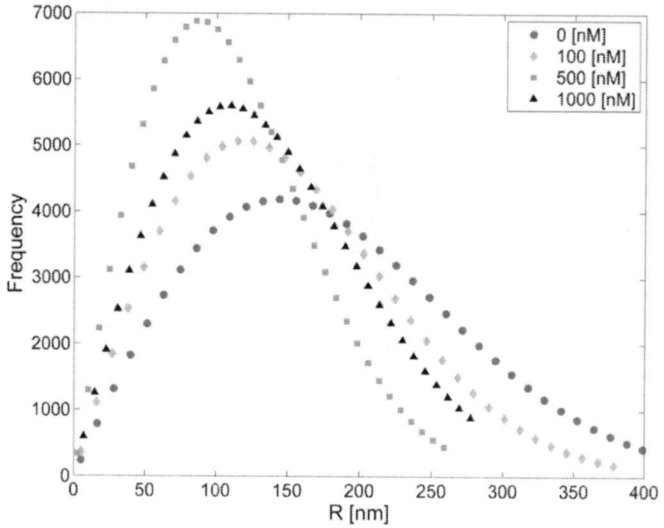

Fig. 12. Comparison of the radial distributions of the bead position for different HU concentrations as measured in a TPM experiment for the same bead. Circles represent DNA without HU proteins and the persistence length is 50 nm. Diamonds are for DNA in a solution with a concentration of 100 nM HU. The distribution is narrower compared to the HU-free DNA and the persistence length is ~39 nm. Squares represent DNA with HU concentration of 500 nM and the distribution is even narrower, with a persistence length of ~26 nm. Triangles represent DNA with HU concentration of 1000 nM and the distribution is now wider than for 500 nM with a persistence length of ~34 nm. Reprinted from Biophysical Journal **100** (2011) (Nir, Lindner *et al.* 2011) with permission from Elsevier.

So far we presented 2D TPM studies. Recently we expanded this method and made it possible to follow the DNA end-to-end distribution in 3D (Lindner, Nir *et al.* 2011). This study provided some powerful insights of the nature of the dynamic axial conformations (perpendicular to the surface) of tethered polymers. It was discovered that while the solution in the *XY* plane follows the normal distribution (Equation 5), the axial end-to-end distribution is different.

It can be described by a 1D random walk in half-space (Chandrasekhar 1943) and the solution is the difference between two Gaussians that centered around $\pm z_0$:

$$P(z)dz \propto \left[\exp\left(-\frac{3(z-z_0)^2}{4l_p L} \right) - \exp\left(-\frac{3(z+z_0)^2}{4l_p L} \right) \right] dz \tag{14}$$

where z_0 is some length parameter, between the width of a DNA (2nm) and its persistence length (50nm) , and it has a negligible effect on the solution. Using the 3D TPM, we measured this distribution, which is similar to the Rayleigh distribution.

Nevertheless, it is clear from the distribution that the DNA free end is repealed from surfaces. This effect may play an important role in experiments on DNA translocation through nanopores and nuclear pores, and should affect DNA dynamics in systems where it is tethered, such as the nucleus (where it is often attached to lamins).

3.2 Studying molecular motors with single molecule techniques

As mentioned above, single-molecule force-techniques enables one to study the mechanochemical properties of specific enzymes, and more specifically also the torque and twist of the DNA. Using these methods, an upgraded magnetic tweezers setup was built for studying the twist induced by *E. Coli* DNA gyrase in DNA (Gore, Bryant *et al.* 2006). Tension was generated in a single dsDNA by pulling a magnetic microsphere attached at one end of the tethered DNA, while a 'rotor' bead (Figure 13) is attached to the center of the DNA just below an engineered single strand nick, which acts as a free swivel. The angle of the rotor bead then reflects changes in twist of the lower DNA segment, and the angular velocity of the bead is proportional to the torque in this segment. Applying tension to the DNA causes changes in linking number to partition into DNA twist, resulting in a torque on the rotor bead. An enzymatic process that changes the linking number by two will cause the rotor bead to spin around twice as the DNA returns to its equilibrium conformation. Thus, the DNA construct serves as a self-regenerating substrate for DNA gyrase. Adding *E. Coli* DNA gyrase and 1 mM ATP results in bursts of directional rotation of the rotor bead, where each burst is an even number of rotations as predicted for type II Topoisomerase, when a single catalytic cycle changes the linking number by two (Brown & Cozzarelli 1979). In order to dissect the different mechanochemical steps of the supercoiling reaction, tension was applied in the range of 0.35-1.3 pN. It was found that the supercoiling velocity doesn't vary significantly, however the processivity and initiation rate have strong dependency to template tension. As the tension increased, bursts length decreased (processivity decreased) and waiting time between bursts increased (initiation rate decreased).

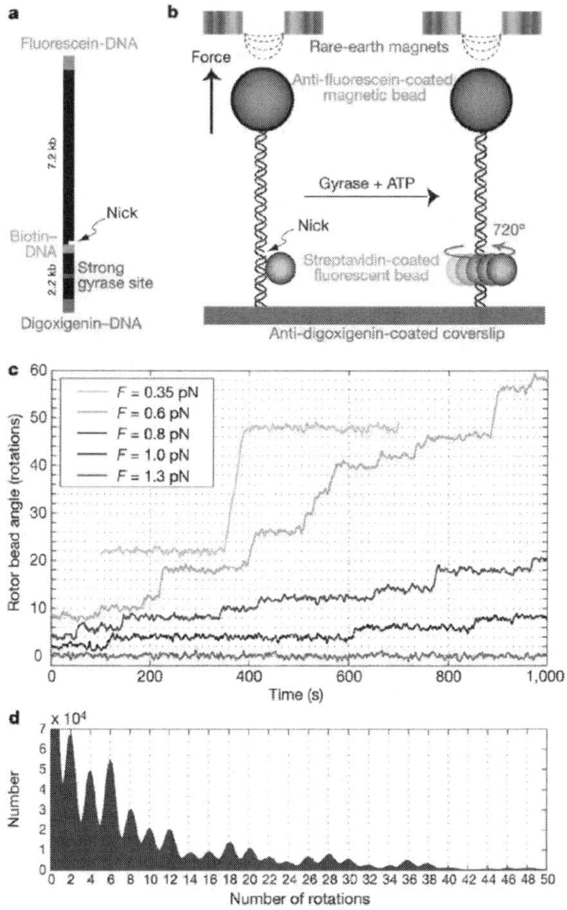

Fig. 13. Experimental design and single-molecule observations of gyrase activity. **a**, The molecular construct contains three distinct attachment sites and a site-specific nick, which acts as a swivel. A strong gyrase site was engineered into the lower DNA segment. **b**, Molecule and bead assemblies were constructed in parallel in a flow chamber and assayed by using an inverted microscope equipped with permanent magnets. Each molecule was stretched between the glass coverslip and a 1-µm magnetic bead, and a 530-nm diameter fluorescent rotor bead was attached to the central biotinylated patch. **c**, Plot of the rotor bead angle as a function of time (averaged over a 2-s window), showing bursts of activity due to diffusional encounters of individual gyrase enzymes. The activity of the enzyme is strongly dependent on tension. **d**, Histogram of the pairwise difference distribution function summed over 11 traces of 15–20 min (averaged over a 4-s window) at forces of 0.6–0.8 pN. The spacing of the peaks indicates that each catalytic cycle of the enzyme corresponds to two full rotations of the rotor bead, as expected for a type II topoisomerase such as DNA gyrase. Reprinted by permission from Macmillan Publishers Ltd: [Nature](Gore, Bryant *et al.* 2006), copyright (2006).

The measured data indicating of untwisting events caused by single DNA gyrase, allowed the researchers to build a physical model for the gyrase-DNA complex kinetics. Such a model cannot be deduced unless the single-molecule data is known.

3.3 DNA replication studies

Another study exploited the ability to apply tension on single DNA tethers using optical tweezers to investigate the conformational dynamics of the intramolecular DNA primer transfer during the processive replicative activity of the Φ29 DNA polymerase and two of its mutants (Ibarra, Chemla *et al.* 2009). Φ29 DNA polymerase has a catalytic unit as well as an exonuclease unit, allowing it to replicate DNA and fix base-pair mismatching at the same time.

The authors used optical tweezers to apply mechanical tension between two beads attached to the ends of an 8-kb dsDNA molecule with a ~400 nucleotide single-stranded gap in the middle (Figure 14A). They monitored the change in the end-to-end distance of the DNA (Δx) at constant force as the single-stranded template is replicated to dsDNA by Φ29 DNA polymerase (Figure 14B). The number of nucleotides incorporated as a function of time was obtained by dividing the observed distance change (Δx) by the expected change at a given force accompanying the conversion of one single-stranded nucleotide into its double-stranded counterpart. They also detected pause events as shown in Figure 14C.

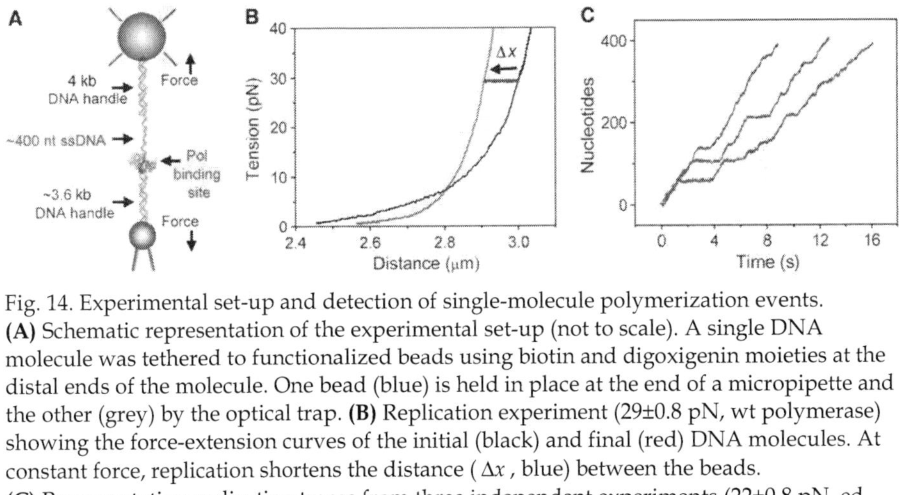

Fig. 14. Experimental set-up and detection of single-molecule polymerization events.
(A) Schematic representation of the experimental set-up (not to scale). A single DNA molecule was tethered to functionalized beads using biotin and digoxigenin moieties at the distal ends of the molecule. One bead (blue) is held in place at the end of a micropipette and the other (grey) by the optical trap. **(B)** Replication experiment (29±0.8 pN, wt polymerase) showing the force-extension curves of the initial (black) and final (red) DNA molecules. At constant force, replication shortens the distance (Δx, blue) between the beads.
(C) Representative replication traces from three independent experiments (22±0.8 pN, ed mutant). Reprinted by permission from Macmillan Publishers Ltd: [EMBO J] (Ibarra, Chemla *et al.* 2009),copyright (2009).

The authors observed an initial sharp increase in the relative pause occupancy and rationalized that it indicates that access to this intermediate from the initial pol1 (initial pol cycle) is force-sensitive and the ensuing saturation requires that the equilibrium between the intermediate and the paused state, Kp, be insensitive to the template tension. Importantly, this new intermediate must be a moving or polymerization-competent cycle (that the authors call pol2), as direct access to a non-active state in a tension-sensitive manner would lead to a continuous exponential increase in the relative pause occupancy, which was not observed.

3.4 Protein refolding studies

Understanding the dynamics of protein folding and unfolding is an ongoing effort that is presumed to benefit a lot from force spectroscopy. Single-molecule force spectroscopy techniques allow the detailed examination of the free-energy surface over which a protein diffuses in response to a mechanical perturbation (Schuler, Lipman *et al.* 2002; Rhoades, Gussakovsky *et al.* 2003). It is possible to pull a protein with an AFM tip and unfold it, a reversible process according to (Rief, Gautel *et al.* 1997; Carrion-Vazquez, Oberhauser *et al.* 1999) and upon reducing the pulling force, the unfolded protein begins to fold from a highly extended conformation that is rare or nonexistent in solution, even in the presence of denaturants. For example, at a typical force of 110 pN, mechanically unfolded ubiquitin proteins extend by >80% of their contour length (~20 nm) (Schlierf, Li *et al.* 2004). By contrast, ubiquitin proteins unfolded chemically in solution by 6 M guanidinium chloride stay compact, with a radius of gyration of only ~2.6 nm (Jacob, Krantz *et al.* 2004; Kohn, Millett *et al.* 2004).

Garcia-Manyes *et al.* studied the collapse and re-folding trajectories of ubiquitin polyproteins (Garcia-Manyes, Bruji *et al.* 2007). A chain of ubiquitin polyproteins was engineered and adsorbed to a surface while its other end was pulled by a force of 110 pN to unfold the polyproteins chain (Garcia-Manyes, Dougan *et al.* 2009). Unfolding events were observed as 20 nm steps (for each protein unfolding), followed by force-quenching to 10 pN. The time spent at the quenched state, Δt, was changed between 0.2-15 seconds before pulling again. The quenching leads to collapsed state which is mechanically unstable that is followed by folding of the protein to the native state. Therefore, changing the duration of the quenched state allows to probe the folding duration. For short durations (Δt =100-200 ms), the unfolding trajectory unravels rapidly and the 20 nm stepwise mechanism is not observed. It indicates that the proteins were not able to fold to their native state. Increasing Δt to 500 ms leads to the detection of steps when re-applying force and they predominant at Δt =3 seconds (Figure 15A). In figure 15B, a two-state process is observed, a fast initial extension that unravels the collapsed states in a stepwise manner featuring different lengths and followed by a much slower staircase of 20-nm steps, characteristic of fully refolded ubiquitin. The two states can be described by a bi-exponential fit with two rate constants, k_1 for the fast stage and k_2 for the slow state, both showing no dependency in Δt, suggesting that the protein does not gradually progress from higher to lower energy states, but populates two distinct conformational states.

The authors adopted the two-state model stating that the fast phase is a mechanically-weak state composed of a number of possible conformations with an equal distance from the transition state corresponding to the unfolding of the native state. With that knowledge, the authors tried to address the question whether these structures represent necessary folding precursors or unproductive kinetic traps in the folding energy landscape. They therefore devised a protocol to disrupt these collapsed conformations by interrupting the folding trajectories with a brief (100 ms) pulse to a higher force of 60 pN. During such a brief pulse,

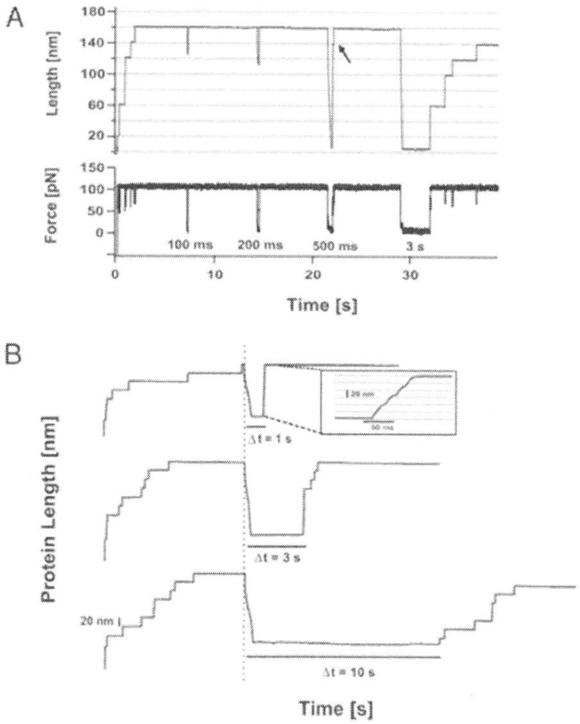

Fig. 15. Identification of a weakly stable ensemble of collapsed conformations in the folding of ubiquitin. **(A)** The authors repeatedly unfold and extend a ubiquitin polyprotein at 110 pN and then reduce the force to 10 pN for a varying amount of time, Δt, to trigger folding. First the polyprotein elongates in well defined steps of 20 nm, because each protein in the chain unfolds at a high force. Upon quenching the force the extended protein collapses. The state of the collapsed polypeptide was probed by raising the force back to 110 pN and measuring the kinetics of the protein elongation. **(B)** After full collapse the protein becomes segregated into 2 distinct ensembles: The first is identified by a fast heterogeneous elongation made of multiple sized steps (Inset); the second corresponds to well defined steps of 20 nm that identify fully folded proteins. The ratio between these 2 states of the protein depends on Δt and longer values of Δt favor the native ensemble. (Garcia-Manyes, Dougan *et al.* 2009). Reprinted with permission of PNAS.

native ubiquitin has a very low probability of unfolding. If the set of mechanically-unstable collapsed conformations are a prerequisite to folding, their disruption would cause a delay in the recovery of mechanical stability as compared with the unperturbed trajectories. By contrast, if the collapsed states represent unproductive traps, then unraveling them would accelerate the rate of folding. The authors showed that an average unfolding trajectory after 5 s of folding has a higher content of folded proteins than the same trajectory with the mechanical interruption. They concluded that the collapsed conformations are necessary precursors of the folded state.

3.5 Studying the unwinding of DNA by hepatitis C virus NS3 helicase

In this final example, the unwinding of DNA is demonstrated by using smFRET. In hepatitis C virus (HCV), nonstructural protein 3 (NS3) is an essential component of the viral replication complex that works with the polymerase NS5B and other protein cofactors (such as NS4A, NS5A, and NS2) to ensure effective copying of the virus. Myong *et al.* used single-molecule FRET to resolve the individual steps of DNA unwinding, catalyzed by NS3 in the absence of applied force (Myong, Bruno *et al.* 2007). Two DNA substrates were prepared. Both consisted of 18 bp and a 3'-ssDNA tail (20 nt) to create a double stranded - single stranded junction as an anchoring position to the helicase and the DNA tail was anchored to the surface. One DNA fragment had a donor and an acceptor fluorophores (cy3 and cy5 respectively) attached to the two different DNA strands at the junction through aminodeoxythymidine (figure 16A). The other DNA fragment had the dyes attached 9 bp away from the junction so that FRET signal is sensitive only to the final 9-bp unwinding

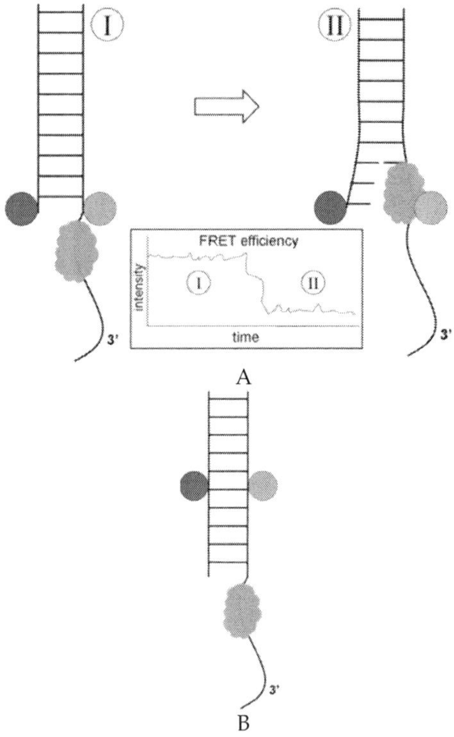

Fig. 16. DNA template for smFRET. **(A)** NS3 helicase translocates on a single dsDNA with a ssDNA tail. A green donor and a red acceptor dyes are attached at the end of the dsDNA template. At state I, the two dyes are close and the FRET efficiency (inset) is high. At state II, the initially dsDNA is partially unwound increasing the distance between the two dyes and resulting in low FRET efficiency. **(B)** In this setup the dyes are attached to the DNA in the middle of the dsDNA.

(figure 16B). Addition of ATP resulted in a decrease of the FRET efficiency as a result of strand separation due to the helicase unwinding. 6 and 3 steps were detected for the 18 bp dsDNA and 9 bp dsDNA respectively, indicating of strand separation in 3 bp steps. If hydrolysis of a single ATP results in a 3-bp step than the dwell time histogram of the steps would follow a single-exponential decay. However the observation revealed non-exponential dwell time histograms.

The authors derived a model suggesting that domains I&II of the helicase move forward one bp at a time and at the third step, the spring-loaded domain 3 moves forward in a burst motion, unzipping 3 bp as a consequence.

4. Conclusions

Single-molecule techniques are an essential tool that opened the door for high temporal and spatial studies, providing the ability to manipulate or passively observe single molecules. We presented some of the major single molecules methods and few applications of these methods for studying key biochemical processes such as DNA replication, protein folding and DNA remodeling. Table 1 summarizes the methods presented here and compares them.

	TPM	**Magnetic Tweezers**	**Optical Tweezers**	**AFM**	**smFRET**
Relies on	Probe's Brownian motion	Application of magnetic force	Momentum change of incident light	Stretching a single molecule with an AFM cantilever	Resonant Energy transfer
Probe	Nanometric gold bead	Polystyrene microsphere	Polystyrene microsphere	Cantilever	Fluorescent dyes
Range of forces (pN)	0	10^{-3} - 10^{2}	10^{-1} - 10^{2}	10 - 10^{4}	0
Data produced	Scattering plot (Fig.4)	Force-extension curve (Fig.7)	Force-extension curve (Fig. 14b)	Force-extension curve (Fig.16)	Energy transfer efficiency

Table 1. A comparison of single-molecule techniques reviewed here.

TPM and smFRET allows to measure without applying force and hence are good candidates to describe DNA-protein interactions in their native forms. On the other hand, force-based techniques allows to manipulate single-molecules and enable to study these interactions in extreme conditions, introducing higher energy states that are usually not observed. They reveal the mechanochemical properties of single enzymes. We showed that using TPM we were able to observe end-to-end fluctuations of single DNA molecules and measure their persistence length dynamic variations induced by HU protein. We proved that the bimodal effect of HU exists even without the use of force.

We showed how magnetic tweezers can be implemented to study the probability of DNA gyrase to achieve a productive catalytic cycle. Optical tweezers were demonstrated as a tool for discovering the proofreading activity of DNA polymerase, revealing the different steps comprising the transition from polymerase activity to exonuclease activity within the same enzyme. The use of AFM was demonstrated for catalyzing the unfolding of single proteins and then the refolding dynamics leading to the collapsed followed by the native state. Finally, smFRET was reviewed for studying helicase motor enzyme mechanisms.

All together these techniques span a broad spectrum of capabilities that can unravel different properties of single DNA-protein interactions and provide unprecedented details that are not visible by ensemble techniques. These properties includes rate constants measured directly on a single molecule, motor enzymes mechanisms, elastic and chemical properties of single molecules and high-energy states.

5. Near-future capabilities

The methodology of single molecule techniques is rapidly growing. Although the achievements that were demonstrated here, and many others, indicate on the usefulness of the existing methods, there are still important challenges that call for further improvements. High spatial resolution and high temporal resolution are necessary, especially with optical methods that can be applied to single molecules. These may be achieved with the newly developed super-resolution techniques such as stimulated depletion emission (STED), 4-pi microscopy, Photoactivated localization microscopy (PALM), stochastic optical reconstruction microscopy (STORM) and others.

Another promising direction relies on nano-optics. Due to improvements in lithography methods, it is now possible to design and fabricate nano-sized devices such as metal-based structures that use plasmonic effects. It was already shown that plasmonic nano-antennas are capable of concentrating an intense light to a sub-diffracted volume (Grigorenko, Roberts *et al.* 2008; Righini, Volpe *et al.* 2008; Huang, Maerkl *et al.* 2009; Juan, Righini *et al.* 2011) and trap dielectric particles and even *E. Coli* bacteria. These plasmonic traps do not require bulk optics (such as optical tweezers) and restrict the trapped object to a smaller volume.

Another intriguing capability is the emerging of different illumination techniques. For instance, the ability to manipulate single molecules under white light and observe fluorescent-conjugated enzymes translocating on single DNAs is fascinating and provide us with further information of the reactions nature since it enables us to follow the dynamics of both a trapped biomolecule and an enzyme translocating on it. One example is the "fleezers", fluorescent tweezers (Comstock, Ha *et al.* 2011). In this setup a dual optical trap integrated with a confocal microscope to illuminate fluorescent molecules was implemented to observe individual single fluorophore-labeled DNA oligonucleotides binding and unbinding to a complementary DNA suspended between two trapped beads.

We will finish by mentioning solid-state nanopores. Nanometer-sized holes in a thin synthetic membrane are a versatile tool for the detection and manipulation of charged biomolecules. For example, a single DNA molecule that translocates through the nanopore will have a unique signature that is attributed to its sequence. That can be done by applying an external electric field which drives a biomolecule through the nanopore, producing a characteristic transient change in the trans-pore ionic current (Heng, Ho *et al.* 2004; Garaj, Hubbard *et al.* 2010; Stefan, Alexander *et al.* 2011).

6. Acknowledgements

This work was supported in part by the Israel Science Foundation grants 985/08, 1729/08, 1793/07, and 25/07.

7. References

Arluison, V., S. Hohng, R. Roy, O. Pellegrini, P. Regnier & T. Ha (2007). Spectroscopic Observation of RNA Chaperone Activities of Hfq in Post-Transcriptional Regulation by a Small Non-Coding RNA. *Nucl. Acids Res*, 353, pp. 999-1006

Binnig, G., C. F. Quate & C. Gerber (1986). Atomic Force Microscope. *Phys. Rev. Lett*, 569, pp. 930

Bouchiat, C., M. D. Wang, J.-F. Allemand, T. Strick, S. M. Block & V. Croquette (1999). Estimating the Persistence Length of a Worm-Like Chain Molecule from Force-Extension Measurements. *Biophys. J*, 76, pp. 409-413

Brown, P. O. & N. R. Cozzarelli (1979). A Sign Inversion Mechanism for Enzymatic Supercoiling of DNA. *Science*, 2064422, pp. 1081-1083

Bustamante, C., Z. Bryant & S. B. Smith (2003). Ten Years of Tension: Single-Molecule DNA Mechanics. *Nature*, 4216921, pp. 423-427

Bustamante, C., Y. R. Chemla, N. R. Forde & D. Izhaky (2004). Mechanical Processes in Biochemistry. *Annu. Rev. Biochem*, 731, pp. 705-748

Bustamante, C., J. C. Macosko & G. J. L. Wuite (2000). Grabbing the Cat by the Tail: Manipulating Molecules One by One. *Nat. Rev. Mol. Cell Biol*, 1, pp. 131-136

Bustamante, C., S. B. Smith, J. Liphardt & D. Smith (2000). Single-Molecule Studies of DNA Mechanics. *Curr. Opin. Struct. Biol*, 103, pp. 279-285

Carrion-Vazquez, M., A. F. Oberhauser, S. B. Fowler, P. E. Marszalek, S. E. Broedel, J. Clarke & J. M. Fernandez (1999). Mechanical and Chemical Unfolding of a Single Protein: A Comparison. *Proc. Natl. Acad. Sci. U S A*, 967, pp. 3694-3699

Chandrasekhar, S. (1943). Stochastic Problems in Physics and Astronomy. *Rev. Mod. Phys*, 151, pp. 1

Comstock, M. J., T. Ha & Y. R. Chemla (2011). Ultrahigh-Resolution Optical Trap with Single-Fluorophore Sensitivity. *Nat. Meth*, 84, pp. 335-340

Destainville, N. & L. Salomé (2006). Quantification and Correction of Systematic Errors Due to Detector Time-Averaging in Single-Molecule Tracking Experiments *Biophys. J*, 902, pp. L17-L19

Dixit, S., M. Singh-Zocchi, J. Hanne & G. Zocchi (2005). Mechanics of Binding of a Single Integration-Host-Factor Protein to DNA. *Phys. Rev. Lett*, 9411, pp. 118101

Finer, J. T., R. M. Simmons & J. A. Spudich (1994). Single Myosin Molecule Mechanics: Piconewton Forces and Nanometre Steps. *Nature*, 3686467, pp. 113-119

Fisher, T. E., A. F. Oberhauser, M. Carrion-Vazquez, P. E. Marszalek & J. M. Fernandez (1999). The Study of Protein Mechanics with the Atomic Force Microscope. *Trends Biochem. Sci*, 2410, pp. 379-384

Garaj, S., W. Hubbard, A. Reina, J. Kong, D. Branton & J. A. Golovchenko (2010). Graphene as a Subnanometre Trans-Electrode Membrane. *Nature*, 4677312, pp. 190-193

Garcia-Manyes, S., Bruji, cacute, Jasna, C. L. Badilla & J. M. Fernandez (2007). Force-Clamp Spectroscopy of Single-Protein Monomers Reveals the Individual Unfolding and Folding Pathways of I27 and Ubiquitin. *Biophys. J*, 937, pp. 2436-2446

Garcia-Manyes, S., L. Dougan, C. L. Badilla, J. Bruji & J. M. Fernandez (2009). Direct Observation of an Ensemble of Stable Collapsed States in the Mechanical Folding of Ubiquitin. *Proc. Natl. Acad. Sci. U S A*,

Gore, J., Z. Bryant, M. D. Stone, M. Nollmann, N. R. Cozzarelli & C. Bustamante (2006). Mechanochemical Analysis of DNA Gyrase Using Rotor Bead Tracking. *Nature*, 4397072, pp. 100-104

Grigorenko, A. N., N. W. Roberts, M. R. Dickinson & Y. Zhang (2008). Nanometric Optical Tweezers Based on Nanostructured Substrates. *Nat. Photon*, 26, pp. 365-370

Ha, T. (2001). Single-Molecule Fluorescence Resonance Energy Transfer. *Methods*, 25, pp. 78-86

Ha, T., T. Enderle, D. F. Ogletree, D. S. Chemla, P. R. Selvin & S. Weiss (1996). Probing the Interaction between Two Single Molecules: Fluorescence Resonance Energy Transfer between a Single Donor and a Single Acceptor. *Proc. Natl. Acad. Sci. U S A*, 93, pp. 6264–6268

Haber, C. & D. Wirtz (2000). Magnetic Tweezers for DNA Micromanipulation. *Rev. Sci. Instrum.*, 7112, pp. 4561

Harlepp, S., J. Robert, N. C. Darnton & D. Chatenay (2004). Subnanometric Measurements of Evanescent Wave Penetration Depth Using Total Internal Reflection Microscopy Combined with Fluorescent Correlation Spectroscopy. *Applied Physcis Letters*, 85, pp. 3917-3919

Heng, J. B., C. Ho, T. Kim, R. Timp, A. Aksimentiev, Y. V. Grinkova, S. Sligar, K. Schulten & G. Timp (2004). Sizing DNA Using a Nanometer-Diameter Pore. *Biophys. J*, 874, pp. 2905-2911

Huang, L., S. J. Maerkl & O. J. Martin (2009). Integration of Plasmonic Trapping in a Microfluidic Environment. *Opt. Express*, 178, pp. 6018-6024

Ibarra, B., Y. R. Chemla, S. Plyasunov, S. B. Smith, J. M. Lazaro, M. Salas & C. Bustamante (2009). Proofreading Dynamics of a Processive DNA Polymerase. *EMBO J*, 2818, pp. 2794-2802

Jacob, J., B. Krantz, R. S. Dothager, P. Thiyagarajan & T. R. Sosnick (2004). Early Collapse Is Not an Obligate Step in Protein Folding. *J. Mol. Biol*, 3382, pp. 369-382

Jeon, J.-H. & R. Metzler (2010). Fractional Brownian Motion and Motion Governed by the Fractional Langevin Equation in Confined Geometries. *Phys. Rev. E*, 812, pp. 021103

Johnson, R. C., L. M. Johnson, J. W. Schmidt & J. F. Garder (2005). Major Nucleoid Proteins in the Structure and Function of the *Escherichia Coli* Chromosome. In: *The Bacterial Chromosome*. N. P. Higgins. 1: 65-132, American Society for Microbiology. Washington, DC.

Juan, M. L., M. Righini & R. Quidant (2011). Plasmon Nano-Optical Tweezers. *Nat. Photon*, 56, pp. 349-356

Kohn, J. E., I. S. Millett, *et al.* (2004). Random-Coil Behavior and the Dimensions of Chemically Unfolded Proteins. *Proc. Natl. Acad. Sci. U S A*, 10134, pp. 12491-12496

Krawiec, S. & M. Riley (1990). Organization of the Bacterial Chromosome. *Microbiol. Rev.*, 544, pp. 502-539

Li, H., A. F. Oberhauser, S. B. Fowler, J. Clarke & J. M. Fernandez (2000). Atomic Force Microscopy Reveals the Mechanical Design of a Modular Protein. *Proc. Natl. Acad. Sci. U S A*, 9712, pp. 6527-6531

Li, L., S. Wetzel, A. Pluckthun & J. M. Fernandez (2006). Stepwise Unfolding of Ankyrin Repeats in a Single Protein Revealed by Atomic Force Microscopy. *Biophys. J*, 904, pp. L30-L32

Lindner, M., G. Nir, H. R. C. Dietrich, Ian T. Young, E. Tauber, I. Bronshtein, L. Altman & Y. Garini (2009). Studies of Single Molecules in Their Natural Form. *Isr. J. Chem*, 493-4, pp. 283-291

Lindner, M., G. Nir, S. Medalion, H. R. C. Dietrich, Y. Rabin & Y. Garini (2011). Force-Free Measurements of the Conformations of DNA Molecules Tethered to a Wall. *Phys. Rev. E*, 831, pp. 011916

Luijsterburg, M. S., M. C. Noom, G. J. L. Wuite & R. T. Dame (2006). The Architectural Role of Nucleoid-Associated Proteins in the Organization of Bacterial Chromatin: A Molecular Perspective. *J. Struct. Biol*, 1562, pp. 262-272

Manosas, M., T. Lionnet, E. Praly, D. Fangyuan, J.-F. Allemand, D. Bensimon, V. Croquette & V. Rivasseau Studies of DNA-Replication at the Single Molecule Level Using Magnetic Tweezers. *Biological Physics*, 60, pp. 89-122

Martin, Y., C. C. Williams & H. K. Wickramasinghe (1987). Atomic Force Microscope - Force Mapping and Profiling on a Sub 100 Å Scale. *J. Appl. Phys*, 6110, pp. 4723-4729

McKinney, S. A., A.-C. Declais, D. M. J. Lilley & T. Ha (2003). Structural Dynamics of Individual Holliday Junctions. *Nat. Struct. Mol. Biol*, 102, pp. 93-97

Myong, S., M. M. Bruno, A. M. Pyle & T. Ha (2007). Spring-Loaded Mechanism of DNA Unwinding by Hepatitis C Virus Ns3 Helicase. *Science*, 3175837, pp. 513-516

Myong, S., I. Rasnik, C. Joo, T. M. Lohman & T. Ha (2005). Repetitive Shuttling of a Motor Protein on DNA. *Nature*, 4377063, pp. 1321-1325

Nelson, P. C., C. Zurla, D. Brogioli, J. F. Beausang, L. Finzi & D. Dunlap (2006). Tethered Particle Motion as a Diagnostic of DNA Tether Length. *J. Phys. Chem. B*, 11034, pp. 17260-17267

Neuman, K. C., E. H. Chadd, G. F. Liou, K. Bergman & S. M. Block (1999). Characterization of Photodamage to Escherichia Coli in Optical Traps. *Biophys. J*, 775, pp. 2856-2863

Neuman, K. C. & A. Nagy (2008). Single-Molecule Force Spectroscopy: Optical Tweezers, Magnetic Tweezers and Atomic Force Microscopy. *Nat. Meth*, 56, pp. 491-505

Nir, G., M. Lindner, H. R. C. Dietrich, O. Girshevitz, C. E. Vorgias & Y. Garini (2011). HU Protein Induces Incoherent DNA Persistence Length. *Biophys. J*, 100, pp. 784-90

Oberhauser, A. F., P. E. Marszalek, H. P. Erickson & J. M. Fernandez (1998). The Molecular Elasticity of the Extracellular Matrix Protein Tenascin. *Nature*, 3936681, pp. 181-185

Perez-Jimenez, R., S. Garcia-Manyes, S. R. K. Ainavarapu & J. M. Fernandez (2006). Mechanical Unfolding Pathways of the Enhanced Yellow Fluorescent Protein Revealed by Single Molecule Force Spectroscopy. *J. Biol. Chem*, 28152, pp. 40010-40014

Pinson, V., M. Takahashi & J. Rouviere-Yaniv (1999). Differential Binding of the *Escherichia Coli* HU, Homodimeric Forms and Heterodimeric Form to Linear, Gapped and Cruciform DNA. *J. Mol. Biol*, 287, pp. 485-497

Rappaport, S. M. & Y. Rabin (2008). Model of DNA Bending by Cooperative Binding of Proteins. *Phys. Rev. Lett*, 1013, pp. 038101

Rhoades, E., E. Gussakovsky & G. Haran (2003). Watching Proteins Fold One Molecule at a Time. *Proc. Natl. Acad. Sci. U S A*, 1006, pp. 3197-3202

Rief, M., M. Gautel, F. Oesterhelt, J. M. Fernandez & H. E. Gaub (1997). Reversible Unfolding of Individual Titin Immunoglobulin Domains by AFM. *Science*, 2765315, pp. 1109-1112

Righini, M., G. Volpe, C. Girard, P. Dimitri & R. Quidant (2008). Surface Plasmon Optical Tweezers: Tunable Optical Manipulation in the Femtonewton Range. *Phys. Rev. Lett*, 10018, pp. 186804

Rohrbach, A. & E. AU - Stelzer (2002). Three-Dimensional Position Detection of Optically Trapped Dielectric Particles. *J. Appl. Phys.*, 918,

Rouvière-Yaniv, J., M. Yaniv & J.-E. Germond (1979). E. Coli DNA Binding Protein HU Forms Nucleosome-Like Structure with Circular Double-Stranded DNA. *Cell*, 172, pp. 265-274

Rouviere-Yaniv, K. D. a. J. (1987). Histonelike Proteins of Bacteria. *Microbiological review*, 513, pp. 19

Rubinstein, M. & R. H. Colby (2003). Polymer Physics, Oxford University Press.

Sagi, D., N. Friedman, C. Vorgias, A. B. Oppenheim & J. Stavans (2004). Modulation of DNA Conformations through the Formation of Alternative High-Order HU-DNA Complexes. *J. Mol. Biol*, 3412, pp. 419-428

Schafer, D. A., J. Gelles, M. P. Sheetz & R. Landick (1991). Transcription by Single Molecules of RNA Polymerase Observed by Light Microscopy. *Nature*, 352, pp. 444-448

Schlierf, M., H. Li & J. M. Fernandez (2004). The Unfolding Kinetics of Ubiquitin Captured with Single-Molecule Force-Clamp Techniques. *Proc. Natl. Acad. Sci. U S A*, 10119, pp. 7299-7304

Schuler, B., E. A. Lipman & W. A. Eaton (2002). Probing the Free-Energy Surface for Protein Folding with Single-Molecule Fluorescence Spectroscopy. *Nature*, 4196908, pp. 743-747

Segall, D. E., P. C. Nelson & R. Phillips (2006). Volume-Exclusion Effects in Tethered-Particle Experiments: Bead Size Matters. *Phys. Rev. Lett*, 96, pp. 0883061-4

Selvin, P. R. & T. Ha (2008). Single-Molecule Techniques: A Laboratory Manual, Cold Spring Harbor Laboratory Press.

Shin, J.-H., T. J. Santangelo, Y. Xie, J. N. Reeve & Z. Kelman (2007). Archaeal Minichromosome Maintenance (Mcm) Helicase Can Unwind DNA Bound by Archaeal Histones and Transcription Factors. *J. Biol. Chem*, 2827, pp. 4908-4915

Skoko, D., D. Yoo, H. Bai, B. Schnurr, J. Yan, S. M. McLeod, J. F. Marko & R. C. Johnson (2006). Mechanism of Chromosome Compaction and Looping by the Escherichia Coli Nucleoid Protein Fis. *J. Mol. Biol*, 3644, pp. 777-798

Slutsky, M. (2005). Diffusion in a Half-Space: From Lord Kelvin to Path Integrals. *Am. J. Phys*, 734, pp. 308-314

Stavans, J. & A. Oppenheim (2006). DNA-Protein Interactions and Bacterial Chromosome Architecture. *Phys. Biol*, 34, pp. R1-10

Kowalczyk, S. W., Y. A. Grosberg, Y. Rabin & C. Dekker (2011). Modeling the Conductance and DNA Blockade of Solid-State Nanopores. *Nanotechnology*, 2231, pp. 315101

Strick, T., J.-F. Allemand, V. Croquette & D. Bensimon Twisting and Stretching Single DNA Molecules. *Prog. Biophys. Mol. Biol*, 741-2, pp. 115-140

Thanbichler, M., S. C. Wang & L. Shapiro (2005). The Bacterial Nucleoid: A Highly Organized and Dynamic Structure. *J. Cell. Biochem*, 963, pp. 506-521

van Noort, J., S. Verbrugge, N. Goosen, C. Dekker & R. T. Dame (2004). Dual Architectural Roles of HU: Formation of Flexible Hinges and Rigid Filaments. *Proc. Natl. Acad. Sci. U S A*, 10118, pp. 6969-6974

Volpe, G., T. Brettschneider, L. Helden & C. Bechinger (2009). Novel Perspectives for the Application of Total Internal Reflection Microscopy. *Opt. Express*, 1726, pp. 23975-23985

Walther, K. A., J. Bruji, H. Li & J. M. Fernandez (2006). Sub-Angstrom Conformational Changes of a Single Molecule Captured by Afm Variance Analysis. *Biophys. J*, 9010, pp. 3806-3812

Wang, J. C. (1998). Moving One DNA Double Helix through Another by a Type Ii DNA Topoisomerase: The Story of a Simple Molecular Machine. *Q. Rev. Biophys*, 3102, pp. 107-144

Williams, M. C. (2002). Optical Tweezers: Measuring Piconewton Forces, Biophysical Society.

Wong, W. P. & K. Halvorsen (2006). The Effect of Integration Time on Fluctuation Measurements: Calibrating an Optical Trap in the Presence of Motion Blur. *Opt. Express*, 1425, pp. 12517-12531

Xiao, B., R. C. Johnson & J. F. Marko (2010). Modulation of Hu-DNA Interactions by Salt Concentration and Applied Force. *Nucl. Acids Res*, 3818, pp. 6176-6185

Yildiz, A., J. N. Forkey, S. A. McKinney, T. Ha, Y. E. Goldman & P. R. Selvin (2003). Myosin V Walks Hand-over-Hand: Single Fluorophore Imaging with 1.5-Nm Localization. *Science*, 3005628, pp. 2061-2065

Yildiz, A., M. Tomishige, R. D. Vale & P. R. Selvin (2004). Kinesin Walks Hand-over-Hand. *Science*, 3035658, pp. 676-678

Yin, H., R. Landick & J. Gelles (1994). Tethered Particle Motion Method for Studying Transcript Elongation by a Single RNA Polymerase Molecule. *Biophys. J*, 67, pp. 2468-2478

Zimmermann, J. L., T. Nicolaus, G. Neuert & K. Blank (2010). Thiol-Based, Site-Specific and Covalent Immobilization of Biomolecules for Single-Molecule Experiments. *Nat. Protocols*, 56, pp. 975-985

Zurla, C., A. Franzini, G. Galli, D. D. Dunlap, D. E. A. Lewis, S. Adhya & L. Finzi (2006). Novel Tethered Particle Motion Analysis of Ci Protein-Mediated DNA Looping in the Regulation of Bacteriophage Lambda. *J. Phys.: Condens. Matter*, 18, pp. S225-S234

20

Characterization of Protein-Protein Interactions via Static and Dynamic Light Scattering

Daniel Some and Sophia Kenrick
Wyatt Technology Corp.
USA

1. Introduction

Light scattering in its various flavors constitutes a label-free, non-destructive probe of macromolecular interactions in solution, providing a direct indication of the formation or dissociation of complexes by measuring changes in the average molar mass or molecular radius as a function of solution composition and time. It is a first-principles technique, thoroughly grounded in thermodynamics, permitting quantitative analysis of key properties such as stoichiometry, equilibrium association constants, and reaction rate parameters.

In the past, light scattering experiments on interacting protein solutions have been labor intensive and tedious, requiring large volumes of sample, and hence impeding widespread adoption by protein researchers. Recent advances in instrumentation and technique hold the promise for simplifying and automating measurements, as well as reducing sample requirements, thus broadening the appeal of these methods to the wider community of analytical biochemistry, biophysics, and molecular biology research. Pioneering work in automating and applying these measurements to equilibrium protein-protein interactions appeared in 2005-2006 (Attri & Minton, 2005a, 2005b; Kameyama & Minton, 2006). This chapter deals primarily with such automated methods.

2. Theory of light scattering from biomacromolecules in solution

2.1 Static light scattering

Static light scattering (SLS) measurements quantify the "excess Rayleigh ratio" R, which describes the fraction of incident light scattered by the macromolecules per unit volume of solution. Knowledge of R vs. scattering angle (θ) and concentration c may determine molar mass, size and self-interactions of the sample, while $R(t)$ will describe the kinetics of self-association or dissociation, via time-dependent changes in the average molar mass or size. Likewise, characterization of R vs. the composition ($[A]$, $[B]$, ...) of a multi-component system, such as hetero-associating proteins, may determine the stoichiometry and equilibrium binding affinity of such a system, as well as binding or dissociation kinetics.

The basic theory of static light scattering from macromolecules in solution is available in myriad publications, including elementary textbooks dealing with physical chemistry (e.g., van Holde et al., 1998; Teraoka, 2002) or essential references on polymers (e.g., Young, 1981). We will cite from these without further reference. More rigorous publications, particularly those dealing with non-ideal, multi-component solutions are found in the scientific literature (e.g., Blanco et al., 2011, and references therein).

2.1.1 Static light scattering in the ideal limit

Macromolecules in solution are subject to correlations arising from intermolecular potentials, which in turn affect the magnitude of scattered light. However, if the particles are few and far between, and the potentials between them sufficiently short-ranged, these correlations may be ignored, leading to what is known as "the ideal limit": essentially, the particles behave like an ideal ensemble of point particles.

The simplest picture of scattering from proteins in solution invokes the ideal limit, i.e., point-like particles with no interactions, much like the more commonly known ideal gas law for pressure and temperature. In this case, the scattering from particles much smaller than the wavelength of incident light, with detectors placed in the plane perpendicular to the incident polarization, can be described by Eq. (1):

$$R = \frac{4\pi^2 n_0^2}{N_A \lambda_0^4} \left(\frac{dn}{dc} \right)^2 Mc = K^* Mc \tag{1}$$

In Eq. (1), N_A represents Avogadro's number, *dn/dc* is the protein's refractive increment, M the protein's molar mass, n_0 the solvent refractive index, λ_0 the wavelength in vacuum, and c the protein concentration in units of mass/volume. K^* incorporates the constants n_0, N_A, λ_0 and *dn/dc*.

The protein refractive increment *dn/dc* describes the change in refractive index of a solution relative to pure buffer, due to a mass/volume protein concentration c; this parameter may be readily measured by means of a common instrument known as a differential refractometer and is, fortuitously, nearly invariable for most proteins in standard aqueous buffers at any given wavelength (*dn/dc*=0.187 mL/g at λ=660 nm). High concentrations of excipient will affect *dn/dc*; adding for example arginine, which has a refractive index higher than that of most proteins, can even reduce *dn/dc* to zero such that no scattering occurs!

For a solution consisting of multiple macromolecular species, e.g., monomer + oligomers or A+B+AB complex, the total light scattered is the sum of intensities scattered by each species:

$$R = K^* \sum_i M_i c_i = K^* M_w c \tag{2}$$

Here M_i and c_i refer to the molar mass and concentration of each species i, M_w is the weight-averaged molar mass and c the total protein concentrations. We have assumed that all species have the same refractive increment and non-ideality may be ignored.

Upon inspection of Eq. (2) it becomes clear that given knowledge of the measurement conditions (solution refractive index, scattering wavelength), sample parameters (*dn/dc*), sample concentration (e.g., by means of a UV absorption or differential refractive index concentration detector), and excess Rayleigh ratio R, it is possible to determine the weight-averaged molar mass of macromolecules in the solution. If the solution is monodisperse (as is often the case in the course of chromatographic fractionation), then the molar mass of the solvated macromolecule may be determined.

Eq. (2) contains the reason that light scattering is famously sensitive to small quantities of dust or other particulates: if the mass if the dust particle is a million times that of the protein, only one-millionth the concentration of dust particles produces the same scattering intensity as the protein.

Generalization of Eq. (2) to species with different refractive increments is obvious, but for reasons of simplicity we will assume henceforth equal *dn/dc* for all proteins. **This "ideal gas law for light scattering" is generally applicable to characterization of specific protein-protein binding with equilibrium dissociation constants $K_D \leq$ 1-10 μM.**

Angular dependence of the scattered intensity comes into play for larger particles such as protein aggregates whose radii exceed ~λ/50. In the limit of transparent particles with radii below ~40 nm, this dependence is described by the Rayleigh-Ganz-Debye (RGD) equation:

$$R(\theta) = K^* McP(\theta); \quad P(\theta) = 1 - \frac{16\pi^2}{3\lambda_0^2}\langle r_g 2\rangle \sin^2(\theta/2) + \ldots \tag{3}$$

For this reason SLS is often referred to as multi-angle light scattering (MALS). Here θ is the angle between the incident and scattered light rays within the plane perpendicular to the incident polarization, r_g is the radius of gyration, and higher order terms in $P(\theta)$ have been ignored. For globular proteins, r_g will be ~80% the average geometrical radius. When r_g < 8-12 nm, $P(\theta)$ ~ 1, angular dependence is eliminated, and the molecules are considered isotropic scatterers. Since the dimensions of most proteins and complexes are below 20 nm, for the remainder of this chapter we will assume isotropic scattering, i.e., $P(\theta) = 1$.

2.1.2 Analysis of protein complexes via static light scattering coupled to online chromatographic separation

Analytical size-exclusion chromatography (SEC) is often an unreliable measure of molar mass, particularly if the standards used to calibrate column elution times do not represent the sample well in terms of shape or column interactions. Because it does not need to make any assumptions regarding separation models or column calibration standards, flow-mode MALS is an invaluable extension of analytical SEC (SEC-MALS) or asymmetric-flow field flow fractionation (AF4-MALS). The analysis almost invariably occurs at concentrations well below 1 mg/mL, low enough to fall squarely within the ideal limit. Figure 1 shows a typical SEC-MALS experimental layout, combining a MALS detector with concentration analysis by means of UV/Vis absorption or differential refractometry (dRI).

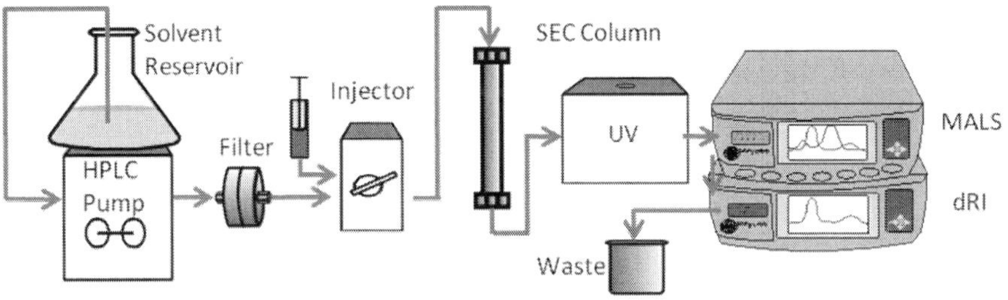

Fig. 1. SEC-MALS instrumentation.

While normally applied to characterize polydisperse ensembles, irreversible oligomers or other tightly-bound complexes, SEC-MALS may be used to assess reversible protein interactions, especially self-association (Bajaj et al., 2007). In this approach, the excess Rayleigh ratio and concentration is measured at multiple points along an eluting peak, and these data are fit to equations representing mass conservation, mass action and ideal light scattering identical to those in Section 2.1.3. Since the ratio of monomer to oligomer in a reversibly associating system is concentration-dependent, the change in weight-average molar mass across the peak should indicate dissociation of oligomers upon dilution in the column. The measurement may be repeated at different initial sample concentrations to enhance the analysis and establish whether the dissociation kinetics are fast or slow.

While this analysis can provide a good semi-quantitative characterization of reversible association, it is subject to certain systematic errors. The analysis must assume either rapid or very slow equilibration. As the sample proceeds through the column and detectors, it dilutes continuously; if the equilibration is neither very fast nor very slow compared to the elution time, the ratio of complex to monomer will not represent equilibrium conditions. Also, band-broadening between detectors means that the concentration measured in the UV or dRI concentration detector is somewhat different than that in the MALS cell, hence systematic errors in the analysis will arise. An integrated UV-SLS cell can eliminate the latter source of error (Bajaj et. al., 2004).

The advantages of analyzing reversible complexes via SEC-MALS are low sample quantity and clean data with little noise due to particulates if the size exclusion column and HPLC system are very clean. The column will separate any dust or aggregates from the sample.

2.1.3 Quantifying specific, reversible protein-protein binding via composition-gradient static light scattering

The use of stop-flow injections with well-defined concentrations permits true equilibrium analysis. In some instances the kinetics of association or dissociation may be analyzed as well. Composition-gradient light scattering apparatus, described in section 3, is more generally useful than SEC-MALS for studying protein-protein interactions. This section presents the principles of this approach.

The analysis of specific, reversible protein-protein interactions in the ideal limit, via light scattering measurements from a series of compositions, has been presented concisely by Attri & Minton (Attri & Minton, 2005b). The equations, with a minor change in notation, include the ideal light scattering law (4), mass action (5) and conservation of mass (6), shown below assuming up to two constituent monomeric species A and B:

$$\frac{R}{K^*} = \sum_{i,j} (iM_A + jM_B)^2 \left[A_i B_j\right] \tag{4}$$

$$K_{ij} = \frac{\left[A_i B_j\right]}{[A]^i [B]^j} \tag{5}$$

$$[A]_{total} = \sum_{i,j} i\left[A_i B_j\right], \quad [B]_{total} = \sum_{i,j} j\left[A_i B_j\right] \tag{6}$$

M_A and M_B represent the molar masses of constituent monomers A and B, respectively; i and j represent the stoichiometric numbers of A and B in the A_iB_j complex, with A_1B_0 and A_0B_1 representing the monomers of A and B; $[A_iB_j]$ represents the molar concentration of the A_iB_j complex; $[A]_{total}$ and $[B]_{total}$ represent the total molar concentration of A and B in solution; and $K_{i,j}$ is the equilibrium association constant relating equilibrium molar concentrations of the A_iB_j complex and free monomer. Light scattering and concentration data acquired over a series of compositions—multiple concentrations, in the case of self-association of a single species, or a series of A:B composition ratios in the case of hetero-association—are fit to Eq. (4) by means of a standard least-squares nonlinear curve fitting procedure. This technique is known as composition-gradient multi-angle static light scattering, CG-MALS or CG-SLS. Beyond the usual curve fitting algorithms, there is an added complication of solving first at each composition and fitted parameter iteration, the nonlinear system of Eqs. (5) (one for each complex present in equilibrium with monomers) + Eqs. (6). Examples of the system of equations to be fit are presented in Section 4.4.1.

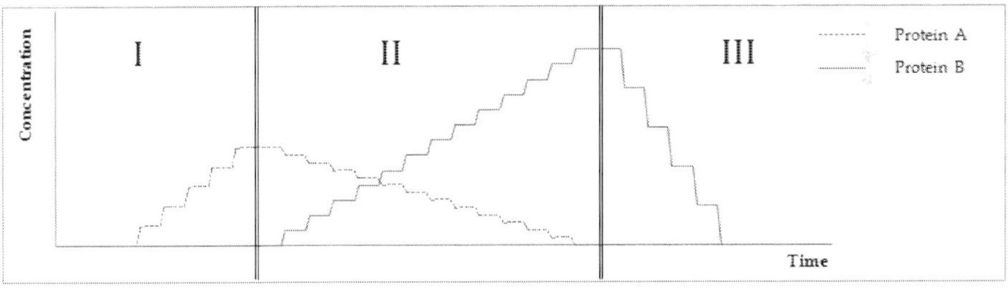

Fig. 2. Composition gradient for characterization of self- and hetero-association.

A typical composition gradient for the analysis of combined self-association and hetero-association, shown in Figure 2, includes three segments: I and III are concentration series in A and B individually to determine any self-association, while II is a "crossover gradient" stepping through a series of A:B composition ratios to characterize hetero-association. Figure 3a simulates LS signals for homodimer and homotrimer association, showing how the appropriate association model is well-determined by LS.

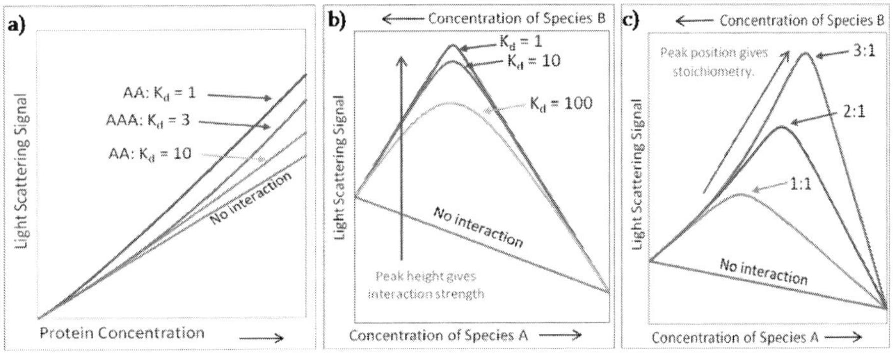

Fig. 3. Simulated CG-MALS signals. a) self association; b) and c) hetero-association.

The crossover gradient, though perhaps not intuitive to those accustomed to sigmoidal titration curves, is in fact quite efficient for analyzing stoichiometry and binding affinity even in the presence of a complex interaction that may include simultaneous self- and hetero-association. Figure 3b and Figure 3c depict a qualitative interpretation of the behavior of the light scattering signal for such a gradient: the position of the peak along the composition gradient axis indicates the stoichiometric ratio of the complex, while the height and shape of the peak indicate both the binding affinity and the true stoichiometry, i.e., discrimination between the formation of 1:1 or 2:2 complexes.

2.1.4 Non-ideal static light scattering for characterization of nonspecific protein interactions

Nonspecific protein interactions arise from various sources such as hard-core molecular repulsion, net charge, dipoles, hydrophobic patches, van der Waals interactions, hydration forces, etc. These are generally weak in relation to specific binding and so only become important at concentrations exceeding about 1 mg/mL. In contrast to site-specific lock-and-key binding, nonspecific interactions do not exhibit well-defined stoichiometry and do not generally saturate at some sufficiently high concentration. The dominant interaction may vary from attractive to repulsive or vice-versa as the concentration increases. The ill-defined, multi-sourced nature of nonspecific interactions lends itself to thermodynamic analysis in terms of small deviations from the ideal case described by Eq. (1), hence the designation "thermodynamic non-ideality." These interactions are of interest since they generally correlate to solubility, viscosity, and propensity for aggregation, and are also key to understanding crowded biomolecular environments such as the intracellular environment.

Akin to the virial expansion of the osmotic pressure as a power series in concentration, light scattering of dilute solutions may be analyzed in terms of a virial expansion, which actually uses the same pre-factors (virial coefficients A_2, A_3, or B_{22}, B_{222}, etc.) as the osmotic pressure equation though with a slightly different functional form:

$$\frac{R}{K^*}=\frac{Mc}{1+2A_2Mc+3A_3Mc^2+\ldots} \quad or \quad \frac{K^*c}{R}=\frac{1}{M}+2A_2c+3A_3c^2+\ldots \tag{7}$$

In many cases a protein's nonspecific interactions may be modeled as those of hard, impenetrable spheres, albeit with an effective specific volume v_{eff} different from that of the molecule's actual specific volume, resulting in a virial expansion containing only one independent parameter describing the non-ideality (Minton & Edelhoch, 1982):

$$\frac{R}{K^*}=\frac{Mc}{1+8v_{eff}c+30\left(v_{eff}c\right)^2+\ldots} \tag{8}$$

Each virial coefficient may be expressed in terms of the effective volume. As the lowest-order correction to ideal light scattering, $A_2=4v_{eff}/M$ tends to be of greatest interest for characterizing nonspecific interactions. Unscreened, long-range charge-charge interactions are not fit well by the effective hard sphere model.

An interesting and counterintuitive feature of purely (or primarily) repulsive interactions is that the LS intensity is not monotonic with concentration. Rather it initially rises as expected from the ideal LS equation, then plateaus, and eventually declines with higher concentration (see high-concentration case study, Section 5.5). Many non-associating proteins exhibit light scattering behavior which is fit well by the effective hard sphere assumption including scattering that eventually decreases with concentration (Fernández & Minton, 2008). The same work described CG-SLS apparatus suitable for high concentrations.

2.1.5 Quantifying repulsive and weakly attractive protein interactions via composition-gradient static light scattering at high concentration

Attractive nonspecific interactions, though weak, will at high enough concentrations invariably lead to the formation of transient clusters which can be analyzed in terms of specific reversible oligomerization, rather than in terms of virial coefficients. The same is not true of repulsive interactions. Hence it is possible to segregate the repulsive interactions into the virial coefficients and treat the attractive interactions as specific self-association.

The analysis is further simplified by assuming not only that the monomers and oligomers behave as effective hard spheres but also that all species have the same effective specific volume v_{eff}. An algorithm has been developed to include the effect of varying thermodynamic activity on the apparent association constants (Minton, 2007). This approach has been shown to work well for enzymes (Fernández & Minton, 2009) as well as antibodies (Scherer et al., 2010). While not quite as rigorous, we have found that a reduced analysis based on Eq. (9) serves to reproduce the essential behavior at high protein concentration in terms of self-associating oligomers subject to an effective hard sphere repulsion:

$$\frac{R}{K^*} = \frac{\sum_i iMc_i}{1 + 8v_{eff}c + 30\left(v_{eff}c\right)^2 + \ldots} \tag{9}$$

2.2 Dynamic light scattering

Rather than measuring the time-averaged intensity of scattered light, dynamic light scattering (DLS) measures the intensity fluctuations which arise from Brownian motion of the scattering molecules. The fluctuations are mathematically processed to produce an autocorrelation spectrum, which is then fit to appropriate functional forms to assess the translational diffusion constant D_t. D_t can be related via the Stokes-Einstein equation to a characteristic dimension, the hydrodynamic radius r_h, which is just the radius of a sphere with the diffusion constant D_t. The theory of DLS is covered in myriad sources, from textbooks (e.g., Teraoka, 2002) to peer-reviewed scientific literature.

DLS has certain practical advantages over SLS. In particular, DLS enjoys a relative immunity to stray light, which permits robust measurements in small volumes with free surfaces, such as a microwell plate. On the other hand, DLS is not as sensitive as SLS and so requires higher concentrations, limiting the range of binding affinities it can measure.

2.2.1 Analyzing protein-protein binding via composition-gradient dynamic light scattering in the ideal limit

The same nonspecific interactions leading to thermodynamic non-ideality in SLS do affect the diffusion constant and apparent r_h measured by DLS. In the ideal limit corresponding to a sufficiently dilute solution, this may be ignored. A solution consisting of multiple species

A_iB_j will exhibit a diffusion constant which is the z-average of the diffusion constants $D_{t,ij}$ of the individual species, leading to an average hydrodynamic radius as shown in Eq. (10):

$$\frac{1}{r_h^{avg}} = \sum_{i,j} \frac{1}{r_{h,ij}} M_{ij}^2 \left[A_iB_j\right] \Big/ \sum_{i,j} M_{ij}^2 \left[A_iB_j\right] \tag{10}$$

Upon measuring a series of concentrations or compositions like that shown in Figure 2, one can determine stoichiometry and equilibrium association constants by analyzing the apparent diffusion constants in terms of Eq. (10) in combination with Eq. (5) and Eq. (6). This technique is known as composition-gradient dynamic light scattering (CG-DLS).

Unlike the molar mass measured by static light scattering, it is not obvious or straightforward to construct the hydrodynamic radius or diffusion constant of an associating complex, given the hydrodynamic radii of its constituent species. This is especially true for stoichiometries higher than 1:1 where the geometry could vary from compact to extended, leading to significantly different diffusion constants. Composition-gradient DLS data has been shown to successfully analyze binding of globular proteins to an A_iB_j complex by assuming a power law relationship (Hanlon et. al, 2010):

$$r_{h,ij} = \left(i r_{h,A}^{\alpha} + j r_{h,B}^{\alpha}\right)^{\frac{1}{\alpha}} \tag{11}$$

The best fits for different associating systems resulted in α values ranging from ~2.8 for homodimers and 1:1 complexes, to ~ 2.0 for a 2:1 stoichimetry. In this work, CG-DLS in a microwell plate reader provided important benefits, including low sample consumption and the ability to measure the same samples at multiple temperatures in order to obtain enthalpy and entropy of the interaction via van 't Hoff analysis.

2.2.2 Analyzing nonspecific interactions via dynamic light scattering

Nonspecific interactions give rise to thermodynamic non-ideality in DLS as well as MALS. The first-order correction to the translational diffusion constant incorporates a combination of parameters: the second osmotic virial coefficient A_2 (B_{22}), the specific volume of the molecule v_{sp}, and the first-order correction to the molecule's friction coefficient due to concentration ζ_1, as presented in Eq. (12) (Teraoka, 2002).

$$D_t = D_0\left(1 + k_D c + \ldots\right); \quad k_D = 2A_2 - 2v_{sp} - \zeta_1 \tag{12}$$

We can expect v_{sp} to remain approximately constant for a given protein in different buffer systems, while ζ_1 actually includes additional A_2 dependence. A measurement of k_D for a series of monoclonal antibodies in different buffers exhibits excellent correlation with A_2 (Lehermayer et al., 2011).

The sample concentrations needed to measure k_D are comparable to those needed to measure A_2, but the volumes can be much smaller. Hence, in order to track trends in nonspecific interactions with buffer modifications such as pH or ionic strength, k_D (particularly as measured in a DLS plate reader) can be a low-volume, high-throughput surrogate for CG-MALS A_2 analysis. Unlike CG-MALS, however, currently the high-concentration behavior of CG-DLS is insufficiently understood to interpret data in the 10-200 mg/mL range.

3. Composition-gradient light scattering instrumentation

3.1 Detectors

Light scattering instrumentation for solutions has evolved considerably in the past two decades, resulting in systems that make characterization of protein interactions via CG-MALS and CG-DLS both feasible and accessible. Current top-of-the-line commercial MALS instruments provide a dynamic range covering protein solutions from tens of ng/mL up to hundreds of mg/mL (HELEOS, Wyatt Technology Corp.), accessing interactions with K_D from sub-nM to mM. Closed systems employing low-stray-light flow cells are important for low through moderately high concentrations, but are not suitable for the most concentrated protein samples that tend to be viscous. Easily-cleaned, microcuvette-based systems are better suited to the latter measurements.

A large selection of microcuvette-based DLS detectors is commercially available (Zetasizer series, Malvern Instruments; DynaPro series, Wyatt Technology; etc.). The lowest protein concentration range that these can analyze is on the order of 10-100 µg/mL, accessing an interaction range from tens of nM to tens of µM. Of particular note is the DynaPro DLS plate reader which can be integrated with standard liquid handling robotics to prepare automatically low-volume composition gradients.

Some instruments provide simultaneous SLS and DLS detection (HELEOS+QELS (flow cell or microcuvette), NanoStar (microcuvette), Wyatt Technology Corp; Zetasizer µV (flow cell or microcuvette), Malvern Instruments).

3.2 Automated composition-gradient delivery systems

An automated composition-delivery system for CG-MALS or CG-DLS is similar in many ways to standard stop-flow apparatus: two or more solutions are mixed by pumping via syringe pumps through a static mixer, an aliquot is delivered to the detector, and the flow stopped. The syringe pumps are operated at different flow ratios in order to obtain different compositions.

The most significant added requirement for light scattering is good in-line filtration of the solutions in order to eliminate large aggregates and particles generated by airborne dust, mechanical motion of syringes and valves, or protein films sloughing off surfaces. The pore size should be on the order of 0.1 µm or less. One approach (Attri & Minton, 2005a) is to add an in-line filter after the mixing point, illustrated in Figure 4. The key disadvantage of this

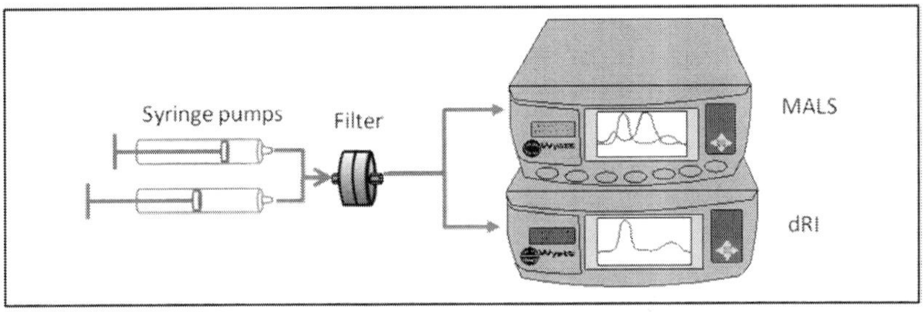

Fig. 4. Single in-line filter, parallel detectors CG-MALS setup, after Attri & Minton 2005a.

single-filter architecture is the changing chemical environment on the filter membrane: proteins will adhere to and release from the filter unpredictably as the environment changes. This is of particular concern in a tight-binding hetero-association analysis carried out at low protein concentrations. An in-line concentration detector is crucial.

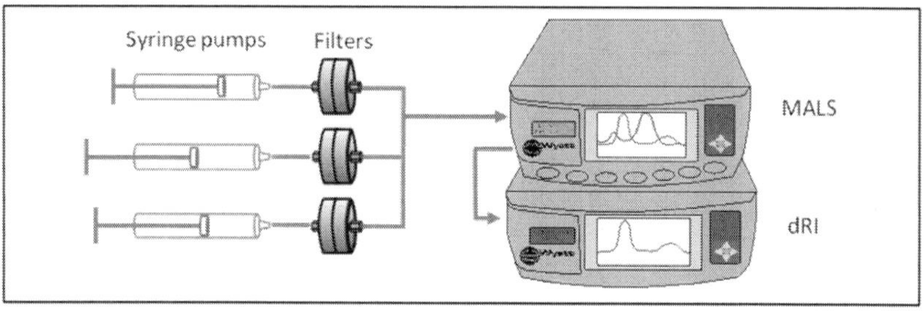

Fig. 5. Multiple in-line filters, serial detectors CG-MALS setup (Calypso, Wyatt Technology Corp.). UV or dRI concentration detector is optional.

Another approach is to flow each solution through a dedicated filter, as shown in Figure 5. In preparation for the gradient, each solution is pumped through its own filter and associated lines until saturation is reached so that, in the course of a subsequent composition gradient, well-defined compositions are produced reliably. This setup potentially eliminates the need for an in-line concentration detector if the stock protein solution concentrations are known prior to loading.

CG-MALS systems typically include light scattering and concentration detectors. The setup of Figure 5 shows a common approach, serially connected detectors, much like that of SEC-MALS. In order to achieve accurate correspondence between the concentration in the MALS flow cell and that in the concentration detector, both cells must be fully flushed with each composition. This can require relatively large sample volume. An alternative approach is to split the flow between the two detectors, as shown in Figure 4. Careful calibration of the flow resistance and delay between the two detectors is required to match the concentrations at the end of each injection. Additional care must be taken to ensure that laboratory temperature fluctuations, clogged capillaries or viscosity changes do not alter the split ratio between the detectors. The parallel detector configuration affords a smaller injection volume per composition.

Automation would not be complete without control and analysis software. Currently the only commercially available hardware/software package integrating syringe pump control, MALS data acquisition, and data analysis of equilibrium and kinetic macromolecular interactions is the Calypso CG-MALS system (Wyatt Technology, Santa Barbara).

4. Practical challenges of composition-gradient light scattering

4.1 Sample and buffer preparation

Due to the sensitivity of light scattering to the presence of just a few large particles, and especially in the absence of a separation step like SEC or FFF, particle-free solutions are essential in CG-MALS. Even though the composition-gradient apparatus provides in-line filtration, samples and solvents must be pre-filtered to the smallest practical pore size into very clean glassware or sterile, disposable containers. Solvents and buffers are generally filtered via bottle-top vacuum filters or large syringe-tip filters with pore sizes of 0.1-0.2 μm. Samples should be diluted to a bit above the appropriate working concentration in *filtered* solvent and then filtered with a syringe-tip filter to the smallest allowable pore size, most commonly 0.02-μm (e.g., Anotop filter, Whatman). All filters should be flushed to wash out any particles prior to introducing sample or final buffers.

4.2 Maintaining clean experimental apparatus

Regular cleaning and maintenance of the LS detectors and sample delivery apparatus are imperative for reliable, reproducible data. As a general rule, after each experiment, the instruments should be flushed with a buffer in which the sample is soluble before changing to storage or cleaning solutions. Common detergents for removing protein and polymer residue from glass and plastic surfaces include 5% v/v Contrad 70 and 1% w/v Tergazyme. Other methods useful for cleaning a dirty system include flushing with a high-salt (0.5-1.0 M NaCl) solution, 20-30% alcohol in water, or 10% nitric acid, as well as manual disassembly and cleaning. Salt and protein residues may be removed from syringes or valves by sonication.

Cleanliness of the instruments and buffers should be verified by observing noise in the MALS signals as the solutions flow through the system.

4.3 Designing optimal methods

The key parameters in CG-MALS experiment design are: 1) stock concentrations; 2) number and spacing of composition steps; 3) injected volume per step; and 4) equilibration time.

4.3.1 Determining optimal concentrations

Since molecular interactions are generally concentration-dependent, it is important to estimate the right concentration range that will, on the one hand, be high enough to produce a significant amount of complex, but on the other, be low enough so as not to saturate the complex leaving no free monomer. A general rule-of-thumb, assuming one has an estimate of K_D, is to prepare stock solution concentrations at 5-10x K_D. A more sophisticated approach is to perform a series of CG-MALS simulations assuming different concentrations, K_D values and even association schemes (e.g. 1:1 or 2:1), selecting composition gradients that best discriminate between reasonably feasible models.

For self-association, a concentration gradient should include concentrations low enough that essentially no self-association occurs, as shown at the low-concentration end of Figure 3a where the no-interaction signal coincides with the associating signal. The gradient should also include concentrations high enough that at least 20-30% of the LS intensity arises from oligomers, as shown on the high-concentration end of Figure 3a.

For heteroassociation, the optimal A:B stock concentration ratio is not necessarily the stoichiometric ratio, but depends on the molar masses of the molecular species. For good contrast, the total LS signal at 100% A (right side of Figure 3b) and 100% B (left side of Figure 3b) will be nearly equal, i.e., $M_A c_A \sim M_B c_B$ where c_A and c_B refer to the stock concentrations of A and B. This should be balanced against centering the LS peak close to the center of the crossover gradient. In particular, juggling these competing considerations can be tricky when the molar masses differ by a factor of 3 or more. If the mass ratio is large it may be better to perform a titration-like gradient where each injection includes a fixed concentration of the larger molecule, but varies the concentration of the smaller molecule.

Once the concentration ratio has been selected, the overall concentrations of A and B in the heteroassociation gradient should be chosen to discriminate well between K_D values within a reasonably expected range. For example, the conditions of Figure 3b discriminate well between K_D values of 1, 10 and 100, but would not be conclusive if the actual K_D is 0.1 or 1000.

An initial CG-MALS analysis may yield multiple association models that fit the data well. Simulations of light scattering from new composition gradients can assist in judiciously designing a follow-on experiment to refine the analysis by eliminating some of the first-round models. Such simulations are incorporated into the Calypso software.

The concentrations required to measure nonspecific interactions characterized by the second virial coefficient (A_2) typically range from 2-20 mg/mL. For proteins, an initial estimate of A_2 may be calculated by assuming a hard sphere of the same molecular weight (M) and hydrodynamic radius (r). The stock concentration needed to achieve a 15% contribution to the total scattered intensity from the A_2 term (see Eq. (7)) can be calculated as per Eq. (13):

$$c_{stock} = \frac{0.15}{2A_2^{sphere}M}; \quad A_2^{sphere} = \frac{16\pi N_A r^3}{3M^2} \tag{13}$$

4.3.2 Composition steps

An adequate number of compositions must be evaluated for proper fitting of CG-MALS/DLS data to an appropriate interaction model. For nonspecific interactions or simple homodimerization, at least five non-zero concentrations are recommended. Likewise, at least 5 composition steps are required for 1:1 binding. More complicated interactions forming larger numbers of species in equilibrium typically require 8-10 different compositions or more.

Sometimes the composition steps, rather than being distributed evenly across a gradient, can be focused in a specific region in order to make best use of the available sample, as shown in Figure 6.

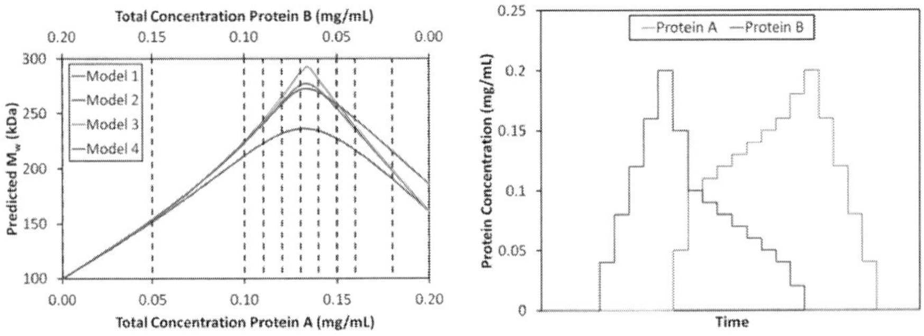

Fig. 6. Simulated interaction data for four possible interaction models (left) and corresponding CG-MALS method (right) focusing compositions around region of interest. Dashed vertical lines indicate plateau compositions.

4.3.3 Step volumes and equilibration time

The volume of sample introduced to the detectors at each composition must suffice to flush out completely the previous contents of the cell. At low sample concentration, or for particularly "sticky" samples, adsorption onto surfaces (especially the in-line filter of Figure 4) may necessitate increased step volumes. The required injection size may also vary with flow rate as well as detector configuration, and so should be determined experimentally for any set of conditions. The proper step volume may be assessed by running an ascending gradient followed by a descending gradient at a series of injection volumes: as the volume increases, the signals will match more closely.

After each injection, flow is stopped and the sample given time to reach equilibrium. Often the time scale for the reaction is faster than the dead time (the time between mixing and reaching the flow cell), but when this is not the case, ample time should be allotted after each injection for equilibration.

Where sample volume is scarce or where high concentration or viscosity prevents performing stop-flow experiments, CG-MALS analyses may be performed using a microcuvette. Stock solutions for each composition are prepared in advance. The light scattering intensity from each sample is measured using a calibrated cuvette and analyzed as for a flow system. Microcuvettes must be carefully cleaned and dried between samples.

4.4 Data analysis

CG-MALS data analysis protocols include two distinct segments: pre-processing and model fitting. The former comprises basic steps common to many measurements: baseline subtraction, application of proportionality (calibration or conversion) factors, smoothing, and selecting the data points for analysis. Specific to multi-angle light scattering are despiking and detector selection, since the main source of noise is foreign particles that primarily affect lower scattering angles and always generate positive signals. For equilibrium analysis, data should be selected after equilibration at each step, and usually a range of data points from each composition step are averaged to provide a single value from each detector.

4.4.1 Equilibrium models: Fitting and interpretation

The essential parameters in a CG-MALS model are monomer molar mass M_A and M_B; association stoichiometries ij; association constants $K_{A,ij}$; and incompetent (inactive) fractions $f_{incomp,A}$, $f_{incomp,B}$. The latter refer to protein molecules in the stock solution that are incompetent to participate in the interaction due to mutation, misfolding, aggregation, etc. In the course of fitting the data, a set of stoichiometries ij must be selected. Parameters such as monomer molar mass and incompetent fractions may be constrained to known values or floated to be optimized in the fit. Additionally, constraints may be imposed on the association constants, e.g., models of equivalent binding sites or isodesmic association confer specific relationships between the $K_{A,ij}$, as described in Table 1.

Standard iterative non-linear curve fitting algorithms, such as Levenberg-Marquardt, are implemented. For each composition, the total concentration of each monomer species is known either from precise dilution or by measurement with a UV or dRI detector. At each iteration of the free parameters, the equations for mass action and mass conservation (Eqs.

(5) and (6)) are solved; then the light scattering is computed (Eq. (4)) and compared to the measured value, the difference thereof serving as the minimization function. The result of fitting the data to a particular model will provide association constants plus any other free parameters, as well as a measure of goodness of fit, such as χ^2.

A broad range of useful equilibrium association models may be implemented, including any combination of self-interactions (formation of oligomers) and hetero-associations (stoichiometries of 1:1, 2:2, 1:n, etc.). Several common association models for proteins are presented in Table 1. Examples of these and more complex association schemes are discussed in Section 5.

Although useful, appropriate fitting of CG-MALS data does not require *a priori* knowledge of the interaction stoichiometry or system constraints. For a well-designed experiment, the best fit of the data should naturally converge on a single solution. This is illustrated in Figure 7 where incorrect models are applied to LS data for 1:1 and 1:2 interactions. In Figure 7A, which depicts a 1:2 interaction with equivalent binding sites as for antibody-antigen binding, a first guess of 1:1 interaction creates a fitted curve that does not peak at the correct stoichiometric ratio and clearly does not fit the data. Similarly, applying combined 1:1 and 1:2 stoichiometries, unconstrained for equivalent and independent binding sites, to a system with only one binding site results in the fitting algorithm eliminating the contribution from the 1:2 species (LS contribution from AB_2 = 0 for all compositions, Figure 7B).

In many instances, the expected interaction scheme fits the data well, resulting in low χ^2 and random residuals. Otherwise different stoichiometric models should be tested until the measured LS behavior is well matched. If the experiment design was far from optimal for the true system behavior or the interaction is particularly complex, more than one model may fit the data equally well. Several strategies may be brought to bear on selecting the most appropriate scheme, including Occam's razor (i.e., the simplest model that fits the data) and information from other techniques such as crystallographic structure or NMR analysis of binding sites. Simulation tools are useful in designing follow-up experiments to discriminate between multiple possibilities.

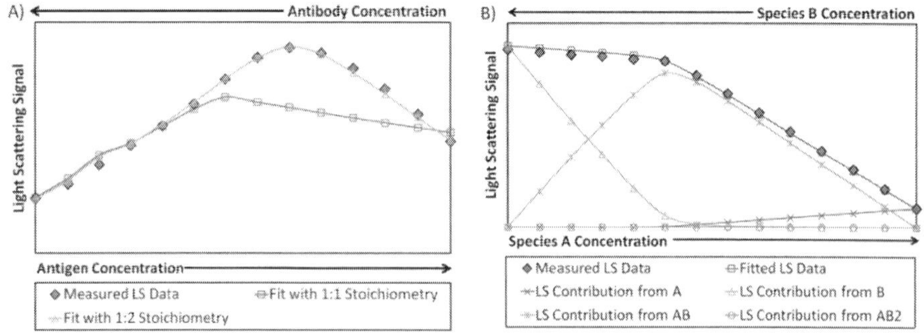

Fig. 7. Proper fitting of CG-MALS data requires the correct association model. A) Fitting of 1:2 interaction by 1:1 or 1:2 stoichiometry. B) Best fit of interaction between chymotrypsin (A) and bovine pancreatic trypsin inhibitor (B) from crossover gradient in Section 5.2.1 requires only 1:1 interaction.

Interaction Model	Unknowns	Governing Equations
Dimerization (e.g., Section 5.1.1)	$K_{20}=\frac{[A_2]}{[A]^2}$	$R/K^*=M_A^2\left([A]+4K_{20}[A]^2\right)$ $[A]_{total}=[A]+2K_{20}[A]^2$
1:1 hetero-association (e.g., Section 5.2.1)	$K_{11}=\frac{[AB]}{[A][B]}$	$R/K^*=M_A^2[A]+M_B^2[B]+(M_A+M_B)^2K_{11}[A][B]$ $[A]_{total}=[A]+K_{11}[A][B]$ $[B]_{total}=[B]+K_{11}[A][B]$
1:2 hetero-association, equivalent binding sites (e.g., Section 5.2.2)	$K_{11}=\frac{[AB]}{[A][B]}$ $K_{12}=\frac{[AB_2]}{[A][B]^2}=\left(\frac{K_{11}}{2}\right)^2$	$R/K^*=M_A^2[A]+M_B^2[B]+(M_A+M_B)^2K_{11}[A][B]$ $+\frac{1}{4}(M_A+2M_B)^2K_{11}^2[A][B]^2$ $[A]_{total}=[A]+K_{11}[A][B]+\frac{1}{4}K_{11}^2[A][B]^2$ $[B]_{total}=[B]+K_{11}[A][B]+\frac{1}{2}K_{11}^2[A][B]^2$
Isodesmic self-association to n-mers (e.g., Section 5.1.2; see Kameyama & Minton 2006)	$K_{20}=\frac{[A_2]}{[A]^2}$ $K_{n0}=\frac{[A_n]}{[A]^n}=K_{20}^{n-1}$	$R/K^*=M_A^2\sum_{i=1}^{n}i^2K_{20}^{i-1}[A]^i$ $[A]_{total}=\sum_{i=1}^{n}iK_{20}^{i-1}[A]^i$

Table 1. Common equilibrium association models that can be quantified by CG-MALS.

4.4.2 Kinetics models: Fitting and interpretation

Reaction kinetics for reversible and irreversible associations can be observed and quantified by light scattering to provide a direct measure of association, dissociation, or aggregation via the evolution of M_w. Quantifying characteristic rate constants from CG-MALS data requires knowledge of the final stoichiometry and, in the case of reversible associations, the appropriate equilibrium association constants. For example, LS data for covalent inhibition of an enzyme by an inhibitor may be fit at varying inhibitor concentrations to yield a second-order rate constant for the interaction, k_a. In the case of irreversible dissociation, the apparent first-order kinetics can be described by an exponential function, and the apparent dissociation rate constant, k, can be related to applicable biomolecular constants:

$$(R/K^*)\propto[A_2]=[A_2]_0\,e^{-kt} \tag{14}$$

More complex analyses, such as the association of two proteins into an equilibrium complex, involve solving the rate equations that govern the system of interest. The equilibrium association constant K_A and final stoichiometry must be determined in addition to the time-dependent change in light scattering. For the simplest heteroassociation $A+B\leftrightarrow AB$, Eq. (15) relates the CG-MALS data to the second order association rate constant $k_a=K_A\cdot k_d$:

$$\frac{1}{K^*}\frac{dR}{dt} = 2M_A M_B \frac{d[AB]}{dt}; \qquad \frac{d[AB]}{dt} = k_a\left\{([A]_{total} - [AB])([B]_{total} - [AB]) - \frac{[AB]}{K_A}\right\} \tag{15}$$

5. CG-MALS examples

5.1 Self-association

5.1.1 Dimerization of chymotrypsin

Dimerization has been observed by CG-SLS for the enzyme α-chymotrypsin with pH-dependent affinity (Kameyama & Minton, 2006; Fernández & Minton, 2009). Figure 8 presents dependence of the reaction on ionic strength (Hanlon & Some, 2007), closely matching results obtained via sedimentation equilibrium (Aune et al., 1971).

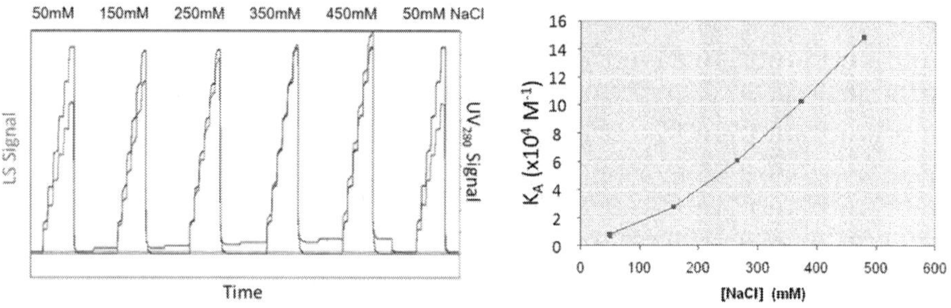

(A) (B)

Fig. 8. Self-association of chymotrypsin forming dimers vs. ionic strength. (A) LS and UV_{280} concentration data over a series of concentration gradients (B) K_A vs. [NaCl].

5.1.2 Isodesmic self-association

Some proteins tend to self assemble into chains, fibrils, or other large oligomers, such as amyloid-β plaques in Alzheimer's disease and α-synuclein aggregates in the Lewy bodies of Parkinson's disease. A model of isodesmic self-association, i.e., the assumption that each protein monomer binds to the growing chain with equal affinity, can often be used to describe such an interaction, especially in the early nucleation phase of the assembly.

Insulin changes its self-association state as a function of pH and the presence of zinc ions (Attri et al., 2010a, 2010b, and references therein). At physiological conditions in the presence of Zn^{2+}, insulin exists as a hexamer that further associates isodesmically to higher order oligomers—dimers of hexamers (12-mers), trimers of hexamers (18-mers), etc. (Attri et al., 2010b). This interaction was studied using both static and dynamic light scattering. Based on the reported equilibrium and diffusion constants, M_w, D_t, and the molar composition of insulin oligomers could be reproduced (Figure 9).

In contrast, in the absence of Zn^{2+}, insulin monomers exist in isodesmic equilibrium with dimers, trimers, and higher order complexes with pH-dependent affinity (Figure 10). Rather than constraining the maximum oligomerization state as in Table 1, both studies considered the possibility of infinitely large oligomers.

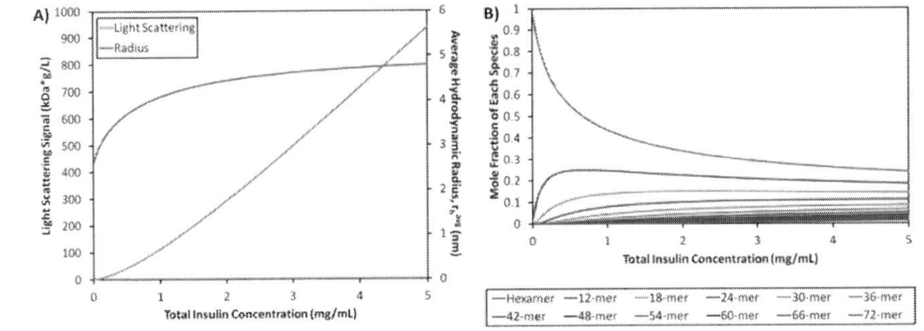

Fig. 9. Infinite self-association of insulin hexamers at neutral pH in the presence of Zn^{2+}. A) LS signal and r_h^{avg} vs. protein concentration, calculated per K_A and D_t in Attri et al., 2010b. B) Calculated molar distribution of species.

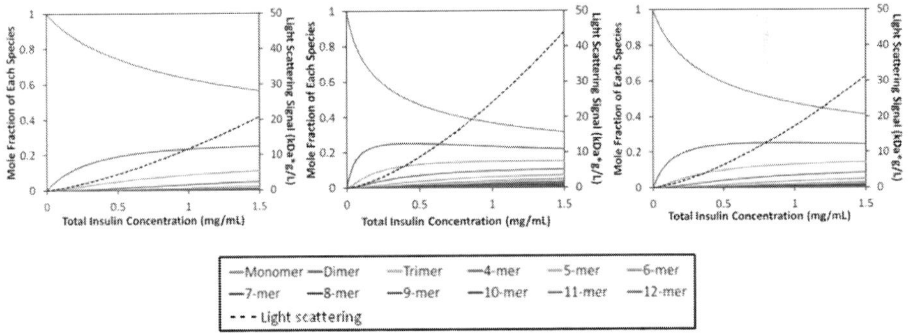

Fig. 10. Molar distribution of insulin self-association products and light scattering signal in the absence of Zn^{2+} at pH 3 (left), 7.2 (middle), and 8 (right), calculated per Attri et al., 2010a.

5.2 Hetero-association

5.2.1 Reversible enzyme-inhibitor binding with 1:1 stoichiometry

Following Kameyama & Minton 2006, we characterized a standard 1:1 reversible association between α-chymotrypsin (CT) and bovine pancreatic trypsin inhibitor (BPTI). A CG-MALS experiment consisting of self-association gradients for each binding partner CT and BPTI and a crossover hetero-association gradient was performed as per Figure 2. The self-association gradients yield the molecular weight for each monomer and confirm the lack of self-association for CT and BPTI at neutral pH. Fitting the LS data in Figure 11A as a function of composition (Figure 7B) results in an equilibrium dissociation K_D = 119 nM (K_A = 8.5×10^6 M^{-1}), consistent with measurements by other techniques (referenced in Kameyama & Minton, 2006). The LS contribution from each species is then transformed to a concentration, giving the species distribution shown in Figure 11B. As expected for a 1:1 interaction, the plateau with the highest amount of complex formation occurs at a molar ratio of CT:BPTI ~1:1 (~11 μM each CT and BPTI). Since the experiment was performed at concentrations >10x K_D, nearly all available free monomer is consumed in the (CT)(BPTI) complex. This is evident in Figure 11B where the mole fraction of CT is ~0 for all compositions [CT]<[BPTI], and the mole fraction of BPTI is ~0 for all compositions [BPTI]<[CT].

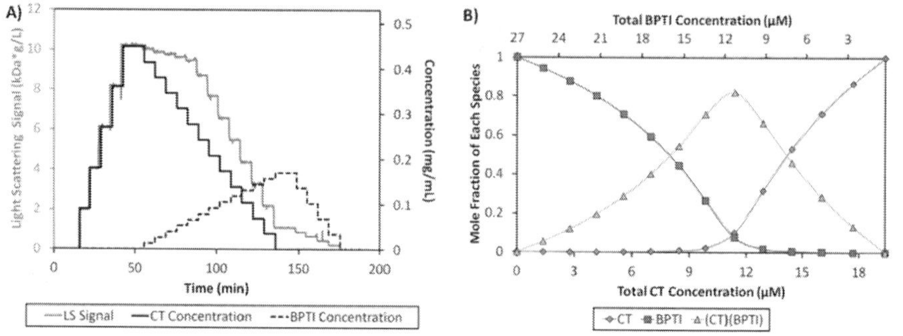

Fig. 11. CG-MALS quantifies binding of CT and BPTI.

Under acidic conditions, the affinity of CT for BPTI decreases, and CT can form reversible dimers, as in Section 5.1.1. At pH 4.4, the K_D for the association of CT and BPTI is of the same order as for the dimerization of CT—K_D = 10 µM and 50 µM respectively (Kameyama & Minton, 2006). Based on these results, we can simulate the expected LS signals for simultaneous self and hetero-associations (Figure 12). Discrimination between 1:1 binding only, and self + heteroassociation, is readily evident where [CT]>[BPTI] (Figure 12A). Despite the additional self-association, the fraction of hetero-association product still peaks at a molar ratio of CT:BPTI ~1:1 (Figure 12B).

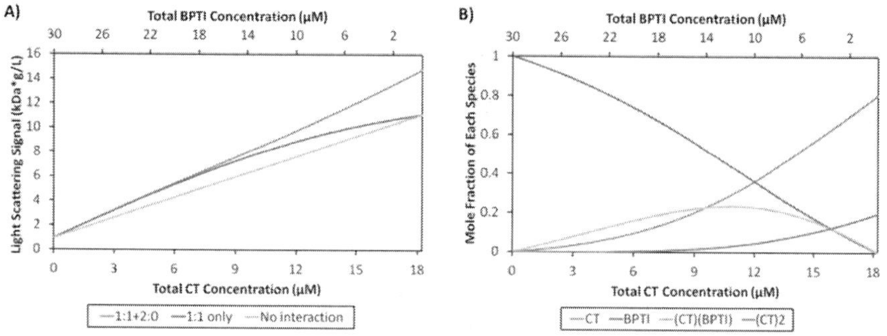

Fig. 12. CT-BPTI interaction at pH 4.4. A) Predicted LS for simultaneous CT self-association and CT-BPTI hetero-association (blue) compared to CT-BPTI hetero-association alone with K_D = 10 µM (red) or no interaction (green). B) Molar distribution of species for simultaneous self- and hetero-association model, based on Kameyama & Minton, 2006.

5.2.2 Antibody-antigen binding with 1:2 stoichiometry, two equivalent binding sites

The power of CG-MALS lies in its ability to identify multiple stoichiometries in solution. For example, a single multivalent receptor A may bind multiple protein ligands B, leading to the simultaneous presence of AB, AB_2, AB_3, etc. The increasing prevalence of therapeutic antibodies brings this type of multivalent binding to the forefront of biotechnology. Moreover, CG-MALS is able to characterize this type of interaction with affinities as low as K_D~0.1 nM, typical of antibody-antigen interactions. Our antibody-antigen binding data (Figure 13) indicates the presence of four species in solution: free antibody (Ab), free antigen (Ag), the 1:1 complex (Ab)(Ag), and the 1:2 complex $(Ab)(Ag)_2$. The CG-MALS K_D value of 10 nM agrees well with the literature value.

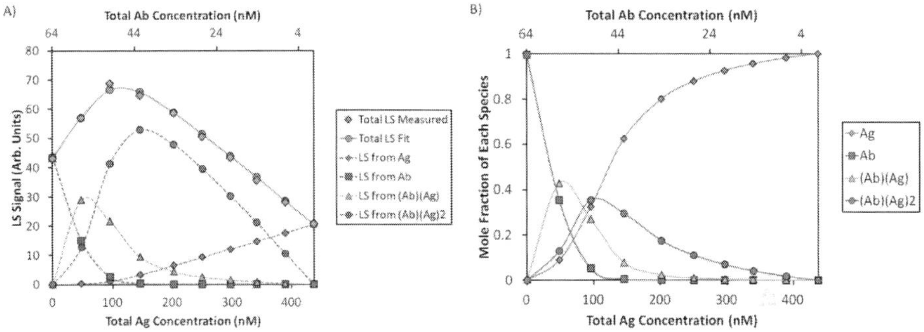

Fig. 13. Light scattering (A) and composition (B) distributions for crossover gradient between an antibody and monovalent antigen, K_D~10 nM.

Conversely, Some et al. (2008a) found that CG-MALS data for a dimeric Fcγ receptor (FcγR) binding to the Fc of a recombinant human Ab (rhumAb), shown in Figure 14, is only fit well by a model assuming two equivalent binding sites on each FcγR dimer (B) for rhumAb (A), producing equilibrium between monomers (A and B), AB, and A_2B (Figure 14C). CG-MALS data do not support other binding models, including 1 A : 1 B association alone (Figure 14A) and 1 A : 2 B with equivalent binding sites (Figure 14B). The calculated single-site affinity of 50 nM agrees closely with surface plasmon resonance (SPR) analysis.

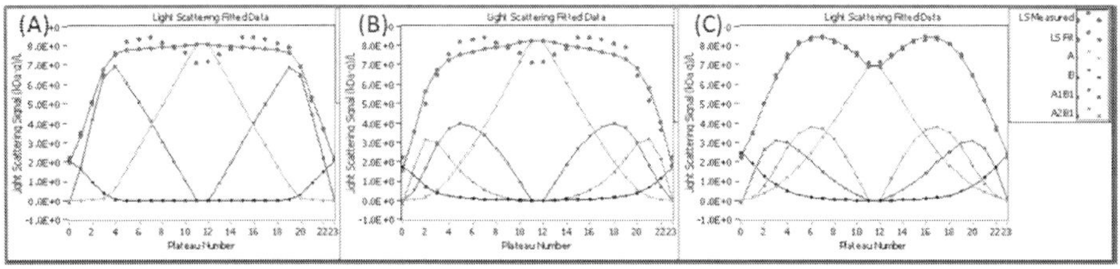

Fig. 14. Best fits (red lines) of measured CG-MALS data (blue circles) to different association models, IgG (rhumAb) : dimeric receptor (FcγR). Stoichiometry: (A) - 1:1; (B) - 1:2; (C) - 2:1. Only the {2 mAb per receptor} model fits the data.

5.2.3 Association of multivalent protein complexes

Combinations of multivalent binding partners can lead to the formation of metacomplexes in solution that are not identified by other techniques. As a homo-tetramer, streptavidin (SA) is composed of four identical binding sites capable of binding either of two Fab domains of an anti-streptavidin IgG. As we have observed, the combination of multivalent proteins enables higher-order stoichiometries to present themselves in solution, including multiple IgG molecules binding a single SA molecule and self-assemblies of IgG-SA complexes (Figure 15). Indeed, the LS signal measured for such a system by CG-MALS is nearly twice the value expected for a simple 1:2 association (Figure 15). Careful analysis of the data indicates that the solution is best described as 1:1 (IgG)(SA) complexes that self-associate (Figure 16). The infinite self-association (ISA) model employed here assumes that each base unit—(IgG)(SA) complex—assembles with other base units with the same affinity; however, this may differ from the binding-site affinity (K_D) for a single IgG-SA interaction. The binding-site K_D for one SA molecule binding one IgG was determined as 22 nM, while these 1:1 base units assemble with an average affinity K_D = 50 nM.

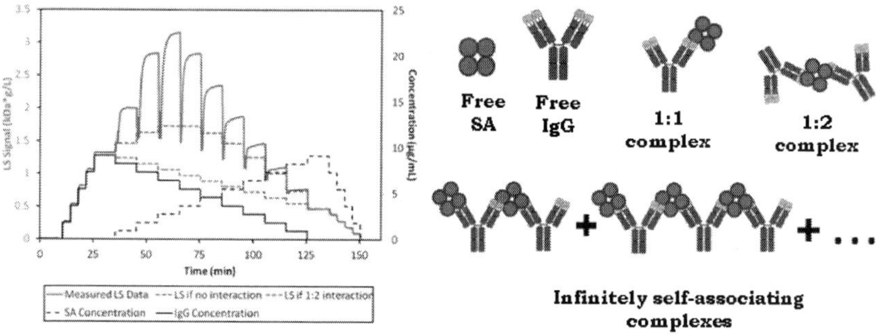

Fig. 15. Light scattering and concentration data for association of SA and anti-SA IgG. Theoretical LS plateaus are indicated for the case of no IgG-SA interaction and a 1:2 equivalent binding site model (Section 5.2.2). Additional stoichiometries that contribute to the measured LS signal, including infinitely self-associating 1:1 complexes, are shown.

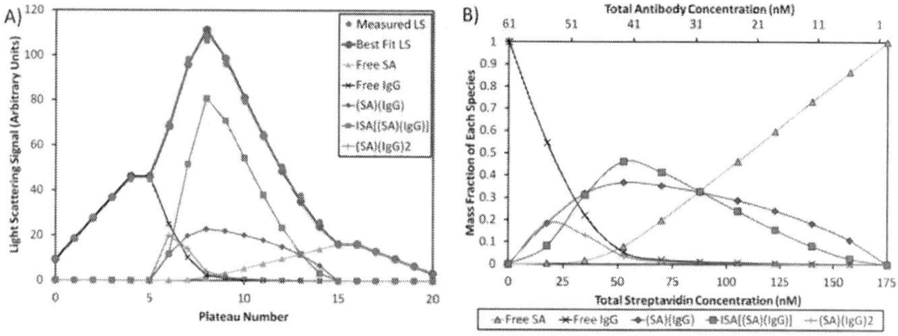

Fig. 16. A) Best fit of LS data for SA + anti-SA IgG includes infinite self assembly (ISA) of 1:1 metacomplexes B) Concentration distribution for hetero-association plateaus (#5-15).

5.3 Dissociation kinetics induced by a small molecule inhibitor

Although other techniques, such as SPR and FRET-based methods, are capable of quantifying association and dissociation kinetics, many require modification of the protein of interest, i.e., immobilization in the case of SPR and labeling with fluorescent tags for

FRET. In contrast, CG-MALS enables real-time observation of reaction kinetics in solution without protein modification. For example, chymotrypsin self-association at low pH is inhibited by the small molecule 4-(2-aminoethyl) benzenesulfonyl fluoride (AEBSF). When introduced to a chymotrypsin solution, AEBSF covalently binds the monomer active site and prevents dimerization. Varying concentrations of AEBSF were mixed with a constant stock solution of chymotrypsin, and the resulting dissociation kinetics quantified with a model of an irreversible dissociation (Some & Hanlon, 2010). For each composition, the solution was allowed to react for >1 hr while observing the decrease in weight-average molar mass of the solution. The characteristic reaction time τ ($1/k$ in Eq. (14)) varies inversely with the AEBSF concentration, consistent with the rate models defined for the system (Figure 17), indicating a rate constant of 0.064 $M^{-1}s^{-1}$.

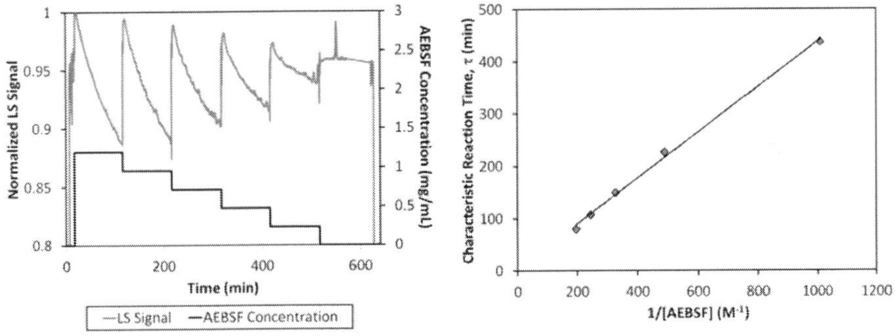

Fig. 17. Decrease in LS signal (left) and change in characteristic reaction time (right) corresponding to irreversible dissociation of chymotrypsin dimers in the presence of AEBSF.

5.4 Nonspecific interactions of non-self-associating proteins

5.4.1 Nonspecific self-interactions

As discussed in Section 2.1.4, all macromolecules at high concentrations exhibit some degree of nonspecific interactions, quantified by the second virial coefficient, A_2. This property is of particular interest in the development of pharmaceutical formulations where A_2 is one metric for the stability of a formulation and the propensity of biomolecular therapeutics to aggregate in solution. Formulations that may appear stable at moderate concentrations (~10 mg/mL or less) may indeed form self-association products at relevant formulation concentrations of 100 mg/mL or more (see Section 5.5). For a well-formulated protein, however, repulsive interactions should dominate for all concentrations of interest. BSA, for example, exhibits nonspecific repulsion even at 100 mg/mL in PBS, as shown in Fig. 18. Long-range interactions are well-screened in this buffer, resulting in an A_2=1.0x10^{-4} mol*mL/g^2, consistent with a hard-sphere of radius 3.5 nm and M_w = 67 kDa.

5.4.2 Nonspecific attraction quantified by the cross-virial coefficient

Carrier proteins, drug delivery vehicles, and other polymers attract their biomolecular targets via nonspecific interactions (e.g., Dong et al., 2011) which cannot be described by an equilibrium association constant. A virial expansion may be employed to quantify nonspecific attraction or repulsion between molecules of the same species or different species. In the example below, the net negative charge of BSA, in PBS with 50 mM NaCl at pH 7, yields repulsion between BSA molecules. Lysozyme exhibits a slight positive charge with a net self-attraction as evidenced by the negative A_2. The charge-mediated attraction between BSA and lysozyme molecules is evident in Figure 19 as the increase in LS when BSA and lysozyme are mixed together. The data are best fit by a model of *nonspecific* attraction, quantified by the cross-virial coefficient A_{11}. The results can be normalized to a unitless value as per Sahin et al., 2010: $a_2 = \left(A_2^{meas} - A_2^{exc}\right)/A_2^{exc}$.

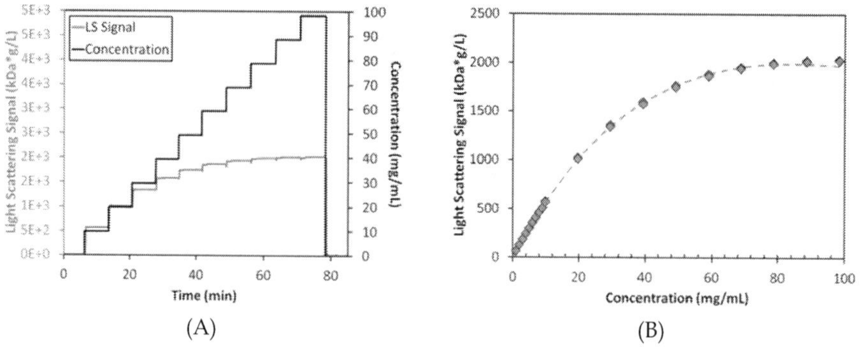

Fig. 18. BSA behaves as an effective hard sphere with $A_2 = 1.0\text{x}10^{-4}$ mol*mL/g² for all concentrations studied. (A) CG-MALS data (B) fit to effective hard sphere model.

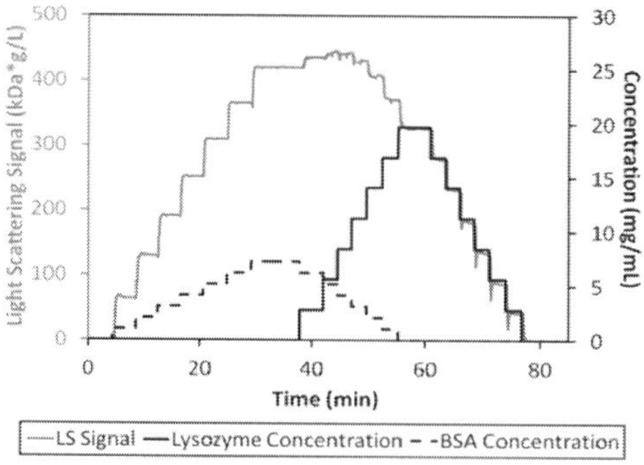

BSA self-virial coefficient	
A_2 (mol*mL/g²)	+1.1x10^{-4}
a_2	+0.2
Lysozyme self-virial coefficient	
A_2 (mol*mL/g²)	-3.6x10^{-4}
a_2	-2.1
Cross-viral coefficient	
A_{11} (mol*mL/g²)	-4.0x10^{-4}
a_{11}	-2.7

Fig. 19. Determination of self-and cross-virial coefficients for nonspecific interactions in BSA-lysozyme solution. Normalized virial coefficients are also presented.

5.5 Interactions of monoclonal antibodies formulated at high concentration

Recently, CG-MALS was applied to investigate interactions between IgG1 monoclonal antibodies at concentrations up to ~300 mg/mL (Scherer et al., 2010). Although the two mAbs studied here were identical except for the CDR sequence, their self-association properties were remarkably different. MAb2 forms dimers with $K_A \leq \sim 10^3$ M^{-1} ($K_D \geq \sim 1$ mM), whereas mAb1 associates into dimers with K_A ~10^3-10^4 M^{-1} (K_D ~0.1-1 mM) and appears to further associate into higher order oligomers of stoichiometry 4-6. The dependence of association properties on ionic strength also differs dramatically between mAb1 and mAb2: while the affinity of the mAb2 homodimer increases with [NaCl], that of mAb1 homodimers is essentially constant. Most significantly, the higher oligomer order of mAb1 decreases from 6 to 4 as [NaCl] increased from 40 to 600 mM.

Based on these results, we reproduce in Figure 20 the relative LS signal for mAbs1 and 2 and the fraction of each oligomer present in solution. Each calculation includes the appropriate correction for non-specific repulsion using v_{eff} = 1.8 cm^3/g (r_{eff} = 4.6 nm) for mAb1 and v_{eff} = 1.4 cm^3/g (r_{eff} = 4.3 nm) for mAb2 (Scherer et al., 2010). Although the antibody molecules continue to self-associate into higher molecular weight species, the LS signal is not monotonically increasing, as would be expected from ideal scattering (Eq. (1)); instead, the LS intensity reaches a maximum at ~100 mg/mL (Figure 20A). Only by accounting for both nonspecific repulsion and specific oligomerization can the light scattering data be fully described for these high-concentration solutions.

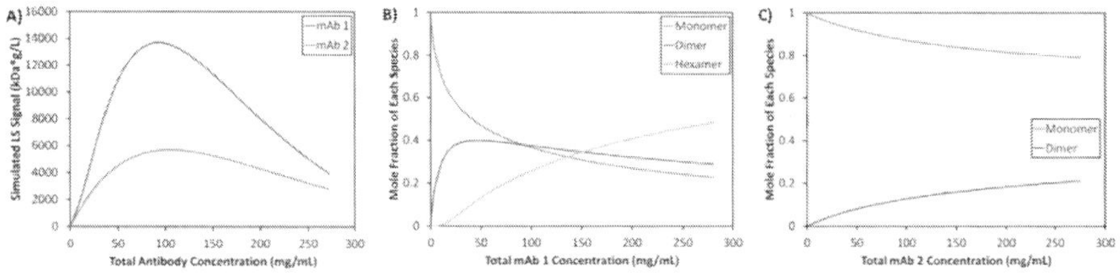

Fig. 20. A) LS signals for mAbs 1 and 2 in buffer containing 75 mM NaCl, calculated to represent results of Scherer et al., 2010. B) and C) Corresponding distribution of oligomers.

6. Conclusion

The power of light scattering, CG-MALS and CG-DLS, for investigating protein interactions lies in their great versatility. These techniques quantify a wide range of protein-protein phenomena in solution and without labeling. Both equilibrium and kinetics may be addressed directly since light scattering provides, from first principles, the molar mass and size of complexes, rather than an indirect indicator such as fluorescence. Hence light scattering is particularly well suited to analyzing higher-order complexes, multiple stoichiometries, and simultaneous self- and hetero-association. The fundamental thermodynamic nature of static light scattering provides a critical window into interactions at high concentration. The development of automation and advanced instrumentation suggests that common use of CG-MALS and CG-DLS is feasible, and hence these are important additions to the protein scientist's toolbox.

7. Acknowledgement

The authors would like to thank Allen P. Minton for many helpful discussions and collaboration in the development of automated CG-MALS; Shawn Cao (Amgen), Joey Pollastrini (Amgen), and Jihong Yang (Genentech) for contributing antibody samples; and the entire team at Wyatt Technology Corp. We are also indebted to the many early adopters of the Calypso CG-MALS system for their support and sharing samples.

8. References

Attri, A.K., Fernández, C., & Minton, A.P. (2010a). pH-dependent self-association of zinc-free insulin characterized by concentration-gradient static light scattering. *Biophysical Chemistry*. Vol. 148, No. 1-3, (May 2010), pp. 28-33, ISSN 0301-4622

Attri, A.K., Fernández, C., & Minton, A.P. (2010b). Self-association of Zn-insulin at neutral pH: investigation by concentration-gradient static and dynamic light scattering. *Biophyisical Chemistry*. Vol. 148, No. 1-3, (May 2010), pp. 23-27, ISSN 0304-4622

Attri, A.K. & Minton, A.P. (2005a). New Methods for Measuring Macromolecular Interactions in Solution via Static Light Scattering: Basic Methodology and Application to Nonassociating and Self-Associating Proteins. *Anal. Biochem.* Vol.337, No.1, (February 2005), pp. 103-110, ISSN 0003-2967

Attri, A.K. & Minton, A.P. (2005b). Composition Gradient Static Light Scattering (CG-SLS): A New Technique for Rapid Detection and Quantitative Characterization of Reversible Macromolecular Hetero-Associations in Solution. *Anal. Biochem.* Vol.337, No.1, (November 2005), pp. 103-110, ISSN 0003-2967

Aune, K.C., Goldsmith, L.C., Timasheff, S.N. (1971) Dimerization of alpha-Chymotrypsin. II. Ionic Strength and Temperature Dependence. *Biochemistry* Vol.10, No.9 (April 1971), pp. 1617-21, ISSN 0006-2960

Bajaj, H., Sharma, V.K., & Kalonia, D. (2004) Determination of Second Virial Coefficient of Proteins Using a Dual-Detector Cell for Simultaneous Measurement of Scattered Light Intensity and Concentration in SEC-HPLC. *Biophys. J.* Vol.87, No.6, (December 2004), pp. 4048-55, ISSN 0006-3495

Bajaj, H., Sharma, V.K., & Kalonia, D. (2007) A High-Throughput Method for Detection of Protein Self-Association and Second Virial Coefficient Using Size-Exclusion Chromatography Through Simultaneous Measurement of Concentration and Scattered Light Intensity. *Pharmaceutical Research* Vol.24, No.11, (November 2007), pp. 2071-83, ISSN 0724-8741

Blanco, M.A., Sahin, E., Li, Y., & Roberts, C.J. (2011) Reexamining Protein-Protein and Protein-Solvent Interactions from Kirkwood-Buff Analysis of Light Scattering in Multi-Component Solutions. *J. Chem. Phys.* Vol.134, No.22, (June 2011) pp. 225103 1-12, ISSN 0021-9606

Fernández, C. & Minton, A.P. (2008) Automated Measurement of the Static Light Scattering of Macromolecular Solutions over a Broad Range of Concentrations. *Anal. Biochem.* Vol.381, No.2, (Oct. 2008) pp. 254-7, ISSN 0003-2967

Fernández, C. & Minton, A.P. (2009) Static Light Scattering from Concentrated Protein Solutions II: Experimental Test of Theory for Protein Mixtures and Weakly Self-Associating Proteins. *Biophys. J.* Vol.96, No.5, (March 2009) pp. 1992-8, ISSN 0006-3495

Hanlon, A.D., Larkin, M.I., & Reddick, R.M. (2010) Free-Solution, Label-Free Protein-Protein Interactions Characterized by Dynamic Light Scattering. *Biophys. J.* Vol.98, No.2, (Jan 2010), p. 297-304, ISSN 0006-3495

Hanlon, A.D., and Some, D. (2007). CG-MALS for Characterization of Protein Self Association and Inhibition Kinetics. International Light Scattering Colloquium 2007, Santa Barbara, October 2007

Kameyama, K. & Minton, A.P. (2006) Rapid Quantitative Characterization of Protein Interactions by Composition Gradient Static Light Scattering. *Biophys. J.* Vol.90, No.6, (March 2006), pp. 2164-9, ISSN 0006-3495

Lehermayer, C., Mahler, H.-C., Mäder, K., & Fischer, S. (2011) Assessment of Net Charge and Protein-Protein Interactions of Different Monoclonal Antibodies. *J. Pharm. Sci.* Vol.100, No.7, (July 2011), pp. 2551-62, ISSN 0022-3549

Minton, A.P. (2007) Static Light Scattering from Concentrated Protein Solutions, I: General Theory for Protein Mixtures and Application to Self-Associating Proteins. *Biophys. J.* Vol.93, No.4, (August 2007), pp. 1321-1328, ISSN 0006-3495

Minton, A.P. and Edelhoch H. (1982) Light Scattering of Bovine Serum Albumin Solutions: Extension of the Hard Particle Model to Allow for Electrostatic Repulsion. *Biopolymers.* Vol.21, No.2, (February 1982), pp. 451-458, ISSN 0006-3525

Scherer, T.M., Liu, J., Shire, S.J. & Minton, A.P. (2010) Intermolecular Interactions of IgG1 Monoclonal Antibodies at High Concentrations Characterized by Light Scattering. *J. Phys. Chem. B.* Vol.114, No.40, (October 2010), pp. 12948-57, ISSN 1089-5647

Sahin, E., Grillo, A.O., Perkins, M.D., Roberts, C.J. Comparative Effects of pH and Ionic Strength on Protein-Protein Interactions, Unfolding, and Aggregation for AgG1 Antibodies. *J. Pharm. Sci.* Vol.99, No.12, (December 2010), pp.4830-48, ISSN 1520-6017

Some, D., Berges, A., Hitchner, E., & Yang, J. (2008a). CG-MALS Characterization of Antibody-Antigen Interactions. International Light Scattering Colloquium 2008, Santa Barbara, October 2008

Some, D. & Hanlon, A. (2010). Characterizing Protein-Protein Interactions via Static Light Scattering: Inhibition Kinetics And Dissociation. *American Biotechnology Laboratory.* Vol. 28, No. 1 (January/February 2010), pp. 9-12, ISSN 0749-3223

Some, D., Hanlon, A., & Sockolov, K. (2008b). Characterizing Protein-Protein Interactions via Static Light Scattering: Reversible Hetero-Association. *American Biotechnology Laboratory.* Vol. 26, No. 4 (March 2008), pp. 18-19, ISSN 0749-3223

Teraoka, I. (2002) *Polymer Solutions: An Introduction to Physical Properties,* John Wiley & Sons, Inc. ISBN 0-471-38929-3, New York, NY, USA van Holde, E.; Johnson, W.C. & Ho, P. S. (1998). *Principles of Physical Biochemistry,* Prentice Hall, ISBN 0-13-720459-0, Upper Saddle River, NJ, USA

Young, R.J. (1981). *Introduction to Polymers,* Chapman and Hall, ISBN 0-412-22170-5, London, UK

21

Site-Directed Spin Labeling and Electron Paramagnetic Resonance (EPR) Spectroscopy: A Versatile Tool to Study Protein-Protein Interactions

Johann P. Klare
Physics Department, University of Osnabrück, Osnabrück
Germany

1. Introduction

The function of a living cell, independent of we are talking about a prokaryotic single-cellular organism or a cell in the context of an complex organism like a human, depends on intricate and balanced interaction between its components. Proteins are playing a central role in this complex cellular interaction network: Proteins interact with nucleic acids, with membranes of all cellular compartments, and, what will be in the focus of this article, with other proteins. Proteins interact to form functional units, to transmit signals for example perceived at the surface of the cell to cytoplasmic or nuclear components, or to target them to specific locations. Thus, the study of protein-protein interactions on the molecular level provides insights into the basic functional concepts of living cells and emerged as a wide field of intense research, steadily developing with the introduction of new and refined biochemical and biophysical methods.

Nowadays there is a vast of methods available to study the interaction between proteins. On the biochemical level mutational studies, crosslinking experiments and chromatographic techniques provide means to identify and characterize the interfaces on the protein surface where interaction takes place. Biophysical methods include calorimetric techniques, fluorescence spectroscopy and microscopy, and "structural techniques" like X-ray crystallography, (cryo-) electron microscopy, NMR spectroscopy, FRET spectroscopy, and EPR spectroscopy on spin labeled proteins.

Site-directed spin labeling (SDSL) (Altenbach et al., 1989a, 1990) in combination with electron paramagnetic resonance (EPR) spectroscopy has emerged as a powerful tool to investigate the structural and the dynamical aspects of biomolecules, under conditions close to physiological i.e. functional state of the system under exploration. The technique is applicable to soluble molecules and membrane bound proteins either solubilised in detergent or embedded in a lipid bilayer. Therein, the size and the complexity of the system under investigation is almost arbitrary (reviewed in Bordignon & Steinhoff, 2007; Hubbell et al., 1996; Hubbell et al., 1998; Klare & Steinhoff, 2009; Klug & Feix, 2007). Especially with respect to protein-protein interactions SDSL EPR can provide a vast amount of information about almost all aspects of this interaction. Spin labeling approaches can provide detailed information about the binding interface not only on the structural level but also give insights into kinetic and thermodynamic aspects of the interaction. EPR also allows determination of distances between pairs of spin labels in the range from ~ 10-80 Å with accuracies down to less than 1 Å, thereby covering a range of sizes including also large multi-domain proteins and protein complexes.

This chapter will give an introduction into the technique of SDSL EPR spectroscopy exemplified with data from studies on the photoreceptor/transducer-complex *Np*SRII/*Np*HtrII, followed by a number of recent examples from the literature where protein-protein interactions have been studied using this technique.

2. Site-Directed Spin Labeling (SDSL)

2.1 Spin labeling of cysteines

For the modification proteins with spin labels, three different approaches have been established. The most commonly used method utilizes the reactivity of the sulfhydryl group of cysteine residues being engineered into the protein applying site-directed mutagenesis. This approach usually requires that the protein of interest possesses only cysteine residues at the desired sites, and that additional cysteine residues present can be replaced by serines or alanines without impairment of protein functionality. Among the various spin labels available the (1-oxyl-2,2,5,5-tetramethylpyrroline-3-methyl) methanethiosulfonate spin label (MTSSL) (Berliner et al., 1982) is most often used due to its sulfhydryl specificity, and its small molecular volume comparable to that of a tryptophane side chain (Fig. 1A,B). The spin label is bound to the protein by formation of a disulfide bond with the cysteine, and the resulting spin label side chain is commonly abbreviated R1. The linker between the nitroxide ring and the protein backbone renders the R1 side chain flexible (Fig. 1B), thus minimizing disturbances of the native fold and the function of the protein it is attached to. In addition, the unique dynamic properties of this spin-label side chain provide detailed structural information from the shape of its room temperature EPR spectrum. Besides MTSSL, a variety of different nitroxide radical compounds are commercially available, for example the 1-oxyl-2,2,5,5-tetramethyl, 2,5-dihydro-1H-pyrrol-3-carboxylic acid (2-methanethiosulphonyl-ethyl) amide (MTS-4-oxyl) spin label (Fig. 1C), comprising different linkers and/or nitroxide moieties. Also pH sensitive spin probes have been used to label the thiol group, e.g. of a synthetic peptide fragment of the laminin B1 chain (Smirnov et al., 2004).

The widely used methanethiosulfonate spin labels suffer from a significant drawback which is the sensitivity of the disulfide bond towards reducing agents like DTT, leading to immediate release of spin label side chains. If reducing conditions are required for sample preparation and/or stability, acetamide or maleimide-functionalized spin label compounds (Steinhoff et al., 1991; Griffith and McConnell, 1966) (Fig. 1D) can be used alternatively. In this case, the spin label is bound via a CS bond, which is not affected by reducing conditions.

Isotopically labeled nitroxide compounds where ^{14}N is exchanged by ^{15}N are important for specialized applications. The corresponding EPR spectra are characterized by a two line spectrum instead of a three line spectrum of the ^{14}N, and the lines of a ^{15}N spectrum are well separated from the ^{14}N lines so that both labels can be used simultaneously in a single experiment (Steinhoff et al., 1991).

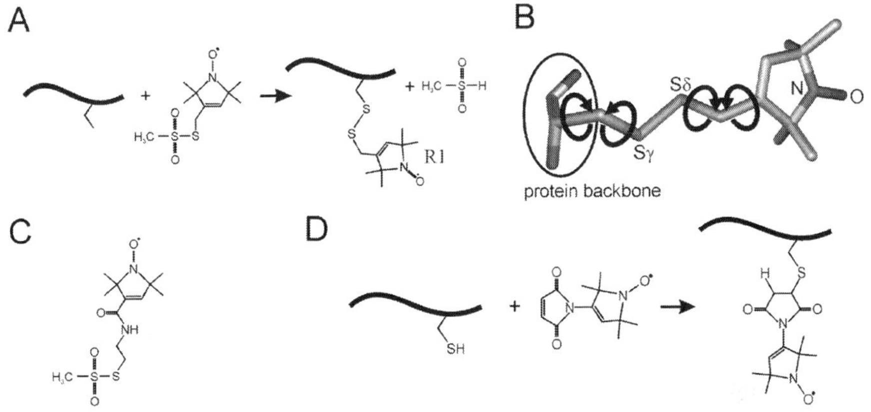

Fig. 1. Spin labeling of cysteines (A) Reaction of the methanethiosulfonate spin label (MTSSL) with a cysteine side chain, generating the spin label side chain R1. (B) Flexible bonds within the R1 side chain are indicated. (C) Chemical structure of the MTS-4-oxyl spin label. (D) Reaction of a maleimide spin label N-(1-oxyl-2,2,6,6-tetramethyl-4-piperidinyl)maleimide with a cysteine side chain.

2.2 Spin labeling by peptide synthesis

A large variety of spin label building blocks for Boc- or Fmoc-based step-by-step peptide synthesis either on a solid support (SPPS) (Merrifield, 1963) or in solution have been synthesized (Barbosa et al., 1999; Elsässer et al., 2005). Being the most popular one, the paramagnetic α-amino acid TOAC (4-amino-1-oxyl-2,2,6,6,-tetramethyl-piperidine-4-carboxylic acid) (Rassat and Rey, 1967) is characterized by only one degree of freedom, the conformation of the six-membered ring (Fig. 2) The nitroxide is rigidly coupled to the peptide backbone, thereby providing the possibility to obtain direct information about the orientation of secondary structure elements, and has for example been used to study the secondary structure of small peptides in liquid solution (Anderson et al., 1999; Hanson et al., 1996; Marsh et al., 2007), and has also been successfully incorporated into the α-melanocyte stimulating hormone without loss of function (Barbosa et al., 1999).

The chemical synthesis of proteins with incorporated unnatural spin labeled amino acids relies on the ability to produce the constituent peptides, typically by SPPS. Although synthesis of polypeptides consisting of more than 160 amino acids (Becker et al., 2003) has become possible through improvements in peptide chemistry, aiming at the incorporation of spin labels into large proteins, esp. membrane proteins, SPPS has to be combined with recombinant techniques. The expressed protein ligation (EPL), also named intein mediated protein ligation (IPL) technique, can be used to semisynthesize proteins from recombinant and synthetic fragments, thereby extending the size and complexity of the protein targets.

The underlying chemical ligation of two polypeptide fragments requires an N-terminal cysteine on one and a C-terminal thioester moiety on the other fragment. After rearrangement through an S→N acyl shift, a native peptide bond is formed. The reaction can be performed also in the presence of other unprotected cysteine residues because of a reversible reaction preceding an irreversible step. Using this methodology, a spin-labeled Ras binding domain has been synthesized, showing a stable paramagnetic center detected by EPR (Becker et al., 2005).

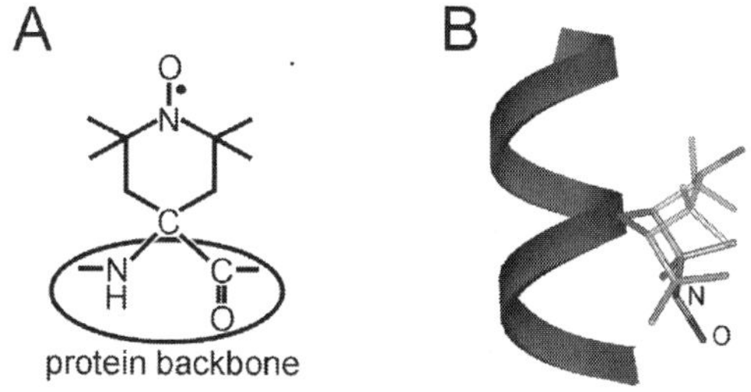

Fig. 2. The TOAC amino acid spin label. (A) chemical structure. (B) three-dimensional structure of the spin label incorporated into an a-helix. The flip of the six-membered ring as the only possible degree of freedom is shown in shaded representation.

2.3 Spin labeling using nonsense suppressor methodology

Spin label amino acids can been introduced into proteins by employing the nonsense suppressor methodology, for example by utilizing the amber suppressor tRNA chemically aminoacylated with the desired spin label amino acid (Cornish et al., 1994). Although this strategy might prove generally applicable in the future using unique transfer RNA(tRNA)/aminoacyl-tRNA-synthetase pairs (Chin et al., 2003), only few laboratories are currently equipped to apply this methodology successfully.

2.4 Spin labeling using nonsense suppressor methodology

As introduced by Kolb, Finn and Sharpless in 2001 (Kolb et al., 2001), the basic concept of "click chemistry" is the highly selective formation of a carbon-heteroatom bond under mild conditions with high yield. Its modular concept renders it a favorable tool for introducing labels into biomolecules. An example is the 1,3-dipolar cycloaddition of organic azides with alkynes in the presence of Cu which has been used to attach fluorescent probes to biomolecules (Deiters and Schultz, 2005). Recently, Tamas and coworkers (Tamas et al., 2009) described the synthesis of nitroxide moieties suitable for click chemistry, thereby opening this approach also for site-directed spin labeling.

3. EPR analysis of spin labeled proteins

In the following, the different experimental techniques of EPR spectroscopy on spin labeled proteins are introduced. Therein, the methods are exemplified with the sensory rhodopsin-transducer complex mediating the photophobic response of the halophilic archaeum

Natronomonas pharaonis. The photophobic response of this organism to green-blue light is mediated by sensory rhodopsin II, *Np*SRII, which is closely related to the light driven proton pump bacteriorhodopsin (for a recent review see (Klare et al., 2007)). *Np*SRII is a seven transmembrane helix (A–G) protein with a retinal chromophore covalently bound via a protonated Schiff base to a conserved lysine residue on helix G. Signal transduction to the intracellular two-component pathway modulating the swimming behavior of the cell takes place via the interaction of *Np*SRII with the tightly bound transducer protein, *Np*HtrII (halobacterial transducer), in a 2:2 complex. A transducer dimer comprising a four-helix transmembrane domain, a linker region consisting of two HAMP domains (Aravind and Ponting, 1999), and a cytoplasmic signaling domain, is flanked by the two SRII receptors.

3.1 Spin label dynamics

The shape of room temperature cw EPR spectra reflects the reorientational motion of the spin label side chain. The influence of spin label dynamics on the spectral shape has been reviewed in detail (Berliner, 1976; Berliner, 1979; Berliner & Reuben, 1989), and the relationship between the dynamics of the spin label side chain and protein structure has been extensively studied for T4 lysozyme (Columbus et al., 2001; Columbus & Hubbell, 2002; Fleissner et al., 2009, 2011; Mchaourab et al., 1996).

In general, the term "mobility" is used to characterize the effects on the EPR spectral features due to the motional rate, amplitude, and anisotropy of the overall reorientational motion of the spin label. Spin labeled sites exposed to the bulk water exhibit weak interaction with the rest of the protein as found for helix surface sites or loop regions and consequently display a high degree of mobility, that is characterized by a small apparent hyperfine splitting and narrow line widths (Fig. 3A & 3B, position 154). If the mobility of the spin label side chain is restricted by interaction with neighboring side chains or backbone atoms of the protein itself or an interaction partner, the line widths and the apparent hyperfine splittings are increased (Fig. 3A & 3B, position 159). Although the relation between the nitroxide dynamics and the EPR spectral line shape is complex, the line width of the centre line, ΔH^0, and the second moment of the spectra, $\langle H^2 \rangle$, have been found to be correlated with the structure of the binding site environment and can therefore be used as mobility parameters (Hubbell et al., 1996; Mchaourab et al., 1996).

The plot of these mobility parameters versus the residue number reveals secondary structure elements through the periodic variation of the mobility as the spin label sequentially samples surface, tertiary, or buried sites. This allows assignment of α-helices, β-strands, or random structures. A more general classification of regions exhibiting buried, surface-exposed, or loop residues can be obtained from the correlation between the inverse of the two mobility parameters, as shown in Figure 3C. Side chains from different topographical regions of a protein can be thereby classified on the basis of the x-ray structures of T4 lysozyme and annexin 12 (Hubbell et al., 1996; Isas et al., 2002; Mchaourab et al., 1996).

For a more quantitative interpretation of the experimental data in terms of dynamic mechanisms and local tertiary interaction, EPR spectra simulations have to be performed. Based on dynamic models developed by Freed and coworkers (Barnes et al., 1999; Borbat et al., 2001; Freed, 1976), excellent agreement of simulations with the corresponding experimental spectra can be obtained. Furthermore, simulations of EPR spectra can be performed on the basis of molecular dynamics (MD) simulations (Beier and Steinhoff, 2006; Budil et al., 2006; Oganesyan, 2007; Sezer et al., 2008; Steinhoff et al., 2000a; Steinhoff and Hubbell, 1996). Thus, a direct link is provided between molecular structure and EPR spectral line shape, thus allowing verification, refinement, or even de-novo prediction (Alexander et al., 2008) of structural models of proteins or protein complexes.

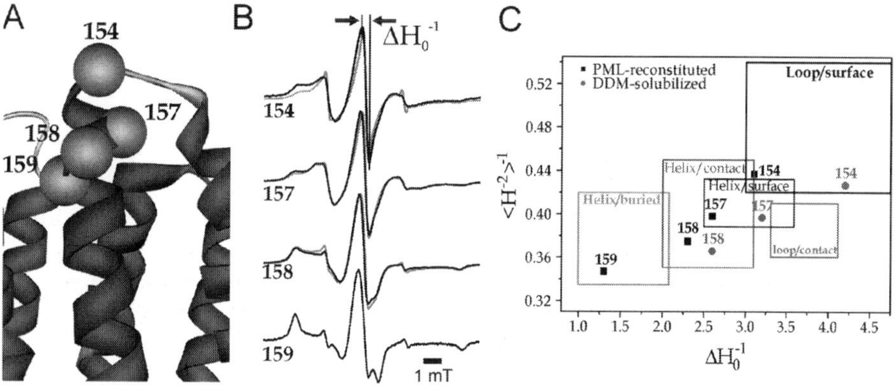

Fig. 3. Mobility analysis of spin labelled proteins. (A) Crystal structure of *Np*SRII (Luecke et al., 2001). The Cα atoms of spin labeled sites are shown as spheres. (B) X band EPR spectra of spin labeled *Np*SRII solubilised in detergent (gray) or reconstituted in purple membrane lipids (black). (C) Two-dimensional mobility plot of the inverse of the second moment *vs.* the inverse of the central linewidth (solubilized: gray circles, reconstituted: black squares), determined from the spectra in B. Boxes indicate the topological regions of proteins according to Isas et al. (2002) and Mchaourab et al. (1996).

The motion of a nitroxide spin label side chain is characterized by three correlation times, the rotational correlation time for the entire protein, the effective correlation time due to the rotational isomerization spin label linker, and the effective correlation time for the segmental motion of the protein backbone. These correlation times can significantly differ in the time scales they occur on. Thus, experimental data for all relevant time scales have to be available to set up an appropriate dynamical model. For this case, correlation times from μs (for the overall protein motion) down to ps (for the rotational isomerization) have to be covered by the experiment. EPR spectra at different microwave frequencies are sensitive to motions on different time scales. EPR at lower frequencies is sensitive to slower motions whereas faster motions are completely averaged out. On the other hand, high-frequency EPR can resolve such fast motions, but slower motions are "frozen" at the high-frequency time scale. Consequently, combining experiments at different microwave frequencies (multifrequency EPR) allows separation of various motional modes in a spin labelled protein according to their different time scales. Most of the work so far has been done using spin-labeled T4 lysozyme as a model system (Liang & Freed, 1999; Liang et al., 2004; Zhang et al., 2010).

Proteins and protein complexes are inherently dynamic structures that can exhibit a number of conformational substates often playing a key role for their function (Cooper, 1973,

Frauenfelder et al., 1988, 1991). A given state of a protein consists of a limited number of such substates with life times in the μs to ms range that can, for example, correspond to "bound" and "unbound" conformations of a protein binding interface. According to the life time of the substates they often can be recognized in room temperature cw spectra of spin labeled proteins if they are characterized by different spin label side chain mobilities due to structural changes in their vicinity. In the past years, Hubbell and co-workers established three experimental techniques to analyze conformational equilibria in proteins and to dissect them from spin label rotameric exchange, namely osmolyte perturbation (Lopez et al., 2009), saturation recovery (Bridges et al., 2010) and high-pressure EPR (McCoy & Hubbell, 2011).

3.2 Spin label solvent accessibilities

Supplementing the motional analysis, the accessibility of the spin label side chain toward paramagnetic probes (exchange reagents), which selectively partition in different environments of the system under investigation, can be used to define the location of spin label with respect to the protein/water/membrane boundaries. The accessibility of the nitroxide spin label side chain is defined by its Heisenberg exchange frequency, W_{ex}, with an exchange reagent diffusing in its environment. Water-soluble metal ion complexes like NiEDDA or chromium oxalate (CrOx) allow to quantify the accessibility from the bulk water phase, whereas molecular oxygen or hydrophobic organic radicals that mainly partition in the hydrophobic part of the lipid bilayer define the accessibility from the lipid phase. The concentration gradients of NiEDDA and molecular oxygen along the membrane normal can be used to characterize the immersion depth of the spin label side chain into the lipid bilayer (Altenbach et al., 1994; Marsh et al., 2006). Two experimental techniques can be used to determine the nitroxide's accessibility toward the paramagnetic probes: Cw power saturation, and saturation recovery.

Most commonly, Heisenberg exchange rates for nitroxide spin label side chains in proteins are measured using cw power saturation. Here, the EPR signal amplitude is monitored as a function of the incident microwave power in the absence and presence of the paramagnetic quencher. From the saturation behaviour of the nitroxide, an accessibility parameter, Π, can be extracted that is proportional to W_{ex} (Altenbach et al., 1989a; Altenbach et al., 2005; Farabakhsh et al., 1992). In Figure 4, the accessibility analysis performed on a 24 amino acid long segment starting at position 78 in the transmembrane region and extending to position 101 in the cytoplasm of the transducer protein *Np*HtrII in complex *Np*SRII is shown as an example for this technique (Bordignon et al., 2005). Figure 4A shows the crystal structure of the transmembrane region of the complex (Gordeliy et al., 2002). Power saturation experiments have been performed with air (21% O_2) and 50 mM CrOx, respectively. The Π values calculated from these experiments are shown in panel B versus residue number. The low Π values for both oxygen and CrOx for residues 78 to 86 indicate their location in a densely packed protein-protein interface. The clear periodicity of 3.6 residues (see inset in panel B) reflects the α-helical structure. For positions 87 to 94 a gradual increase in the Π_{CrOx} and Π_{oxygen} values is observed, indicating that this region is protruding away from the protein-protein interface into the cytoplasm. For positions 92 to 101 the Π_{CrOx} and Π_{oxygen} values observed are typical for water exposed residues. Also here a periodical pattern corresponding to an α-helical structure is observed.

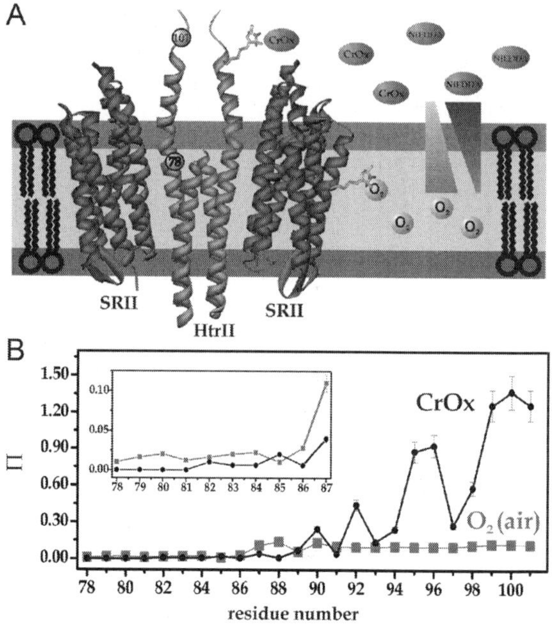

Fig. 4. Spin label accessibilities by power saturation. (A) Structure of the *Np*SRII/*Np*HtrII complex in a lipid bilayer (light gray: hydrophobic region, medium gray: headgroup region). The concentration gradients for water-soluble reagents (CrOx and NiEDDA) and lipid-soluble reagents (O_2) are indicated by shaded triangles. The first (78) and last residue (101) of the region investigated are indicated. (B) Accessibility parameters Π_{CrOx} (black circles) and Π_{Oxygen} (gray squares) vs. residue number. Π_{CroX} values have been obtained with 50 mM CrOx, Π_{Oxygen} values with air (21% O_2). The inset depicts magnified the region from residues 78 to 87 to show the periodicity of 3.6 for Π_{CroX} and Π_{Oxygen}.

Saturation recovery EPR (SR-EPR) allows measuring the spin-lattice relaxation time T_{1e} which is connected to the Heisenberg exchange frequency W_{ex} by

$$W_{ex} = \Delta\left(\frac{1}{T_{1e}}\right)_R \tag{5}$$

directly. In this type of experiment, saturating microwave pulses are applied to the sample in the absence and in the presence of exchange reagents, and the recovery of the *z*-magnetization is monitored as a function of time. Analyses of the recovery curves provide T_{1e} and thus the accessibility for the respective exchange reagent (Altenbach et al., 1989a, 1989b; Nielsen et al., 2004). One major advantage of SR-EPR compared to cw power saturation is that in the presence of multiple spin populations (see chapter 2.1) all corresponding T_{1e} values and accessibilities can be determined by SR-EPR. In contrast, cw power saturation can only provide an average accessibility for all components present in the cw EPR spectrum, and moreover, this average value will be biased towards the most mobile component as it dominates the amplitude of the resonance lines. Moreover, saturation recovery can be used to distinguish between rotamer exchange for the spin label side chain (~ 0.1-1 μs range) and conformational exchange of the protein, which is at least one order of magnitude slower (Bridges et al., 2010).

3.3 Polarity and proticity of the spin label micro-environment

Polarity and proticity of the spin label microenvironment are reflected in the hyperfine component A_{zz} and the g tensor component g_{xx}. A polar environment shifts A_{zz} to higher values, whereas the tensor component g_{xx}, determined from the B-field of the canonical peak position ($g_{xx} = h\nu / \mu_B B$), is decreased. The A_{zz} component can be obtained from cw X-band EPR spectra of spin labeled proteins in frozen samples. The principal g-tensor components and their variation can be determined with high accuracy using high-field EPR techniques due to the enhanced Zeeman resolution (Steinhoff et al., 2000b). In regular secondary structure elements with anisotropic salvation (e.g., surface exposed α-helices), the water density and hence the tensor component values A_{zz} and g_{xx} are a periodic function of residue number. Therefore, similarly to accessibility measurements with water soluble exchange reagents (see 3.2), these data can be used to obtain structural and topological information and the polarity of the spin label environment can reveal detailed information on the protein fold.

The polarity parameter values for the sequence 88 to 94 in the first HAMP domain of *Np*HtrII in complex with *Np*SRII are shown in Figure 5A (Brutlach et al., 2006). It is evident that positions 90 and 93 in *Np*HtrII are located in a more polar, water accessible environment. The same holds for position 154 on the cytoplasmic surface of *Np*SRII. In contrast, positions 88, 89, 91, 92 and 94 reside in a more apolar environment that is less accessible to water. Also evident is a periodical pattern, characterizing the α-helical structure. The exceptional apolar character of position 78 indicates that this side chain is deeply buried in a protein-protein or protein-lipid interface. These results are reflected in a structural model, which has been based in

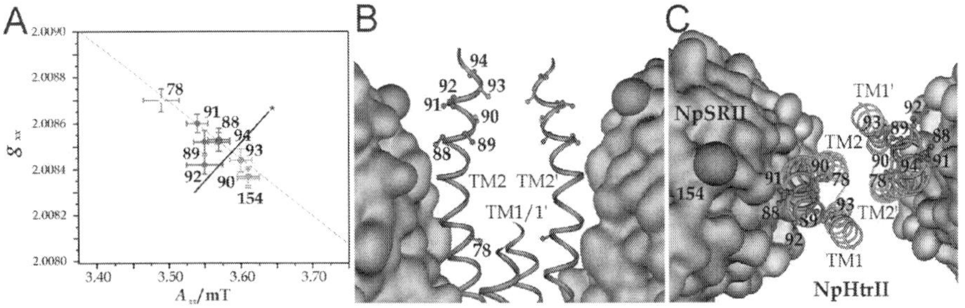

Fig. 5. (A) Plot of g_{xx} versus A_{zz} for positions 88 to 94 in *Np*HtrII according to Brutlach et al., 2006. The plot also includes values for position 78 on the second transmembrane helix (TM2) and for position 154 in *Np*SRII. An arbitrary threshold of g_{xx} / A_{zz} indicated by the diagonal line marked with * classifies the sites into more polar (blue) or more apolar sites (red).
(B) Side view onto *Np*SRII (surface representation) and the four-helix bundle of the transducer (ribbon representation) up to position 96 according to Bordignon et al. (2005).
(C) Cytoplasmic view of the structural model.

addition on mobility, accessibility and distance data (see 2.4) for this region (Figure 5B and C) (Bordignon et al., 2005). Residues 88, 91 and 92 are located in protein-protein interfaces, and positions 90 and 93 are positioned at the opposite side of the transducer helix. Position 78 is, in line with the exceptional low polarity of its micro-environment, buried in the densely packed four helix bundle of the *Np*HtrII dimer.

3.4 Inter-spin label distance measurements

If two spin label side chains are introduced into a biomolecule or two singly labeled molecules are in a stable macromolecular complex, the distance between the two labels can be determined through quantification of their spin–spin interaction, thus providing valuable structural information.

Spin–spin interaction is composed of static dipolar interaction, modulation of the dipolar interaction by the residual motion of the spin-label side chains, and exchange interaction. The static dipolar interaction in an unordered immobilized sample leads to considerable broadening of the cw EPR spectrum if the inter spin distance is less than 2 nm (Figure 6A, C). Distances can be quantified by a detailed line shape analysis of EPR spectra of frozen protein samples or proteins in solutions of high viscosity (Altenbach et al., 2001; Rabenstein and Shin, 1995; Steinhoff et al., 1991; Steinhoff et al., 1997). Pulse EPR techniques expand the accessible distance range up to 8 nm (Borbat and Freed, 1999; Pannier et al., 2000; Martin et al., 1998). Two major protocols have been successfully applied, the 4-pulse DEER or 4-pulse PELDOR (Figure 6D) and the Double Quantum Coherence approaches (for a recent review see (Schiemann and Prisner, 2007)).

The combination of cw and pulse EPR techniques, taking into account borderline effects in the region from 1.6-1.9 nm (Banham et al., 2008; Grote et al., 2008), provide means to determine interspin distances in the range from 1-8 nm, thereby covering the most important distance regime necessary for structural investigations on biomacromolecules. Remarkably, the DEER data can provide, besides the distance information, also information about the orientation of the spin label side chains, their conformational flexibility and the spin density distribution, thereby increasing the amount and quality of data for a setup or verification of structural models.

Based on inter spin distance measurements on 26 different pairs of spin labels introduced into the cytoplasmic regions of NpSRII and *Np*HtrII the arrangement of the transmembrane domains of this complex was modelled (Wegener et al., 2001) (Fig. 7A). Direct comparison of the EPR model with the later determined crystal structure (Gordeliy et al., 2002) (Fig. 7B) shows the consistency of the EPR model with the crystal structure concerning the general topology and the location and relative orientation of the transmembrane helices. Remarkably, also most of the side-chain orientations within the complex coincide quite well in the two models, although for the EPR based model the bacteriorhodopsin structure had to be used as a template for *Np*SRII, since its structure was not known at that time. In a later study it was shown that the structural properties of the HAMP domain as characterized by mobility, solvent accessibility, and intra-transducer-dimer distance data are in agreement with the NMR model of the HAMP domain from *Archaeoglobus fulgidus* (Döbber et al., 2008).

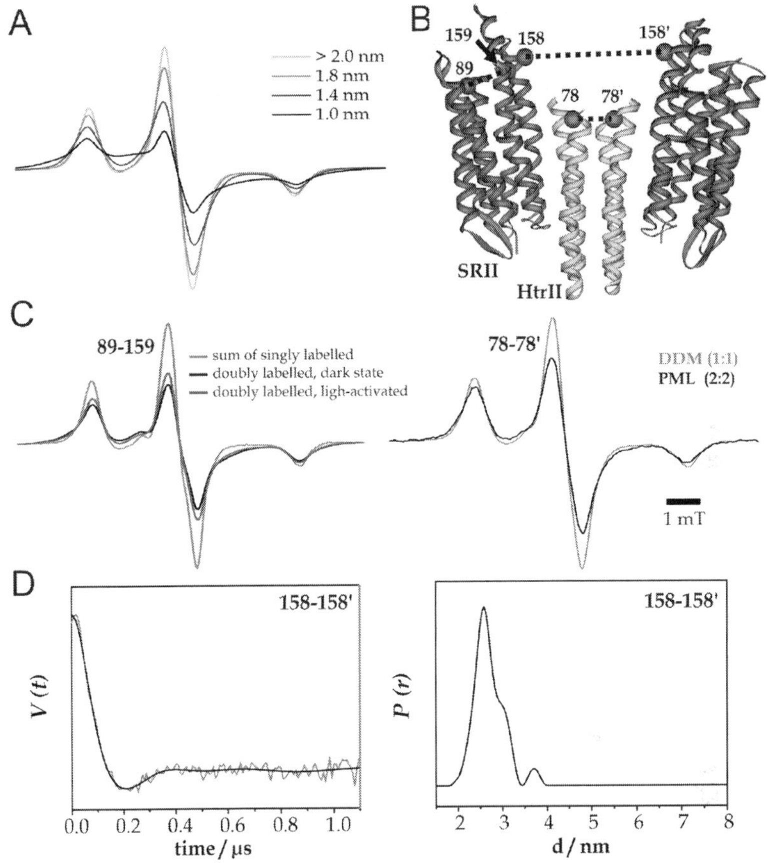

Fig. 6. Interspin distance measurements. (A) Simulated powder spectra (obtained by cw EPR on frozen samples) for different interspin distances. (B) Ribbon representation of the X-ray structure of the 2:2 *Np*SRII–*Np*HtrII complex (PDB 1H2S). Positions L89, S158 and L159 in *Np*SRII, and V78 in *Np*HtrII are shown as spheres. (C) left: cw EPR spectra (T = 160 K) of the double mutant L89R1/L159R1 in the receptor ground state (black) and in the trapped signaling state (M-state, red) compared to the sum of the spectra of the singly labeled samples (gray) reveals line broadening due to dipolar interaction. Interspin distances of 1.1 (±0.2) nm for the ground state and 1.3 (±0.2) nm for the M-trapped have been determined (Bordignon et al., 2007); right: *Np*HtrII–V78R1 solubilized in DDM (gray) or reconstituted in PML (black) in the absence of *Np*SRII. The interspin distance obtained in the reconstituted sample is 1.3 (±0.2) nm (Klare et al., 2006). (D) Left: background corrected DEER time domain for *Np*SRII-S158R1. The distance distribution shown in the right panel shows a mean distance of 2.6 nm between the two spin labels bound to positions 158 in the 2:2 complex that is in good agreement with the distance of 3 nm between the oxygen atoms of the respective serine residues as calculated from the crystal structure.

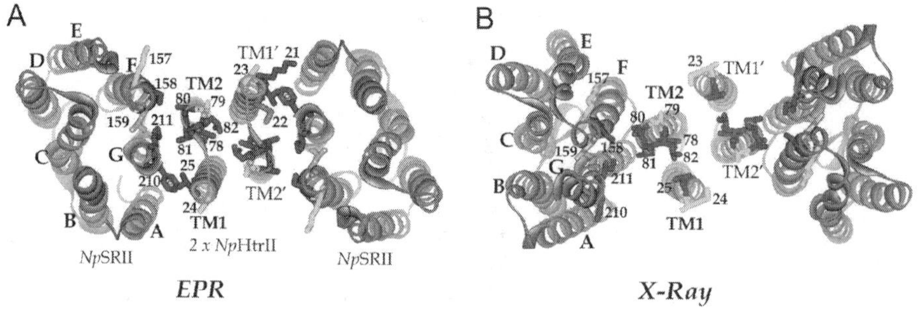

Fig. 7. (A) EPR based model of the transmembrane region of the *Np*SRII/*Np*HtrII complex (viewed from the cytoplasmic side). Side chains where spin labels have been attached are shown in stick representation. The color code represents the strength of the observed dipolar interaction (blue and red: strong; cyan and orange: weak). (B) Crystal structure of *Np*SRII/*Np*HtrII (PDB: 1H2S) (Klare et al., 2004). The color code for the side chains is the same as in (A).

4. SDSL EPR in protein interaction studies

4.1 Structure of the Na^+/H^+ antiporter dimer

An excellent example for the application of DEER spectroscopy to investigate interactions between proteins is the *E. coli* Na^+/H^+ Antiporter NhaA (Hilger et al., 2005, 2007). The protein is responsible for the specific exchange of Na^+ for H^+, and is known to be regulated by pH. From studies on two-dimensional crystals diffracting at 4 Å it had been revealed that NhaA forms a dimer in crystals, and *in vivo* studies as well as cross-linking data suggested that it also works as a dimer *in vivo*. Applying DEER on a NhaA variant labeled at position H225, it could be shown (Hilger et al., 2005) that a pH independent distance of 4.4 nm between spin labeled sites in neighboring molecules can be resolved, and that the degree of dimerization, as judged from the modulation depth of the DEER dipolar evolution data, strongly depends on pH. Thereby, experiments utilizing a singly spin labelled position yielded data strongly supporting the stoichiometry of the functional unit and providing evidences for a mechanistic picture of pH regulation, i.e. that the affinity between the protomers is modulated.

In a later study (Hilger et al., 2007), an extensive distance mapping of the NhaA dimer with nine different spin labeled amino acid positions was carried out. Based on these distance distribution data, explicit modeling of the spin label side chain conformations with a rotamer library approach, and combination with the available X-ray structures, a high-resolution structure of the presumably physiological dimer was determined.

4.2 Structure and function of the tRNA modifying MnmE/GidA complex

The GTP hydrolyzing protein MnmE is, together with the protein GidA, involved in the modification of the wobble position of certain tRNAs (Meyer et al, 2009). It belongs to the expanding class of G proteins activated by nucleotide-dependent dimerization (GADs). The crystal structure shows that MnmE is a multidomain protein with a central helical part in which a Ras-like domain is inserted, and an N-terminal tetrahydrofolate-binding unit. MnmE was predicted to form a dimer in solution, in which the two G domains are separated with a distance of about 50Å between the two P-loops (Fig. 10A). A G domain dimerization had been proposed based on biochemical data and on the crystal structure of the isolated G domains in complex with the GTP hydrolysis transition state mimic GDP·AlF_x.

In a DEER study, distance measurements between spin labels positioned in the MnmE G domain and in the dimerization domain have been carried out (Meyer et al., 2009).The distance distributions for position E287 in the G domains (Fig. 8A) are shown in Figure 8B. In the apo state without any nucleotide bound, the two spin labels exhibit a distances of 55Å for E287R1 and a broad distribution of distances from 25Å to 50Å. GTP (GppNHp) binding induces the formation of additional distances at 27Å (S278R1) / 37Å (E287R1), contributing about 30% to the distance distributions. Upon GTP hydrolysis (GDP · AlF_x state), the distance distribution shows a single population maximum at 28Å (S278R1) / 36Å (E287R1). Thus, an equilibrium between an open conformation with distant G domains and a closed conformation with the G domains being in close proximity, exists when GTP is bound, and is shifted completely towards the closed state upon hydrolysis. A schematic representation of the proposed conformational changes of the MnmE G domains based on the EPR data is depicted in Figure 8C.

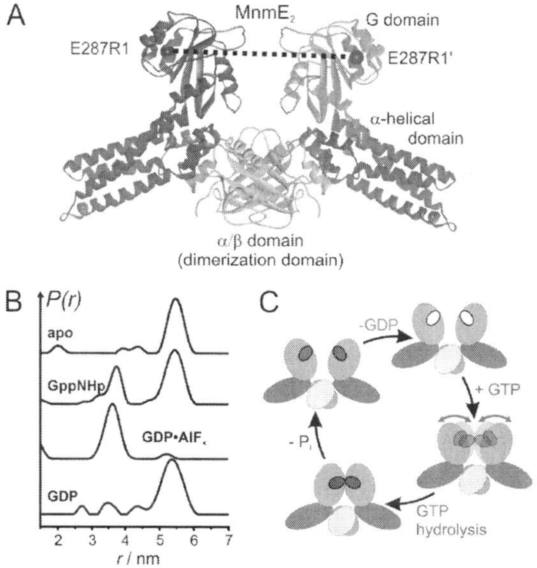

Fig. 8. G domain dimerization of MnmE monitored by DEER. (A) Structural model of the MnmE-dimer. Position 287 in the G domains, which was spin labeled is indicated by black spheres. (B) DEER distance distributions obtained by Tikhonov regularization using the program DeerAnalysis 2008. (C) Schematic representation of the G domain conformational states during the GTPase cycle.

In addition, spin labels attached at position 105 in the dimerization domain showed no significant distance changes during the GTPase cycle, indicating that the initial dimerization interface is largely preserved despite the large G domain movements. In addition, a dependency of the GTPase activity and consequently of the G domain motions on the presence of specific cations could be fully corroborated by the DEER analysis performed in this study. In a subsequent study, the influence of binding of GidA to MnmE in a 2:2 complex was investigated, showing that the interaction of GidA with MnmE partly abolishes the previously observed cation dependency.

4.3 Subunit binding in the chaperone/usher pathway of pilus biogenesis

Type 1 pili are adhesive multisubunit fibres in Gram-negative bacteria. During pilus assembly, subunits dock as chaperone-bound complexes to an usher, which catalyses their polymerization and pilus translocation across the membrane. In the background of the recently determined crystal structure of the full-length FimD usher bound to the FimC–FimH chaperone–adhesin complex, SDSL EPR was used to show that subsequent subunits bind to the usher c-terminal domains after undergoing so-called donor-strand exchange (Phan et al., 2011).

DEER spectroscopy was carried out on spin label pairs introduced into Fim proteins and the results ware compared to calculated distance distributions obtained from alternative models of the complex. The pair residue 74 of FimC and residue 756 of FimD in the complex revealed that the distance distribution obtained experimentally overlaps with that predicted when FimC–FimG (FimG is the adjacent subunit within the pilus) locates at the c-terminal domains. For the FimC-Q74R1/FimD-S774R1 the experimental distance distributions compared with those calculated for the crystal structure of FimD-C-H, assuming that the position of FimC-G, is similar to the previously bound chaperone-subunit complex FimC-H, are in good agreement, further supporting the assumption that FimC-G in the FimD-C-G-H complex locates at the c-terminal domains and is bound to c-terminal domain 2 (Phan et al., 2011).

4.4 Conformation of peptides bound to TAP

The ATP-binding cassette transporter associated with antigen processing (TAP) is involved in the adaptive immune defense against infected or malignantly transformed cells. It translocates proteasomal degradation products, i.e. peptides of 8 to 40 residues, into the lumen of the endoplasmic reticulum for loading onto MHC class I molecules. EPR spectroscopy and simulations based on a rotamer library were used to reveal conformational details about the bound peptides (Herget et al., 2011).

The authors used two different spin label side chains, namely the PROXYL spin label to monitor side-chain dynamics and environmental, and TOAC-labeled peptides (see chapter 2.2) to detect backbone properties. For different locations of the spin label on the peptide, striking differences in affinity, dynamics, and polarity were found. The mobility of the spin labels was found to be strongly restricted at the ends of the peptide. In contrast, the central region was flexible, suggesting a central peptide bulge. Furthermore, DEER spectroscopy was used for the determination of intrapeptide distances in doubly labeled peptides bound to TAP. Comparison with calculated distance distributions based on a rotamer library led the authors to the conclusion that peptides bind to TAP in an extended kinked structure, analogous to those bound to MHC class I proteins (Herget et al., 2011).

5. Acknowledgement

Part of this work was supported by the Deutsche Forschungsgemeinschaft.

6. References

Alexander, N., Al-Mestarihi, A., Bortolus, M., Mchaourab, H.S. & Meiler, J. (2008) De Novo High-Resolution Protein Structure Determination from Sparse Spin-Labeling EPR Data, *Structure*, Vol. 16, No. 2, (February 2008), pp. 181-195, ISSN 0969-2126

Altenbach, C., Flitsch, S.L., Khorana, H.G. & Hubbell, W.L. (1989a). Structural studies on transmembrane proteins, 2: spin labeling of bacteriorhodopsin mutants at unique cysteines. *Biochemistry*, Vol. 28, No. 19, (September 1989), pp. 7806-7812, ISSN 0006-2960

Altenbach, C., Froncisz, W., Hyde, J.S. & Hubbell, W.L. (1989b). Conformation of spin-labeled melittin at membrane surfaces investigated by pulse saturation recovery and continuous wave power saturation electron-paramagnetic resonance. *Biophysical Journal*, Vol. 56, No. 6, (December 1989), pp. 1183-1191, ISSN 1542-0086

Altenbach, C., Marti, T., Khorana, H.G. & Hubbell, W.L. (1990). Transmembrane protein structure: spin labeling of bacteriorhodopsin mutants. *Science*, Vol. 248, No. 4959, (June 1990), pp. 1088-1092, ISSN 1095-9203

Altenbach, C., Greenhalgh, D.A., Khorana, H.G. & Hubbell, W.L. (1994). A collision gradient method to determine the immersion depth of nitroxides in lipid bilayers: application to spin-labeled mutants of bacteriorhodopsin. *Proceedings of the National Academy of Sciences of the USA*, Vol. 91, No. 5, (March 1994), pp. 1667-1671, ISSN 1091-6490

Altenbach, C., Oh, K.J., Trabanino, R.J., Hideg, K. & Hubbell,W.L. (2001). Estimation of Inter-Residue Distances in Spin Labeled Proteins at Physiological Temperatures: Experimental Strategies and Practical Limitations. *Biochemistry*, Vol. 40, No. 51, (December 2011), pp. 15471-15482, ISSN 0006-2960

Altenbach, C., Froncisz, W., Hemker, R., Mchaourab, H.S. & Hubbell, W.L. (2005). Accessibility of nitroxide side chains: absolute Heisenberg exchange rates from power saturation EPR. *Biophysical Journal*, Vol. 89, No. 3, (September 2005), pp. 2103-2112, ISSN 1542-0086

Anderson,D.J., Hanson,P., McNulty,J., Millhauser,G.L., Monaco,V., Formaggio,F. et al. (1999). Solution Structures of TOAC-Labeled Trichogin GA IV Peptides from Allowed (g = 2) and Half-Field Electron Spin Resonance. *Journal of the American Chemical Society*, Vol. 121, No. 29, (July 1999), pp. 6919-6927, ISSN 1520-5126

Aravind, L. & Ponting, C.P. (1999). The cytoplasmic helical linker domain of receptor histidine kinase and methyl-accepting proteins is common to many prokaryotic signalling proteins. *FEMS Microbiology Letters*, Vol. 176, No. 1, (July 1999), pp. 111-116, ISSN 0378-1097

Banham, J.E., Baker, C.M., Ceola, S., Day, I.J., Grant, G.H., Groenen, E.J.J., Rodgers, C.T., Jeschke, G. & Timmel, C.R. (2008). Distance measurements in the borderline region of applicability of CW EPR and DEER: A model study on a homologous series of spin-labelled peptides. *Journal of Magnetic Resonance*, Vol. 191, No. 2, (April 2008), pp. 202-218, ISSN 1090-7807

Barbosa,S.R., Cilli,E.M., Lamy-Freund,M.T., Castrucci,A.M., and Nakaie,C.R. (1999). First synthesis of a fully active spin-labeled peptide hormone. *FEBS Letters*, Vol. 446, No. 1, (March 1999), pp. 45-48, ISSN 0014-5793

Becker, C.F.W., Hunter, C.L., Seidel, R., Kent, .B.H., Goody, R.S. & Engelhard,M. (2003). Total chemical synthesis of a functional interacting protein pair: The protooncogene H-Ras and the Ras-binding domain of its effector c-Raf1. *Proceedings of the National Academy of Sciences of the USA*, Vol. 100, No. 9, (April 2003), pp. 5075-5080, ISSN 1091-6490

Becker, C.F.W., Lausecker, K., Balog, M., Kalai, T., Hideg, K., Steinhoff, H.-J. & Engelhard,M. (2005). Incorporation of spin-labelled amino acids into proteins. Magnetic Resonance in Chemistry, Vol. 43, No. S1, (December 2005), pp. S34-S39, ISSN 1097-458X

Beier, C. & Steinhoff, H.-J. (2006). A structure-based simulation approach for electron paramagnetic resonance spectra using molecular and stochastic dynamics simulations. *Biophysical Journal*, Vol. 91, No. 7, (October 2006), pp. 2647-2664, ISSN 1542-0086

Berliner, L.J., (Ed.). (June 1976). *Spin labeling: theory and applications.* Academic Press, ISBN 978-0120923502, New York

Berliner, L.J., (Ed.). (September 1979). *Spin labeling II: theory and applications.* Academic Press, ISBN 978-0120923526, New York

Berliner,L.J., Grunwald,J., Hankovszky,H.O., and Hideg,K. (1982). A novel reversible thiol-specific spin label: Papain active site labeling and inhibition. *Analytical Biochemistry*, Vol. 119, No. 2, (January 1982), pp. 450-455, ISSN 0003-2697

Berliner, L.J. & Reuben, J., (Eds.). (June 1989). *Spin labeling,* Vol. 8: *Biological magnetic resonance.* Plenum Press, ISBN 978-0306430725, New York

Borbat, P.P. & Freed, J.H. (1999). Multiple-quantum ESR and distance measurements. *Chemical Physics Letters*, Vol. 313, No. 1-2, (November 1999), pp. 145-154, ISSN 0009-2614

Bordignon, E., Klare, J.P., Döbber, M.A., Wegener, A.A., Martell, S., Engelhard, M. & Steinhoff, H.-J. (2005). Structural Analysis of a HAMP domain: The Linker Region of the Phototransducer in Complex with Sensory Rhodopsin II. *Journal of Biological Chemistry*, Vol. 280, No. 46, (November 2005), pp. 38767-38775, ISSN 1083-351X

Bordignon, E. & Steinhoff, H.-J. (February 2007). Membrane protein structure and dynamics studied by site-directed spin labeling ESR., In: *ESR spectroscopy in membrane biophysics*, Hemminga, M.A. & Berliner, L.J. (Eds.), pp. 129-164, Springer Science and Business Media, ISBN 978-0387250663, New York

Bridges, M.D., Hideg, K. & Hubbell, W.L. (2010). Resolving Conformational and Rotameric Exchange in Spin-Labeled Proteins Using Saturation Recovery EPR. *Applied Magnetic Resonance*, Vol. 37, No. 1, (January 2010), pp. 363-390, ISSN 0937-9347

Brutlach, H., Bordignon, E., Urban, L., Klare, J.P., Reyher, H.-J., Engelhard, M. & Steinhoff, H.-J. (2006). High-Field EPR and Site-Directed Spin Labeling Reveal a Periodical Polarity Profile: The Sequence 88 to 94 of the Phototransducer, NpHtrII, in Complex with Sensory Rhodopsin, NpSRII. *Applied Magnetic Resonance*, Vol. 30, No. 3-4, (June 2006), pp. 359-372, ISSN 0937-9347

Budil, D.E., Sale, K.L., Khairy, K. & Fajer, P.G. (2006). Calculating slow-motional electron paramagnetic resonance spectra from molecular dynamics using a diffusion operator approach. *Journal of Physical Chemistry A*, Vol. 110, No. 10, (March 2006), pp. 3703-3713, ISSN 1520-5215

Chin, J.W., Cropp, T.A., Anderson, J.C., Mukherji, M., Zhang, Z. & Schultz,P.G. (2003). An Expanded Eukaryotic Genetic Code. *Science*, Vol. 301, No. 5635, (August 2003), pp. 964-967, ISSN 1095-9203

Columbus, L., Kalai, T., Jekö, J., Hideg, K. & Hubbell, W.L. (2001). Molecular motion of spin labelled side chains in α-helices: analysis by variation of side chain structure. *Biochemistry*, Vol. 40, No. 13, (April 2001), pp. 3828-3846, ISSN 0006-2960

Columbus, L. & Hubbell, W.L. (2002). A new spin on protein dynamics. *Trends in Biochemical Sciences*, Vol. 27, No. 6, (June 2002), pp. 288-295, ISSN 0968-0004

Cooper, A. (1976). Thermodynamic fluctuations in protein molecules. *Proceedings of the National Academy of Sciences of the USA*, Vol. 73, No. 8, (August 1976), pp. 2740-2741, ISSN 1091-6490

Cornish, V.W., Benson, D.R., Altenbach, C., Hideg, K., Hubbell, W.L. & Schultz, P.G. (1994). Site-Specific incorporation of biophysical probes into proteins. *Proceedings of the National Academy of Sciences of the USA*, Vol. 91, No. 8, (April 1994), pp. 2910-2914, ISSN 1091-6490

Deiters, A. & Schultz, P. G. (2005). In vivo incorporation of an alkyne into proteins in Escherichia coli. *Bioorganic & Medicinal Chemistry Letters*, Vol. 15, No. 5, (March 2005), pp. 1521-1524, ISSN 0960-894X

Elsässer,C., Monien,B., Haehnel,W. & Bittl,R. (2005). Orientation of spin labels in de novo peptides. Magnetic Resonance in Chemistry, Vol. 43, No. S1, (December 2005), pp. S26-S33, ISSN 1097-458X

Farahbakhsh, Z.Z., Altenbach, C. & Hubbell, W.L. (1992). Spin labeled cysteines as sensors for protein lipid interaction and conformation in rhodopsin. *Photochemistry and Photobiology*, Vol. 56, No. 6, (December 1992), pp. 1019-1033, ISSN 1751-1097

Fleissner, M.R., Cascio, D. & Hubbell, W.L. (2009). Structural origin of weakly ordered nitroxide motion in spin-labeled proteins. *Protein Science*, Vol. 18, No. 5, (May 2009), pp. 893-908, ISSN 1469-896X

Fleissner, M.R., Bridges, M.D., Brooks, E.K., Cascio, D., Kalai, T., Hideg, K. & Hubbell, W.L. (2011). Structure and dynamics of a conformationally constrained nitroxide side chain and applications in EPR spectroscopy. *Proceedings of the National Academy of Sciences of the USA*, Vol. 108, No. 39, (September 2011), pp. 16241-16246, ISSN 1091-6490

Frauenfelder, H., Parak, F.G. & Young, R.D. (1988). Conformational Substates in Proteins. *Annual Reviews of Biophysics and Biophysical Chemistry*, Vol. 17, No. 1, (June 1988), pp. 451-479, ISSN 0084-6589

Frauenfelder, H., Sligar, S.G. & Wolynes, P.G. (1991). The energy landscapes and motions of proteins. *Science*, Vol. 254, No. 5038, (December 1991), pp. 1598-1603, ISSN 1095-9203

Griffith,O.H., and McConnell,H.M. (1966). A Nitroxide-Maleimide Spin Label. *Proceedings of the National Academy of Sciences of the USA*, Vol. 55, No. 1, (January 1966), pp. 8-11, ISSN 1091-6490

Grote, M., Bordignon, E., Polyhach, Y., Jeschke, G., Steinhoff, H.-J. & Schneider, E. (2008). A Comparative EPR Study of the Nucleotide-binding Domains' Catalytic Cycle in the

Assembled Maltose ABC-Importer. *Biophysical Journal*, Vol. 95, No. 6, (June 2008), pp. 2924-2938, ISSN 1542-0086

Hanson,P., Millhauser,G., Formaggio,F., Crisma,M. & Toniolo, C. (1996) ESR Characterization of Hexameric, Helical Peptides Using Double TOAC Spin Labeling. *Journal of the American Chemical Society*, Vol. 118, No. 32, (August 1996), pp. 7618-7625, ISSN 1520-5126

Herget, M., Baldauf, C., Schoelz, C., Parcej, D., Wiesmueller, K.-H., Tampe, R., Abele, R. & Bordignon, E. (2011). Conformation of peptides bound to the transporter associated with antigen processing (TAP). *Proceedings of the National Academy of Sciences of the USA* , Vol. 108, No. 4, (January 2011), pp. 1349-1354, ISSN 1091-6490

Hilger, D., Jung, H., Padan, E., Wegener, C., Vogel, K.P., Steinhoff, H.-J. & Jeschke, G. (2005). Assessing Oligomerization of Membrane Proteins by Four-Pulse DEER: pH-Dependent Dimerization of NhaA Na+/H+ Antiporter of E. coli, *Biophysical Journal*, Vol. 89, No. 2, (August 2005), pp. 1328-1338, ISSN 1542-0086

Hilger, D., Polyhach, Y., Padan, E., Jung, H. & Jeschke, G. (2007). High-Resolution Structure of a Na+/H+ Antiporter Dimer Obtained by Pulsed Electron Paramagnetic Resonance Distance Measurements, *Biophysical Journal*, Vol. 93, No. 10, (November 2007), pp. 3675-3683, ISSN 1542-0086

Hubbell, W.L., Mchaourab, H.S., Altenbach, C., & Lietzow, M.A. (1996). Watching proteins move using site-directed spin labeling. *Structure*, Vol. 4, No. 7, (July 1996), pp. 779-783, ISSN 0969-2126

Hubbell, W.L., Gross, A., Langen, R. & Lietzow, M.A. (1998). Recent advances in site-directed spin labeling of proteins. *Current Opinion in Structural Biology*, Vol. 8, No. 5, (October 1998), pp. 649-656, ISSN 0959-440X

Isas, J.M., Langen, R., Haigler, H.T. & Hubbell, W.L. (2002). Structure and dynamics of a helical hairpin and loop region in annexin 12: a site-directed spin labeling study. *Biochemistry*, Vol. 41, No. 5, (February 2002), pp. 1464-1473, ISSN 0006-2960

Kalai, T., Hubbell, W.L. & Hideg, K. (2009). Click Reactions with Nitroxides, *Synthesis*, Vol. 2009, No. 8, (April 2009), pp. 1336-1340, ISSN 1437-210X

Klare, J.P., Bordignon, E., Engelhard, M. & Steinhoff, H.-J. (2004). Sensory rhodopsin II and bacteriorhodopsin, light activated helix F movement. *Photochemical and Photobiolological Sciences*, Vol. 3, No. 6, (June 2004), pp. 543–547, ISSN 1474-9092

Klare, J.P., Bordignon, E., Döbber, M.A., Fitter, J., Kriegsmann, J., Chizhov, I. et al. (2006). Effects of solubilization on the structure and function of the sensory rhodopsin II/transducer complex. *Journal of Molecular Biology*, Vol. 356, No. 5, (March 2006), pp. 1207–1221, ISSN 0022-2836

Klare, J.P., Chizhov, I. & Engelhard,M. (2007). Microbial Rhodopsins: Scaffolds for Ion Pumps, Channels, and Sensors. Results and Problems in Cell Differentiation, Vol. 45, pp. 73-122, ISSN 1861-0412

Klare, J.P. & Steinhoff, H.-J. (2009). Spin labeling EPR. *Photosynthesis Research*, Vol. 102, No. 2-3, (December 2009), pp. 377-390, ISSN 0166-8595

Klug, C.S. & Feix, J.B. (November 2007). Methods and applications of site-directed spin labeling EPR spectroscopy., In: *Methods in cell biology. Biophysical tools for biologists, volume one: in vitro techniques*, Correia, J.J. & Detrich, H.W. (Eds.), pp. 617-658, Academic Press, ISBN 978-0123725202, New York

Kolb, H., Finn, M. G. & Sharpless, K. B. (2001). Click Chemistry: Diverse Chemical Function from a Few Good Reactions, *Angewandte Chemie International Edition*, Vol. 40, No. 11, (June 2001), pp. 2004-2021, ISSN 1521-3773

Liang, Z. & Freed, J.H. (1999). An assessment of the Applicability of Multifrequency ESR to Study the Complex Dynamics of Biomolecules. *Journal of Physical Chemistry B*, Vol. 103, No. 30, (July 1999), pp. 6384-6396, ISSN 1520-6106

Liang, Z., Lou, Y., Freed, J.H., Columbus, L. & Hubbell, W.L. (2004). A Multifrequency Electron Spin Resonance Study of T4 Lysozyme Dynamics Using the Slowly Relaxing Local Structure Model. *Journal of Physical Chemistry B*, Vol. 108, No. 45, (November 2004), pp. 17649-17659, ISSN 1520-6106

Lopez, C.J., Fleissner, M.R., Guo, Z., Kusnetzow, A.N. & Hubbell, W.L. (2009). Osmolyte perturbation reveals conformational equilibria in spin-labeled proteins. *Protein Science*, Vol. 18, No. 8, (June 2009), pp. 1637-1652, ISSN 1469-896X

Luecke, H., Schobert, B., Lanyi, J.K., Spudich, E.N. & Spudich, J.L. (2001). Crystal structure of sensory rhodopsin II at 2.4 Å: insights into color tuning and transducer interaction. *Science*, Vol. 293, No. 5534, (August 2001), pp. 1499-1503, ISSN 1095-9203

Marsh, D., Dzikovski, B.G. & Livshits, V.A. (2006). Oxygen profiles in membranes. *Biophysical Journal*, Vol. 90, No. 7, (April 2006), pp. L49-L51, ISSN 1542-0086

Marsh, D., Jost, M., Peggion, & Toniolo, C. (2007). TOAC Spin Labels in the Backbone of Alamethicin: EPR Studies in Lipid Membranes. Biophysical Journal, Vol. 92, No. 2, (January 2007), pp. 473-481, ISSN 1542-0086

Martin, R. E., Pannier, M., Diederich, F., Gramlich, V., Hubrich, M. & Spiess, H. W. (1998). Determination of End-to-End Distances in a Series of TEMPO Diradicals of up to 2.8 nm Length with a New Four-Pulse Double Electron Electron Resonance Experiment, *Angewandte Chemie International Edition*, Vol. 37, No. 20, (December 1998), pp. 2833-2837, ISSN 1521-3773

Mchaourab, H.S., Lietzow, M.A., Hideg, K. & Hubbell, W.L. (1996). Motion of spin-labeled side chains in T4 lysozyme. Correlation with protein structure and dynamics. *Biochemistry*, Vol. 35, No. 24, (June 1996), pp. 7692-7704, ISSN 0006-2960

McCoy, J. & Hubbell, W.L. (2011). High-pressure EPR reveals conformational equilibria and volumetric properties of spin-labeled proteins. *Proceedings of the National Academy of Sciences of the USA*, Vol. 108, No. 4, (January 2011), pp. 1331-1336, ISSN 1091-6490

Merrifield, B. (1963). Solid Phase Peptide Synthesis. I. The Synthesis of a Tetrapeptide. *Journal of the American Chemical Society*, Vol. 85, No. 14, (July 1963), pp. 2149-2154, ISSN 1520-5126

Nielsen, R.D., Canaan, S., Gladden, J.A., Gelb, M.H., Mailer, C. & Robinson, B.H. (2004). Comparing continuous wave progressive power saturation EPR and time domain saturation recovery EPR over the entire motional range of nitroxides psin labels. *Journal of Magnetic Resonance*, Vol. 169, No. 1, (July 2004), pp. 129-163, ISSN 1090-7807

Oganesyan, V.S. (2007). A novel approach to the simulation of nitroxide spin label EPR sdpectra from a single truncated dynamical trajectory. *Journal of Magnetic Resonance*, Vol. 188, No. 2, (October 2007), pp. 196-205, ISSN 1090-7807

Pannier, M., Veit, S., Godt, A., Jeschke, G. & Spiess, H.W. (2000). Dead-Time Free Measurement of Dipole-Dipole Interactions between Electron Spins. *Journal of Magnetic Resonance*, Vol. 142, No. 2, (February 2000), pp. 331-340, ISSN 1090-7807

Phan, G., Remaut, H., Wang, T., Allen, W.J., Pirker, K.F. et al. (2011). Crystal structure of the FiumD usher bound to its cognate FimC-FimH substrate. *Nature*, Vol. 474, No. 7349, (June 2011), pp. 49-53, ISSN 0028-0836

Rabenstein, M.D. & Shin, Y.K. (1995). Determination of the distance between 2 spin labels attached to a macromolecule. *Proceedings of the National Academy of Sciences of the USA*, Vol. 92, No. 18, (August 1995), pp. 8239–8243, ISSN 1091-6490

Rassat,A. & Rey, P. (1967). Nitroxides, 23: preparation of amino-acid free radicals and their complex salts. *Bulletin De La Societe Chimique De France*, Vol. 3, (March 1967), pp. 815-818, ISSN 0037-8968

Schiemann, O. & Prisner, T.F. (2007). Long-range distance determinations in biomacromolecules by EPR spectroscopy. *Quarterly Reviews of Biophysics*, Vol. 40, No. 1, (June 2007), pp. 1-53, ISSN 1469-8994

Sezer, D., Freed, J.H. & Roux, B. (2008). Simulating electron spin resonance spectra of nitroxide spin labels from molecular dynamics and stochastic trajectories. *Journal of Chemical Physics*, Vol. 128, No. 16, (April 2008), pp. 165106-165116, ISSN 1089-7690

Smirnov,A.F., Ruuge,A., Reznikov,V.A., Voinov,M.A., and Grigor'ev,I.A. (2004). Site-Directed Electrostatic Measurements with a Thiol-Specific pH-Sensitive Nitroxide: Differentiating Local pK and Polarity Effects by High-Field EPR. *Journal of the American Chemical Society*, Vol. 126, No. 29, (July 2004), pp. 8872-8873, ISSN 1520-5126

Steinhoff, H.-J., Dombrowsky, O., Karim, C., & Schneiderhahn, C. (1991). Two dimensional diffusion of small molecules on protein surfaces: an EPR study of the restricted translational diffusion of protein-bound spin labels. *European Biophysics Journal*, Vol. 20, No. 5, (December 1991), pp. 293-303, ISSN 0175-7571

Steinhoff, H.-J. & Hubbell, W.L. (1996). Calculation of electron paramagnetic resonance spectra from Brownian dynamics trajectories: application to nitroxide side chains in proteins. *Biophysical Journal*, Vol. 71, No. 4, (October 1996), pp. 2201-2212, ISSN 1542-0086

Steinhoff, H.-J., Radzwill, N., Thevis, W., Lenz, V., Brandenburg, D., Antson, A., Dodson, G.G. & Wollmer, A. (1997). Determination of interspin distances between spin labels attached to insulin: comparison of electron paramagnetic resonance data with the X- ray structure. *Biophysical Journal*, Vol. 73, No. 6, (December 1997), pp. 3287-3298, ISSN 1542-0086

Steinhoff, H.-J., Müller, M., Beier, C. & Pfeiffer, M. (2000a). Molecular dynamics simulation and EPR spectroscopy of nitroxide side chains in bacteriorhodopsin. *Journal of Molecular Liquids*, Vol. 84, No. 1, (January 2000), pp. 17-27, ISSN 0167-7322

Steinhoff, H.J., Savitsky, A., Wegener, C., Pfeiffer, M., Plato, M. & Möbius, K. (2000b) High-field EPR studies of the structure and conformational changes of site-directed spin labeled bacteriorhodopsin. *Biochimica et Biophysica Acta*, Vol. 1457, No. 3, (April 2000), pp.253–262, ISSN 0005-2728

Wegener, A.A., Klare, J.P., Engelhard, M. & Steinhoff, H.-J. (2001). Structural insights into the early steps of receptor-transducer signal transfer in archaeal phototaxis. *EMBO Journal*, Vol. 20, No. 19, (October 2001), pp. 5312-5319, ISSN 1460-2075

Zhang, Z., Fleissner, M.R., Tipikin, D.S., Liang, Z., Moscicki, J.K., Earle, K.A., Hubbell, W.L. & Freed, J.H. (2010). Multifrequency Electron Spin Resonance Study of the Dynamics of Spin Labeled T4 Lysozyme. *Journal of Physical Chemistry B*, Vol. 114, No. 16, (April 2010), pp. 5503-5521, ISSN 1520-6106

22

Modification, Development, Application and Prospects of Tandem Affinity Purification Method

Xiaoli Xu[1,2,*], Xueyong Li[1,*],
Hua Zhang[2] and Lizhe An[2]
[1]*Department of Burn and Plastic Surgery, Tangdu Hospital,*
Fourth Military Medical University, Xi'an,
[2]*Key Laboratory of Arid and Grassland Agroecology of Ministry of Education,*
School of Life Sciences, Lanzhou University, Lanzhou,
[1,2]*China*

1. Introduction

Since the completion of genome sequences of several organisms, attention has been focused on the analysis of the function and functional network of proteins. Most cell type-specific functions and phenotypes are mediated and regulated by the activities of multiprotein complexes as well as other types of protein–protein interactions and posttranslational modifications. Accordingly, the formation and function of macromolecular protein complexes support the whole of cell processes. Consequently, analysis of the variations of protein complex composition in different cell and tissue types is essential to understand the relationship between gene products and cellular functions in diverse physiological contexts (Alberts, 1998; Cusick et al., 2005).

With the development of research strategies, many large-scaleprotein-protein interaction studies have been performed in model organisms, especially the budding yeast Saccharomyces cervisiae. Genome-wide yeast two-hybrid screens (Fromont-Racine et al., 1997; Ito et al., 2001; Uetz et al., 2000) and protein chip-based methods (Zhu et al., 2001) allow broader insight into the interaction networks and afford the possibility of high-throughput analysis of function and functional network of proteins. While the former approach provides information relating to interactions between two proteins, typically of binary nature, and has the potential for false-positive and false-negative results, the latter approach is time consuming and labor intensive. These defects may limit their application in large scale protein complex purification.

A novel protein complex purification strategy, named tandem affinity purification (TAP) (Puig et al., 2001; Rigaut et al., 1999), in cooperation with mass spectrometry allows identification of interaction partners and purification of protein complexes. This strategy was originally developed in yeast and has been tested in many cells and organisms.

---

[*] These authors contributed equally to this work

2. TAP method: A brief overview

The basic principle of TAP is similar to the epitope tagging strategy but different on the utilization of two sequential tags instead of one. Rigaut et al. (Rigaut et al., 1999) compared several tag combinations aiming at high recovery rates without hampering protein functions and developed the standard TAP tag. The TAP method requires fusing a TAP tag to the target protein. The TAP tag consist of two IgG-binding domains of protein A of *Staphylococcus aureus* (ProtA) and a calmodulin-binding domain (CBP), separated by a cleavage site for the tobacco etch virus (TEV) protease (Rigaut et al., 1999). In addition to the C-terminal TAP tag, an N- terminal TAP tag (Puig et al., 2001), which is a reverse orientation of the C-terminal TAP tag, was also generated (Fig. 1A).

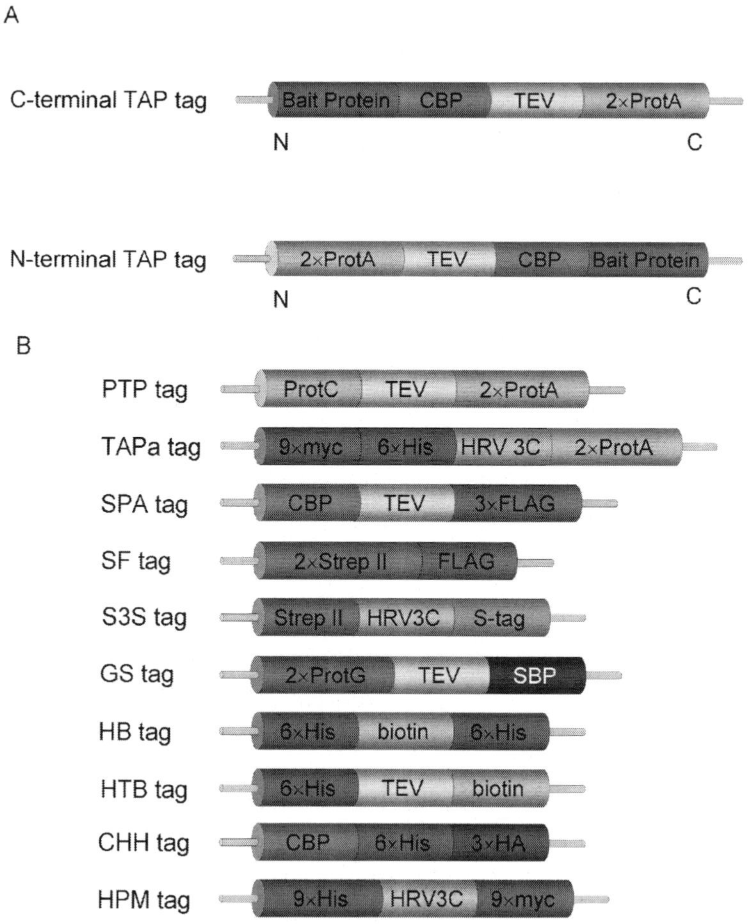

Fig. 1. Diagrammatic sketch of the TAP tag. (A) The original C- and N-terminal TAP tag. (B) Variation of TAP tags developed over the past few years.

The TAP method requires the fusion of the TAP tag to proteins of interest, either at the C- or N-terminus, and the transformation of the construct into appropriate host organisms. The TAP-tagged protein is expressed in host cells at close to physiological concentrations to form a complex with endogenous components. Extracts prepared from cells expressing TAP-tagged proteins are subjected to two sequential purification steps (Fig. 2).

It is well known that the TAP system is very useful for the identification of relatively stable protein complexes, and have helped in the discovery of novel interactions. The TAP method has many advantages: first, the TAP system allows rapid purification of protein complexes without the knowledge of their function or structure. Second, the TAP method enables protein complex purification under native conditions. Third, the tandem purification steps provide highly specific and reduce the high background caused by contaminants substantially. Finally, all protein complex purification can be processed under the same conditions, thus the results are reproducible and comparable, which is significant in large-scale systematic proteome researches. Due to these advantages, the TAP method has been successfully applied in the research of protein-protein interactions in prokaryotic and eukaryotic cells.

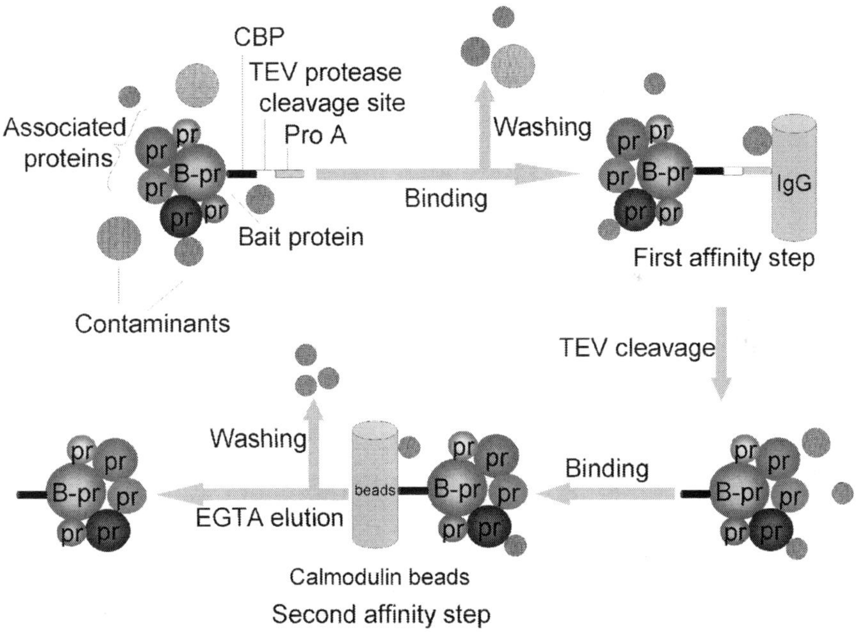

Fig. 2. Schematic of the original TAP method. In the first step, the protein complex, which contains the tagged target protein, combines with an IgG matrix by the ProtA fraction. The protein complex is then eluted using TEV protease under native conditions. In the second step, the elution fraction of the first purification step is incubated with beads coated by calmodulin in the presence of calcium. Subsequently, contaminants and the remainder of TEV protease used in the first step are eliminated through washing. Ultimately, the target protein complex is obtained by elution using EGTA. Adapted from (Xu et al., 2010).

3. The development of TAP tags

Although the TAP system was originally developed in yeast, it has been proven to successfully work in a broad range of organisms. The classic ProtA-TEV-CBP tag may be inefficient to purify all given protein complexes. Therefore several variations of the TAP tag based on other affinity tags have been developed that offer advantages in specific cases (Fig. 1B). The properties of these basic affinity tags (Li, 2010; Lichty et al., 2005; Stevens, 2000; Terpe, 2003) are summarized (Xu et al., 2010) to highlight the advantages and disadvantages of corresponding recombinant tags.

The CBP tag could not always recover protein complexes with high efficient, especially where EGTA may irreversibly interfere with the metal-binding protein function. Consequently, one major type of variation is the replacement of the CBP tag. For example, when purifying protein complexes from mammalian cells growing in monolayer cultures, a biotinylation tag is used as the second affinity tag, taking advantage of the high biotin-avidin binding affinity and resulting in an increased yield of the fusion protein (Drakas et al., 2005). Another example is that the CBP tag has been replaced with a protein C epitope (ProtC) resulting in a new TAP tag, designated PTP (Mani et al., 2011; Schimanski et al., 2005). The advantage of this is that ProtC shows more efficiency and allows the elution either by EGTA or by the ProtC peptide. With respect to isolation of active metal-binding proteins, another replacement of the CBP moiety has been a 9×myc with a 6×His sequence (Rubio et al., 2005). This tag is known as TAPa tag and also contains a human rhinovirus 3C protease cleavage site (HRV 3C) instead of the original TEV site. In contrast to TEV protease, 3C protease still has enzymatic activity at 4℃. These modifications are thought to be beneficial to keep the stabilization of protein complex structures and activities.

Another type of variation is a series TAP tags with smaller size. The original TAP tag is as large as approximately 21 kDa, and this size might impair the function of the tagged protein or interfere with protein complex formation. Because of this, many affinity tags, which range in size from 5-51 amino acids, can be used to replace of CBP or ProtA moiety (Terpe, 2003). One example of a smaller TAP tag is SPA tag, made by substituting 3×FLAG for ProtA (Zeghouf et al., 2004). Replacement of the CBP with a spacer and a single FLAG sequence constitute another smaller tag for TAP (Knuesel et al., 2003). The combination of a streptavidin-binding peptide (SBP) and a CBP has been verified in human cells (Ahmed et al., 2010; Colpitts et al., 2011). Recently, use of another alternative tandem affinity tag, composed of two Strep-tag II and a FLAG-tag (SF), has been published (Gloeckner et al., 2007). This SF tag reduced the size of the TAP tag to 4.6 kDa. This smaller size is less likely to disturb protein activity and structure. Because both tags can be eluted under native conditions, the SF-TAP strategy allows purification of protein complexes in less than 2.5 h. Another similar tandem combination of FLAG-Strep tag II has been developed to purify protein complexes efficiently from *Thiocapsa roseopersicina* (Fodor et al., 2004). And a tandem SBP-FLAG tag has been used to uncover the interacting proteins from HEK293 cells (Zhao et al., 2011). These FLAG-containing combination tags may take advantage of shorter length of the tag and result in higher purity of fusion proteins, while the disadvantage of FLAG tag is the relatively high cost during purification. Lehmann et al. (Lehmann et al., 2009) developed a novel S3S tag comprising a S-tag, a HRV 3C and a Strep-tag II. The S3S tag with a size of 4.2 kDa fulfils the requirements of specificity, high yield and no adverse effects on protein function. Nevertheless, it is doubtful as to whether large tags actually disturb the function of tagged proteins. It would appear that the majority of proteins tagged with the original protA-TEV-CBP tag remain functional, and even small proteins such as acyl-carrier protein (< 10 kDa) (Gully et al., 2003) and thioredoxin (~12 kDa) (Kumar et al., 2004) can be used as bait to purify protein complexes.

In addition to those described above, there are a variety of other TAP tags that are largely different from the classic TAP tag. Bürckstümmer et al. (Bürckstümmer et al., 2006) designed a new TAP tag, designated as GS tag. This tag comprised two copies of IgG binding units of protein G from *Streptococcus* sp. (ProtG) and a SBP. The GS tag was able to purify recombinant proteins with high efficiency and purity, however, the size of the GS tag, at approximately 19 kDa, might be the obvious disadvantage. In a recently published paper, a new tandem affinity tag, the HB tag (Guerrero et al., 2006), consisting of two 6×His motifs and a biotinylation signal peptide has been developed. The HB tag is compatible with *in vivo* cross-linking to purify protein complexes under fully denaturing conditions, which may be beneficial to detect transient and weak protein-protein interactions. A useful derivative of the HB tag is the HTB tag, which includes a TEV cleavage site allowing for protease-driven elution from streptavidin resins (Tagwerker et al., 2006). A CHH tag consisting of a CBP, 6×His residues and three copies of the hemagglutinin (3×HA) has been designed (Honey et al., 2001). However, in fact, the 3xHA peptide is usually used to detect the expression levels of tagged proteins rather than act as the third purification step. In practice, the elution buffer for the calmodulin resin is incompatible with binding to the Ni^{2+} resin. Although buffer exchange may solve this problem, it results in a significant loss of yield. For this reason, the combination of CBP and His tags is generally not recommended. As for purification of associated proteins from *Drosophila* tissues, the 3xFLAG-6xHis tag provided significant higher yields than the tranditional tag (Yang et al., 2006). At the same time, a similar combination of His and FLAG epitope was constructed to isolate protein complexes from pathogenic fungus (Kaneko et al., 2004). The HPM tag, another bipartite affinity tag, consisting of 9×His, 9×myc epitope and two copies of HRV 3C inserted was successfully applied in yeast (Graumann et al., 2004).

4. Application of the TAP method

With the development of the TAP approach over the past decade, this method has been employed in the analysis of protein-protein interactions and protein complexes in many different organisms, including yeast, mammals, plants, *Drosophila* and bacteria (Table 1) (Chang, 2006; Xu et al., 2010).

4.1 TAP in yeast

The TAP method was originally developed for analysis of protein complexes in yeast at near-physiological conditions. Gavin et al. (Gavin et al., 2006; Gavin et al., 2002) and Krogan er al. (Krogan et al., 2006) utilized TAP in the large-scale analysis of multi-protein complexes in *Saccharomyces cerevisiae,* in which hundreds to thousands tagged proteins were successfully purified and the associated proteins and involved protein-protein interactions were identified. These results give the possibility to intensive study the functional and organizational network of proteins in yeast.

As to a given protein, the TAP system could also provide opportunity to investigate protein interaction (Graumann et al., 2004; Guerrero et al., 2006; Honey et al., 2001; Krogan et al., 2002). For example, TAP analysis revealed more than one hundred previously known and possible interacting proteins for 21 tagged proteins, which are involved in transcription and progression during mitosis (Graumann et al., 2004). In addition, an active Clb2-Cdc28 kinase complex was purified from yeast cell lysate by TAP (Honey et al., 2001), and four proteins were identified by mass spectrometry to be associated with this complex.

The application of TAP protocol was successful not only in *S. cerevisiae,* but also in *Schizosaccharomyces pombe* and *Candida albicans.* A large number of researches have carry out the TAP strategy to isolate protein complexes and associated partners (Cipak et al., 2009; Gould et al., 2004; Kaneko et al., 2004; Tasto et al., 2001).

Organism	Protein complex	Bait protein	Tandem tag	Proteins identified	Functional pathway
Yeast	RNA polymerase II elongation factors	15 proteins	Protein A, CBP	25	Transcriptional elongation
	Cyclin-dependent kinase complex	Cdc2p	Protein A, CBP	3	Orchestrating cell cycle
	Arp2/3 complex	Arp2p	Protein A, CBP	6	Nucleation
	Swi 5 contaning complex	Swi 5	Protein A, CBP	4	Mating-type switching and homologous recombination regulation
	Spetin complex	CaCDC11	His, FLAG	4	Cytokinesis, virulence
Human cells	DNA-dependent protein kinase	Ku70, Ku80	Protein G, SBP	Many	Multiple
	Human cytomegalovirus protein kinase	UL97	Protein A, Protein C	1	Human cytomegalovirus replication
	ADAP	ADAP	S tag, Strep II	Many	Integrin adhesion regulation
	Many	Spartin	Protein A, CBP	94	Multiple
	Parkin-associated complex	Parkin	streptavidin-binding peptide, CBP	14	Regulation mitochondrial activity
	Many	32 proteins	Protein A, CBP	Many	mRNA formation
	MCM complex	MCM-BP	Protein A, CBP	5	Initiation of DNA replication
	IRF-4 containing complex	IRF-4	Protein A, CBP	1	Activiates IL-2 and IL-4 promoters
	Flavivirus capsid containing complex	Flavivirus capsid	SBP, CBP	4	Disruption of nucleosome formation

Organism	Protein complex	Bait protein	Tandem tag	Proteins identified	Functional pathway
	HDGF	HDGF	SBP, FLAG	106	Multiple
	HP1 containing complex	HP1α, β, γ	Protein A, CBP	Many	Multiple
	Gβγ-Rap1-Radil complex	$G\beta_2$, $G\gamma_2$	Strep, CBP	2	Cell-matrix adhesion
Nicotiana benthamiana	Resistance protein N containing complex	Resistance protein N	ProteinA, Myc, His	1	Plant defense
Arabidopsis	Many	6 proteins	Protein A, CBP	42	Cell division signaling
Rice	Many	45 proteins	Protein A, CBP	Many	Multiple
Drosophila	Notch signaling complex	4 proteins	Protein A, CBP	400	Signaling pathway
Escherichia coli	Thioredoxin associated complex	Thioredoxin	Protein A, CBP	80	Multiple
	RecQ complex	RecQ	Protein A, CBP	3	DNA unwinding
	RNP	SrmB	Protein A, CBP	2	Ribosome assembly
Dictyostelium	Microtubule-associated protein	DdEB1, DdCP224	Protein A, CBP	Many	Multiple
	Arp2/3 complex	pArc-34	Protein A, CBP	6	Nucleation
Trypanosome brucei	RNA polymerase I	RPB6z	Protein A, Protein C	3	Transcription
	Alba containing complex	Alba1, 2, 3, 4	Protein A, Protein C	Many	Regulation of translation
Plasmodium falciparum	Translation elongation factor	PfEF-1β	Protein A, Protein C	4	Translation elongation

Table 1. Representative applications of the TAP method.

4.2 TAP in mammalian systems

The application of TAP method has made considerable progress in mammalian systems (Bürckstümmer et al., 2006; Davison et al., 2009; Drakas et al., 2005; Gottlieb and Jackson, 1993; Holowaty et al., 2003; Jeronimo et al., 2007; Kamil and Coen, 2007; Knuesel et al., 2003; Lehmann et al., 2009; Milewska et al., 2009; Sakwe et al., 2007). For instance, human active SMAD3 protein complex was purified from cell lysates through TAP method, and HSP70 was identified as a novel combination partner of SMAD3 (Knuesel et al., 2003). However, in this research the TAP system took a FLAG tag for the second purification step, the elution conditions of which was incompatible with the liquid chromatography-MS/MS sequence application. Therefore, an additional purification step might be reqired to resolve this trouble, which would be time consuming and lead to more sample loss. Sakwe et al. identified a new form of the minichromosome maintenance (MCM) complex in human cells (Sakwe et al., 2007). In another study using the TAP process, Holowaty et al. (Holowaty et al., 2003) expressed Epstein-Barr nuclear antigen-1 (EBNA1) protein in fusion with a TAP tag at the C-terminus in human 293T cells. Several specific cellular protein interactions and some important regulating proteins were discovered. The TAP method could also be used in analysis nuclear protein interaction. The specific association of interferon regulatory factor (IRF)-4 with c-Rel was revealed from human HUT102 cells (Shindo et al., 2011). A directed proteomic analysis of heterochromatin protein 1 (HP1) isotypes interacting partners were identified by the TAP approach (Rosnoblet et al., 2011). Using TAP strategy, more than one hundred proteins were found to interact with hepatoma-derived growth factor (HDGF) in human HEK293 cells (Zhao et al., 2011). The relationship between HDGF and associated proteins suggests that DHGF as a multifunctional molecular might be involved in many cellular activities. In human HEK293 cells, Gβγ subunits of heterotrimeric G proteins were identified to enter in a protein complex with Rap1a and its effector Radil (Ahmed et al., 2010). This result sggested that the Gβγ -Rap1-Radil complex played an important role in cell adhesion.Besides, TAP also allows for purification of protein complexes from mouse fibroblast cells growing in monolayer cultures and mouse embryonic stem cells (Drakas et al., 2005; Mak et al., 2010). It is meaningful that the TAP system could be used to analysis the interaction of virus and host cells during the infection procedure (Colpitts et al., 2011).

4.3 TAP in plants

Recent studies have shown that the TAP strategy is useful in plant protein complex analysis. The first report of the purification of protein complexes from plant tissue by the TAP method was published in 2004 by Rohila et al (Rohila et al., 2004). By using a TAP-tagged hybrid transcription factor as bait, HSP70 and HSP60 were co-purified. This result was verified by former reports (Dittmar et al., 1997; Stancato et al., 1996). Through the TAP strategy, Liu et al. demonstrated that Hsp90 associated with the plant resistance protein N (Liu et al., 2004), which meant that Hsp90 plays an important role in plant defense (Kanzaki et al., 2003; Takahashi et al., 2003). In another study, the Cf-9 protein function in initiating defense signaling was also investigated by TAP (Rivas et al., 2002). The TAP applications described above were all carried out in a transient expression system of *Nicotiana benthamiana*.

On the other hand, the TAP system was utilized to purify a protein complex in stable expression system of *Arabidopsis thaliana*, for the first time in 2005 (Rubio et al., 2005). The components of the target protein complex were all co-purified with the tagged bait. The superiority of this TAP strategy is based on a constitutive promoter, which allowed for over-expression of TAP fusion proteins. The strength of this method is that over-expression increases incorporation of the tagged protein into a protein complex, when the tagged protein is the core component of a complex or a mutant and suppressed expression for the target protein is harmful to cells. In 2006, Brown et al. (Brown et al., 2006) utilized TAP tagged fatty acid synthase components to investigate protein interactions *in vivo* from stably transfected *A. thaliana.* In addition to the application of TAP to a whole, *A. thaliana* cell suspension culture is ideal for investigating protein-protein interactions involved in cell cycle (Van Leene et al., 2007).

Purification of protein complexes by TAP was demonstrated to be effective in rice (Rohila et al., 2006; Rohila et al., 2009), suggesting that the TAP method could be utilized in cereal crops.

4.4 TAP in Drosophila

In 2003 Forler et al. (Forler et al., 2003) successfully expressed TAP-tagged human proteins and purified their *Drosophila melanogaster* (Dm) binding partners in Dm Schneider cells. The critical advantage in this system is the introduction of RNA interference (RNAi), which can suppress the expression of the corresponding endogenous proteins, thereby avoiding competition from them during protein complex assembly. But the complexes purified through this system consisted of two different source proteins, human bait protein and Dm binding partners, therefore the reliability of the interaction needed validation with other experimental strategies. Both in *Drosophila* cultured cells and embryos, several components of the Notch signaling pathway were tagged with a TAP tag and many novel interactions were uncovered (Veraksa et al., 2005). Throughout the TAP progress, Hsc70 and Hsp83 were validated as cofactors of the *Drosophila* nuclear receptor protein for the first time (Yang et al., 2006).

4.5 TAP in bacteria

In recent years, with the development of the TAP procedure, the application of TAP was extended to purification of protein complexes from bacteria. Gully et al. (Gully et al., 2003) first used the TAP protocol in *E. coli* to isolate native protein complexes. Kumar et al. (Kumar et al., 2004) have identified 80 proteins associated with thioredoxin in *E. coli* suggesting multifunction of thioredoxin. Shereda et al. (Shereda et al., 2007) employed the TAP approach to purify the RecQ complexes, and three heterologous proteins were identified. On the basis of the amount of these three binding proteins, these interactions were classed as direct or indirect. This may imply a new application aspect of TAP in interaction identification. SrmB, one of the five *E. coli* DEAD-box proteins was discovered to form a specific ribonucleoprotein (RNP) complex with r-proteins L4, L24 and the 5′ region of 23S rRNA using the TAP procedure (Trubetskoy et al., 2009). Similar to the application of the TAP method in global protein complexes analysis in yeast, a large-scale analysis of protein complexes, which revealed a novel protein interaction network in *E. coli*, was reported (Butland et al., 2005). Besides the application of the TAP method in *E. coli*, TAP was also carried out in *Thiocapsa roseopersicina* (Fodor et al., 2004) and *Bacillus subtilis* (Yang et al., 2008).

4.6 TAP in other organisms

The efficiency of the TAP method in purification of protein complexes and identification of interactions was also tested in other organisms, including *Dictyostelium* (Koch et al., 2006; Meima et al., 2007), *Trypanosoma bruce* (Mani et al., 2011; Nguyen et al., 2007; Palfi et al., 2005; Schimanski et al., 2005; Walgraffe et al., 2005) and *Plasmodium falciparum* (Takebe et al., 2007).

5. Problems and future prospects

The TAP method has been successfully used for purification and identification of protein complexes and interacting components both in prokaryotic and eukaryotic organisms. However, in practice, the application effects of the method may have been influenced by its inherentvice. Gavin et al. (Gavin et al., 2002) found that in their large-scale analysis of yeast proteome, not all of the tagged proteins could be purified and not all of the purified tagged proteins could interact with other proteins. They ascribed this failure to the intrinsic quality of the TAP tag. The TAP tag fused to a target protein may interfere with protein function, location and interactions (Mak et al., 2010). One of the possible solutions is to add the tag at the other terminus of the ORFs or to replace the original tag with another one. The CBP affinity step has been proved to be problematic in that case where many endogenous proteins of mammalian cells interact with calmodulin in a calcium-dependent manner (Agell et al., 2002; Head, 1992). A simple alternative solution is replacing the CBP tag with other affinity tags, such as the FLAG sequence (Gloeckner et al., 2007; Knuesel et al., 2003), ProtC (Schimanski et al., 2005) and biotinylation tag (Drakas et al., 2005). The main challenge of the TAP strategy comes from the competition of endogenous proteins with the tagged protein in protein complex assembly. This can be resolved by using RNAi to reduce the endogenous expression level (Forler et al., 2003).In some cases, when the target protein is essential and a mutant of it might be harmful and lethal, the over-expression strategy is a perfect strategy to obtain a protein complex containing the tagged target protein (Ho et al., 2002; Rohila et al., 2006; Rubio et al., 2005). However, bait overexpression possibly cause the formation of nonbiological interactions. In addition, overespression may affect cell viability or cellular activity (e.g. negative regulators of cell metabolism). On this occasion, an inducible promoter is a viable choice, which allows experimental modulation of target protein expression, both in terms of amount and timing.

It is thought that the TAP approach is not a powerful tool to detect transient interactions. Therefore, an *in vivo* cross-linking step is added to freeze both weak and transient interactions taking place in intact cells before lysis (Guerrero et al., 2006; Rohila et al., 2004). The cross-linking method has been widely used in the investigation of protein-DNA and protein-protein interactions (Hall and Struhl, 2002; Jackson, 1999; Kuo and Allis, 1999; Orlando et al., 1997; Otsu et al., 1994; Schmitt-Ulms et al., 2004; Schmitt-Ulms et al., 2001; Vasilescu et al., 2004).

Although the two sequential purification steps of the TAP method largely reduce the background resulting from non-specific protein binding compared to a single purification step, these contaminants cannot be removed completely. Collins et al. (Collins et al., 2007) have compared the results from the two large-scale studies of protein complexes in yeast (Gavin et al., 2006; Krogan et al., 2006) and found the two datasets shared very low degrees of overlap. The major difference between the two datasets was mainly caused by non-specific interactions. The problem of non-specifically interacting proteins can be overcome by comparing several interaction datasets (Ewing et al., 2007), using stable-isotope labelling by amino acids in cell culture (Blagoev et al., 2003; Mann, 2006) or isotope-coded affinity tag (Ranish et al., 2003), thereby completely eliminating false-positive interactions.

The TAP system is considered to be inefficient in identifying interactions occurring only in special physiological states or those which occur for a short period. Whether the TAP tag impairs protein function and complex assembly also remains largely unknown and speculative. These disadvantages may affect its application in such instances.

6. Conclusion

Understanding protein function is a major goal in biology. Although the TAP method has some inherent shortcomings, it is undoubtedly a reasonable system for use in purification of protein complexes and identification of protein-protein interactions. In addition to identifying interactions between proteins, the TAP method could be used to characterize and verify interactions between protein and DNA or between protein and RNA (Hogg and Collins, 2007; Zhao et al., 2011). For protein-DNA/RNA interaction analysis, the use of benzonase must be avoided, and RNase inhibitors should be added to protect RNA intact. At the same time, the TAP method can also be used to analyze the effect of mutants on protein interaction and association, possibly resulting in the discovery of binding sites. Protein purification under near-physiological conditions through the TAP strategy is compatible with functional studies and this advantage allows for mapping of large-scale functional interaction networks. As the procedures and conditions used during the TAP process do not vary greatly among different proteins, the results that are generated by this method should be compiled in a database in order to provide comparable and detailed information on the potential and confirmed functions of proteins, as well as the composition of protein complexes and even the structure and activity of protein complexes.

7. Abbreviations used

TAP, tandem affinity purification; ProtA, IgG-binding units of protein A of *Staphylococcus aureus*; CBP, calmodulin-binding domain; TEV, tobacco etch virus; ProtC, protein C epitope; HRV 3C, human rhinovirus 3C protease cleavage site; SBP, streptavidin-binding peptide; ProtG, IgG binding units of protein G from *Streptococcus* sp.; HA, hemagglutinin; ADAP, adhesion and degranulation promoting adaptor protein; MCM, minichromosome maintenance; EBNA1, Epstein-Barr nuclear antigen-1; IRF, interferon regulatory factor; HP1, heterochromatin protein 1; HDGF, hepatoma-derived growth factor; Dm, *Drosophila melanogaster*; RNAi, RNA interference; RNP, ribonucleoprotein.

8. References

Agell, N., Bachs, O., Rocamora, N.&Villalonga, P. (2002). Modulation of the Ras/Raf/MEK/ERK pathway by Ca(2+), and calmodulin. *Cell. Signal.*, Vol. 14, No. 8, pp. 649-654.

Ahmed, S. M., Daulat, A. M., Meunier, A.&Angers, S. (2010). G protein betagamma subunits regulate cell adhesion through Rap1a and its effector Radil. *J. Biol. Chem.*, Vol. 285, No. 9, pp. 6538-6551.

Alberts, B. (1998). The cell as a collection of protein machines: preparing the next generation of molecular biologists. *Cell*, Vol. 92, No. 3, pp. 291-294.

Bürckstümmer, T., Bennett, K. L., Preradovic, A., Schutze, G., Hantschel, O., Superti-Furga, G.&Bauch, A. (2006). An efficient tandem affinity purification procedure for interaction proteomics in mammalian cells. *Nat. Methods*, Vol. 3, No. 12, pp. 1013-1019.

Blagoev, B., Kratchmarova, I., Ong, S. E., Nielsen, M., Foster, L. J.&Mann, M. (2003). A proteomics strategy to elucidate functional protein-protein interactions applied to EGF signaling. *Nat. Biotechnol.*, Vol. 21, No. 3, pp. 315-318.

Brown, A. P., Affleck, V., Fawcett, T.&Slabas, A. R. (2006). Tandem affinity purification tagging of fatty acid biosynthetic enzymes in Synechocystis sp. PCC6803 and Arabidopsis thaliana. *J. Exp. Bot.*, Vol. 57, No. 7, pp. 1563-1571.

Butland, G., Peregrin-Alvarez, J. M., Li, J., Yang, W., Yang, X., Canadien, V., Starostine, A., Richards, D., Beattie, B., Krogan, N., Davey, M., Parkinson, J., Greenblatt, J.&Emili, A. (2005). Interaction network containing conserved and essential protein complexes in Escherichia coli. *Nature*, Vol. 433, No. 7025, pp. 531-537.

Chang, I. F. (2006). Mass spectrometry-based proteomic analysis of the epitope-tag affinity purified protein complexes in eukaryotes. *Proteomics*, Vol. 6, No. 23, pp. 6158-6166.

Cipak, L., Spirek, M., Novatchkova, M., Chen, Z., Rumpf, C., Lugmayr, W., Mechtler, K., Ammerer, G., Csaszar, E.&Gregan, J. (2009). An improved strategy for tandem affinity purification-tagging of Schizosaccharomyces pombe genes. *Proteomics*, Vol. 9, No. 20, pp. 4825-4828.

Collins, S. R., Kemmeren, P., Zhao, X. C., Greenblatt, J. F., Spencer, F., Holstege, F. C., Weissman, J. S.&Krogan, N. J. (2007). Toward a comprehensive atlas of the physical interactome of Saccharomyces cerevisiae. *Mol. Cell. Proteomics*, Vol. 6, No. 3, pp. 439-450.

Colpitts, T. M., Barthel, S., Wang, P.&Fikrig, E. (2011). Dengue virus capsid protein binds core histones and inhibits nucleosome formation in human liver cells. *PLoS One*, Vol. 6, No. 9, pp. e24365.

Cusick, M. E., Klitgord, N., Vidal, M.&Hill, D. E. (2005). Interactome: gateway into systems biology. *Hum. Mol. Genet.*, Vol. 14 No. pp. 171-181.

Davison, E. J., Pennington, K., Hung, C. C., Peng, J., Rafiq, R., Ostareck-Lederer, A., Ostareck, D. H., Ardley, H. C., Banks, R. E.&Robinson, P. A. (2009). Proteomic analysis of increased Parkin expression and its interactants provides evidence for a role in modulation of mitochondrial function. *Proteomics*, Vol. 9, No. 18, pp. 4284-4297.

Dittmar, K. D., Demady, D. R., Stancato, L. F., Krishna, P.&Pratt, W. B. (1997). Folding of the glucocorticoid receptor by the heat shock protein (hsp) 90-based chaperone machinery. The role of p23 is to stabilize receptor.hsp90 heterocomplexes formed by hsp90.p60.hsp70. *J. Biol. Chem.*, Vol. 272, No. 34, pp. 21213-21220.

Drakas, R., Prisco, M.&Baserga, R. (2005). A modified tandem affinity purification tag technique for the purification of protein complexes in mammalian cells. *Proteomics*, Vol. 5, No. 1, pp. 132-137.

Ewing, R. M., Chu, P., Elisma, F., Li, H., Taylor, P., Climie, S., McBroom-Cerajewski, L., Robinson, M. D., O'Connor, L., Li, M., Taylor, R., Dharsee, M., Ho, Y., Heilbut, A., Moore, L., Zhang, S., Ornatsky, O., Bukhman, Y. V., Ethier, M., Sheng, Y., Vasilescu, J., Abu-Farha, M., Lambert, J. P., Duewel, H. S., Stewart, II, Kuehl, B., Hogue, K., Colwill, K., Gladwish, K., Muskat, B., Kinach, R., Adams, S. L., Moran, M. F., Morin, G. B., Topaloglou, T.&Figeys, D. (2007). Large-scale mapping of human protein-protein interactions by mass spectrometry. *Mol. Syst. Biol.*, Vol. 3, No. pp. 89-105.

Fodor, B. D., Kovacs, A. T., Csaki, R., Hunyadi-Gulyas, E., Klement, E., Maroti, G., Meszaros, L. S., Medzihradszky, K. F., Rakhely, G.&Kovacs, K. L. (2004). Modular broad-host-range expression vectors for single-protein and protein complex purification. *Appl. Environ. Microbiol.*, Vol. 70, No. 2, pp. 712-721.

Forler, D., Kocher, T., Rode, M., Gentzel, M., Izaurralde, E.&Wilm, M. (2003). An efficient protein complex purification method for functional proteomics in higher eukaryotes. *Nat. Biotechnol.*, Vol. 21, No. 1, pp. 89-92.

Fromont-Racine, M., Rain, J. C.&Legrain, P. (1997). Toward a functional analysis of the yeast genome through exhaustive two-hybrid screens. *Nat. Genet.*, Vol. 16, No. 3, pp. 277-282.

Gavin, A. C., Aloy, P., Grandi, P., Krause, R., Boesche, M., Marzioch, M., Rau, C., Jensen, L. J., Bastuck, S., Dumpelfeld, B., Edelmann, A., Heurtier, M. A., Hoffman, V., Hoefert, C., Klein, K., Hudak, M., Michon, A. M., Schelder, M., Schirle, M., Remor, M., Rudi, T., Hooper, S., Bauer, A., Bouwmeester, T., Casari, G., Drewes, G., Neubauer, G., Rick, J. M., Kuster, B., Bork, P., Russell, R. B.&Superti-Furga, G. (2006). Proteome survey reveals modularity of the yeast cell machinery. *Nature*, Vol. 440, No. 7084, pp. 631-636.

Gavin, A. C., Bosche, M., Krause, R., Grandi, P., Marzioch, M., Bauer, A., Schultz, J., Rick, J. M., Michon, A. M., Cruciat, C. M., Remor, M., Hofert, C., Schelder, M., Brajenovic, M., Ruffner, H., Merino, A., Klein, K., Hudak, M., Dickson, D., Rudi, T., Gnau, V., Bauch, A., Bastuck, S., Huhse, B., Leutwein, C., Heurtier, M. A., Copley, R. R., Edelmann, A., Querfurth, E., Rybin, V., Drewes, G., Raida, M., Bouwmeester, T., Bork, P., Seraphin, B., Kuster, B., Neubauer, G.&Superti-Furga, G. (2002). Functional organization of the yeast proteome by systematic analysis of protein complexes. *Nature*, Vol. 415, No. 6868, pp. 141-147.

Gloeckner, C. J., Boldt, K., Schumacher, A., Roepman, R.&Ueffing, M. (2007). A novel tandem affinity purification strategy for the efficient isolation and characterisation of native protein complexes. *Proteomics*, Vol. 7, No. 23, pp. 4228-4234.

Gottlieb, T. M.&Jackson, S. P. (1993). The DNA-dependent protein kinase: requirement for DNA ends and association with Ku antigen. *Cell*, Vol. 72, No. 1, pp. 131-142.

Gould, K. L., Ren, L., Feoktistova, A. S., Jennings, J. L.&Link, A. J. (2004). Tandem affinity purification and identification of protein complex components. *Methods*, Vol. 33, No. 3, pp. 239-244.

Graumann, J., Dunipace, L. A., Seol, J. H., McDonald, W. H., Yates, J. R., 3rd, Wold, B. J.&Deshaies, R. J. (2004). Applicability of tandem affinity purification MudPIT to pathway proteomics in yeast. *Mol. Cell. Proteomics*, Vol. 3, No. 3, pp. 226-237.

Guerrero, C., Tagwerker, C., Kaiser, P.&Huang, L. (2006). An integrated mass spectrometry-based proteomic approach: quantitative analysis of tandem affinity-purified in vivo cross-linked protein complexes (QTAX) to decipher the 26 S proteasome-interacting network. *Mol. Cell. Proteomics*, Vol. 5, No. 2, pp. 366-378.

Gully, D., Moinier, D., Loiseau, L.&Bouveret, E. (2003). New partners of acyl carrier protein detected in Escherichia coli by tandem affinity purification. *FEBS Lett.*, Vol. 548, No. 1-3, pp. 90-96.

Hall, D. B.&Struhl, K. (2002). The VP16 activation domain interacts with multiple transcriptional components as determined by protein-protein cross-linking in vivo. *J. Biol. Chem.*, Vol. 277, No. 48, pp. 46043-46050.

Head, J. F. (1992). A better grip on calmodulin. *Curr. Biol.*, Vol. 2, No. 11, pp. 609-611.

Ho, Y., Gruhler, A., Heilbut, A., Bader, G. D., Moore, L., Adams, S. L., Millar, A., Taylor, P., Bennett, K., Boutilier, K., Yang, L., Wolting, C., Donaldson, I., Schandorff, S., Shewnarane, J., Vo, M., Taggart, J., Goudreault, M., Muskat, B., Alfarano, C., Dewar, D., Lin, Z., Michalickova, K., Willems, A. R., Sassi, H., Nielsen, P. A., Rasmussen, K. J., Andersen, J. R., Johansen, L. E., Hansen, L. H., Jespersen, H., Podtelejnikov, A., Nielsen, E., Crawford, J., Poulsen, V., Sorensen, B. D.,

Matthiesen, J., Hendrickson, R. C., Gleeson, F., Pawson, T., Moran, M. F., Durocher, D., Mann, M., Hogue, C. W., Figeys, D.&Tyers, M. (2002). Systematic identification of protein complexes in Saccharomyces cerevisiae by mass spectrometry. *Nature*, Vol. 415, No. 6868, pp. 180-183.

Hogg, J. R.&Collins, K. (2007). RNA-based affinity purification reveals 7SK RNPs with distinct composition and regulation. *RNA*, Vol. 13, No. 6, pp. 868-880.

Holowaty, M. N., Zeghouf, M., Wu, H., Tellam, J., Athanasopoulos, V., Greenblatt, J.&Frappier, L. (2003). Protein profiling with Epstein-Barr nuclear antigen-1 reveals an interaction with the herpesvirus-associated ubiquitin-specific protease HAUSP/USP7. *J. Biol. Chem.*, Vol. 278, No. 32, pp. 29987-29994.

Honey, S., Schneider, B. L., Schieltz, D. M., Yates, J. R.&Futcher, B. (2001). A novel multiple affinity purification tag and its use in identification of proteins associated with a cyclin-CDK complex. *Nucleic Acids Res.*, Vol. 29, No. 4, pp. 24-32.

Ito, T., Chiba, T., Ozawa, R., Yoshida, M., Hattori, M.&Sakaki, Y. (2001). A comprehensive two-hybrid analysis to explore the yeast protein interactome. *Proc. Natl. Acad. Sci. USA*, Vol. 98, No. 8, pp. 4569-4574.

Jackson, V. (1999). Formaldehyde cross-linking for studying nucleosomal dynamics. *Methods*, Vol. 17, No. 2, pp. 125-139.

Jeronimo, C., Forget, D., Bouchard, A., Li, Q., Chua, G., Poitras, C., Therien, C., Bergeron, D., Bourassa, S., Greenblatt, J., Chabot, B., Poirier, G. G., Hughes, T. R., Blanchette, M., Price, D. H.&Coulombe, B. (2007). Systematic analysis of the protein interaction network for the human transcription machinery reveals the identity of the 7SK capping enzyme. *Mol. Cell*, Vol. 27, No. 2, pp. 262-274.

Kamil, J. P.&Coen, D. M. (2007). Human cytomegalovirus protein kinase UL97 forms a complex with the tegument phosphoprotein pp65. *J. Virol.*, Vol. 81, No. 19, pp. 10659-10668.

Kaneko, A., Umeyama, T., Hanaoka, N., Monk, B. C., Uehara, Y.&Niimi, M. (2004). Tandem affinity purification of the Candida albicans septin protein complex. *Yeast*, Vol. 21, No. 12, pp. 1025-1033.

Kanzaki, H., Saitoh, H., Ito, A., Fujisawa, S., Kamoun, S., Katou, S., Yoshioka, H.&Terauchi, R. (2003). Cytosolic HSP90 and HSP70 are essential components of INF1-mediated hypersensitive response and non-host resistance to Pseudomonas cichorii in Nicotiana benthamiana. *Mol Plant Pathol*, Vol. 4, No. 5, pp. 383-391.

Knuesel, M., Wan, Y., Xiao, Z., Holinger, E., Lowe, N., Wang, W.&Liu, X. (2003). Identification of novel protein-protein interactions using a versatile mammalian tandem affinity purification expression system. *Mol. Cell. Proteomics*, Vol. 2, No. 11, pp. 1225-1233.

Koch, K. V., Reinders, Y., Ho, T. H., Sickmann, A.&Graf, R. (2006). Identification and isolation of Dictyostelium microtubule-associated protein interactors by tandem affinity purification. *Eur. J. Cell Biol.*, Vol. 85, No. 9-10, pp. 1079-1090.

Krogan, N. J., Cagney, G., Yu, H., Zhong, G., Guo, X., Ignatchenko, A., Li, J., Pu, S., Datta, N., Tikuisis, A. P., Punna, T., Peregrin-Alvarez, J. M., Shales, M., Zhang, X., Davey, M., Robinson, M. D., Paccanaro, A., Bray, J. E., Sheung, A., Beattie, B., Richards, D. P., Canadien, V., Lalev, A., Mena, F., Wong, P., Starostine, A., Canete, M. M., Vlasblom, J., Wu, S., Orsi, C., Collins, S. R., Chandran, S., Haw, R., Rilstone, J. J., Gandi, K., Thompson, N. J., Musso, G., St Onge, P., Ghanny, S., Lam, M. H.,

Butland, G., Altaf-Ul, A. M., Kanaya, S., Shilatifard, A., O'Shea, E., Weissman, J. S., Ingles, C. J., Hughes, T. R., Parkinson, J., Gerstein, M., Wodak, S. J., Emili, A.&Greenblatt, J. F. (2006). Global landscape of protein complexes in the yeast Saccharomyces cerevisiae. *Nature*, Vol. 440, No. 7084, pp. 637-643.

Krogan, N. J., Kim, M., Ahn, S. H., Zhong, G., Kobor, M. S., Cagney, G., Emili, A., Shilatifard, A., Buratowski, S.&Greenblatt, J. F. (2002). RNA polymerase II elongation factors of Saccharomyces cerevisiae: a targeted proteomics approach. *Mol. Cell. Biol.*, Vol. 22, No. 20, pp. 6979-6992.

Kumar, J. K., Tabor, S.&Richardson, C. C. (2004). Proteomic analysis of thioredoxin-targeted proteins in Escherichia coli. *Proc. Natl. Acad. Sci. USA*, Vol. 101, No. 11, pp. 3759-3764.

Kuo, M. H.&Allis, C. D. (1999). In vivo cross-linking and immunoprecipitation for studying dynamic Protein:DNA associations in a chromatin environment. *Methods*, Vol. 19, No. 3, pp. 425-433.

Lehmann, R., Meyer, J., Schuemann, M., Krause, E.&Freund, C. (2009). A novel S3S-TAP-tag for the isolation of T-cell interaction partners of adhesion and degranulation promoting adaptor protein. *Proteomics*, Vol. 9, No. 23, pp. 5288-5295.

Li, Y. F. (2010). commonly used tag combinations for tandem affinity purification. *Biotechnol. Appl. Biochem.*, Vol. 55, No. pp. 73-83.

Lichty, J. J., Malecki, J. L., Agnew, H. D., Michelson-Horowitz, D. J.&Tan, S. (2005). Comparison of affinity tags for protein purification. *Protein Expr. Purif.*, Vol. 41, No. 1, pp. 98-105.

Liu, Y., Burch-Smith, T., Schiff, M., Feng, S.&Dinesh-Kumar, S. P. (2004). Molecular chaperone Hsp90 associates with resistance protein N and its signaling proteins SGT1 and Rar1 to modulate an innate immune response in plants. *J. Biol. Chem.*, Vol. 279, No. 3, pp. 2101-2108.

Mak, A. B., Ni, Z., Hewel, J. A., Chen, G. I., Zhong, G., Karamboulas, K., Blakely, K., Smiley, S., Marcon, E., Roudeva, D., Li, J., Olsen, J. B., Wan, C., Punna, T., Isserlin, R., Chetyrkin, S., Gingras, A. C., Emili, A., Greenblatt, J.&Moffat, J. (2010). A lentiviral functional proteomics approach identifies chromatin remodeling complexes important for the induction of pluripotency. *Mol. Cell. Proteomics*, Vol. 9, No. 5, pp. 811-823.

Mani, J., Guttinger, A., Schimanski, B., Heller, M., Acosta-Serrano, A., Pescher, P., Spath, G.&Roditi, I. (2011). Alba-domain proteins of Trypanosoma brucei are cytoplasmic RNA-binding proteins that interact with the translation machinery. *PLoS One*, Vol. 6, No. 7, pp. e22463.

Mann, M. (2006). Functional and quantitative proteomics using SILAC. *Nat. Rev. Mol. Cell Biol.*, Vol. 7, No. 12, pp. 952-958.

Meima, M. E., Weening, K. E.&Schaap, P. (2007). Vectors for expression of proteins with single or combinatorial fluorescent protein and tandem affinity purification tags in Dictyostelium. *Protein Expr. Purif.*, Vol. 53, No. 2, pp. 283-288.

Milewska, M., McRedmond, J.&Byrne, P. C. (2009). Identification of novel spartin-interactors shows spartin is a multifunctional protein. *J. Neurochem.*, Vol. 111, No. 4, pp. 1022-1030.

Nguyen, T. N., Schimanski, B.&Gunzl, A. (2007). Active RNA polymerase I of Trypanosoma brucei harbors a novel subunit essential for transcription. *Mol. Cell Biol.*, Vol. 27, No. 17, pp. 6254-6263.

Orlando, V., Strutt, H.&Paro, R. (1997). Analysis of chromatin structure by in vivo formaldehyde cross-linking. *Methods*, Vol. 11, No. 2, pp. 205-214.

Otsu, M., Omura, F., Yoshimori, T.&Kikuchi, M. (1994). Protein disulfide isomerase associates with misfolded human lysozyme in vivo. *J. Biol. Chem.*, Vol. 269, No. 9, pp. 6874-6877.

Palfi, Z., Schimanski, B., Gunzl, A., Lucke, S.&Bindereif, A. (2005). U1 small nuclear RNP from Trypanosoma brucei: a minimal U1 snRNA with unusual protein components. *Nucleic Acids Res.*, Vol. 33, No. 8, pp. 2493-2503.

Puig, O., Caspary, F., Rigaut, G., Rutz, B., Bouveret, E., Bragado-Nilsson, E., Wilm, M.&Seraphin, B. (2001). The tandem affinity purification (TAP) method: a general procedure of protein complex purification. *Methods*, Vol. 24, No. 3, pp. 218-229.

Ranish, J. A., Yi, E. C., Leslie, D. M., Purvine, S. O., Goodlett, D. R., Eng, J.&Aebersold, R. (2003). The study of macromolecular complexes by quantitative proteomics. *Nat. Genet.*, Vol. 33, No. 3, pp. 349-355.

Rigaut, G., Shevchenko, A., Rutz, B., Wilm, M., Mann, M.&Seraphin, B. (1999). A generic protein purification method for protein complex characterization and proteome exploration. *Nat. Biotechnol.*, Vol. 17, No. 10, pp. 1030-1032.

Rivas, S., Romeis, T.&Jones, J. D. (2002). The Cf-9 disease resistance protein is present in an approximately 420-kilodalton heteromultimeric membrane-associated complex at one molecule per complex. *Plant Cell*, Vol. 14, No. 3, pp. 689-702.

Rohila, J. S., Chen, M., Cerny, R.&Fromm, M. E. (2004). Improved tandem affinity purification tag and methods for isolation of protein heterocomplexes from plants. *Plant J.*, Vol. 38, No. 1, pp. 172-181.

Rohila, J. S., Chen, M., Chen, S., Chen, J., Cerny, R., Dardick, C., Canlas, P., Xu, X., Gribskov, M., Kanrar, S., Zhu, J. K., Ronald, P.&Fromm, M. E. (2006). Protein-protein interactions of tandem affinity purification-tagged protein kinases in rice. *Plant J.*, Vol. 46, No. 1, pp. 1-13.

Rohila, J. S., Chen, M., Chen, S., Chen, J., Cerny, R. L., Dardick, C., Canlas, P., Fujii, H., Gribskov, M., Kanrar, S., Knoflicek, L., Stevenson, B., Xie, M., Xu, X., Zheng, X., Zhu, J. K., Ronald, P.&Fromm, M. E. (2009). Protein-protein interactions of tandem affinity purified protein kinases from rice. *PLoS One*, Vol. 4, No. 8, pp. e6685.

Rosnoblet, C., Vandamme, J., Volkel, P.&Angrand, P. O. (2011). Analysis of the human HP1 interactome reveals novel binding partners. *Biochem. Biophys. Res. Commun.*, Vol. 413, No. 2, pp. 206-211.

Rubio, V., Shen, Y., Saijo, Y., Liu, Y., Gusmaroli, G., Dinesh-Kumar, S. P.&Deng, X. W. (2005). An alternative tandem affinity purification strategy applied to *Arabidopsis* protein complex isolation. *Plant J.*, Vol. 41, No. 5, pp. 767-778.

Sakwe, A. M., Nguyen, T., Athanasopoulos, V., Shire, K.&Frappier, L. (2007). Identification and characterization of a novel component of the human minichromosome maintenance complex. *Mol. Cell Biol.*, Vol. 27, No. 8, pp. 3044-3055.

Schimanski, B., Nguyen, T. N.&Gunzl, A. (2005). Characterization of a multisubunit transcription factor complex essential for spliced-leader RNA gene transcription in Trypanosoma brucei. *Mol. Cell. Biol.*, Vol. 25, No. 16, pp. 7303-7313.

Schimanski, B., Nguyen, T. N.&Gunzl, A. (2005). Highly efficient tandem affinity purification of trypanosome protein complexes based on a novel epitope combination. *Eukaryot Cell*, Vol. 4, No. 11, pp. 1942-1950.

Schmitt-Ulms, G., Hansen, K., Liu, J., Cowdrey, C., Yang, J., DeArmond, S. J., Cohen, F. E., Prusiner, S. B.&Baldwin, M. A. (2004). Time-controlled transcardiac perfusion cross-linking for the study of protein interactions in complex tissues. *Nat. Biotechnol.*, Vol. 22, No. 6, pp. 724-731.

Schmitt-Ulms, G., Legname, G., Baldwin, M. A., Ball, H. L., Bradon, N., Bosque, P. J., Crossin, K. L., Edelman, G. M., DeArmond, S. J., Cohen, F. E.&Prusiner, S. B. (2001). Binding of neural cell adhesion molecules (N-CAMs) to the cellular prion protein. *J. Mol. Biol.*, Vol. 314, No. 5, pp. 1209-1225.

Shereda, R. D., Bernstein, D. A.&Keck, J. L. (2007). A central role for SSB in Escherichia coli RecQ DNA helicase function. *J. Biol. Chem.*, Vol. 282, No. 26, pp. 19247-19258.

Shindo, H., Yasui, K., Yamamoto, K., Honma, K., Yui, K., Kohno, T., Ma, Y., Chua, K. J., Kubo, Y., Aihara, H., Ito, T., Nagayasu, T., Matsuyama, T.&Hayashi, H. (2011). Interferon regulatory factor-4 activates IL-2 and IL-4 promoters in cooperation with c-Rel. *Cytokine*, Vol. No. pp.

Stancato, L. F., Hutchison, K. A., Krishna, P.&Pratt, W. B. (1996). Animal and plant cell lysates share a conserved chaperone system that assembles the glucocorticoid receptor into a functional heterocomplex with hsp90. *Biochemistry*, Vol. 35, No. 2, pp. 554-561.

Stevens, R. C. (2000). Design of high-throughput methods of protein production for structural biology. *Structure*, Vol. 8, No. 9, pp. 177-185.

Tagwerker, C., Zhang, H., Wang, X., Larsen, L. S., Lathrop, R. H., Hatfield, G. W., Auer, B., Huang, L.&Kaiser, P. (2006). HB tag modules for PCR-based gene tagging and tandem affinity purification in Saccharomyces cerevisiae. *Yeast*, Vol. 23, No. 8, pp. 623-632.

Takahashi, A., Casais, C., Ichimura, K.&Shirasu, K. (2003). HSP90 interacts with RAR1 and SGT1 and is essential for RPS2-mediated disease resistance in Arabidopsis. *Proc. Natl. Acad. Sci. USA*, Vol. 100, No. 20, pp. 11777-11782.

Takebe, S., Witola, W. H., Schimanski, B., Gunzl, A.&Ben Mamoun, C. (2007). Purification of components of the translation elongation factor complex of Plasmodium falciparum by tandem affinity purification. *Eukaryot Cell*, Vol. 6, No. 4, pp. 584-591.

Tasto, J. J., Carnahan, R. H., McDonald, W. H.&Gould, K. L. (2001). Vectors and gene targeting modules for tandem affinity purification in Schizosaccharomyces pombe. *Yeast*, Vol. 18, No. 7, pp. 657-662.

Terpe, K. (2003). Overview of tag protein fusions: from molecular and biochemical fundamentals to commercial systems. *Appl. Microbiol. Biotechnol.*, Vol. 60, No. 5, pp. 523-533.

Trubetskoy, D., Proux, F., Allemand, F., Dreyfus, M.&Iost, I. (2009). SrmB, a DEAD-box helicase involved in Escherichia coli ribosome assembly, is specifically targeted to 23S rRNA in vivo. *Nucleic Acids Res.*, Vol. 37, No. 19, pp. 6540-6549.

Uetz, P., Giot, L., Cagney, G., Mansfield, T. A., Judson, R. S., Knight, J. R., Lockshon, D., Narayan, V., Srinivasan, M., Pochart, P., Qureshi-Emili, A., Li, Y., Godwin, B., Conover, D., Kalbfleisch, T., Vijayadamodar, G., Yang, M., Johnston, M., Fields,

S.&Rothberg, J. M. (2000). A comprehensive analysis of protein-protein interactions in Saccharomyces cerevisiae. *Nature*, Vol. 403, No. 6770, pp. 623-627.

Van Leene, J., Stals, H., Eeckhout, D., Persiau, G., Van De Slijke, E., Van Isterdael, G., De Clercq, A., Bonnet, E., Laukens, K., Remmerie, N., Henderickx, K., De Vijlder, T., Abdelkrim, A., Pharazyn, A., Van Onckelen, H., Inze, D., Witters, E.&De Jaeger, G. (2007). A tandem affinity purification-based technology platform to study the cell cycle interactome in Arabidopsis thaliana. *Mol. Cell. Proteomics*, Vol. 6, No. 7, pp. 1226-1238.

Vasilescu, J., Guo, X.&Kast, J. (2004). Identification of protein-protein interactions using in vivo cross-linking and mass spectrometry. *Proteomics*, Vol. 4, No. 12, pp. 3845-3854.

Veraksa, A., Bauer, A.&Artavanis-Tsakonas, S. (2005). Analyzing protein complexes in Drosophila with tandem affinity purification-mass spectrometry. *Dev. Dyn.*, Vol. 232, No. 3, pp. 827-834.

Walgraffe, D., Devaux, S., Lecordier, L., Dierick, J. F., Dieu, M., Van den Abbeele, J., Pays, E.&Vanhamme, L. (2005). Characterization of subunits of the RNA polymerase I complex in Trypanosoma brucei. *Mol. Biochem. Parasitol.*, Vol. 139, No. 2, pp. 249-260.

Xu, X., Song, Y., Li, Y., Chang, J., Zhang, H.&An, L. (2010). The tandem affinity purification method: an efficient system for protein complex purification and protein interaction identification. *Protein Expr. Purif.*, Vol. 72, No. 2, pp. 149-156.

Yang, P., Sampson, H. M.&Krause, H. M. (2006). A modified tandem affinity purification strategy identifies cofactors of the Drosophila nuclear receptor dHNF4. *Proteomics*, Vol. 6, No. 3, pp. 927-935.

Yang, X., Doherty, G. P.&Lewis, P. J. (2008). Tandem affinity purification vectors for use in gram positive bacteria. *Plasmid*, Vol. 59, No. 1, pp. 54-62.

Zeghouf, M., Li, J., Butland, G., Borkowska, A., Canadien, V., Richards, D., Beattie, B., Emili, A.&Greenblatt, J. F. (2004). Sequential Peptide Affinity (SPA) system for the identification of mammalian and bacterial protein complexes. *J. Proteome Res.*, Vol. 3, No. 3, pp. 463-468.

Zhao, J., Yu, H., Lin, L., Tu, J., Cai, L., Chen, Y., Zhong, F., Lin, C., He, F.&Yang, P. (2011). Interactome study suggests multiple cellular functions of hepatoma-derived growth factor (HDGF). *J Proteomics*, Vol. No. pp.

Zhu, H., Bilgin, M., Bangham, R., Hall, D., Casamayor, A., Bertone, P., Lan, N., Jansen, R., Bidlingmaier, S., Houfek, T., Mitchell, T., Miller, P., Dean, R. A., Gerstein, M.&Snyder, M. (2001). Global analysis of protein activities using proteome chips. *Science*, Vol. 293, No. 5537, pp. 2101-2105.